Soil Ecology and Management

THE LIBRARY
WRITTLE COLLEGE
CHELMSFORD CM1 3RR

Soil Ecology and Management

Joann K. Whalen

McGill University,
Québec, Canada

Luis Sampedro

Centro de Investigación Forestal de Lourizán–Unidad Asociada MBG-CSIC,
Galicia, Spain

CABI is a trading name of CAB International

CABI Head Office
Nosworthy Way
Wallingford
Oxfordshire OX10 8DE
UK

Tel: +44 (0)1491 832111
Fax: +44 (0)1491 833508
E-mail: cabi@cabi.org
Website: www.cabi.org

CABI North American Office
875 Massachusetts Avenue
7th Floor
Cambridge, MA 02139
USA

Tel: +1 617 395 4056
Fax: +1 617 354 6875
E-mail: cabi-nao@cabi.org

©CAB International 2010. All rights reserved. No part of this publication may be reproduced in any form or by any means, electronically, mechanically, by photocopying, recording or otherwise, without the prior permission of the copyright owners.

A catalogue record for this book is available from the British Library, London, UK.

Library of Congress Cataloging-in-Publication Data

Whalen, Joann K.
Soil ecology and management / by Joann K. Whalen, Luis Sampedro.
p. cm.
Includes bibliographical references and index.
ISBN 978-1-84593-563-4 (alk. paper)
1. Soil ecology. 2. Soil management. I. Sampedro, Luis. II. Title.

QH541.5.S6W45 2009
577.5'7--dc22

2009016841

ISBN: 978 1 84593 563 4

Typeset by SPi, Pondicherry, India.
Printed and bound in the UK by Cambridge University Press, Cambridge.

Contents

The colour plate section can be found following p. 214

Preface

Soil ecology is the study of interactions between the soil physico-chemical components (minerals and soil solution) and the soil biota. These creatures range in size from microscopic to those visible to the naked eye. They live within the pore spaces between solid particles and contribute vitally to plant nutrition, in symbiotic association with roots or by releasing essential nutrients into the soil solution around roots. The breakdown of plant and animal residues, as well as many synthetic chemicals, is catalysed by soil microorganisms and facilitated by soil animals. Soil pedogenesis involves the mixing of organic material and mineral particles by soil fauna such as collembola, enchytraeids, mites and earthworms. Soil bacteria and fungi produce binding substances that stabilize aggregates, building a soil structure that allows for root penetration, water infiltration, chemical buffering and gas exchange.

Humans are highly reliant on the soil ecosystem, which provides food, fibre, fuel and other ecological services such as water filtration/purification and the recycling of atmospheric gases. Yet, the development of deep soil profiles underlying most terrestrial ecosystems takes thousands of years or longer. Since soils form so slowly, they are considered to be a non-renewable resource that merits careful attention and protection. At the present time, we are experiencing unprecedented global change, largely induced by anthropogenic activities such as fossil fuel burning and land use change. Food and fuel insecurity, as well as environmental pollution and ecological degradation, are tremendous societal challenges for the 21st century. As scientists, we believe that understanding the functions and dynamic nature of the soil ecosystem is crucial for predicting and mitigating the long-term consequences of present-day actions. In this way, we can make recommendations and decisions today that will sustain our soils for the future.

This modular textbook is intended to provide an overview of the soil ecosystem, soil organisms and their functions, as well as the impact of human-induced management and global change. It is meant for advanced undergraduates, beginning graduate students and practising scientists from diverse disciplines. Some of the scientific literature and Internet sources relevant to soil ecology are provided for further reading.

The first part of the book provides the context for understanding the soil ecosystem. Physical, chemical and biological components and their interactions are reviewed, with a view towards a holistic approach and interpretation of soil functions. Part II examines the diversity of life found within our soils, from microscopic to macroscopic organisms. Methods for collecting and enumerating soil organisms, including emerging molecular approaches, are introduced. The concept of the soil foodweb is introduced and trophic interactions (feeding relationships) are discussed. Part III gives an overview of the major ecological and pedological functions of soil organisms, including primary production, decomposition and carbon cycling, nutrient cycling (nitrogen, phosphorus and sulfur), soil structure and biological control. This section is intended to give the reader a sense of what organisms do in soils, and at what temporal and spatial scales their activities affect ecosystem processes. Part IV discusses the impact of human-induced management on the soil foodweb in agricultural, grassland and forest ecosystems. Finally, we consider the effect of global changes on the diversity and function of the soil foodweb, as well as the consequences for sustainable soil use.

Acknowledgements

This book could not have been completed without support from many people. The authors particularly thank Hicham and Janet, and Pilar and Luisiño for their extraordinary patience during this time.

Many topics in the text were first presented to undergraduate and graduate students at McGill University, and benefited from the feedback and ideas arising from student discussions.

أشكر عائلة بنسليم شكرا جزيلا لترحيبهم وكرمهم اللذين ساعداني على كتابة الجزء الأول من هذا الكتاب

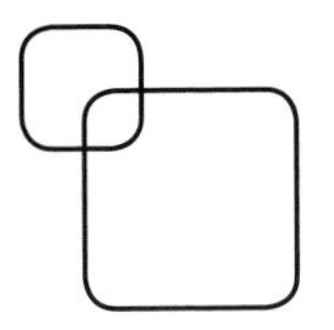

PART I

Introductory Concepts

1 Fundamental Properties of the Soil Ecosystem

Most terrestrial ecosystems on Earth are based upon a thin layer of finely fragmented rocks and decomposed organic materials that we refer to as soil. This material is distinguished for its ability to support plant growth and thus sustain human and animal life on our planet. Soil is also home to a myriad of creatures, many microscopic and hidden from human sight, known as the soil biota. The soil biota and their importance to soil health, plant growth, nutrient cycling and soil structure are central topics in this book.

Humans have been actively managing the soils on our planet for millennia, developing specialized methods to use the soils in various regions. We are keenly aware that many soils in the world are degraded or at risk for degradation, which diminishes our ability to produce food, fibre and fuel. When we look beyond the agricultural goods that are harvested from soils, we begin to have a clear view of the importance of soils for the Earth's ecosystems (Table 1.1).

The health and integrity of terrestrial, and even planetary, ecosystems absolutely depends on chemical and biochemical transformations of materials in soil. For instance, water that enters the soil as rainfall, snowfall or as irrigation water generally contains soluble salts and may contain inorganic and organic pollutants. Many of these substances are adsorbed to soil surfaces and thus removed from the water as it moves vertically through the soil profile to recharge aquifers beneath soil and bedrock. Chemical buffering capacity describes the equilibrium between compounds adsorbed to soil surfaces and released into the soil solution. Plant nutrition is affected by the rates of ion adsorption and desorption from soil surfaces. Decomposition and nutrient cycling are equally essential to sustaining life on this planet. Photosynthesis and nutrient uptake by green plants are fundamental processes on Earth. A part of foodwebs in terrestrial ecosystems survives by withdrawing energy from living plants, but an even larger part of the foodweb relies on the energy contained in dead materials, essentially dead vegetal tissues. When terrestrial plants and animals die, a portion of the carbon and nutrients in their dead biomass is stored in the soil and the rest is recycled by soil organisms, thus supporting the next generation of plants and animals. There is more carbon stored in the top 1 m of soil than in any other terrestrial pool, and thus soils are an integral part of the global carbon cycle. Managing the soil in a way that builds up its carbon content will counterbalance excessive carbon dioxide emitted to the atmosphere from anthropogenic sources

The production of food, fibre and fuel essential for homes and industries is highly dependent on maintaining and protecting fertile soils. Agricultural and forest bioproducts, as well as fast-growing crops, can be used as biofuels, thus reducing our reliance on fossil fuel energy. The soil environment is home to a diverse array of organisms that enhance nutrient transfers from the soil to terrestrial plants and are a source of nutrients and energy when consumed by insects, birds, amphibians and mammals. In addition, soil microorganisms produce a wide array of novel chemical compounds and antibiotics that are effective against pathogens that cause disease in plants, animals and humans.

Soils also serve as a repository of our cultural heritage. The burial of ancient towns and settlements under layers of soil and desert sands preserves the history of past civilizations. Soil is the foundation of modern civilizations, supporting the buildings, infrastructure and transportation systems of our towns and cities.

To summarize, soils are at the base of terrestrial ecosystems. They provide a wide array of essential ecosystem services, buffer climate change and support the growth of diverse plant communities that are the habitat and food of creatures ranging from the tiniest microorganisms to the largest terrestrial vertebrates inhabiting the Earth's emerged land.

©CAB International 2010. *Soil Ecology and Management*
(J.K. Whalen and L. Sampedro)

Table 1.1. Major ecological functions provided by soils.

Transformations	Production	Aesthetic
Water filtration	Food production	Cultural heritage
Chemical buffering capacity	Raw materials for homes and industries (fibre, fuel)	Physical support for homes and industries
Organic matter decomposition	Biodiversity	Maintaining landscapes
Nutrient cycling		

Focus Box 1.1. Soil use and the development of civilizations.

A change in soil use due to the exploitation of natural resources by human populations leaves a lasting imprint in ecosystems. Charcoal buried in the soils of the European steppes and the North American prairies comes from the ancient practice of burning these grasslands to remove the surface residues and stimulate plant growth, which would be favourable for wild and domestic grazing animals. Since Neolithic times, humans have used fire to remove unwanted trees or grasses from an area before planting agricultural crops or using the land for grazing animals, a practice known as slash and burn agriculture, that continues to this day in some regions.

In the Amazon basin, pre-Columbian inhabitants burned wood and other wastes to generate low-temperature charcoal and deliberately mixed it with the soil. The Terra Preta soils in Amazonia were formed between 450 BC and AD 950 and still have greater concentrations of essential plant nutrients and support higher crop yields than the surrounding low-fertility soils (Acrisols, Ferrasols and Arenosols). They also have the ability to bind carbon from newly incorporated organic residues and thus may be beneficial in soil carbon sequestration. Local residents claim that Terra Preta soils are self-regenerating, which explains their persistence for centuries and millennia. Smaller areas with charcoal-containing black earth soils from postmodern human activities are also found in West Africa (Benin, Liberia), South African savannahs and South America (Ecuador, Peru, Guyana).

Unfortunately, not all anthropogenic soil use is equally sustainable. For instance, decades of intensive agricultural production in the upland fields of Central and South America, Madagascar and Vietnam led to soil erosion, which reduces the concentration of essential plant nutrients, water availability and results in a subsequent decline in crop yields. In tropical regions, slash and burn agriculture with short fallow periods is responsible for nutrient exhaustion, soil compaction, increased water runoff, soil erosion and land degradation. These problems are further exacerbated when slash and burn is practised on steep hillsides and in regions with intense rainfall, such as Honduras, Ecuador, Bolivia, as well as rainforests in West Africa and Indonesia. In some cases, farmers are forced to abandon their fields because they are no longer productive and economically viable.

Considerable time and effort is needed to rehabilitate land that has suffered soil loss due to erosion. Yet, this is an essential investment for future generations. History shows us that the decline of many civilizations and cultures was due, in part, to poor soil management that eventually caused food shortages, malnutrition, disease and starvation. The decline of the great Maya civilization and repeated declines in human population to the point of collapse on Rapa Nui (Easter Island; Fig. 1.1), both within the past ten centuries, should serve as a stark reminder of the importance of soils in guaranteeing our food supply. Similarly, uncontrolled human activities following occidental colonization on some Caribbean islands such as Haití have led to severe land degradation in less than 500 years.

Not all regions of the world exhibit such dramatic and rapid soil degradation. Yet, large-scale soil loss occurs every year as a consequence of the cumulative human activities through hundreds to thousands of years in the landscape. Galicia is a province in north-west Spain with a warm-temperate climate (2000 mm rain/year; 14°C mean temperature), hilly topography and high organic matter content but low-fertility soils. Humans have been logging and extensively burning the climactic broadleaf forests to clear pastures for grazing animals for the past 5000 years. Most agricultural fields were abandoned during the second half of the 20th century. Today, the landscape is extremely fragmented into many small cropped fields and grasslands, interdispersed among small stands of non-exploited mixed secondary forests, and larger areas with managed forest plantations (pine and eucalyptus). The anthropogenic practices of ancient times have a strong influence on the current land use and soil fertility in this region.

Continued

Focus Box 1.1. *Continued*

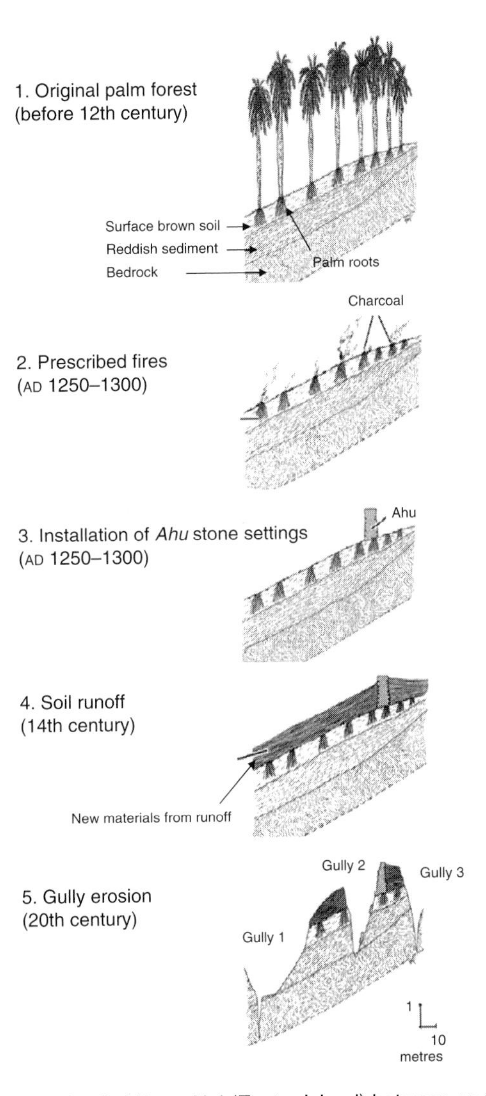

Fig. 1.1. Polynesians reached the island of Rapa Nui (Easter Island) between AD 300 and 800. They introduced new crops that grew exceedingly well in the fertile volcanic soils. They progressively cleared the woodland of indigenous palm trees and shrubs, creating an efficient food production system that supported population growth and a rich new culture. Agricultural intensification peaked around AD 1300 to 1500, when the woodland was practically cleared. Wood resources became scarce, soil erosion and a decline in soil fertility led to famine and subsequent social conflict. During this time, the population declined from 8000 inhabitants to fewer than 2000. Raids by slave traders and a smallpox epidemic further decimated the population, and there were only about 100 indigenous people left by the late 1800s. Although the population has grown since then, much of the cultural memory was lost. Intensive sheep production in the past century has caused further damage and led to the impressive soil erosion and land degradation that we see nowadays. (Source: reprinted from Mieth, A. and Bork, H.-R. (2005) 'History, origin and extent of soil erosion on Easter Island (Rapa Nui)', *Catena* 63, 244–260, Copyright (2005), with permission from Elsevier.)

1.1 Historical and Current Perspectives in Soil Biology

The ecological foundations of soil science date back to research conducted in the 1800s. The earliest textbooks of relevance to soil ecologists tended to focus on specific organisms or groups of organisms. Some relevant works were the *Monograph of the Collembola and Thysanura* published by John Lubbock (1873) and *The Formation of Vegetable Mould Through the Action of Worms* by Charles Darwin (1881). The first text on soil microbiology was *Handbuch der Landwirtschaftliche Bakteriologie* published by the German scientist Felix Löhnis in 1910. This textbook was translated into four languages and had 19 editions. Another text of interest to soil microbiologists was M.C. Rayner's *Mycorrhiza: an Account of Non-pathogenic Infection by Fungi in Vascular Plants and Bryophytes*, published in 1927.

From the 1900s to the 1960s, researchers working on soil biology devoted much time to studies that described the interactions between climate, soils and plant communities (Fig. 1.2). This pioneering work contributed to a deeper understanding of ecosystem ecology, succession and community dynamics in terrestrial ecosystems. Predator–prey studies provided new insight into the relationship between pests, pathogens and plants, which is fundamental for the identification of biological control agents. The use of radioactive and stable tracers in field experiments since the 1950s has allowed researchers to quantify the transfer of energy and nutrients (especially carbon, nitrogen, sulfur and oxygen) within the boundaries of an ecosystem, and between systems (e.g. from terrestrial to aquatic systems). These studies are at the heart of complex landscape, regional and global-scale models that can help us understand how basic processes such as primary production and nutrient cycling are affected by land management, thus leading to predictions about world food production, greenhouse gas emissions and eutrophication of waterways. Questions about the importance and role of species diversity led to texts such as Wallwork's 1976 book *The Distribution and Diversity of Soil Fauna* and the development of the soil foodweb concept, a framework for describing the interactions among plants, soil microorganisms and larger soil fauna. The importance of the soil biota and how they may help to provide ecological services in terrestrial ecosystems and buffer climate change is a major preoccupation of many soil ecologists at present. While much of the early work on soil ecology focused on natural ecosystems such as forests and grasslands, contemporary research in soil ecology does not differentiate between natural ecosystems and those that are managed intensively by humans, such as urban and agricultural ecosystems.

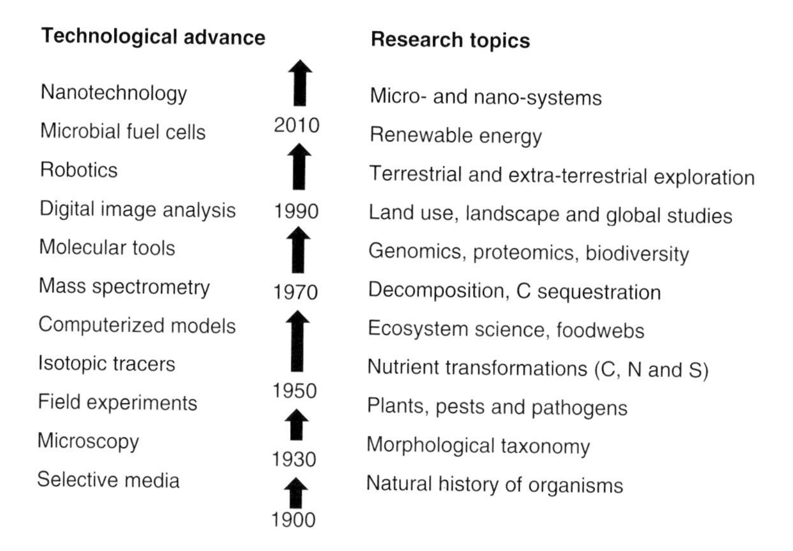

Fig. 1.2. A brief history of the technological advances that have led to significant breakthroughs in soil ecology research since the 1900s. (Modified from Wall *et al.*, 2005.)

Soil ecology is distinguished from other fields of soil science because soil ecologists attempt to understand the interactions among soil physical, chemical and biological components that permit the soil to support the ecological services described in Table 1.1. Soil ecology is the study of the interactions among soil organisms, and between the biotic (living) and abiotic (non-living) components in the soil environment. It is particularly concerned with how living creatures influence decomposition and nutrient cycling, soil aggregate formation and soil porosity, the dispersal and vitality of pathogens, and the diversity of biological communities. Soil ecologists strive to make connections between the above-ground system and the below-ground system containing the soil environment, soil organisms and transient animal residents, as well as plant roots.

There are many challenges in assessing the function of the below-ground system because soils are heterogeneous and dynamic. If we excavate soils for analysis and to collect the creatures living within the soil, we disrupt the soil structure and porosity, which introduces bias to our measurements. If we consider the soil as a 'black box' and simply measure inputs and outputs to an undisturbed soil, we may not be able to gather information on changes in the soil environment and the population dynamics of soil organisms or changes in the soil foodweb community during the measurement period. Devising techniques to account for the spatial heterogeneity in the soil environment, which occurs vertically within the soil profile and horizontally through ecosystems and landscapes, and evaluating the temporal changes in soils related to the activity of soil organisms and plant roots remains a constant preoccupation of soil ecologists.

1.2 Soil-forming Factors

Understanding and making predictions about the ecological functions of soils in any terrestrial ecosystem requires that researchers are able to describe the soils that they study. The fundamentals of soil genesis, which describes how soil is formed from parent material (rock), and soil classification were established by the Russian scientist V.V. Dokuchaev in the 1870s. Dokuchaev devised the genetic classification of soils, which was further refined by Russian, European and American scientists, and identified the five major factors that control soil formation, namely:

1. The nature of the parent material (chemical and mineralogical composition).
2. The climate (especially temperature and precipitation).
3. The influences of organisms – flora, fauna and humans.
4. The topography of the area (relief).
5. The length of time over which soil formation had occurred.

The importance of the soil-forming factors was emphasized by H. Jenny (1941), who stated that 'soils are dynamic natural bodies having properties derived from the combined effect of *climate* and *biotic activities*, as modified by *topography*, acting on *parent materials* over periods of *time*'. Most terrestrial ecosystems have mineral soils, which arise from the weathering (disintegration and solubilization) of rocks and minerals, so we will begin by evaluating the importance of the parent material in soil formation.

Soil formation from sedentary parent material

If we examined the soils in a particular climatic zone, such as those under tallgrass prairies in the central USA, we would observe soils in various stages of development (Fig. 1.3). When bare rock (solum) is weathered, the first soils formed are called Entisols, which possess a thin A horizon overlying the bedrock (C horizons). Wind-dispersed seeds and organisms such as bacterial and fungal spores are deposited on the surface and begin to grow. With time, the continued weathering of the parent material and decomposition of organic matter from the vegetation leads to the development of Mollisols and Alfisols. These have three to four distinct layers, the A, B and C horizons. If these soils formed in a forested area, they may also possess an organic O horizon.

The chemical and mineralogical composition of the parent material exerts an important influence on the soil texture and the depth of the soil horizons. Sandstones with a high quartz content will produce sandy soils, while soils developed from limestone and shale will have a finer texture. Granite and rhyolite are igneous rocks with the same chemical composition, but rhyolite has smaller mineral grains because it was cooled more rapidly after extrusion. Thus, rhyolite will weather more slowly and produce soil with a finer texture than one produced from granite.

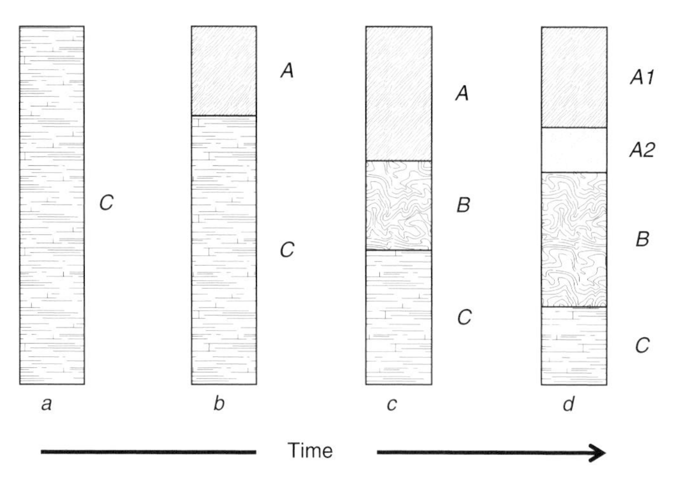

a. Parent material	*b.* Young soil (Entisol)	*c.* Mature soil (Mollisol)	*d.* Old soil (Alfisol)
Original material before soil development begins.	Thin A horizon, from which carbonates have been leached. Organic matter accumulation in A horizon. Minimal weathering and eluviation.	High organic matter content in A horizon. Moderate clay accumulation in the B horizon. Slightly acidic. Potential for high agricultural production.	Moderately acidic, heavily weathered and much less organic matter than mature soil. Leached topsoil layer (A2 horizon). Clay accumulation in B horizon can form a clay pan.

Fig. 1.3. A summary of the stages in the development of soils in the central USA under tallgrass vegetation. The Entisols are azonal soils because they do not have a fully developed profile. The Mollisols and Alfisols are zonal soils with a full profile (A, B and C horizons). (Adapted from Foth, 1990.)

Soil formation from transported parent material

A large geographic area with a zonal soil may possess small pockets of other soil types, known as interzonal soils. The presence of interzonal soils may indicate that there is some variation in the geology of the sedentary parent material or soil-forming factors in a particular part of the landscape if the interzonal soil developed from the underlying rock. It is also possible that the interzonal soil was formed from transported materials. There are four types of transported materials that can contribute to soil formation, namely: (i) colluvial debris; (ii) water-deposited (alluvial) material; (iii) glacier-deposited material; and (iv) wind-transported material.

Colluvial debris is transported by gravity, such as rock fragments that fall from the mountain slopes and cliff faces due to frost action. Avalanches generally carry a considerable quantity of colluvial debris. This material is often coarse and stony, although some medium- to fine-textured minerals (loess) are found at the base of some slopes. Soils of colluvial origin represent a small fraction of soils formed from transported materials.

Alluvial soils are formed from sediments deposited during floods. Sediments may originate from the edge of a waterway or be soil particles that were transported to streams and rivers during a heavy rainfall or snowmelt event. Wind erosion can also carry particles far from their point of origin, and some may enter waterways. When a flooding river overflows its banks, the flow velocity of the suspended particles declines and sediments are deposited. Coarse sands and gravels settle along the banks of the river, while silts and clays are carried further inland and deposited in the flood plain.

Materials that are carried to the river outlet, whether it be a lake, gulf or other water body, tend to be deposited in a delta. When drained, many river deltas serve as highly productive agricultural land, as seen in the deltas of the Nile, Po, Tigrus, Euphrates and Mississippi rivers. Another type of alluvial soil is that formed by marine sediments. These are generally sandy, but may also be clays that were weathered and eroded, swept into the ocean and weathered further by tidal action before they were deposited above sea level.

The glaciations of northern Europe, Asia, North America and some parts of South America, New Zealand and Australia during the Pleistocene Ice Age contributed to the formation of certain soil types. The movement of continental ice sheets pushed against the soil surface, gathering large rock fragments that scraped and ground against the rock floor as they passed. When the ice melted, all of the debris within the melting ice sheet fell to the ground. In some areas, we find an unsorted heterogeneous mass of boulders, rocks, sand, silt and clay called till, which was deposited in ridges or hills called moraines. Coarse-textured sediments, mostly rocks and sand, were transported by the melting ice into outwash plains. Finer silt and clay particles were carried further from the melting ice, settling into the lacustrine plains (Fig. 1.4).

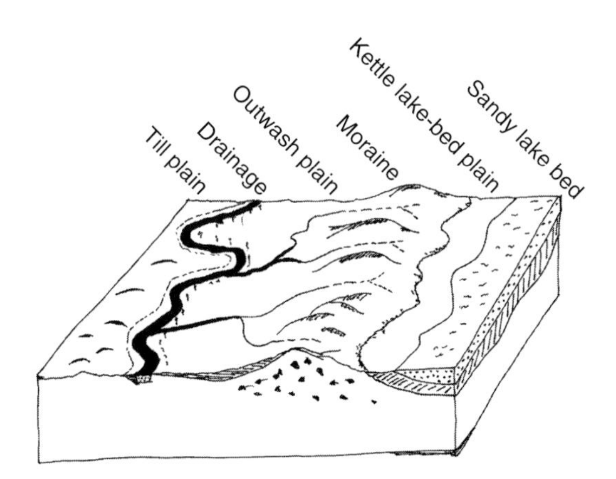

Fig. 1.4. A generalized view of physiographic features in a glaciated area. The till plain is separated from the moraine by an outwash plain. A border drainage developed while the ice front was at the moraine line and melted rapidly as it advanced to deposit the morainic material. To the right, a glacial lake (kettle lake) has receded to leave a plain of heavy sediment (till) partially covered with lacustrine sands. (Adapted from Foth, 1990.)

Aeolian soils are those formed from materials that were transported by wind, including: (i) fine sand that may be collected on low swells or steep ridges, like the sand dunes along the coastline of the Atlantic Ocean or the sand dunes of the Sahara desert; (ii) volcanic ash; and (iii) loess, a silty material with a characteristic yellowish buff colour, which originates from quartz, feldspar, mica, hornblende and augite minerals. The source of this material is often a glacial outwash deposit containing finely ground rocks, although some loess comes from non-glaciated rocks in arid climates (e.g. deep loess deposits in southern China originate from deserts in northern China). When dry, the deposit is highly susceptible to wind erosion and is blown some distance downwind from the source.

The soils formed from sedentary and transported parent material can be seen in Fig. 1.5. The landscape is near the prairie–forest transition and trees occur mostly in the protected, steep-sided valleys on Downs and Fayette soils. These soils are formed on a loess of glacial origin on land with slight to steep slopes (2–15%) and are stabilized by forest vegetation. Erosion has removed loess from the steeper, non-forested slopes so the soils at the margins of the Tama soil series are developing from the underlying glacial till rather than the loess. The Garwin soil is developing in a depression where runoff water from adjacent areas accumulates and will be the most heavily leached soil because the water table is far below the surface. The Wabash Judson complex contains alluvial soils and colluvial debris that has fallen from the valley walls into this depression. The spatial variability of soil characteristics in this landscape is affected by variation in the parent material, topography, drainage and vegetation.

Climate

Climatic conditions are perhaps the most influential of soil-forming factors, because temperature and precipitation affect the mechanical and chemical processes that lead to soil formation and the development of the soil profile. The thin mineral soil of an Entisol found in an arid desert is distinct from the deep, organic soil formed under grassland (Mollisol), an acidic, partially leached forest soil profile (Podzol) or a heavily weathered soil from the humid tropics (Oxisol).

Climate has a profound influence on the weathering of rock into smaller fragments and eventually

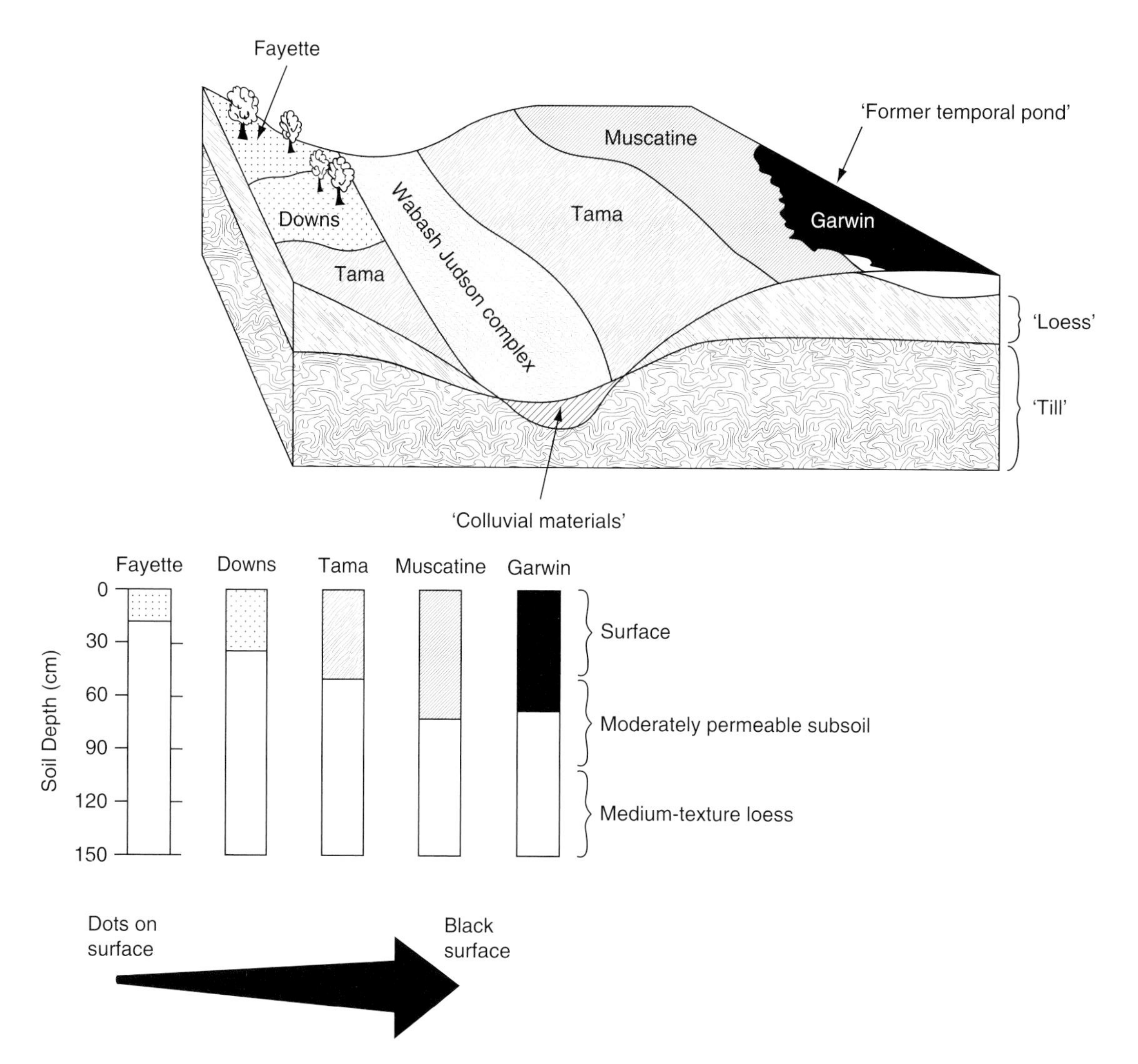

Fig. 1.5. Association of soils in Iowa. Note the relationship of soil type to parent material, vegetation, topography and drainage. Two Alfisols (Fayette and Downs) and three Mollisols (Tama, Muscatine and Garwin) are shown. (Adapted from Brady and Weil, 2008.)

into soil minerals. Mechanical weathering, which causes disintegration of the rock, is affected by: (i) temperature – expansion, exfoliation and frost action; (ii) erosion and soil deposition by water, ice and wind; and (iii) plants and animals.

Temperature fluctuations cause heating and cooling, which exposes rocks to differential stresses and eventually forms cracks and rifts in the rock. If the external surface of the rock heats more dramatically than the internal rock, the differential stress causes the surface layer to peel away from the parent mass, a process known as exfoliation. When water moves into these cracks and freezes, it creates a great deal of pressure, which widens cracks and dislodges mineral grains from rock fragments.

Rain falls on the land with a force sufficient to disrupt and dislodge small particles from the sedentary rock mass. Water moving from terrestrial ecosystems into streams and rivers will inevitably pick up and transport small rocks, minerals and soil particles. In running waters, these particles are mixed and pounded against one another. If you have ever picked up a stone from the bottom of a stream or examined a boulder at the seaside, you will notice how rounded and smooth these rocks are, due to the constant movement of water and impact of abrasive particles contained within the water. Similarly, ice and wind are also important agents that abrade, erode and deposit soil materials from their place of origin to other parts of the landscape.

Some mechanical weathering is attributed to plants and animals. Plant roots enter fissures and disrupt the structural integrity of the rock, while burrowing animals move and reorganize loose rocks and stones. However, the influence of plants and animals on soil formation is minor when compared with the forces of temperature, water, ice and wind. The distribution of plants and animals is strongly influenced by climate, as discussed below.

Chemical weathering involves a series of reactions that decompose small rock fragments into the constituent minerals, including quartz sand, silicate clays, iron and aluminium oxides and soluble ions such as K^+ and Ca^{2+}. Water falling on the soil surface and leaching through the profile accelerates chemical weathering. Carbonic acids such as HCO_3^- are formed when water reacts with calcite ($CaCO_3$). Other acids found in the soil solution, including dilute nitric (HNO_3) and sulfuric (H_2SO_4) acids, come from acidic rainfall. Plant roots and microbial cells also secrete H^+ ions as well as dilute organic acids such as oxalic acid and citric acid. An example of the reaction of H^+ ions with a feldspar mineral (anorthite) shows the displacement of Ca^{2+}, creating an unstable acid silicate that is prone to further dissolution and subsequent reactions that eventually form silicate clays:

$$\underset{\text{Anorthite (solid)}}{CaAl_2Si_2O_8} + 2\,H^+ \rightarrow \underset{\text{Acid silicate (solid)}}{H_2Al_2Si_2O_8} + Ca^{2+} \quad [1.1]$$

Oxidation is another important chemical reaction. The conversion of reduced ferrous (Fe^{2+}) oxide to the ferric (Fe^{3+}) form is illustrated below:

$$\underset{\text{Olivine}}{3\,MgFeSiO_4} + 2\,H_2O \rightarrow \underset{\text{Serpentine}}{H_4Mg_3Si_2O_9} + SiO_2 + \underset{\text{Ferrous oxide}}{3\,FeO} \quad [1.2]$$

$$\underset{\text{Ferrous oxide}}{4\,FeO} + O_2 + 2\,H_2O \rightarrow \underset{\text{Goethite}}{FeOOH} \quad [1.3]$$

In these reactions, the hydration of olivine produces serpentine, quartz (SiO_2) and ferrous oxide, which is rapidly oxidized to ferric oxide (goethite).

Organisms

Plants are often considered to have a predominant influence on soil formation. In fact, plant communities often depend on soil microorganisms for survival. If we examine bare rock and desert sands, we find that they are colonized by a complex group of mosses, lichens and cyanobacteria, referred to as 'soil crusts'. Lichens are a symbiotic association between fungi and algae, both considered to be soil microorganisms. Cyanobacteria are another group of soil microorganisms capable of photosynthesis and N_2 fixation, which is beneficial to the mosses and lichens within the soil crust. Soil formation may begin when dust from aeolian deposits is captured by this biological community, binding to extracellular polysaccharides secreted by living organisms or released from dead biomass. These organisms also produce organic acids that contribute to the chemical weathering of rocks.

In semi-arid climates, the dominant vegetation is generally grasses, although some trees will be found in protected microsites, such as valleys where there is more water. Many grasses interact with an important group of nonsymbiotic N_2-fixing bacteria called *Azotobacter*. Together, this plant–microbial association leads to considerable primary production and deposition of organic residues on the surface and within the soil profile. Grassland soils generally have more organic matter, especially in the subsurface horizons, than forest soils. Consequently, grassland soils are characterized by their dark colour, high cation exchange capacity and water-holding capacity, and structural stability.

Soils developing under forests tend to have shallower horizons, and the organic matter tends to accumulate closer to the soil surface, but the type of soil depends on the vegetation and organisms present. In well-drained soils with sufficient calcium in the parent material, the litter of deciduous trees falls to the surface and is thoroughly mixed with the underlying mineral soil by earthworms and other soil fauna. Early soil ecologists such as P.E. Muller (1879) and Charles Darwin (1881) recognized that soil organisms were responsible for the burial and decomposition of leaf litter, the process in which dead organic residues are degraded, releasing CO_2 and leaving a stable fraction called humus. Earthworms are notable for their ability to mix and incorporate the surface litter into the mineral soil horizons. When earthworms are present, mineral particles become coated with humus, water leaches more readily through the soil profile and soils may develop into Alfisols or other types of 'mull soils' (Fig. 1.6). In forest sites with low calcium, the leaf litter does not become mixed with the mineral layer but remains in a mat on the surface. The L/F horizon includes intact leaves, twigs and woody material (L materials) and partially decomposed litter that has been colonized by filamentous fungi (F materials). The H horizon is decomposed organic matter whose origin cannot be distinguished (Fig. 1.6). The acidic soils developed in these forests are Spodosols, also referred to as 'mor soils'. As shown in Fig. 1.6, the soil foodweb is dominated by fungi, mites and collembola in a mor soil.

Human activities should not be overlooked as factors influencing soil formation. The removal of natural vegetation from forest clearing or the cultivation of grasslands changes the soil-forming factors abruptly. Likewise, irrigation of semi-arid or arid lands, or agricultural activities such as fertilization and liming may alter the course of soil formation in an area. Humans have contributed to nitrogen deposition, acid rain, global warming and climate change since the Industrial Revolution, which is expected to have a significant impact on soil-forming processes in some areas.

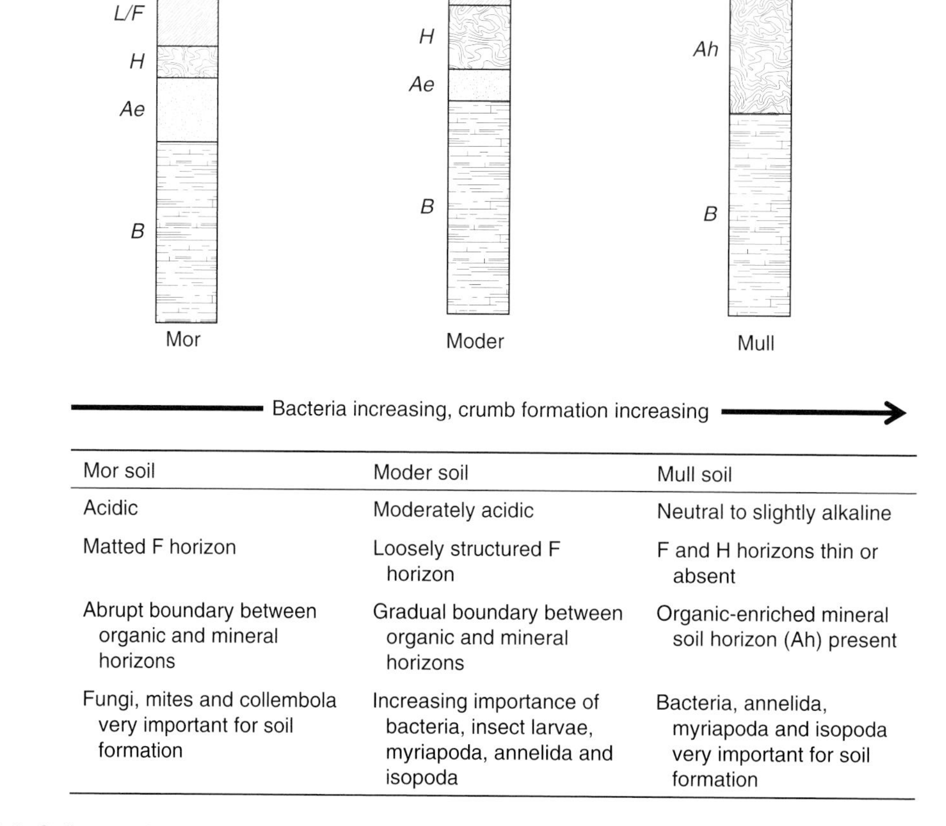

Mor soil	Moder soil	Mull soil
Acidic	Moderately acidic	Neutral to slightly alkaline
Matted F horizon	Loosely structured F horizon	F and H horizons thin or absent
Abrupt boundary between organic and mineral horizons	Gradual boundary between organic and mineral horizons	Organic-enriched mineral soil horizon (Ah) present
Fungi, mites and collembola very important for soil formation	Increasing importance of bacteria, insect larvae, myriapoda, annelida and isopoda	Bacteria, annelida, myriapoda and isopoda very important for soil formation

Fig. 1.6. Soil categories, pH and soil organisms responsible for soil formation in forests. (Adapted from Foth, 1990.)

Focus Box 1.2. Soil protection and soil conservation policies.

Soils play an essential role in regulating the global carbon cycle, support food production, water purification and other ecosystem services. However, land use change and soil degradation is on the rise worldwide and is becoming critical in some regions (Fig. 1.7). The root of the problem seems to be an idea that soils can resist disruptions caused by human activities. Yet, this is not the case. Cultural and historical choices about land use, combined with growing human population and mechanization, have led to our current landscape configuration. Within these landscapes, soils may be susceptible to degradation due to soil erosion, salinization, compaction and contamination from agricultural and industrial activities. Careful management is necessary to sustain or improve a soil's capacity to produce crops, support biodiversity, accumulate atmospheric carbon dioxide and degrade pollutants, but we must not lose sight of the fact that soils are a non-renewable resource. There are a limited number of highly fertile soils and many fragile soils in the world that must be protected.

Some soil scientists have proposed that soils be protected in the same manner as endangered species. The first step would be to classify soils as rare, endangered or protected, depending on the risks to soil health, their rarity and their ecological and economic value. For example, Gelisols are under threat because global warming could lead to the disappearance of the permafrost layer, which covers approximately 9% of the Earth's ice-free land area, within 1000 years. This would result in the extinction of an

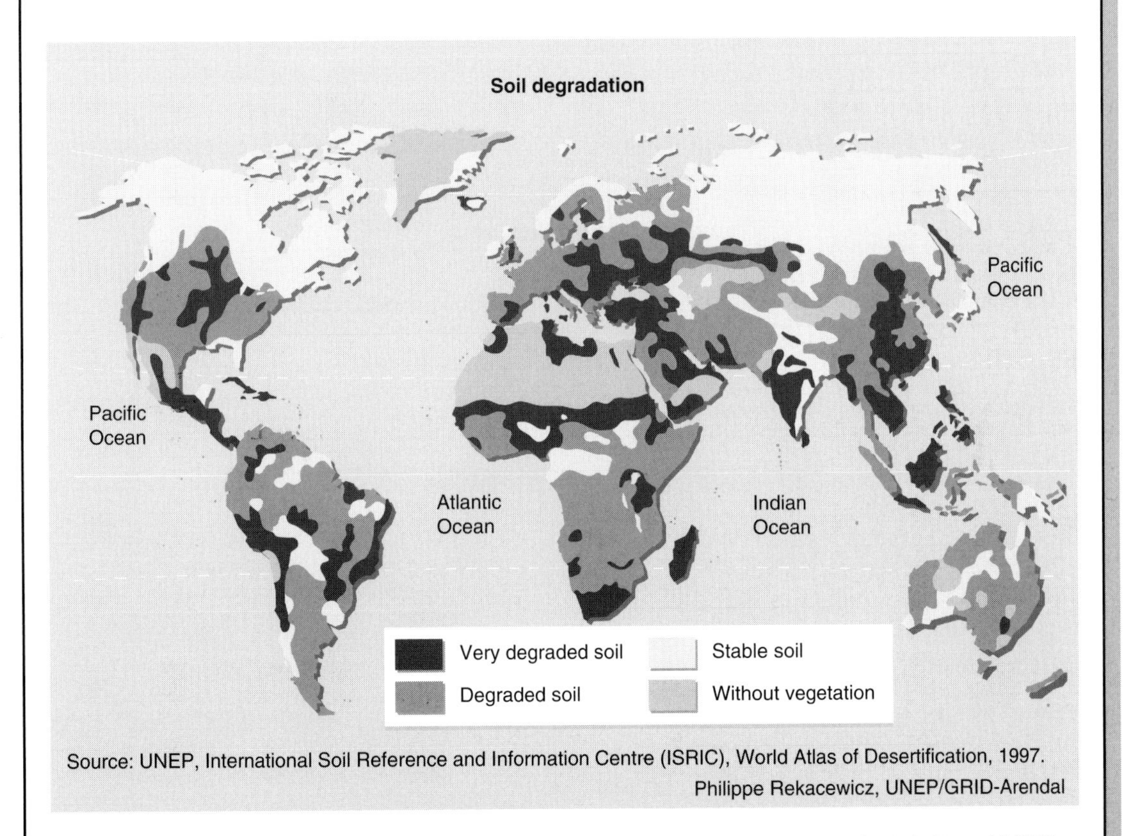

Fig. 1.7. Global map of soil degradation according to the *World Atlas of Desertification* (1997). (From UNEP/GRID-Arendal, Degraded soils, UNEP/GRID-Arendal Maps and Graphics Library, author Philippe Rekacewicz: http://maps.grida.no/go/graphic/degraded-soils/ (accessed 4 February 2009).)

Continued

Focus Box 1.2. *Continued*

entire soil order from soil taxonomy and its affiliated biotic assemblage (microorganisms, plants and wildlife), biological and metabolic biodiversity.

In spite of the global magnitude of the problem of soil degradation, there is no international policy framework to guide sustainable soil management. Lindsay Stringer (2008) wrote an article titled 'Can the UN Convention to Combat Desertification guide sustainable use of world's soils?' (Stringer, 2008), where she argues that the United Nations should take leadership and legislating practices that will improve soils' ability to conserve biodiversity, store atmospheric carbon, produce agricultural crops and support human well-being. Delaying to take action now will increase the cost for future generations to restore these and other essential soil services (Table 1.2).

Today, the major challenge is to develop soil conservation policies that are consistent with modern-day needs for fuel, fibre and food. Soil degradation is a global problem, but it is also a complex problem because the underlying reasons for soil degradation vary in every part of the world. The climate, geomorphology, soil fertility, cultural practices, land use management and human decisions in a region affect the health of our soils. International agreement about the importance and value of soils would be an excellent starting point that would lead to support for local soil conservation agencies working to solve regional soil degradation problems (Table 1.3).

Table 1.2. Consequences of soil degradation due to contamination, erosion, declining organic matter, salinization and landslides. (Based on the European Commission, 2006.)

Property depreciation	Loss of fertility and crop yields
Insurance costs	Higher agricultural production costs
Cost of installing protective barriers/infrastructure	Cost of increased food safety controls
Flood damage to property and infrastructure	Increased cost related to the release of greenhouse gases related to loss of organic carbon
Degradation-induced income loss (e.g. loss of trade in agricultural produce, less tourism)	Increase in health-care costs
Sediment removal, treatment and disposal costs	Loss of human life and well-being

Table 1.3. Key supranational political initiatives to conserve land and soil (1980s to present). (Adapted from Stringer, 2008.)

International agency/ organization	Year	Strategy	Key aim(s)
FAO/UNEP	1982	World Soil Charter and World Soils Policy	To establish a set of principles for the optimum use of the world's land resources, for the improvement of their productivity and for their conservation for future generations
UNEP	1994	United Nations Convention to Combat Desertification	To combat desertification in countries experiencing serious drought and/or desertification, particularly in Africa, through effective action and an integrated approach consistent with the goals of sustainable development, supported by international cooperation and partnership agreements
FAO/World Bank	1996	Soil Fertility Initiative	To address the widening food gap in developing regions due in part to poor soil quality
Tutzing Initiative (private sector)	1998	Convention on Sustainable Use of Soils	To move toward the sustainable use of all kinds of soils by all states of the Earth to preserve soil functions and uphold the objectives of the Convention on Desertification

Continued

Focus Box 1.2. *Continued*

Table 1.3. Continued

International agency/ organization	Year	Strategy	Key aim(s)
Alpine Convention of 1991	1998	Soil Protocol on the implementation of the Alpine Convention	To preserve soil sustainability and allow it to perform its natural functions
IUCN (World Conservation Union)	2000	Endorsed proposal to investigate the feasibility of a global convention for soil	To evaluate the capability and suitability of existing global conventions to provide adequately for the ecologically sustainable use of soils
IUSS	2002	Global Soils Agenda	IASUS working group established to examine emerging global soil initiatives
EU	2002	Soil Thematic Strategy (approved 2006)	To prevent further degradation of soil, to preserve and restore degraded soil to a level that enables at least its current or intended use, considering the cost of restoration
IUCN/Iceland declaration for a soil protocol	2005	Draft Soil Protocol under the UN Convention on Biological Diversity	The conservation/protection of soil and its sustainable and equitable use.

Source: Stringer (2008)

Topography and spatial heterogeneity

Topography, or relief, affects soil formation by influencing the microclimate, plant growth, hydrology, drainage and transport processes in the landscape (Fig. 1.8). Runoff, evaporation and transpiration are controlled by the presence of knolls, slopes and depressions in the landscape. Rolling topography encourages natural erosion and deposition processes at the soil surface. In wet depressions like the swamp shown in Fig. 1.8, organic matter will accumulate because the decomposition process is retarded by anaerobic conditions associated with poor soil drainage. The swamp will also receive organic matter and soil minerals eroded from adjacent slopes. In such areas, we find organic soils such as mucks and peats, which are classified as Histosols. The mineral soils under well-drained forest and grasslands are probably yellow, red or brown, indicating good aeration and oxidizing conditions, while the mineral soil under a Hisotol is often grey or bluish grey due to chemical reduction (gleyed B horizon).

Soil ecologists recognize that soil spatial heterogeneity, whether it be vertically in a soil profile or horizontally across the landscape, is of great importance. Any general statement that we would like to make about soil functions in terrestrial ecosystems is affected by the depth of the surface soil and characteristics of the underlying soil horizons, as well as the variety of soils found in a geographical area. The realization that soils are spatially heterogeneous helps us to understand why crop growth and water infiltration are not always uniform across an agricultural field or why certain tree species and understorey plants grow in one area of a forest, but not elsewhere. Information about soil characteristics and their distribution in the landscape also helps us to make decisions about land use for the long term (decades, centuries), such as:

- Where should roads, railways and airports be built?
- What land should be left as natural forests and grassland, and which lands have the potential to support agricultural production?
- Which areas are suitable for residential/commercial building sites and waste disposal?

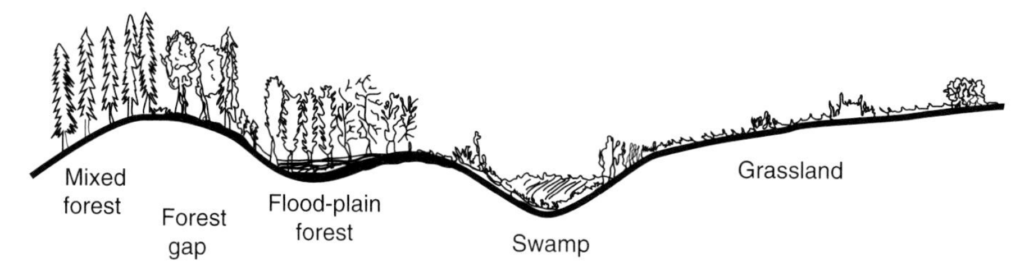

Fig. 1.8. Variation in vegetation and topography affecting soil formation. (Based on Tate, 1992.)

To answer these questions requires information about soil formation processes and soil characteristics at a macroscale (across watersheds and ecosystems). Yet, the management of our soils requires that we understand the short-term processes (daily, monthly, yearly) that occur from the activity of soil biota, which range in size from a few micrometres to perhaps 20 cm. The soil environment controls the growth and survival of soil organisms, and is instrumental in sustaining terrestrial ecosystems.

Focus Box 1.3. Spatial variation in soil properties.

Environmental heterogeneity is a key factor explaining major ecological properties such as biodiversity, productivity and competition (Ettema and Wardle, 2002). When we look at soils, they often appear to be homogeneous, but actually the small-scale spatial heterogeneity in the soil conditions is probably responsible for the diversity and distribution of plant species found in grasslands, forests and other natural ecosystems. Recent studies reveal that soil properties in field experiments usually exhibit a non-random spatial structure to some extent, with more similarity in values that come from sampling points that are close together than those that are far apart. When spatial heterogeneity occurs and near neighbours are more similar than far neighbours, we say that the data are autocorrelated.

This fact is of major relevance for soil scientists because it means that the requirement of data independence in standard parametric statistics is violated. Researchers have been fighting against spatial autocorrelation with specific experimental designs such as complete and incomplete block designs. Sometimes the block design absorbs the spatial variation pattern, but sometimes the spatial autocorrelation persists within blocks, and in those cases the use of conventional statistics for the analysis of spatially correlated observations may be disastrous. The more intuitive way for checking if a variable shows spatial autocorrelation is using semivariograms (Fig. 1.9). Semivariograms are simple figures that synthesize the spatial structure of a set of data in a particular site, representing the amount of variation existing between each pair of observations at different positions (the 'semivariance') as a function of the distance that separates them.

For a randomly distributed soil property that is not autocorrelated we would expect little change in the semivariance with increasing distances, and semivariograms would be essentially flat (Fig. 1.10). For spatially patterned data, we would find that the semivariance tends to be lower at closer proximity, that is closer pairs of data are more similar than distanced pairs. Such a diagnosis should be included in all protocols for field experimentation. In the case of detecting a non-random spatial pattern, the researcher is obliged to use spatial analysis techniques. Nowadays, several computer software applications permit researchers to incorporate the spatial autocorrelation structure of data into the statistical model (SAS, S-plus, ASREML, etc.), to correct the values of the dependent variable by the spatial autocorrelation.

Gallardo *et al.* (2000) found that the spatial distribution of soil organic matter and mineral nitrogen occurred at the same scale as the oak tree canopy, so soil properties were likely associated with biotic processes such as litterfall, root growth and turnover, and soil biological activity (Fig. 1.11). As result of these biotic interactions, the spatial pattern of nutrients cycled through biological processes, such as carbon and nitrogen, would differ from those elements that are recycled through biological and geological processes, such as phosphorus, and those under strict geological control, such as lithium (Gallardo, 2003).

Continued

Focus Box 1.3. *Continued*

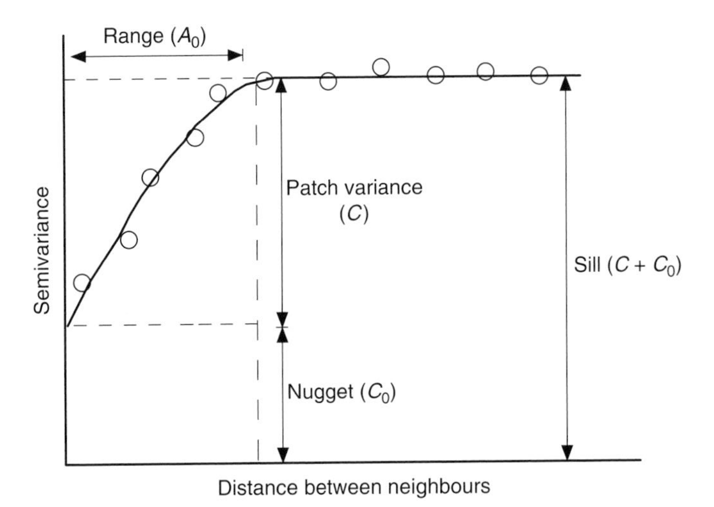

Fig. 1.9. Generalized semivariogram showing the proportion of variance (semivariance) found at increasing distances between paired soil samples (distance between neighbours; open dots) and the modelled function. The semivariogram summarizes the structure of the spatial autocorrelation with three parameters of the function. The range (A_0), the distance to the asymptote, indicates the patch size at which the data are spatially dependent. The sill is an estimate of the total population variance, which comprises a variance component explained by the model (patch variance C), and some variance due to sampling error and/or spatial autocorrelation at scales not accounted for by the model (nugget).

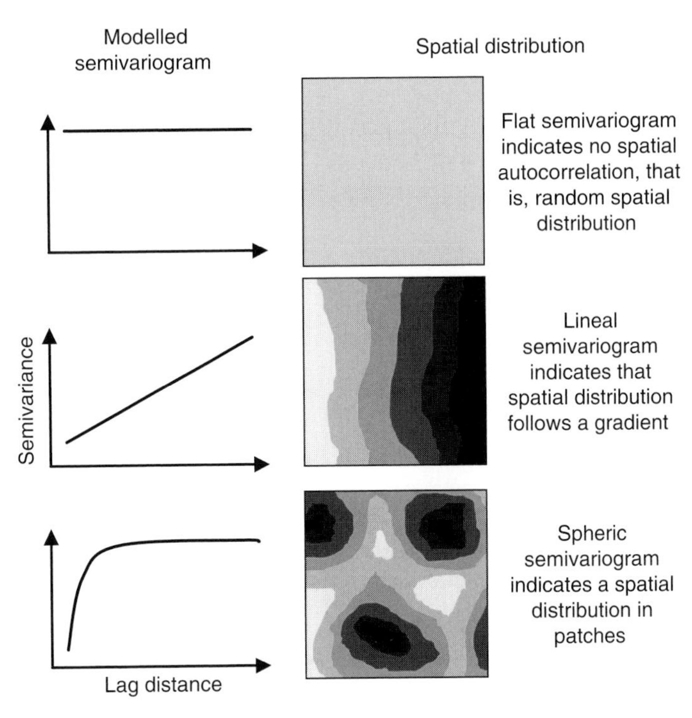

Fig. 1.10. Some spatial distribution patterns usually found in soils, and the corresponding semivariance models. (Based on Ettema and Wardle, 2002; Zas, 2008.)

Continued

Focus Box 1.3. *Continued*

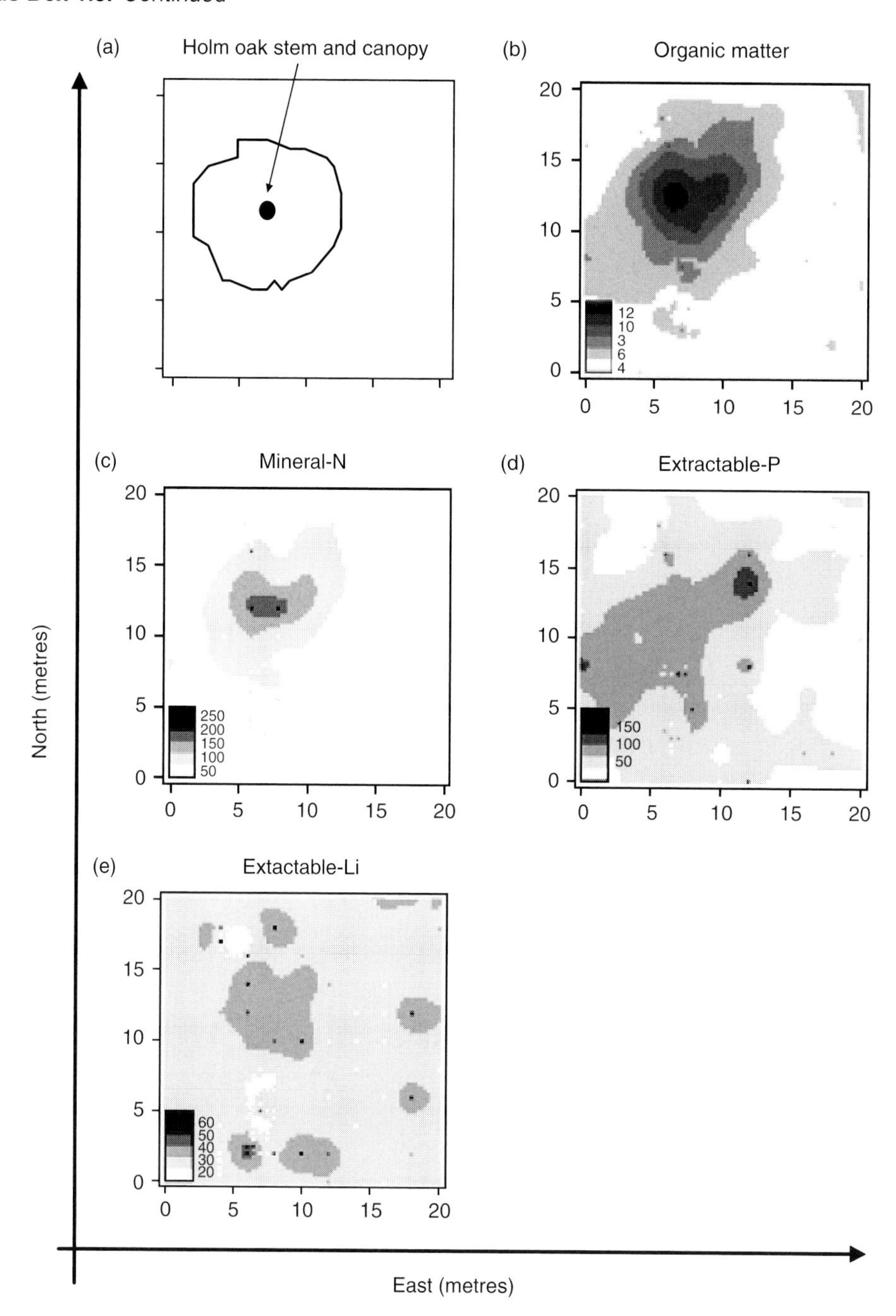

Fig. 1.11. Maps of soil organic matter content (b) and plant-available nutrients, including mineral nitrogen (c), extractable phosphorus (d) and extractable lithium (e), around a holm oak (*Quercus ilex*) in a Spanish Dehesa ecosystem. The maps were obtained by kriging interpolation methods. The approximate position of the oak (a) corresponds to the area with the highest soil organic matter and N content, but was not related strongly to the P and Li distributions. Darker colours indicate greater nutrient concentrations. (Modified from Gallardo, 2003.)

1.3 The Soil Environment

If you dig in the soil, you will see many things – plant roots, dead organic debris, soil crumbs, cracks and crevices. Sometimes small insects or earthworms can be observed. The soil often exudes a rich odour and feels moist. We can think of the soil environment as a system that contains solids, liquids and gases within a three-dimensional matrix. The solid components include rocks of various sizes, minerals and organic materials, both living and dead. The dead organic matter may be visible pieces of leaves, straw or roots, a source of food for soil biota. Soil organisms consume these materials and transform them into an organic material known as humus, which often binds to the soil minerals and forms aggregates. Most plant roots and soil organisms inhabit the pore spaces between dead organic matter, aggregates and soil minerals, although some of the soil bacteria live in tiny micropores within soil aggregates. The solid components could constitute about half the soil volume, and the other half is the pore space.

The pore space is occupied by water and air, and the volume occupied by each of these components is proportional. Thus, when water enters the soil due to rainfall, snowmelt or irrigation, air is forced out of the pores. When water is removed by growing plants, evaporation or drainage, the air-filled pore space increases. We will begin by discussing the soil water, due to the central role of water for all living creatures.

Soil water

It is rare to find a natural soil system with no water, because water molecules are strongly adsorbed to soil particles through electrostatic bonds (positively charged hydrogen atoms in the water are bound to negatively charged clays and organic matter). This is called hygroscopic water, a layer of water several molecules thick that is not available to plants and most soil organisms. The capillary water binds through hydrogen bonds to the hydroscopic water layer (Fig. 1.12). Capillary water constitutes the majority of water in water films around soil particles and contained in micropores (less than 10 μm in diameter). About two-thirds of the capillary water can be accessed by plant roots. The water films that form around soil particles are vital to the survival of soil microorganisms and other tiny animals that live in micropores.

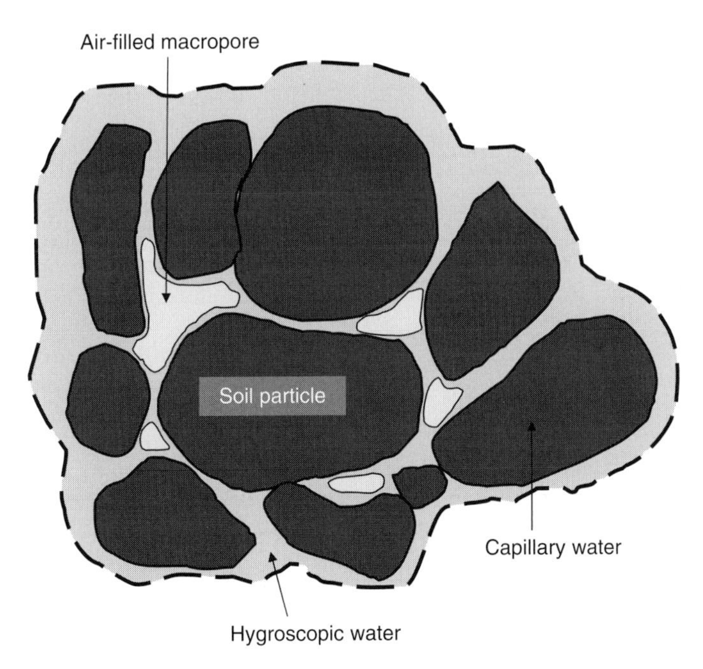

Fig. 1.12. Relationship of hygroscopic and capillary water with respect to soil particles and air-filled macropores. (Redrawn from Foth and Turk, 1972.)

The water in macropores (greater than 10 μm in diameter) is called gravitational water. Typically, gravitational water flows freely, although it can be loosely bound to soil particle surfaces under low tension and drains through the soil profile quickly, due to the force of gravity. When soil pores are completely filled with water, the soil is saturated, but once the free-flowing gravitational water has drained, the soil is unsaturated and at field capacity. The presence of gravitational water excludes air, which is necessary for root respiration and for the metabolic activities of most soil organisms. The effect of soil saturation on millet root growth is shown in Fig. 1.13.

Plant roots, especially the root tips, have a high oxygen requirement and tend to suffocate in saturated soils because oxygen diffusion is about 10,000 times slower through water than through air. Most crop plants do not grow roots in saturated soil because there is not enough oxygen to support their development. Yet, there are some plants that have specialized adaptions to tolerate soil saturation. Paddy rice does well in flooded soils because it absorbs oxygen through the stem and transports it to the root through the aerenchymous tissue, and thus the plant does not rely on soil oxygen for root functions. The aerenchyma also removes ethylene from the root zone, which is advantageous because high concentrations of ethylene interfere with root elongation. Mangroves and bald cypress survive in flooded environments because they produce aerial 'breathing' roots known as pneumatophores that transfer oxygen to submerged parts of the root.

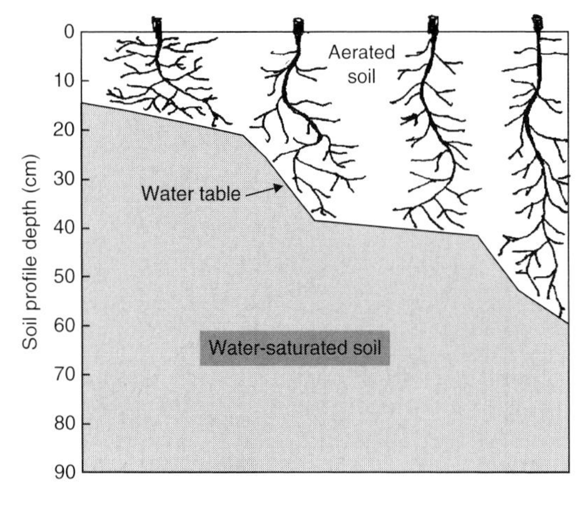

Fig. 1.13. The effect of depth to water table on root length. The roots of most agricultural crops will not enter the saturated soil. (Adapted from Foth and Turk, 1972.)

Similarly, most soil organisms require oxygen and function optimally when soils contain 40–60% water-filled pore space. Yet, some soil microorganisms thrive in the anaerobic conditions of a saturated soil, which can be expressed as the soil redox potential. Soil conditions can range from highly reduced to highly oxidized, corresponding to redox potentials of −300 mV to 900 mV. As the oxygen concentration in the soil water diminishes, other electron acceptors are used by microorganisms, and oxygen-based respiration is substituted by fermentative processes and respiration based on other electron acceptors:

1. At 250 mV, NO_3^- is reduced to N_2O and N_2 gases by denitrifying bacteria.
2. At 200 mV, Mn^{4+} is reduced to Mn^{2+} by manganese-reducing bacteria.
3. At 100 mV, Fe^{3+} is reduced to Fe^{2+} by iron-reducing bacteria.
4. At −150 mV, SO_4^{2-} is reduced to S^{2-} by sulfur-reducing bacteria.
5. At −400 mV, CO_2 is reduced to methane by methanogens.

All of these reactions may occur in soils, although the most common are nitrate, manganese and iron reduction. Sulfur-reducing bacteria are important in salt marshes, bogs and swamps. When all of these electron acceptors have been exhausted, methanogens belonging to the domain Archaea will metabolize hydrogen and fermentation by-products left by other anaerobic microorganisms and produce methane gas.

Soils can be saturated temporarily, after a heavy irrigation or rainfall event, or for prolonged periods of time due to flooding and improper drainage that causes the soil to remain waterlogged. Anaerobic conditions are noticed very soon after the soil becomes saturated because plants and microorganisms will rapidly deplete the O_2 reserves. Anaerobic conditions impair plant growth because:

- Root and microbial respiration increases the CO_2 concentration in soil. There is a negative feedback between the soil CO_2 concentration and root functions. Depending on the plant species, cellular metabolism can be impacted when the CO_2 level reaches 1 to 5% by volume in the soil water and pore space surrounding roots. In anaerobic soils, photosynthesis, respiration and mineral uptake are reduced, and inhibitory compounds (ethylene, acetaldehyde) are produced.

- The O_2 deficiency limits the cellular metabolism of plant roots and obligate aerobic microorganisms that are beneficial for plant growth, such as mycorrhizal fungi and N_2-fixing bacteria.
- Metabolic by-products released by anaerobic bacteria, including CO_2, H_2S, CH_4 and fermentation substances, may be toxic to plants and other soil organisms.

Once the gravitational water has drained from soil pores, the soil is said to be at field capacity (Fig. 1.14). About half of the pore space is filled with water at field capacity, although this varies depending on soil texture. At field capacity, the soil water tension is low and plant roots can easily absorb capillary water. As the soil becomes drier, water is more tightly bound in water films, the soil water tension rises and water flows less readily through the pore space to roots. Eventually, the plant will absorb less water than it loses through transpiration and wilting occurs, unless the plant has some special adaptation that can minimize water loss through transpiration. If watered, a plant that has wilted will often recover from this stress and continue growing. The permanent wilting point is defined as the point that a wilted plant cannot regain turgor when placed in a saturated atmosphere. The wilting point for plants in a loam-textured soil is about −1.5 MPa, corresponding to about 15 bars of tension, 25% of pore space occupied by water and less than 10% gravimetric soil moisture content (Fig. 1.14). These values vary with soil texture, porosity and environmental conditions. Soft-bodied soil organisms, such as earthworms and molluscs, are equally sensitive to fluctuations in soil moisture, but soil protozoa and microorganisms can tolerate drier soil conditions as low as −6 MPa.

The amount of water available for plant growth is related to the soil surface area, because water adheres to soil particles, and the volume of pore space. Thus, the water-holding capacity is related to soil texture as well as soil aggregation. Fine-textured (clay) soils have the greatest water-holding capacity, while coarse-textured (sandy) soils have the lowest.

Soil water contains a variety of dissolved and suspended materials that nourish plants and

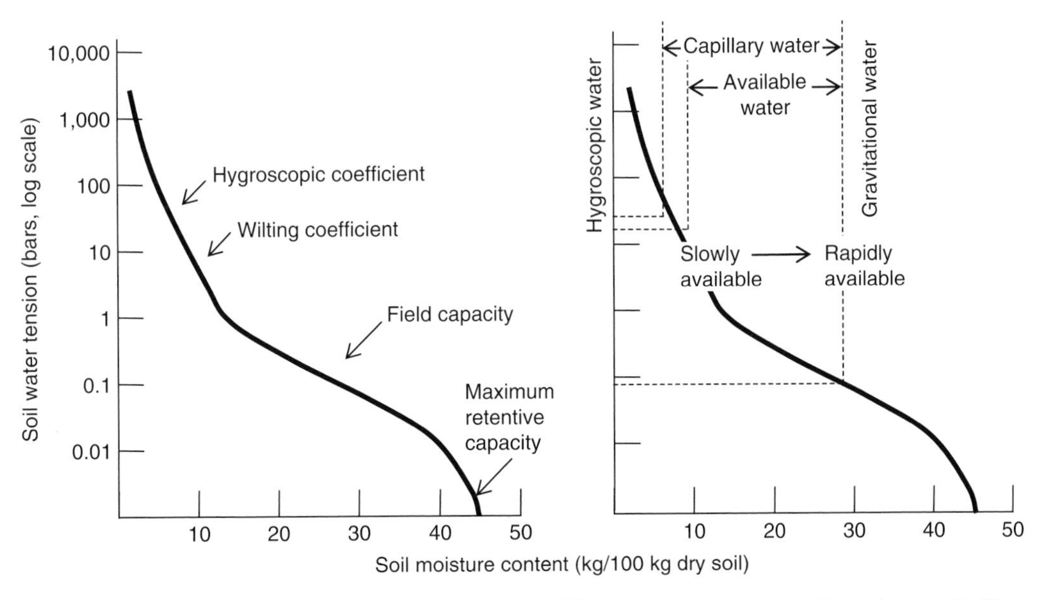

Fig. 1.14. Tension–moisture curve of a loam soil as related to different terms used to describe water in soils. The tension–moisture values corresponding to hygroscopic water, capillary water, available water and gravitational water are approximate. (Adapted from Brady and Weil, 2008.)

soil creatures. Water flowing to plant roots contains dissolved ions that are essential for the growth of living organisms, such as nitrate (NO_3^-) and calcium (Ca^{2+}). Dissolved organic materials, including carbon compounds, small proteins and amino acids, plant growth hormones and vitamins, are found in the soil water. Suspended clays also move through macropores with the gravitational water. Ions, organic compounds and soil microorganisms adsorbed to the surface of suspended clays will also be transported through the soil profile with free-flowing gravitational water. The soil water also contains dissolved gases such as carbon dioxide (CO_2) and oxygen (O_2).

Soil air

Most of the gases within the soil environment are found in the water-free space of soil pores. The soil air is a mixture of gases and water vapour, and the composition is quite distinct from that found in the Earth's atmosphere, as illustrated in Table 1.4.

The importance of O_2 for root growth cannot be overemphasized. In well-aerated soil with good structure and drainage, it is rare that the O_2 content within the soil will limit plant growth. As shown in Table 1.4, the O_2 concentration in a well-aerated soil may be 18.0–20.5%, which is slightly less than the O_2 concentration in the atmosphere.

As the soil bulk density increases (i.e. greater mass of solid particles per unit volume, perhaps due to compaction by heavy machinery), it becomes more difficult for roots to penetrate the soil profile. The O_2 diffusion rate is limited, which further slows root development and growth. Periodic flooding can also reduce the O_2 supply to plant roots, restricting root growth and ion uptake by the plant until the flood waters recede.

The CO_2 concentration in a well-aerated soil may range from 0.3 to 3.0%, which is higher than atmospheric concentrations due to respiration from plant roots and soil organisms. The CO_2 concentration is greater in certain biological 'hot spots', such as near elongating root tips and around fresh organic residues that are being decomposed by soil animals and microorganisms. The CO_2 respired by these organisms diffuses away from areas with high CO_2 content to pores with low CO_2 content, eventually leaving the soil. The flux of CO_2 and other gases (N_2O, CH_4) from soils forms part of the greenhouse gas emissions from terrestrial ecosystems.

Soils containing 5–10% CO_2 or greater CO_2 concentrations are close to anoxic and may not contain enough O_2 to support the metabolism of plant roots, soil fauna and obligate aerobic microorganisms. The rate of O_2 diffusion to living organisms should be equal to or greater than the rate of CO_2 diffusion away from the plant roots and soil biota for aerobic metabolism to continue. In reality, gas diffusion is

Table 1.4. Globally-averaged atmospheric and soil concentrations (% by volume of dry air) of the major gases and some trace gases. (Modified from Coleman and Crossley, 2003.)

Gas	Atmospheric concentration	Range of soil concentrations	Comments
N_2	78.084	78.084	Little variation
N_2O	3.2×10^{-5}	< 0.8	Anaerobic soil
NH_3	4.2×10^{-4}	< 0.7	High concentrations in soils receiving NH_4-based fertilizers
O_2	20.948	18.0–20.5	Aerobic soil
		< 2.0–10.0	Anaerobic soil
O_3	6×10^{-5} to 0.0015	–	Peak concentrations in the stratosphere
CO_2	0.0376	0.3–3.0	Aerobic soil
		5.0– > 10.0	Anaerobic soil
CO	1.2×10^{-5}	7.0×10^{-5}–15×10^{-5}	
CH_4	1.8×10^{-4}	$< 6.4 \times 10^{-4}$	
Ar	0.934	–	
H_2	5×10^{-5}	5×10^{-5}–7×10^{-5}	
SO_2	2×10^{-8}	–	
H_2S	2×10^{-8}	–	

not uniform throughout the soil profile. Some of the factors that affect gas diffusion include:

- Pore size and the number of pores – soil pores may be a few micrometres to a few centimetres in diameter. Fine-textured soils have many small pores, while coarse-textured soils have fewer, large pores.
- Pore distribution – there may be discontinuities in the pore structure due to vehicle traffic (compacted zones) or the activities of burrowing animals, including earthworms.
- Air-filled pore volume – gases move more rapidly through the air-filled pore space and much more slowly through water-filled pores.

Soil permeability describes the speed at which air or water moves through soil pores, ranging from very slow to very rapid (Table 1.5). In a well-drained, aerobic soil, it is possible to find anaerobic microsites where O_2 and CO_2 diffusion is limited. Therefore, biological reactions mediated by aerobic and obligate anaerobic microorganisms may occur simultaneously in the same soil profile.

Table 1.5. Soil permeability and percolation classes, with water infiltration rates.

Classes	Permeability (cm/h)	Percolation (min/cm)
Very slow	< 0.10	> 3050
Slow	0.1–0.5	760–3050
Moderately slow	0.5–2.0	190–760
Moderate	2.0–6.4	60–190
Moderately rapid	6.4–12.7	30–60
Rapid	12.7–25.0	15–30
Very rapid	> 25	< 15

Soil minerals

Soils are formed from the weathering of rocks, which is achieved through physical breakdown and chemical weathering. Water enters cracks within the rocks, expanding and contracting as it freezes and thaws, thus causing the mechanical breakdown of rocks into primary particles – sand, silt and clay. This process is accelerated by acidic radicals found in the soil environment, coming from rainfall and organic acids and released by soil microorganisms and plant roots.

Soil texture is determined by measuring the relative proportions of sand, silt and clay. Many important physical and chemical reactions in soils are governed by the texture, because it determines the surface area upon which these reactions occur. Table 1.6 shows the physical characteristics and surface area of various constituents.

Since the sand separates are relatively large particles with small surface area, their main function is to facilitate the passage of air and water through pore spaces. Silt particles are generally coated with a thin layer of clays, so they exhibit some plasticity, cohesion (stickiness) and adsorption, but much less than found in clays and colloidal clays (Fig. 1.15).

Sand and silt particles are considered to contribute little to chemical reactions due to their low solubility and lack of absorptive surfaces. The majority of soil physical and chemical reactions occur on clay surfaces. Clay particles are small, platy and highly plastic when wetted. Their size ranges from 0.2 to 2.0 μm in diameter. The colloidal clays are those with a diameter smaller than 1 μm. Clays have a greater surface area than other soil

Table 1.6. Some characteristics of soil particles. (From Foth and Turk, 1972.)

Primary particles	Diameter (mm)	Number of particles (per g of soil)	Surface area (cm^2/g of particles)
Very coarse sand	2.0–1.0	90	11
Coarse sand	1.0–0.50	720	23
Medium sand	0.50–0.25	5,700	45
Fine sand	0.25–0.10	46,000	91
Very fine sand	0.10–0.05	722,000	227
Silt	0.05–0.002	5,776,000	454
Clay	< 0.002	90,260,853,000	8,000,000

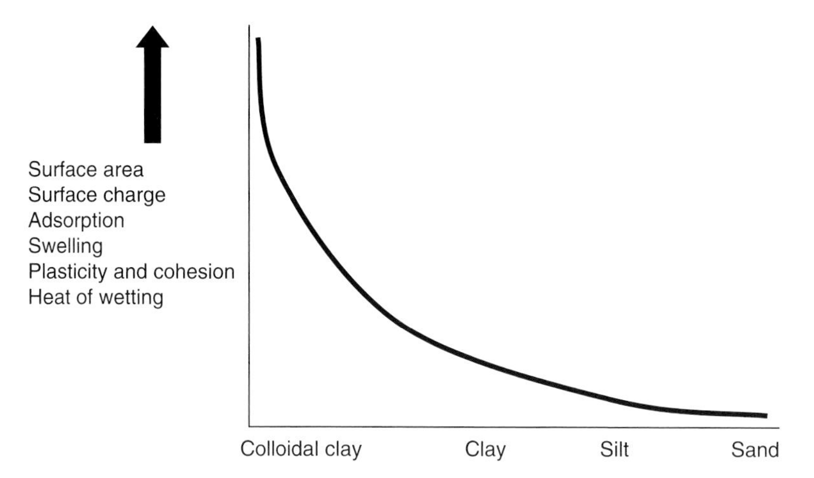

Fig. 1.15. The finer the texture of a soil, the greater the effective surface exposed by its particles. Note that adsorption, swelling and the other physical properties cited follow the same general trend and that their intensities go up rapidly between the clay and colloidal clay fractions. (Modified from Brady, 1984.)

particles, but not all clays are equally reactive. The least reactive clays are the oxide clays, primarily iron oxides such as hematite (Fe_2O_3) and aluminium oxides such as gibbsite ($Al(OH)_3$). These clays predominate in highly weathered soils such as those in the humid tropics. Soils dominated by oxide clays form extremely stable soil aggregates and have a lower degree of plasticity than other clays. In acidic soils, these clays confer a net positive charge, which promotes flocculation but reduces the capacity of the soil to attract and adsorb cations such as H^+, K^+, Ca^{2+} and Mg^{2+}. Consequently, these cations may be susceptible to leaching from acidic, highly weathered soils.

The silicate clays include kaolinite, illite, montmomorillonite, smectite, vermiculite and chlorite. These are known as layer silicates because they are composed of individual layers, consisting of planes of silicon and aluminium (or magnesium) cations sandwiched between oxygen and hydroxyl ion planes. Layers appear as alternating tetrahedral and octahedral sheets, separated by an interlayer of water and adsorbed cations.

The mineralogical organization of silicate clays can be classified into three groups, and is summarized in Table 1.7:

1. 1:1 clays containing one tetrahedral (silica) sheet and one octahedral (alumina) sheet. The lattice is fixed and does not expand when the clays are wetted, so no water or cations enter between the structural layers. Kaolinite is an example of a 1:1 clay.

Table 1.7. Comparative properties of three major types of silicate clays. (Modified from Brady, 1984.)

Property	Kaolinite	Smectite	Illite
Size (μm)	0.1–5.0	0.01–1.0	0.1–2.0
Shape	Hexagonal crystals	Irregular flakes	Irregular flakes
Specific surface area (m^2/g)	5–20	700–800	100–120
External surface	Low	High	Medium
Internal surface	None	Very high	Low
Cohesion, plasticity	Low	High	Medium
Swelling capacity	Low	High	Medium
Cation exchange capacity (cmol (+)/kg)[a]	3–10	80–120	15–40

[a] Centimoles of positive charge/kg of clay

2. 2:1 clays contain one octahedral sheet between two tetrahedral sheets. Smectite and vermiculite are expanding-type minerals while illite is a non-expanding mineral. The expanding-type minerals are noted for their high plasticity and cohesion and their marked shrinkage on drying. Soils with a high proportion of expanding 2:1 clays exhibit marked cracking upon drying and form hard clods that can be difficult to till. Non-expanding minerals become hydrated, swell and shrink, but much less intensely than the expanding-type minerals.
3. 2:1:1 clays are ferro-magnesium silicates that do not absorb water, but they have considerable surface charge that provides cation exchange capacity and surface reactions, similar to illite. Chlorites are common 2:1:1 clays.

Soil structure

Texture is based on the proportions of primary soil minerals (sand, silt and clay) in a given sample, while soil structure refers specifically to how the primary particles are bound together into clusters known as aggregates or peds. The three categories of soil structure we will consider here are:

1. Particulate structure with no interlinking between particles (i.e. a sandy soil).
2. Massive structure with moderately to tightly bound particles. In this structure, silica particles are held together by cementing agents such as iron and aluminium oxides or calcium carbonate. The resulting material is extremely hard and difficult to break. Massive structure is seen in duripans, which are cemented by silica, iron oxide or calcium carbonate, as well as hardpans, ironpans and claypans.
3. Fragmentary structure contains micro- and macro-aggregates created from clay particles coated with silt particles and bound to sandy particles. The aggregates can range in size from less than 0.25 mm to more than 1 cm in diameter.

Most agricultural soils have a fragmentary structure, although hardened layers can be found deeper in the profile. Aggregation is an important characteristic of such soils. Unlike soil texture, which does not change appreciably in our lifetime, aggregation is a dynamic soil characteristic. Aggregates are formed when clay particles bind with organic materials, such as particulate organic matter encrusted with microbial by-products (extracellular polysaccharides) and earthworm mucus. They are broken apart primarily by mechanical forces such as wind, water and machinery. Soil structure is discussed in more detail in Chapter 10.

Soil organic matter

Organic matter has a fundamental role in aggregate formation, but the organic matter is more than a binding agent to hold together the soil mineral particles. Soil organic matter is defined as the plant, animal and microbial residues found in soils, both decomposed and undecomposed. This includes dead microbial cells, organic materials secreted from microbial cells and plant roots, dissolved organic substances in the soil solution, dead root hairs, fungal hyphae and animal corpses. We also consider the living biomass to be part of the soil organic matter, including the soil bacteria and archaea, fungi, nematodes, mites and earthworms (Fig. 1.16).

The contribution of soil organisms to aggregation and soil structure is very important, but their role in nutrient cycling through terrestrial ecosystems is perhaps even more vital. Carbon cycling is controlled by photosynthesis, the process by which plants obtain CO_2 from the atmosphere, and decomposition, the breakdown of dead plants and other organic residues that returns CO_2 to the atmosphere. Many organisms in the soil foodweb are saprophytes that consume dead organic residues to obtain the energy necessary for their survival. During the decomposition process, these organisms also transform organically bound nitrogen, phosphorus and sulfur into inorganic ions that can be absorbed by plants and animals. These topics are discussed further in Chapters 8 and 9.

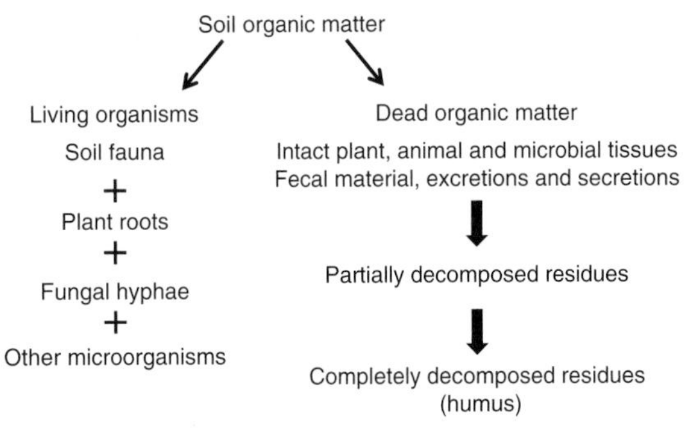

Fig. 1.16. Components of the soil organic matter.

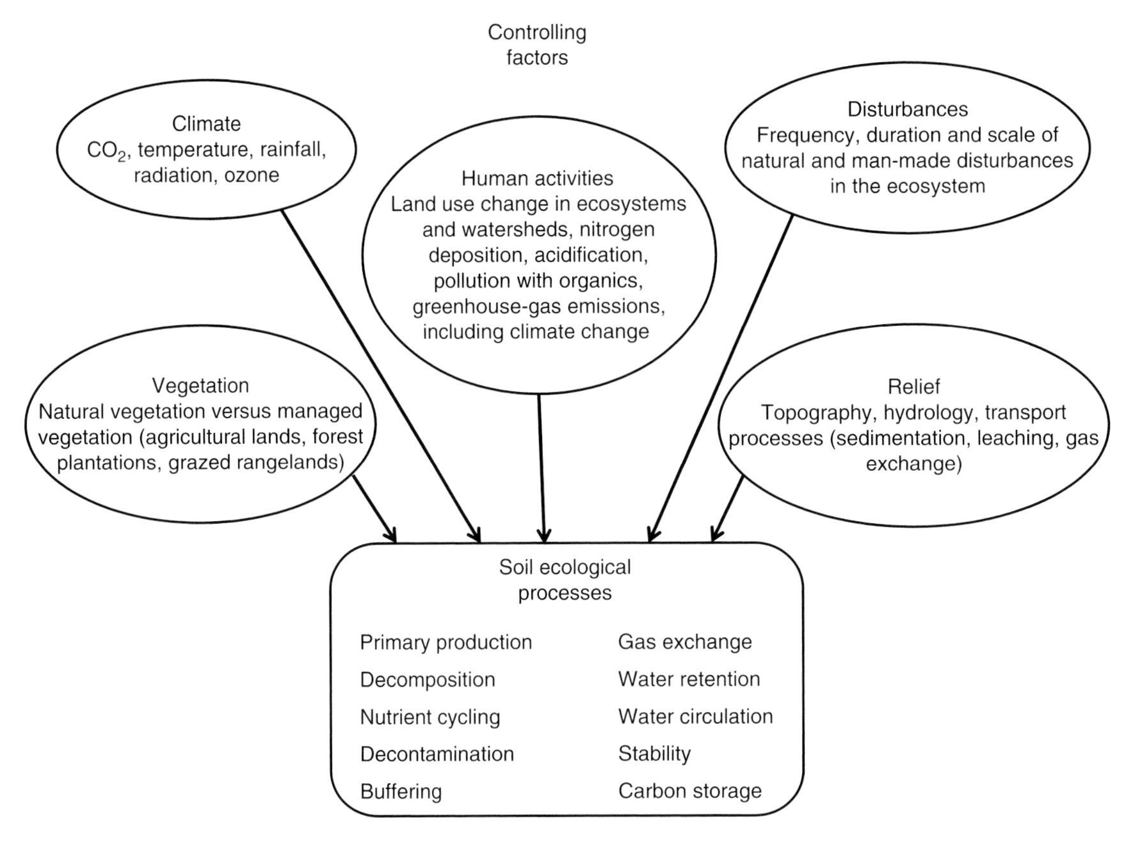

Fig. 1.17. Human activities, climate, vegetation, disturbances and relief are major factors controlling the soil ecological processes that are essential for our survival. It should not be forgotten that human activities are having a profound effect on climate, vegetation and relief and are responsible for much of the disturbance in ecosystems and landscapes. Based on the concept of Jenny (1980).

1.4 Conclusions

Soils provide a broad array of vital ecosystem services, but there is considerable variation in the ability of soils to support primary production, decompose organic materials, adsorb pollutants and filter water. The formation of the diverse soils found around the world has occurred through geological time and is influenced by factors such as parent material, climate, organisms and topography. However, recent changes in land use and climatic conditions induced by human activities are having a profound effect on soil ecological processes (Fig. 1.17). A deeper understanding of how human intervention affects the integrity and functioning of the soil is required to sustain the ecological services of this vital resource.

Further Reading and Web Sites

United States Department of Agriculture – Natural Resources Conservation Service. Available at: http://soils.usda.gov/technical/classification/ (accessed 5 February 2009).

World Resources Report 2008 released in Oslo. GRID-Arendal/UNEP (a collaborating centre of the United Nations Environment Programme (UNEP)): http://www.grida.no/

PART II

Soil Organisms and the Soil Foodweb

Soil organisms range in size from 1–2 μm to 10 cm or more in length. Scientists who study them may be soil microbiologists, if their focus is on microscopic creatures such as viruses, bacteria and fungi, or soil zoologists who are interested in the nematodes, arthropods or annelids that live in the soil. Some of these organisms have received much attention due to their evident importance to agricultural crops, such as the nitrogen-fixing bacteria that form symbiotic associations with leguminous plants. Yet, we have limited knowledge of the natural history and population dynamics of many soil organisms, especially some of the smaller, cryptic insects and non-culturable soil microorganisms.

Soil ecologists are interested in the entire suite of microscopic and macroscopic creatures that are found in the soil foodweb, although they recognize that some are less well studied than others. The diversity of the soil biota, as well as the interactions between soil organisms and their natural environment, lead us to ask many questions – What do they eat? Where do they live? How do they interact with each other and how do they contribute to plant nutrition and soil health?

Before answering these questions, which are central to understanding how soil organisms and the soil foodweb contribute to soil ecological and pedological functioning, we need first to understand the biology and ecology of the major groups of soil biota. The soil organisms can be divided into various groups, depending on their body length and where they live in the soil. The microorganisms are considered to be the Archaea, Bacteria (including some relevant groups such as Cyanobacteria and actinomycetes), fungi and algae found in natural soils. They are associated with clay and silt particles, and secrete extracellular polysaccharides and protein-rich compounds that bind inorganic materials in the soil structure. The microfauna are protists and nematodes that live in water films surrounding soil particles and in small micropores. Mesofauna are soil-dwelling arthropods and annelids with a body size of from 100 μm to 5 mm that inhabit and move through macropores in the soil profiles. The macrofauna range in size from less than 1 mm to more than 50 mm in length. These organisms move through soil fissures, root channels or create their own burrow systems in search of food. The macrofauna include top-level predators in the soil foodweb as well as detrivores that consume dead organic residues and mix these with soil minerals, contributing to decomposition and soil aggregation processes.

2 Soil Microorganisms

2.1 Biology of Soil Microorganisms

The soil microorganisms encompass four general groups: bacteria, archaea, fungi and algae. Viruses can be considered with this group because of their comparable size and interactions with soil microorganisms. New insights into the phylogeny of these groups have come from DNA sequencing and the taxonomy continues to be revised. Although hundreds of genera have already been described, it is likely that many more remain to be discovered. Thus, the description below is designed to give a general overview of this group, with examples of a few genera or species to illustrate key points.

Viruses

Viruses range in size from 0.01 to 1 μm long, and are acellular structures composed of a central nucleic acid core (DNA or RNA) with a protein covering. They are widely distributed in temperate and tropical soils and have been detected in the most extreme environments on Earth, including soils from the cold, dry valleys of Antarctica and from the Namib desert in south-western Africa, which is believed to be the oldest desert on Earth (80 million years old).

Viruses are internal parasites of plants, animals, bacteria and fungi, meaning that they are not active when found outside a living animal. However, some are stabilized and persist in soil until they encounter a host organism. Among the plant viruses, tobacco mosaic virus of the genus *Tobamovirus* is notable for its ability to persist in soil – it can even survive heat-steam sterilization (soil pasteurization in an autoclave). Viruses in infected plants are often transmitted by soil nematodes and fungi. As discussed further in Chapter 3, some nematodes are plant parasites that penetrate the root with a stylet to consume sap and cellular cytoplasm. If a virus is ingested by the nematode, there is a good possibility that it will be transported and injected via the stylet when the nematode parasitizes a non-infected plant.

Land-spreading of animal manure, sewage sludge and wastewater introduces a group of viruses that cause human and mammalian diseases into soils. The enteroviruses belonging to the family Picornaviridae are single-stranded RNA viruses such as poliovirus, coxsackie, echovirus, porcine and bovine enteroviruses. Once they enter soil, enteroviruses are generally inactivated within a period of days to weeks. Livestock should not be allowed in pastures that have received manure, sewage sludge or wastewater for a period of 2 months (cows and horses) to 6 months (goats, sheep and swine). Mixing manure and other wastes into the soil can hasten the inactivation of enteroviruses.

Naturally occurring viruses affect the soil microbial community. Bacteria are vulnerable to viruses called bacteriophages or phages. The most common soil phages are the tailed phages, grouped into Myoviridae with a contractile tail (100–300 nm), Siphoviridae with a long, non-contractile tail (150–200 nm) and Podoviridae with a short, non-contractile tail (less than 50 nm). These viruses have a characteristic shape, illustrated in Fig. 2.1.

In tailed viruses, the icosahedral heads are attached to the tail by a connector. When they encounter a host bacterium, the tail section of the phage binds to receptors on the cell surface and injects the DNA into the cell. Once inside, the genes are expressed from transcripts made by the host's ribosomes. The genome is replicated by the use of concatamers, in which overlapping segments of DNA are generated and assembled to form the whole genome. Bacteriophage infections that cause bacterial mortality (the infected bacterium bursts) are called lytic infections. Some bacteriophage persists inside the bacterial cell, reducing the bacterial function without killing the cell. This

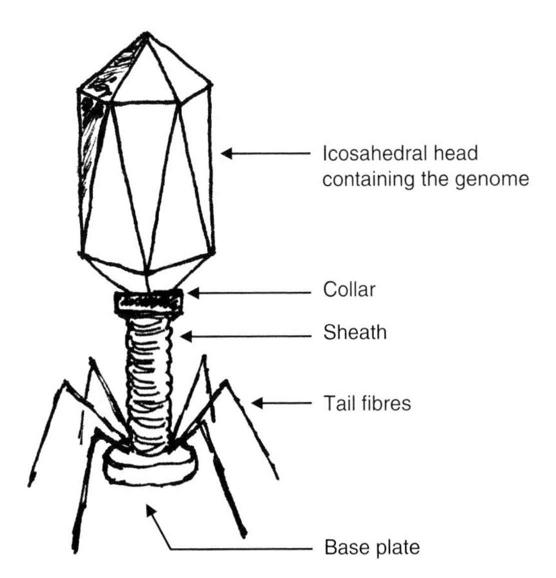

Fig. 2.1. Representative bacteriophage in the Myoviridae morphotype. Phages possess an icosahedral head that contains genetic material. The genetic material of Myoviridae is double-stranded DNA with 1.8–200k base pairs. The size of these bacteriophages ranges from about 20 to 200nm.

is called a lysogenic infection and the infected bacteria are known as lysogens. Eventually the lysogen dies and the cellular contents, including phages that replicated inside the bacterium, are released into the soil solution.

The coexistence of bacteria and phages in soils of all terrestrial ecosystems suggests that they co-evolve. Resistant hosts appear soon after bacteria are exposed to the phages. This leads to an 'arms race' between resistant bacterium and the corresponding phage, in which the bacterium modifies the adsorption sites on its cell surface and the phage must change its binding proteins. In some cases, modification of the surface phage receptors can either confer an evolutionary advantage or reduce the ability of the bacteria to use substrates in their environment. Thus, bacteria–phage interactions can affect the composition and activity of soil biological communities.

In agricultural fields with legumes, a large number of rhizobiphages affecting *Rhizobium* spp. can be found. Although rhizobiphages can reduce nodule formation and yield of leguminous plants in some cases, resistant strains of *Rhizobium* soon appear in the nodules. Rhizobiphages seem to be host-specific at the genus level, rather than species level, meaning that any bacterium in the genus *Rhizobium* could be infected by a rhizobiphage. Soil environmental factors such as nutrient availability, temperature, soil moisture content, pH and soil structure affect phage–bacteria interactions. It is difficult to make generalizations because the effect of the soil environment on virus survival is not completely understood. In addition, soil bacteria are not always very active; encysted cells and bacteria with low metabolic rates are not easily infected by viruses. However, bacteria under nutrient limitation are more susceptible to lytic infections than those with sufficient substrates for growth.

Soil fungi can also be affected by viruses, typically double-stranded RNA viruses belonging to the families Narnaviridae, Hypoviridae, Partitiviridae, Reoviridae and Totiviridae. These mycoviruses are dependent on the fungal host for replication. Since they do not have an extracellular phase in their life cycle, their survival is not directly affected by the soil environment. They can be transmitted between related fungi when cytoplasm is transferred through the hyphal network or when fungal propagules such as conidia are released during sexual reproduction. In some cases, the mycovirus affects the fungal host dramatically; the chestnut blight fungus *Cryphonectria parasitica* had reduced growth, sporulation, enzyme activity and was less damaging to chestnut trees when infected by a virus (Kimura *et al.*, 2008). In other cases, the plant-pathogenic fungi *Nectria radicicola*, which causes ginseng root rot, and *Rhizoctonia solani*, the causative agent for 'damping off', were more virulent when they contained a mycovirus. This phenomenon is called hypervirulence, meaning that the plant disease is more severe when the plant is colonized by a virus-infected fungal pathogen than a pathogenic fungus without the virus. Genetic information from the virus affects biochemical and physiological processes that make the pathogenic fungus more infective (e.g. higher production of enzymes that damage plant cells, more rapid growth and sporulation). There is virtually no information on the effect of mycoviruses on non-pathogenic soil fungi.

Genetic diversity and ecological functions of soil viruses

There is scant information on the genetic diversity of soil bacteriophages and even less about soil mycoviruses, although about 100 million genotypes possibly exist in the world, based on estimates from fluorescence and electron microscopy. Fewer than 1% of these bacteriophages have ever been grown in culture, even though we may find 10^8 to 10^9 bacteriophages/g soil. Modern DNA sequencing of phages has led to the isolation of about 5400 bacteriophages (Kimura *et al.*, 2008). It appears that soil bacteriophages may be much larger and possess more genetic material (> 200 kbp) than was previously detected due to classical procedures for propagating bacteriophages and preparing soil samples for microscopy. A bacteriophage of *Bacillus thuringiensis* that had a similar morphology to the Myoviridae was recently isolated from a dry soil sample, except that its contractile tail was longer (about 480 nm), tail fibres were larger (about 10 nm diameter, compared with 2 nm in other Myoviridae) and it exhibited a different tail configuration to other isolates. Not only was the genome large (221 kbp), but the genes coding for structural components and energy generation were not previously reported for soil bacteriophages. The diversity of soil viruses could be much greater than the current estimates, perhaps because small-scale heterogeneity in the soil environment permits complex co-evolution of viruses and hosts. DNA-based technologies will be invaluable in exploring and cataloguing the biodiversity of soil viruses.

In marine and freshwater environments, viruses are recognized for their importance in biogeochemical nutrient cycles and as genetic reservoirs due to their widespread distribution, large numbers and great diversity. The death of phage-infected bacterial cells accounts for about 25% of bacterial mortality in oceanic waters and 58% of bacterial mortality in coastal waters, inducing the release of carbon, nitrogen, phosphorus and other nutrients that were contained in bacterial cells. Due to the large microbial biomass and the rapid turnover time (days to weeks) of phage-infected bacteria, viruses act as an important driver of the biogeochemical cycle in aquatic systems. Scant information is available for terrestrial ecosystems, although it is believed that soil viruses have a similar ecological role.

Archaea

Archaea are an extremely diverse group of organisms – they are found in every habitat imaginable, from the deepest thermal vents in the ocean to the top of Mount Everest, from the frozen tundra of the Arctic to the dry, cold valleys of Antarctica. Early studies focused on archaea living in extreme environments that would be deadly to most other bacteria and eukaryotes, such as salt marshes, hot sulfur springs, deep-sea vents and the rumen of cattle and other ruminant animals. Initially, this group was presented as extremophiles but nowadays their presence and ecological relevance are increasingly recognized in terrestrial, freshwater and marine ecosystems, where they may be essential actors in carbon and nitrogen cycles. Several species are able to fix CO_2 using the energy from the oxidation of inorganic substances. In this way, they are some of the most relevant primary producers not linked to sunlight on the Earth, such as those found in hot springs and in deep-sea thermal vents. General characteristics of archaea that live in extreme environments (methanogens, extreme halophiles, thermoacidophiles and hyperthermophiles) and the factors influencing their survival are listed in Table 2.1.

Discovered in 1977, the archaea are morphologically similar to bacteria, both probably diverging from a common prokaryotic ancestor billions of years ago (Fig. 2.2). Archaea and bacteria share some genes as a result of their common ancestry; however, the archaea have many capabilities that are completely unique. Archaea have several characteristics in common with Eukaryota, because the eukaryotic branch split off from Archaea, and not from Bacteria. One of the key differences is that archaea do not possess peptidoglycan, a structural compound, in their cell wall. Their rigid cell wall contains proteins and non-cellulosic polysaccharides and the cell membrane is made of branched chain lipids that enable them to survive extreme temperature, pH and salinity. However, they are sensitive to many antibiotics such as kanamycin, chloramphenicol and rifampicin that do not necessarily affect bacteria. At the molecular level, we find that the tRNA nucleotides of archaea do not contain thymine. The promoter structures and ATPases in archaea are also quite different from those of other organisms.

Some halophiles possess a photosynthetic membrane that captures solar energy through a protein pump using pigmented proteins such as bacteriorhodopsin (purple pigment). The protons are converted to chemical energy within the cell. This energy capture system is quite different from chlorophyll-based photosynthesis in bacteria and

Table 2.1. Diversity of archea living in extreme environments.

Archea	Characteristics	Factors affecting survival
Methanogens	Generate CH_4 when they oxidize H_2 gas to produce energy, using CO_2 or other substrates as the terminal electron acceptor	Require anaerobic conditions and CO_2, which is produced by associated microflora in the microbial community
Extreme halophiles	Found near salt lakes and brines. They produce pigments and appear as pink blooms in saltwater ponds	Halophiles exposed to strong light develop a purple-pigmented membrane that captures solar energy
Methane-generating thermophiles	Found near oceanic hydrothermal vents at depths of 2–3 km	Require anaerobic conditions. Survive temperatures of 100°C and 35 MPa pressure
Thermophilic extreme acidophiles	Grow in extremely hot, acid environments	Survive temperatures up to 80°C and acidity as low as pH 2
Sulfur- and sulfate-reducing hyperthermophiles	Obligate anaerobes that use S or SO_4^{2-} as a terminal electron acceptor, generating H_2S	Require anaerobic conditions
Sulfur oxidizers	Oxidize S as an energy source, using O_2 as a terminal electron acceptor to generate H_2SO_4	Require aerobic conditions

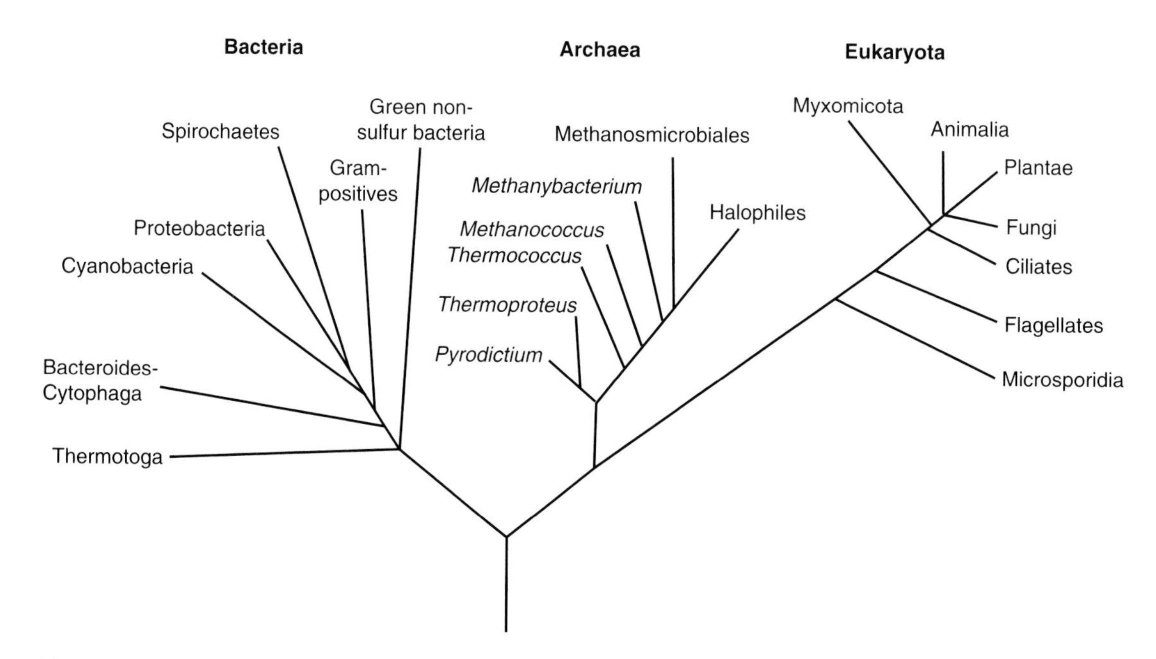

Fig. 2.2. A phylogenetic tree of life based on ribosomic RNA sequencing as proposed by Woese *et al.* (1990). The tree shows the three domains: Archaea, Bacteria and Eukaryota. The archaea are related both to the bacteria and eukaryotes. The names of archaea denote their physiological group (thermo-, pyro-, methano- and halo-). The major branches of bacteria are shown. The eukaryotes include unicellular organisms in Protista as well as multicellular eukaryotes in Plantae, Fungi and Animalia. Other eukaryotes that do not fit into the traditional five-kingdom system of classification are listed. The tips of the branches in the three domains represent different taxonomic levels. Several other phylogenies based on nuclear DNA are congruent with rRNA. Due to the extensive horizontal transfer of genes between some branches of bacteria and archaea, it has been suggested that a simple two-dimensional tree is not the best representation of all the genetic relationships.

plants, which has led some scientists to suggest that photosynthesis evolved at least twice, once in Archaea and once in Bacteria. Methanogens capable of nitrogen fixation were identified in 1984. Although the biochemical reactions that led to nitrogen fixation are the same in these archaea and bacteria, there are some differences in the transcriptional regulation of genes that control protein synthesis and generate the nitrogenase enzyme required for nitrogen fixation.

Genetic diversity and ecological functions of soil archaea

Since the late 1980s, molecular studies have revealed that archaea are found in ordinary environments as well as the extreme environments discussed above. Garden soils, forests, lake sediments and coastal waters are all favourable habitats for archaea. These archaea were not previously isolated with culture methods and thus very little is known about their genetic diversity. However, there is considerable divergence in the rRNA gene sequences that suggests physiological diversity and large populations of archaea, indicating that they make a major contribution to global biogeochemical cycles.

Most of the soil archaea belong to the phylum Crenarchaeota, including the ammonia-oxidizing archaeon that was identified in 2005 and currently referred to as group I.1B. They may respresent as much as 5% of the total prokaryotic community in soils, although it is difficult to know at this time since they are not culturable. Aerobic soils are a habitat for low-temperature methanogens like *Methanosarcina*, while flooded rice paddy soils support methanogens like *Methanobacterium*, *Methanosaeta* and *Methanosarcina*, all members of the phylum Euryarchaeota. The 'Korarchaeota' and 'Nanoarchaeota' have been proposed as separate phyla due to their divergent genetic characteristics, but have not yet been isolated from soils (Fig. 2.3).

The importance of the ammonia-oxidizing archaea of the phylum Crenarchaeota to soil nitrogen cycling is still not fully known, but there is a sense that the nitrification process catalysed by these archaea may convert ammonia to nitrate in a variety of terrestrial ecosystems. In pristine and agricultural soils, the ammonia-oxidizing archaeon produced more than 3000 times more ammonia monoxygenase (AMO)-encoding genes than bacteria, which suggests that they are important ammonia-oxidizers in these ecosystems (Hayatsu *et al.*, 2008).

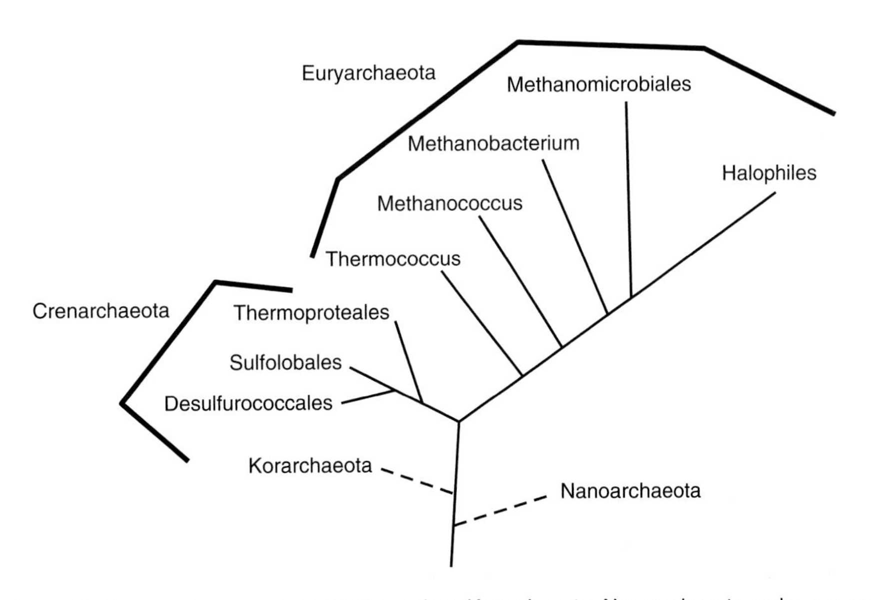

Fig. 2.3. Phylum of Archaea and some representative orders. Korarchaeota, Nanoarchaeota and some ammonia-oxidizer sequences belonging to Crenarchaeota have not yet been cultured. (Modified from Schleper *et al.*, 2005.)

Some Crenarchaeota and Euyarchaeota possess genes that code for nitrite reductase (NirK), the enzyme that catalyses dissimilatory nitrate reduction. This reaction produces nitrogen gases such as nitrous oxide (NO_2), a greenhouse gas, and dinitrogen (N_2).

Further physiological, biochemical and genomic studies are needed to determine the relative contribution of ammonia oxidizers and denitrifying archaea to nitrogen cycling in soils.

Another key group of soil archaea are low-temperature methanogens belonging to the phylum Euyarchaeota. All methanogens require an extremely low redox potential (–300 mV) and thus their habitat is restricted to anaerobic zones. In aerobic soils that normally have sufficient oxygen to inhibit methanogenesis, it is believed that these organisms survive in anaerobic microniches such as micropores that are permanently water-saturated. Biofilms that have a low oxygen concentration due to the action of oxygen-scavenging facultative anaerobes could also support methanogenic archaea.

Soil methanogens can use CO_2/H_2 and a variety of other substrates for methane production (Table 2.2). These reactions lead to organic matter decomposition and carbon loss from soil as methane, significant in the global carbon cycle because methane is a more potent greenhouse gas than carbon dioxide.

Bacteria

The organisms we refer to as bacteria belong to the domain (superkingdom) Bacteria, sometimes called Eubacteria. Like the Archaea, they are single-celled prokaryotes that range in size from 1 to 2 μm long and 0.5–1.0 μm in diameter. These prokaryotes have smaller cells than the more highly evolved eukaryotes and lack a nuclear membrane around the nucleus of their cell. The outer cell wall is composed mainly of mucopeptides. Most reproduce by binary fission. There is a great deal of variety in their metabolic processes, which may explain why they are so broadly distributed in ecosystems of the world. The amount of carbon and nitrogen contained in soil bacterial biomass is thought to be equivalent to the carbon and nitrogen stored in all plants that cover our world. Virtually every biochemical reaction that we know about can be catalysed enzymatically by a prokaryotic organism (bacteria and/or archaea).

Soil bacteria have diverse morphology, ranging from spherical cocci to rod-like bacilli and spiral forms (Fig. 2.4). The bacillus form is the most common among the culturable soil bacteria. They can be divided into two groups based on the origin of their energy source. The autochthonous, or indigenous populations, are uniformly distributed and obtain energy from the oxidation of soil organic and mineral matter. Representative bacteria in this group belong to the genera *Arthobacter* and *Nocardia*. Zymogenous populations rely on external substrates such as plant and animal residues for their metabolism. Since their activity depends on the input of external substrates, the zymogenous bacteria have the ability to produce spores or enter resting states when the food supply runs out. The genera *Pseudomonas* and *Bacillus* are considered to be zymogenous bacteria.

Bacteria exhibit a wide variety of metabolic types. Although these characteristics were traditionally used for taxonomic classification, metabolic

Table 2.2. Methanogensis reactions catalysed by some representative soil archaea.

Reaction	Representative genera
$4\ H_2 + CO_2 \rightarrow CH_4 + 2\ H_2O$	All soil methanogens
4 Formate $\rightarrow CH_4 + 3\ CO_2 + 2\ H_2O$	*Methanobacterium*
Acetate $\rightarrow CH_4 + CO_2$	*Methanosaeta, Methanosarcina*
4 Methanol $\rightarrow 3CH_4 + CO_2 + 2\ H_2O$	*Methanosarcina*
4 2-Propanol $+ CO_2 \rightarrow CH_4 + 4$ Acetone $+ 2\ H_2O$	*Methanobacterium*
4 Methylamine $+ 2\ H_2O \rightarrow 3\ CH_4 + CO_2 + 4\ NH_4^+$	*Methanosarcina*
2 Dimethylamine $+ 2\ H_2O \rightarrow 3\ CH_4 + CO_2 + 2\ NH_4^+$	*Methanosarcina*
4 Methylamine $+ 6\ H_2O \rightarrow 9\ CH_4 + 3\ CO_2 + 4\ NH_4^+$	*Methanosarcina*
$4\ CO + 2\ H_2O \rightarrow CH_4 + 3\ CO_2$	*Methanobacterium, Methanosarcina*

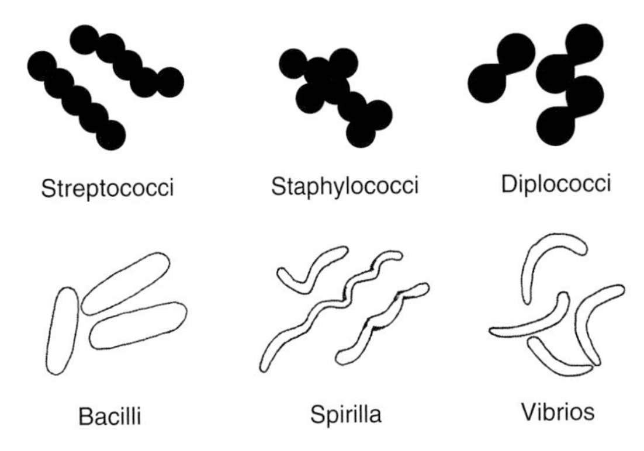

Fig. 2.4. Diversity of major forms of soil bacteria.

traits do not necessarily correspond to modern genetic classifications. The nutritional groups of bacteria are based on the energy source, the carbon source and the electron donors used for growth, as presented in Table 2.3 and in Focus Box 2.1.

The growth of photoautotrophic bacteria is achieved by capturing energy from sunlight and transforming CO_2 into bacterial biomass. Among the photoautotrophs, purple sulfur bacteria of the order Chromatiales capture light energy in bacterochlorophyll *a* and *b* and carotenoid pigments in their cellular membrane. Unlike plants, algae and cyanobacteria, they do not use water as the electron donor and so do not produce oxygen. Instead, they use hydrogen sulfide, which is oxidized to produce elemental sulfur or sulphuric acid. Purple sulfur bacteria are found in anoxic zones of lakes and also in sulfur springs where hydrogen sulfide accumulates. Similarly, green sulfur bacteria in the order Chlorobiales are obligate anaerobes that can photosynthesize using bacterochlorophyll *c*, *d* and *e* and chlorophyll *a* present in chlorosomes attached to their cellular membrane. Generally, they use sulfide ions as the electron donor for the reaction and produce elemental sulfur.

Table 2.3. Nutritional groups of soil bacteria.

Nutritional group	Energy source	Carbon source	Electron donors	Examples
Photoautotroph	Light	CO_2	H_2S, SO_3^-, H_2O	Purple sulfur bacteria, green sulfur bacteria, cyanobacteria
Photoheterotroph	Light	Organic compounds	H_2, S^{2-}	Purple non-sulfur bacteria, *Chloroflexi*
Chemolithoautotroph	Inorganic compounds	CO_2	NH_4^+, NO_2^-, S^{2-}, Fe^{2+}	*Nitrosomonas*, *Nitrobacter*, *Thiobacillus*, *Ferrobacillus*
Chemolithoheterotroph	Organic compounds	Organic compounds	H_2O, organic compounds	Most soil bacteria (also soil fungi and animals)

Focus Box 2.1. Metabolic diversity in the biosphere.

Our biosphere supports a great diversity of metabolic processes. The following illustrates the energy and carbon sources, electron donors and electron acceptors of prokaryotes (Fig. 2.5). Possible reactions in photoautotrophs and photoheterotrophs that function in aerobic and anaerobic environments are shown below. As well, the metabolism of chemolithoautotrophs and chemolithoheterotrophs living in aerobic and anaerobic environments are illustrated. Note that fermentation of organic compounds also provides energy, but this process uses intermediate metabolites for electron transfer and does not rely on redox reactions with external electron acceptors.

Organisms are classified as phototrophic or chemotrophic, depending on whether they use sunlight or energy liberated during exogenous chemical reactions to produce ATP. The carbon source they use for producing biomass leads to their definition as autotrophic (use CO_2) or as heterotrophic (use organic substrates). Depending on the type of reduced electron donor, organisms could be named organotrophic or lithotrophic, indicating the use of organic or inorganic compounds, respectively. It is important to note that all organotrophic organisms are heterotrophic, because they use exogenous organic matter as a source of electrons and carbon simultaneously. On the other hand, most of lithotrophic organisms are autotrophic because they use energy from sunlight or inorganic chemical reactions to fix carbon dioxide, their source of carbon and inorganic sources of electrons.

Continued

Focus Box 2.1. *Continued*

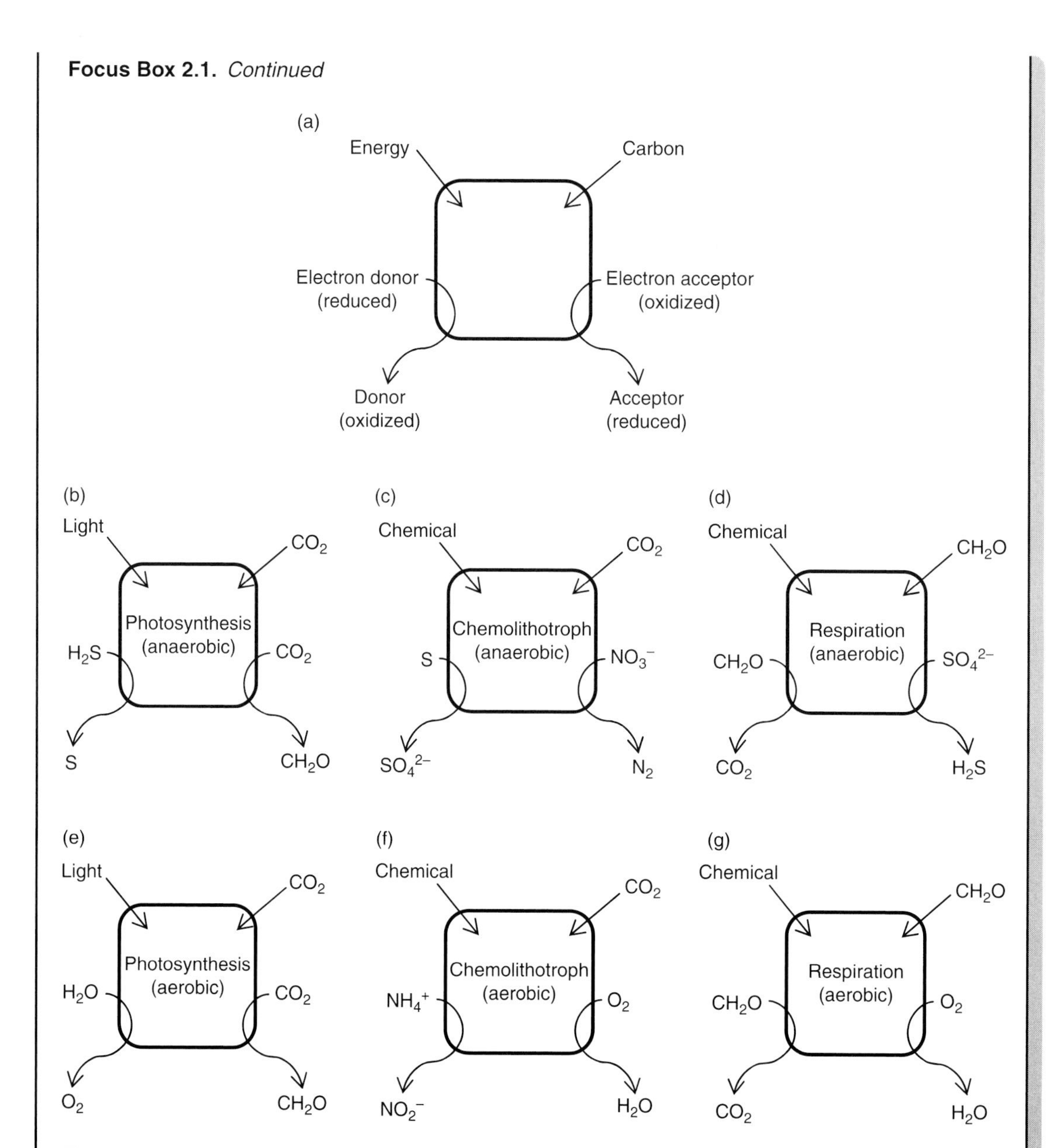

Fig. 2.5. Metabolic processes in prokaryotes involving redox reactions: (a) general scheme, showing energy and carbon sources and redox reactions of electron donors and acceptors; (b) anaerobic photoautotroph (e.g. purple sulfur bacteria); (c) anaerobic chemolithoautotroph (e.g. *Thiobacillus* spp.); (d) anaerobic chemolithoheterotroph (e.g. *Desulfovibrionales* spp.); (e) aerobic photoautotroph (e.g. cyanobacteria); (f) aerobic chemolithoautotroph (e.g. *Nitrosomas*); (g) aerobic chemolithoheterotroph (e.g. *Bacillus* spp.). (Modified from Rodriguez, 1999.)

In the photoheterotroph group are purple non-sulfur bacteria that also capture light energy in bacterochlorophyll and carotenoid pigments, but they use H_2 or other compounds (S^{2-}, $S_2O_3^-$ and S°) as electron donors. However, they cannot fix CO_2 and they depend on organic carbon sources (such as succinate) to produce biomass. Purple non-sulfur bacteria belong to the families Rhidospirillales,

Rhizobiales, Rhodobacteraceae and Rhodocyclaceae. Although they were believed to be close relatives of purple sulfur bacteria, RNA analysis reveals that they are more closely related to other heterotrophs. The similarity of the photosynthetic processes in the purple sulfur bacteria and purple non-sulfur bacteria suggests that they originated from the same source – perhaps a common ancestor or maybe the genes controlling this process were passed from one group to the other by lateral gene transfer.

The *Chloroflexi* were formerly known as green non-sulfur bacteria because they capture light energy in bacterochlorophyll *c* and *d* and chlorophyll *a* present in chlorosomes, using S^{2-} as the electron donor. Phylogenetic analysis reveals that they are distinct from the order Chlorobiales. These filamentous, mobile organisms move through bacterial gliding. Although they are facultatively aerobic, they do not produce oxygen during photosynthesis. Common habitats are anoxic, high-temperature geothermal vents. In alkaline hot springs (pH 5.5 to 10), *Chloroflexus auranticus* grows in an orange mat below a layer of cyanobacteria, suggesting that it uses carbon exudates from cyanobacteria cells for growth.

Among the photosynthetic bacteria, only cyanobacteria are capable of oxygenic and anoxygenic photosynthesis. They were the first prokaryotes responsible for oxygen production, sediment oxidation and subsequent accumulation of oxygen in the atmosphere more than 2000 million years ago. The profound change in the oxidative status of the atmosphere due to the biological activity of cyanobacteria promoted a massive change in the Earth's geochemistry. Thus, cyanobacteria have influenced large-scale processes, both spatially (planetary scale) and temporally (millions of years) (Focus Box 2.3.).

Cyanobacteria are a structurally diverse group of Gram-negative bacteria that have a phycocyanin pigment and chlorophyll in their cellular membranes and a thick gelatinous cell wall. Due to the colour of this pigment and certain similarities to algae, they are often referred to as 'blue-green algae', although they are in fact prokaryotes. Their morphology includes being unicellular and colonies that may be filamentous, sheets or hollow balls. The filamentous colonies can differentiate into different cell types: vegetative cells that are photosynthetic, akinetes that are resistant spores produced during unfavourable conditions or stress, and heterocysts, which contain the enzyme nitrogenase and catalyse nitrogen fixation.

Cyanobacteria are the only group of organisms that are able to reduce nitrogen and carbon in aerobic conditions, which may be a reason for their widespread distribution in aquatic and terrestrial ecosystems. However, nitrogenase is inactivated by contact with oxygen, which is generated during photosynthesis. Cyanobacteria are able to support both photosynthesis and nitrogen fixation because they produce specialized cells called heterocysts (Fig. 2.6) that are capable of reducing or excluding oxygen gas through one or more strategies:

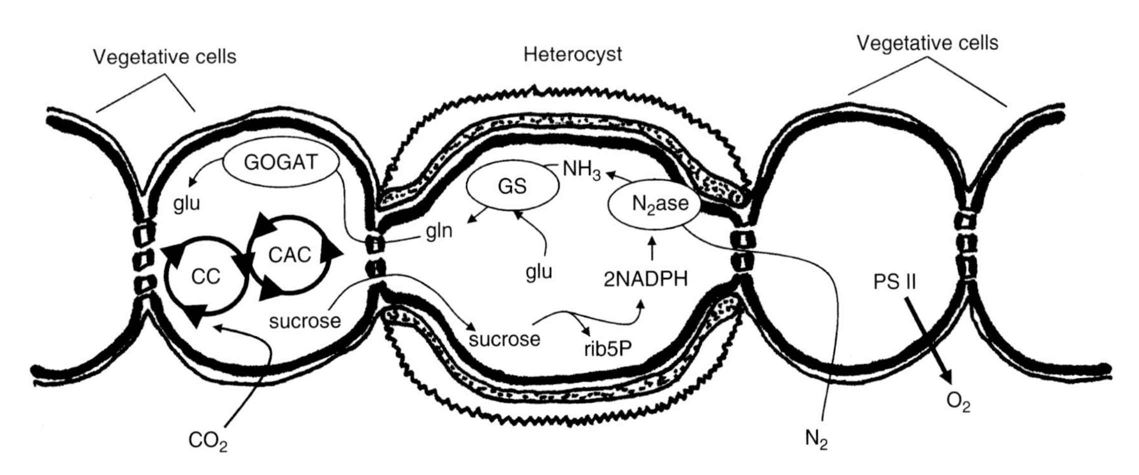

Fig. 2.6. Diagram of carbon and nitrogen transfer between vegetative cells and heterocysts in filamentous cyanobacteria. The enzymes involved in these reactions include glutamine synthetase (GS), glutamine-oxoglutarate amidotransferase (GOGAT) and nitrogenase (N_2ase). PS II, photosystem II; CC, Calvin cycle; CAC, citric acid cycle; glu, glucose; gln, glucosamine. (Redrawn from Sylvia *et al.*, 2005.)

- thick walls containing glycolipids slow O_2 diffusion into the cell;
- high levels of cellular respiration deplete O_2 that may diffuse into the cell; and
- photosystem II, which produces O_2, is eliminated from the heterocyst.

The coupling of photosystems I and II into the vegetative cell for aerobic photosynthesis is a special feature of cyanobacteria. Energy from the vegetative cell is transported to the heterocyst to support nitrogen fixation. In anaerobic environments, they use photosystem I and electron donors such as hydrogen sulfide, thiosulfate or hydrogen in photosynthesis. Like some archaea, they can reduce elemental sulfur by anaerobic respiration in the dark.

The nitrogen-fixing cyanobacteria can provide an important quantity of nitrogen in many ecosystems, including rice fields, deserts and in the tropics. Estimates of nitrogen fixation by *Anabaena* spp. in Asian rice fields range from 5 to 80 kg N/ha/year with an average of about 30 kg N/ha/year (approximately 25% of the nitrogen required for rice production). Cyanobacteria can therefore reduce the amount of manufactured nitrogen fertilizer required for rice production, which occurs on about 150 million ha of agricultural land around the world. Cyanobacteria are the pioneer colonists on newly disturbed soils in arid and semi-arid regions, leading to the growth of biological soil crusts that stabilize the soil surface and contribute to soil pedogenesis. Biological soil crusts are especially important in biomes that are too dry for vascular plants, such as desert soils, xerothermic steppes and tundra soils. Filamentous species of the genus *Microcoleus* are of major importance because their gelatinous fibres bind to soil particles, creating a stable mat at the soil surface that protects against erosion. Photosynthesis and nitrogen fixation in these biological crusts means that they accumulate carbon and nitrogen, perhaps important for the succession of xeric and mesophyllic plants.

The chemolithoautotrophs use inorganic electron donors as an energy source; at the same time they gain carbon from CO_2. The transfer of electrons from the electron donor to a terminal electron acceptor such as oxygen releases energy that can be used to synthesize ATP. This process is illustrated in Focus Box 2.2. This metabolic pathway is very common both in bacteria and archaea. Chemolithoautotrophs can use inorganic substances such as hydrogen, carbon monoxide and reduced metal ions as electron donors. Among the bacteria that have adopted this nutritional strategy are *Nitrosomonas* and *Nitrobacter*. *Nitrosomonas* is a rod-shaped bacterium with aerobic metabolism that obtains energy for growth and cell division by transferring hydrogen ions from ammonia or ammonium to oxygen. *Nitrobacter* is also a rod-shaped bacterium with aerobic metabolism that obtains energy by oxidizing nitrite to nitrate. Other aerobic bacteria in this group include *Thiobacillus*, capable of converting inorganic sulfur compounds to sulfate, and *Ferrobacillus*, which converts ferrous ions (Fe^{2+}) into ferric ions (Fe^{3+}). The activity of these bacteria depends on the availability of their required electron donor, which is strongly related to redox conditions in the soil environment. Other habitats for iron bacteria include terrestrial runoff water channels, hot springs that provide iron and sulfur, and deep-sea smoker chimneys emitting iron sulfides at extremely high temperatures (up to 350°C).

Focus Box 2.2. Bacterial electron transport chain.

Chemolithoautotrophic bacteria cannot capture solar energy to meet their energetic requirements. So, how do they acquire energy? All of the chemolithotrophs possess an electron transport chain that captures chemical energy from electron donors and transfers it to electron acceptors. In Fig. 2.5, we see a series of membrane-bound reactions leading to the transfer of electrons to oxygen, which generates ATP (energy). During this process, hydrogen ions (H^+) are released outside the cell membrane during reactions at coenzyme Q. In the final phosphorylation steps (conversion of ADP to ATP), water is produced and dissociates to produce hydroxyl ions (OH^-), which remain inside the cell membrane (Fig. 2.7). The presence of H^+ outside the membrane and OH^- inside the membrane creates a pH gradient and electrical potential, which allows the ADPase enzyme to convert ADP to ATP. Bacteria can also use this energy for active transport of ions and small molecular weight compounds through their cellular membrane.

Continued

Focus Box 2.2. *Continued*

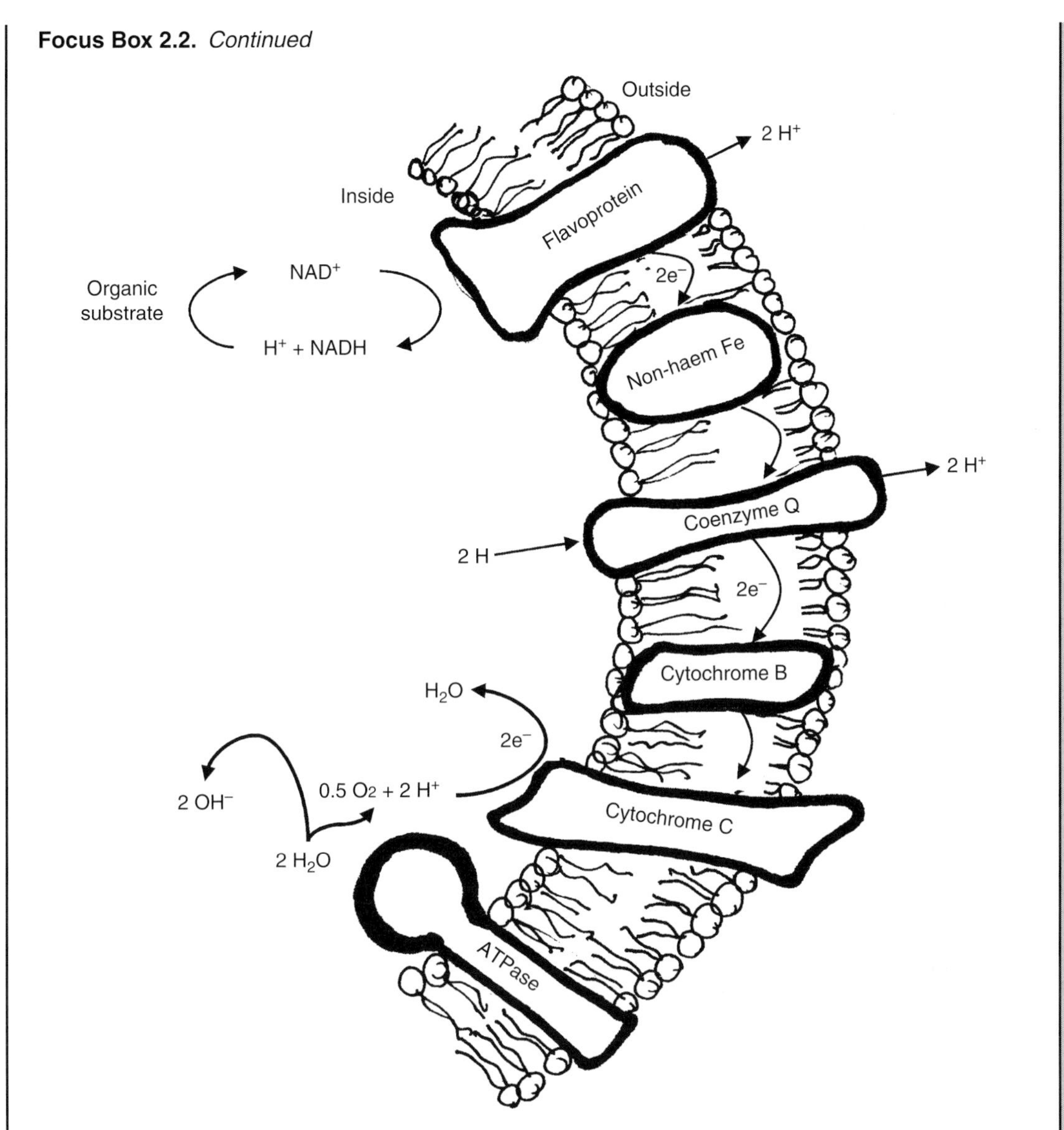

Fig. 2.7. Bacterial electron transport chain, showing the series of membrane-bound reactions that can result in the transfer of electrons to O_2, resulting in the generation of ATP. The proton gradient produced facilitates energy production and the production of H^+ outside the membrane. (Redrawn from Paul and Clark, 1989.)

These examples have highlighted aerobic bacteria that use oxygen as the terminal electron acceptor. Anaerobic bacteria are able to thrive in oxygen-depleted environments by generating energy from electron transport chains using nitrate, sulfur or carbon dioxide as the electron acceptors. For example, denitrifying bacteria such as the autotrophic *Thiobacillus denitrificans* use nitrate as a terminal electron acceptor in the process of anaerobic respiration, producing a variety of reduced, gaseous nitrogen compounds and eventually dinitrogen gas. However, these electron acceptors are not used when oxygen is available, so the bacteria involved in these reactions are either facultative anaerobes or obligate anaerobes (active only when oxygen concentrations are low, or in the absence of oxygen). Other possible electron acceptors for anaerobic bacteria are given in Table 2.4.

Soil chemolithoheterotrophs obtain energy by oxidizing the soil organic matter and using carbon for growth. Many are aerobic bacteria that feed on senescing plant tissues or dead organic matter (from plants, animals and other microorganisms). Others are parasites of plants, animals and other soil biota. They obtain their energy by feeding on the living tissue of another organism. Finally, there is a group of soil bacteria called symbionts. Perhaps the most well-known symbiotic soil bacterium is *Rhizobium*, which enters the root of a host plant and fixes N_2 gas from the atmosphere, in exchange for energy (photosynthates) from the plant.

All of these organisms require energy for growth and reproduction as well as basic metabolic processes. As an example, the oxidation of soil organic matter requires a considerable amount of energy. First, the bacteria must produce extracellular hydrolytic enzymes required to break down complex organic substrates (e.g. protein) into simpler monomeric units (e.g. amino acids). The initial decomposition of proteins, complex carbohydrates and lipids is done outside the bacterial cell, since large molecules cannot be transported across the cellular membrane. Small peptide chains and urea can be absorbed by bacteria and degraded into simpler amino acids and ammonium by intracellular peptidase and urease enzymes. Under aerobic conditions, simple sugars undergo glycolysis to generate ATP and NADH. Short-chain fatty acids are oxidized to acetyl-CoA molecules that enter the TCA cycle, which produces NADH and FADH.

Since soils are not always well-oxygenated and even aerobic soils may contain anaerobic microsites, many bacteria function under anaerobic conditions. Anaerobic respiration must be distinguished from fermentation: in bacteria, anaerobic respiration refers to a membrane-bound biological process that couples the oxidation of electron-donating substances like sugars and organic compounds to the reduction of an external electron acceptor, but not oxygen; fermentation is the oxidation of molecules coupled to the reduction of an internal electron acceptor, usually pyruvate. Facultative anaerobes are well known for their ability to switch from anaerobic respiration to fermentation when their preferred electron acceptors are not available. Fermentation reactions generate simple compounds such as lactate, ethanol, hydrogen and butyric acid, all of which have lower energy than the starting materials, leading to a net energy gain by the bacteria. Common substrates for fermentation reactions include glucose and more complex sugars, as well as compounds like homoacetic acid, propionic acid, acetylene, oxalate and malonate.

Table 2.4. Electron acceptors used by bacteria or archaea during aerobic or anaerobic metabolism. Electron acceptors and typical products are arranged according to their reduction potential, where positive values indicate a greater affinity to gain electrons.

Metabolic process	Terminal electron acceptor	Typical product	Reduction potential (E_h in millivolts)
Aerobic	O_2	H_2O	0.82
Anaerobic	Fe^{3+}	Fe^{2+}	0.76
	NO_3^-	N_2	0.74
	Mn^{4+}	Mn^{2+}	0.48
	NO_3^-	NO_2^-	0.42
	Fumarate	Succinate	0.03
	SO_4^{2-}	H_2S	−0.22
	CO_2	Acetate	−0.28
	CO_2	CH_4	−150
	AsO_4^{2-}	AsO_3^{2-}	−200

Focus Box 2.3. The evolution of our oxygen-rich atmosphere.

The Earth's atmosphere and oceans contain reactive compounds and unstable gases that have been used and modified by biological organisms. We now understand that the present oxygen-rich atmosphere that supports life on our planet developed due to the activity of ancient prokaryotes, and that the evolution of oxygen-producing cyanobacteria was one of the most relevant events after life appeared on the Earth's surface. Reconstruction of atmospheric oxygen concentrations is consistent with several geochemical proxies such as the presence of oxidized metallic sediments, banded iron sedimentary

Continued

Focus Box 2.3. *Continued*

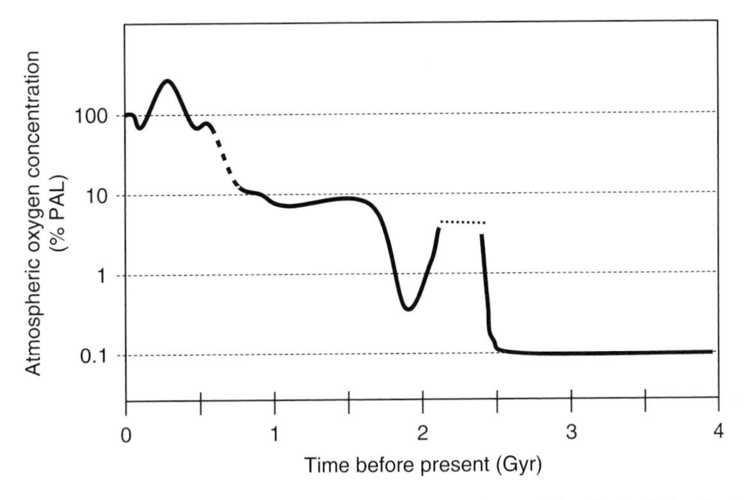

Fig. 2.8. Reconstruction of the concentration of oxygen in the atmosphere of the Earth during the first 4000 million years. Dashed line indicates uncertain periods. PAL: present atmospheric level. (Modified from Canfield, 2005.)

formations and sulfur isotope ratios. There is a general agreement that by 2.45 giga years (Gyr) ago atmospheric oxygen concentrations rose from less than 10^{-5} of the present atmospheric level (PAL) to more than 0.01 PAL, and possibly to more than 0.1 PAL (Fig. 2.8). The 'Great Oxygenation' of the Earth's surface environment was characterized by rising and fluctuating O_2 levels in the atmosphere that reached steady state approximately 0.6 Gyr ago. However, it is broadly accepted that the history of atmospheric O_2 concentrations was very dynamic and complex and is still far from well understood. The presence of oxygenic photosynthesis together with non-linear changes in the chemistry and thermodynamics of Earth's atmosphere probably had a joint role in producing the gaseous components of the atmosphere.

1. The atmosphere and the primordial ocean had moderate reducing potential (Table 2.5). The most striking difference was in the atmosphere, which contained about 98% CO_2 and trace amounts of acidic H_2S. The oceans were saline with high Cl^- and Na^+ concentrations, similar to present-day conditions, and contained acetate and abiotic C compounds such as HCO_3^-.

Table 2.5. Comparison of the Earth's atmosphere with and without life.

	With life	Without life
Nitrogen (%)	78	1.9
Oxygen (%)	21	trace
Carbon dioxide (%)	0.035	98
Surface temperature (°C)	13	290 ± 50

2. Simple procaryotes were able to metabolize acetate and abiotic carbon. These organisms were probably the ancestors of modern-day methanogens and fermenting bacteria.

3. A major change occurred when chemoautotrophs and autotrophs appeared on Earth. These prokaryotes live in an anoxic environment and used H_2S and CO_2 in their metabolism. The abundance of atmospheric CO_2, compared with abiotic carbon, was highly favourable for their growth and evolution.

4. The definitive step in the evolution of our oxygen-rich atmosphere occurred when the capacity for oxygen-producing photosynthesis was developed by primitive cyanobacteria. Living in the ocean, they had an unlimited supply of water as an electron donor and CO_2 gas as a carbon source. They left extensive coral reef-like fossil formations called stromatolites. The oxygen produced by these organisms had a profound effect on the chemical composition at the surface of our planet. The oxygen produced as a result of their metabolic activity was absorbed by oceans and seabeds, then oxidized metal ions in emerged land surface and sediments, followed by O_2 accumulation in the atmosphere. Oxygen concentration increased dramatically from trace levels to the

Continued

Focus Box 2.3. *Continued*

21% in the atmosphere today (Fig. 2.8). The ozone layer was also formed by that time, reducing the energy of incoming solar radiation, filtering the UV light and cooling the Earth's surface. Earth became an oxidizing environment, causing a mass extinction event and the onset of extreme Ice Ages.

5. Around 1.2 to 2 billion years ago, the cellular machinery for chloroplast and mitochondria formation, essential for life in the new oxidative environment, appeared during the evolution of the eukaryotes.

The intimate relationship between oxygen, life and biogeochemistry is broadly recognized; however, there is still debate about how the appearance of oxygenic photosynthesis is coupled to Earth surface oxidation and atmospheric oxygen accumulation in the geological timescale. Recent advances in accurate dating of cyanobacterial evolution are the key to understanding these relationships. Although dating of molecular sequences has shown that aerobic respiration (under marginal concentrations of oxygen) appeared early in the evolution of prokaryotes, the earliest life was extensively dominated by anaerobic metabolism. Genes for sulfate reduction, anoxygenic photosynthesis, fermentation, chemolithoautotrophy and organic matter respiration with Fe oxides are deeply rooted in ancient prokaryote lineages.

Methylhopane biomarkers and lacustrine stromatolites, most likely formed by associations of ancient prokaryotes with oxygenic phototrophic organisms, have been found in 2.7-Gyr-old organic-rich shales in Western Australia. Some scientists have recently questioned this evidence, instead dating the existence of cyanobacteria to fossil remains that are approximately 2.15 Gyr old. Thus, the debate is reopened about what happened during the short time lag between cyanobacterial appearance and the 'Great Oxygenation Event' about 2.45 Gyr ago.

Actinomycetes (Actinobacteria)

Actinomycetes are similar in size and structure to Gram-positive bacteria, with a single cell membrane (no nuclear membrane) and a thick cell wall composed of peptidoglycan layers. Until recently, actinomycetes were thought to be related to the fungi because their morphology is mycelial, like a fungus. However, they possess prokaryotic cells (in contrast to fungi, which have eukaryotic cells), so they must be considered in the domain Bacteria.

Actinomycetes are filamentous bacteria that possess an extensive branched pseudomycelium and aerial hyphae, but their hyphae are typically much smaller than fungal hyphae (perhaps 10–15 mm long and 0.5–2.0 mm in diameter). The hyphae often bear long chains of spores (perhaps more than 50 spores) that can be straight, wavy, hooked or in tight spirals (Fig. 2.9). Spores are the reproductive unit of this organism. Although actinomycetes species can be distinguished by examining spores, genetic analysis using DNA hybridization followed by sequence analysis is a more reliable way of identifying actinomycetes species.

The most abundant soil actinomycetes belong to the genus *Streptomycetes*. They have an unusual life cycle that resembles that of a multicellular organism more than bacteria. Each cell possesses a single chromosome with the genetic material, which is shared among individual cells by conjugation (two cells fuse briefly to share genetic information). This leads to a higher genetic diversity within actinomycetes populations than is found in other bacteria. The life cycle of actinomycetes begins with spore germination, in which a germ tube is formed and grows into a branched mycelial network, bound to

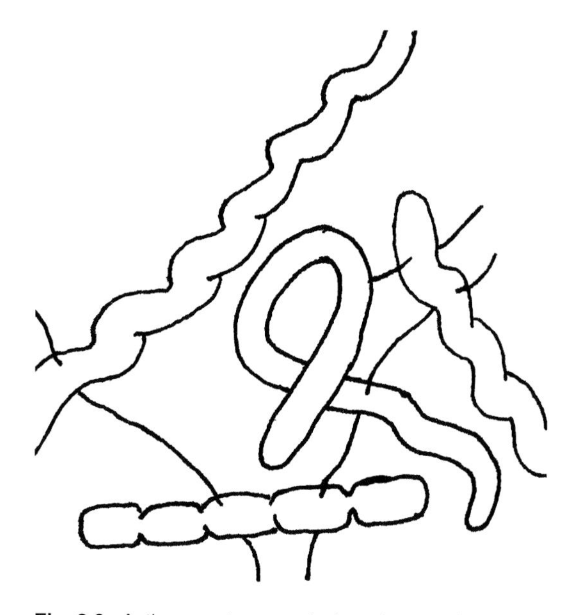

Fig. 2.9. Actinomycetes morphology features long, spiral chains of cells.

the soil matrix. The mycelium is considered the vegetative part of the actinomycete colony. The hyphae that grow from the mycelium are the reproductive part of the colony, and produce chains of spores, protected from desiccation by a hydrophobic outer sheath. These spores are adapted for wind dispersal. In contrast, the sporangia-forming actinomycetes (family Actinoplanaceae) and micromonosporas (family Micromonosporaceae) produce hydrophilic spores that are dispersed by water movement through the soil. Not all hyphae produce spores. When non-sporulated hyphae die, their cellular contents are reassimilated and used for the production of sporulated hyphae.

The majority of soil actinomycetes are free-living saprophytes that are capable of decomposing recalcitrant organic compounds under aerobic conditions, most effectively in neutral to alkaline pH conditions. Their decomposition activities release geosmins that give soils their characteristic earthy smell. Actinomycetes produce a range of extracellular hydrolytical and oxidative enzymes such as cellulases, xylanases and proteases that degrade cellulose, hemicelluloses and proteins originating from plant residues. Extracellular peroxidases produced by the ligninolytic *Streptomycetes* and other actinomycetes such as *Thermomonospora mesophila* contribute to lignin depolymerization. Actinomycetes are not as efficient at lignocellulosic breakdown as fungi, but do make an important contribution to the decomposition and humification processes in soils and compost piles. They could also be exploited as a source of enzymes for lignocellulose bioconversion, which transforms agricultural, industrial and forest residuals into biofuels and other value-added products.

The non-sporulating actinomycetes in the genera *Rhodococcus, Nocardia, Gordonia, Mycobacterium* and *Dietzia* are free-living aerobic bacteria that have broad metabolic capabilities, including the degradation of diverse classes of hydrocarbons, including aliphatics, aromatics and xenobiotic pollutants. They can degrade even the most persistent man-made chemical compounds such as organochlorines, carbamates and organophosphates. In spite of the broad metabolic versatility of this group, none are able to completely break down these complex organic substances to carbon dioxide and water. Normally, corsortia of bacteria are required for complete degradation. Co-metabolism is frequently observed, meaning that two compounds are metabolized simultaneously and the degradation of the second compound (complex organic substrate that does not support microbial growth) depends on the presence of the first compound (primary substrate used for microbial growth). The aerobic co-metabolism of the actinomycete *Rhodococcus corallinus* depends on an alkene monooxygenase enzyme to initiate the oxidation of propylene, the growth-supporting substrate and consequently hydrolyses alkenes and chlorinated alkenes.

Soil actinomycetes in the genus *Frankia* form a symbiotic N_2-fixing association, with about 200 trees and shrubs in 24 genera on all continents except Antarctica. These filamentous soil actinomycetes form root nodules that fix gaseous nitrogen in a symbiotic exchange that resembles that between legumes and their symbionts. *Frankia* receives carbon from the host and supplies from 70 to 100% of the host's nitrogen requirements. The survival of alders (genus *Alnus*, family Casuarinales), primary colonizers of recently disturbed soils and low-fertility soils, is due largely to the actinorhizal symbiosis. The nitrogen fixation occurring in actinorhizal alders in pristine soils ranges from 40 to 300 kg N/ha/year, comparable to nitrogen fixation by *Rhizobium* bacteria associated with clover (up to 160 kg N/ha/year) and lucerne (up to 300 kg N/ha/year). Alder roots are also colonized by ectomycorrhizal fungi such as *Alpova diplophloeus* and *Lactarius obsuratus*, which also contribute to plant nutrition and growth. Ectomycorrhizal fungi effectively extend the root zone of the plant through their fine hyphal networks and acquire water and nutrients that are normally inaccessible to plant roots. Alder seedlings with both *Frankia* and ectomycorrhizal fungi grow more vigorously than with one symbiont or the other. Research is ongoing to determine the process of hypersymbiosis (plant–bacterial–fungal association), which is expected to be energetically costly to the host plant.

The genus *Streptomycetes* is also well known for its ability to produce bioactive compounds such as antibiotics. In fact, 75–80% of the commercially and medicinally useful antibiotics currently in use were derived from this genus, including streptomycin, neomycin, chloramphenicol and tetracyclines. In natural environments, *Streptomycetes* probably uses antibiotics to kill or reduce the populations of other microorganisms that are competition for food, water, oxygen and other resources necessary for its survival. The production of antibiotics is elicited by environmental stresses that trigger regulatory genes in the organism. Actinomycetes populations are present in virtually all soils, with larger populations in soils with near-neutral pH conditions as well as in soils under water stress and heat stress.

A few actinomycetes are pathogens of mammals and plants. In tropical soils, agricultural workers can acquire *Streptomyces somaliensis*, which causes a chronic subcutaneous infection called mycetoma. Three *Streptomycetes* species, including *Streptomycetes scabies*, are economically important pathogens of potaoes and other root crops such as beets, turnips and carrots. The deep lesions and scabs formed on the potato peel do not affect the integrity of the tuber, but they mar the appearance and reduce the market value of the crop.

Fungi

The soil fungi are numerically dominant micro-organisms, with an estimated 1,500,000 species worldwide. This group appeared more recently than the prokaryotes, less than 1000 million years ago. Fungi are obligate aerobes and most species are found in terrestrial ecosystems, probably because they co-evolved with vascular plants. These eukaryotic organisms are characterized by a distinctive, multinucleate, vegetative thallus (body) called the mycelium. The mycelium is composed of long, filamentous, branched cells called hyphae. Hyphae are from 2 to 10 μm in diameter, but much more extensive than those produced by actinomycetes. Fungal hyphae may extend from millimetres to centimetres in the soil, although we can find examples in old-growth forests where the fungal hyphae extend for many metres from the host tree and colonize multiple trees. Hyphae may also group as thick, cord-like rhizomorphs that look like roots.

In most fungi, the hyphae are divided into cells by internal membranes called septa. The septa are porous, which allows cytoplasmic fluid, ribosomes, mitochondria and even nuclei to move between cells. Some fungi have non-septate hyphae, meaning that no internal membranes are found in the hyphae. Fungal growth occurs through extention of the hyphae, which can branch through bifurcation of the growing tip or when a new apical tip emerges from the hyphae. The intracellular organelle responsible for the production of new hypha is called the spitzenkörper, which functions as a supply centre for wall-building vesicles and controls the direction of hyphal growth.

The importance of hyphae for fungal nutrition is illustrated in Fig. 2.10. Hydrolytic enzymes are released from a fungal hypha into the soil solution, where they bind with suitable substrates and break complex organic substrates (e.g. cellulose) into sim-

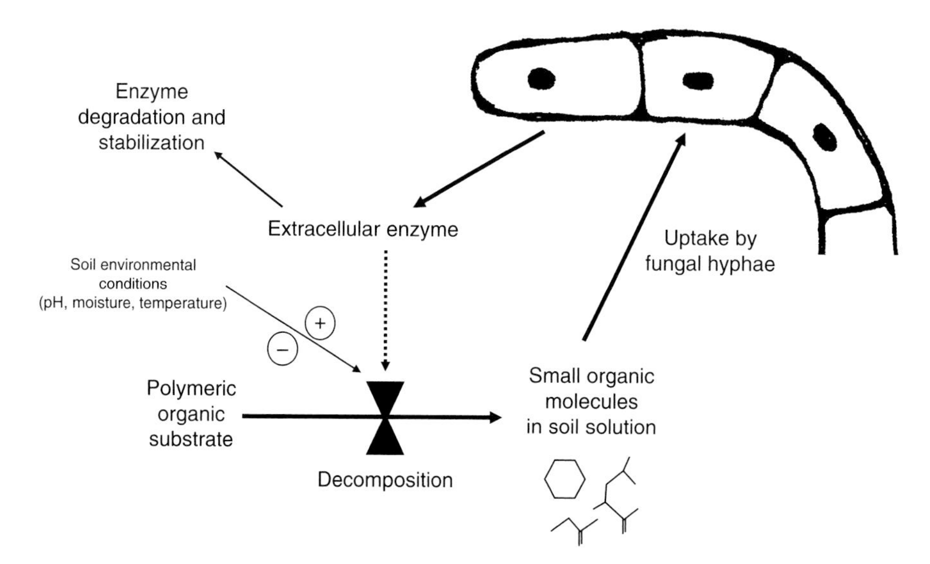

Fig. 2.10. Extracellular enzymes secreted from fungal hyphae enter the soil solution or decomposing organic residues, where the substrates are degraded to inorganic forms and low-molecular-weight molecules that can be absorbed by the hypha and transported through the mycelium by facilitated diffusion and active transport. The fungus illustrated here has a septate hypha, meaning that it is divided into cells by internal membranes called septa.

pler monomeric units (e.g. cellobiose, glucose) that can be absorbed through the cell wall. The growth of the vegetative thallus into a decomposing substrate (e.g. a dead leaf) increases the chance that fungi will intercept cellobiose, small peptides and inorganic substances needed for their survival.

Fungal cell walls are composed of chitin, cellulose and glycoproteins, which distinguishes them from prokaryotes. The composition of the cell wall differs among taxonomic groups and is one of the morphological characteristics examined when distinguishing species. Traditionally, fungal taxonomy was based on the cell wall composition, morphology of the vegetative thallus and reproductive strategies (Table 2.6). The classical taxonomy system grouped the soil fungi into the following groups: zygomycetes, ascomycetes, basidiomycetes and deuteromycetes (Fungi Imperfecti). The deuteromycetes are an artificial group that includes the asexual stages of ascomycetes and basidiomycetes, as well as fungi that lack sexual structures (reproduce by asexual means only). The development of rapid DNA sequencing methods since the 1990s has allowed mycologists to remove this artificial group and properly classify fungi according to their phylogenetic identity. The former deuteromycetes are now referred to as anamorphic fungi, meaning that they produce asexual spores through mitosis (Table 2.7). Soil fungi with sexual stages, known as teleomorphs, possess morphological structures where meiosis occurs and sexual spores are produced and are still grouped into zygomycetes, basidiomycetes and ascomycetes.

Soil fungi are aerobic chemolithoheterotrophs that exhibit considerable diversity in their metabolism and preferred substrates. This group includes free-living organisms, root symbionts, plant pathogens and animal parasites. Some of the common trophic groups in the soil fungi are:

- saprotrophs, like the sugar fungi (e.g. *Mucor* and *Rhizopus*) that consume simple sugars and by-products from cellulose decomposition;
- cellulose- and lignin-decomposing fungi belonging to the ascomycetes and the basidiomycetes;
- plant symbionts, which obtain their energy from plants and provide nutrients and water to the plant (e.g. mycorrhizal fungi, discussed further in Part 3);
- plant pathogens, which obtain their energy from plants while causing diseases (e.g. *Fusarium* species that cause damping off or wilt);
- animal pathogens, which obtain energy from the animal host;
- oligotrophs, which survive under low nutrient conditions;
- coprophilous fungi, which feed on animal manure; and
- 'carnivorous' fungi, such as nematode-trapping fungi, which build hyphal webs that can trap passing nematodes; the fungi then excrete enzymes to digest the nematode.

Table 2.6. Soil fungi groups based on classical taxonomy.

Group	Distinguishing characteristics	Asexual reproduction	Sexual reproduction	Examples
Zygomycetes	Multicellular, coenocytic mycelia	Asexual spores develop in sporangia on the tips of aerial hyphae	Sexual spores (zygospores) are persistent in the environment	Sugar fungi (e.g. *Mucor*) and mycorrhizal fungi
Basidiomycetes	Multicellular, uninucleated mycelia	Generally absent	Sexual spores (basidiospores) are found on a club-shaped structure on hyphal tips	Brown rot and white rot fungi (e.g. *Fomes*, *Polysporus*)
Ascomycetes	Unicellular and multicellular with septate hyphae	Budding, conidiophores	Asci form on specialized structures	Cellular slime moulds (e.g. *Dictylostelium*, *Physarum*)
Deuteromycetes	Unicellular or multicellular, septate hyphae	Budding	Absent	Fungi Imperfecti (e.g. *Aspergillus*, *Penicillium*, *Fusarium*)

Table 2.7. Fungi with asexual reproduction stages, grouped as anamorphic fungi based on DNA sequence analysis. There are about 16,000 known fungi belonging to this taxonomic group, distributed among 2900 genera. (Adapted from Gams and Seifert, 2008.)

Group	Characteristics	Examples
Anamorphs of Eurotiales, Trichocomaceae	Prolific spore production, chains of dry, wind-dispersed conidia. Many species produce antiobiotics and mycotoxins	*Penicillium*, *Aspergillus*
Anamorphs of Onygenales	Unicellular translucent (hyaline) conidia. Some have multicellular macroconidia	*Chrysosporium*, dermatophytes
Anamorphs of Hypocreales	Slimy, hooked multicellular conidia. Many species are plant pathogens	*Fusarium*, *Gliocladium*, *Clonostachys*, *Trichoderma*
Anamorphs of Clavicipitaceae	Colourful mixture of mononematous and conidiomatal spores, or sclerotia. Group includes insect and arachnid pathogens and endophytes	*Beauveria*, *Hirsutella*, *Metarhizium*, *Tolypocladium*
Anamorphs of ophiostomatoid fungi	Conidia produced on a dark, spherical, hollow fruiting body similar to an ascus. Group includes plant pathogens and mild human pathogens	*Pesotum*, *Leptographium*, *Sporothrix*
Anamorphs of Sordariales	Pigmented conidiophores producing slimy unicellular conidia. Found on dung, in soil or in wood	*Chaetosphaeria*, *Chrysonilia*
Anamorphs of Chaetothyriomycetidae and Dothideomycetes	Olive-brown or pigmented hyphomycetes with conidia that bud prolifically in moist environments before converting to hyphal growth (black yeasts) or produce large, multicellular dry conidia. Phylogenetically heterogeneous group includes saprophytes and plant pathogens	Loculoascomycetes and bitunicate ascomycetes, cercosporoid and phycnidial genera of the Dothideales, alternarioid fungi and Helminthosporia
Anamorphs of basidiomycetes	Refers to non-sporulating stages in the life cycle of basidiomycetes, especially mycelia and sclerotial forms. Plant pathogens, saprophytes, mycorrhizal fungi	*Rhizoctonia*, *Pleurotus*, *Coprinopsis*
Nematophagous fungi	Sickle-like conidia are ingested by nematodes, while slimy conidia adhere to nematode surfaces. Some are parasites, others capture and feed upon nematodes	Parasites: *Haptocillium* and *Hirsutella* Nematode-trapping fungi: *Orbilia*
Ingoldian and aeroaquatic hyphomycetes	Branched, radial conidia produce waterborne propagules. Found in highly aerated streams and rivers, generally on leaves that have fallen into the stream	Ingoldian

Fungi are sensitive to environmental conditions. When soil water is limiting, fungi produce resistant spores or enter into a resting state to conserve energy. When environmental conditions are not favourable for fungal survival, they can produce structures that enable them to survive, such as sclerotia, rhizomorphs and mycelial cords. Another survival strategy is the translocation of living cytoplasm towards the young, growing mycelial tips.

Fungi disperse when spores are displaced by wind and water or transported from the point of origin by soil fauna and other animals. The growth of new mycelium and translocation of living cytoplasm towards the growing ends constitutes another mechanism for fungal movement.

Focus Box 2.4. Co-evolution of angiosperms and fungi was linked to lignin degradation and soil organic carbon reserves.

The rates of organic carbon burial through the Earth's history can be estimated from the composition and masses of sedimentary rock during different ages, and also from the isotopic record of marine carbonates (Fig. 2.11). These data show a peak of organic carbon burial during the coal age of the late Paleozoic (300 million years ago), with rates twofold greater than modern values. Carbon burial at that time was accompanied by depletion in the atmospheric CO_2 concentration and a peak of 35% O_2 content by volume in the atmosphere, which dropped to the present value (about 21% O_2 content) around 200 million years ago. Such high O_2 availability is consistent with the existence of giant insects; but also suggests an extremely high risk of extensive wildfires due to the oxidizing conditions.

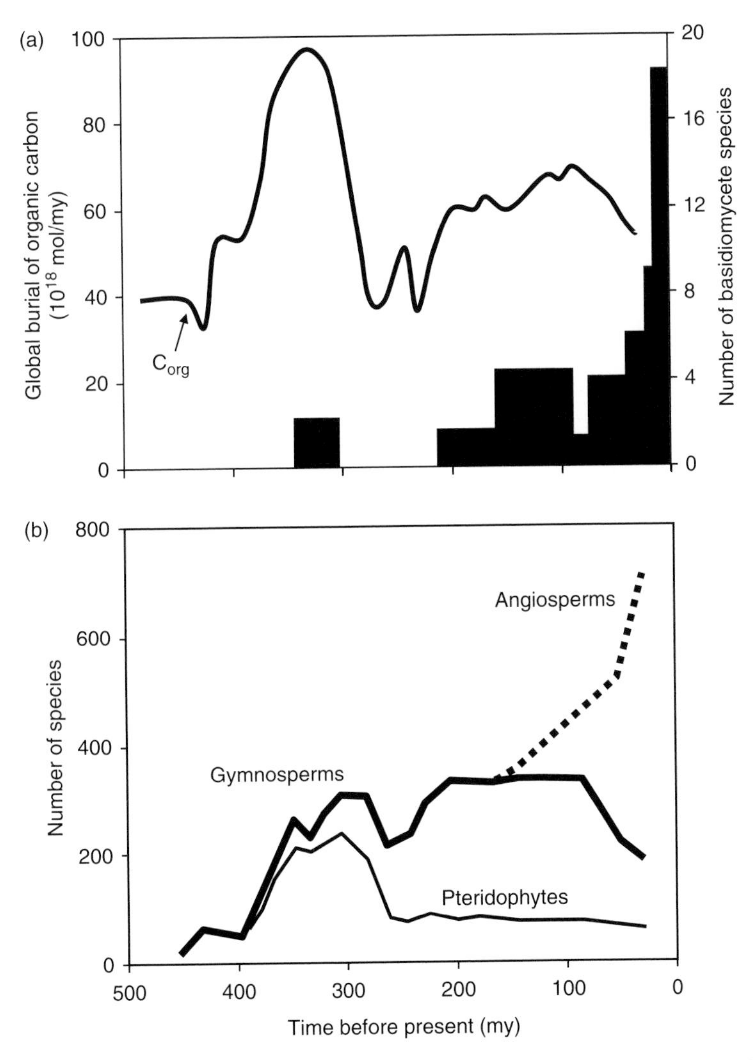

Fig. 2.11. (a) Reconstructed chronologies for rate of burial of sedimentary organic C, estimated number of basidiomycete species identified in the fossil record; and (b) major trends in plant radiation in the last 500 million years. (Redrawn from Robinson, 1990.)

Continued

Focus Box 2.4. *Continued*

The Paleozoic period was also characterized by the extensive colonization of marine ecosystems and the development of terrestrial ecosystems on emerged land masses. Besides the famous Cambrian explosion of the animal kingdom, two new kingdoms (plants and fungi) appeared, undergoing extensive radiation and colonizing most of the land surface. Plants had to evolve to adapt to terrestrial conditions, developing mechanisms to maintain and optimize the water balance, CO_2 assimilation, nutrient uptake and to tolerate excessive solar radiation. In only 50 million years, evolution changed the body plan of plants, which now exhibited roots, leaves, cuticule, stomata, seeds, secondary xylem and phloem. The growth of a thick periderm (bark) to protect the plant organelles also supported the growth of trees as tall as 20 m and led to the development of forests in terrestrial ecosystems. Soil organic matter accumulation began and most of the primary production was routed through detritivorous foodwebs, which dominate the fossil record of the Carboniferous.

The first terrestrial forests were dominated by ancient lycopsids and pteridophytes. Periderm and cortex were the main mechanical support for these dominant trees, and their extremely thick ancient bark represented as much as 80% of body mass, leading to about 45% of lignin content in above-ground biomass. During the Mesozoic period, ancient gymnosperms replaced lycopsids and ferns as the dominant trees, possibly due to lower energetic requirements to produce their mechanical support (wood rather than periderm) and reduced lignin production in their tissues. In the late Cretaceous period, angiosperms and pinaceous flora of gymnosperms, known to possess lower contents of lignin and lignocellulosic mechanical tissues, replaced the ancient gymnosperms.

It is believed that the accumulation of lignin-rich plant detritus on emerged land during the Paleozoic period led to the very high rates of organic carbon burial in sediments. The high productivity and extent of coverage of the Earth's terrestrial ecosystems with the ancient lycopsid and pteridophyte forests, coupled with the low degradability of lignin, probably explains the subsequent perturbation of the atmospheric concentrations of CO_2 (depleted) and O_2 (increased) during the Paleozoic. The abrupt decline in the atmospheric O_2 concentration from 35% to 21% O_2 content about 200 million years ago is thought to have signalled, a new equilibrium in the co-evolution of Plantae, anabolizers of decay-resistent compounds, and Fungi, which act as catabolizers of decay-resistent compounds.

The most important fungi for lignin decomposition are the basidiomycetes (white-rot fungi). Due to the complexity of lignin, it must be broken into simpler subunits before it can be absorbed and metabolized by the fungus. The basidiomycetes produce and secrete the extracellular enzymes needed for lignin breakdown, but extracellular ligninolysis produces very little energy gain for most white-rot fungi. The major energy gain comes from having access to cellulose, hemicelluloses, proteins and other nutrients embedded within and protected by the complex net of phenolic chains. It has been proposed that lignin-degrading ability, which involves several complex enzymes, did not evolve until available plant detritus had lower lignin content and greater concentrations of other energy-rich substrates. Although fungi were present in the early Paleozoic, it is presumed that most were not decomposers. By the end of the Paleozoic era, all modern classes of fungi had appeared in the fossil record. This illustrates a 'megasymbiosis' that links the co-evolution of plants (lignin producers) and fungi (lignin decomposers) to stored soil organic carbon reserves and the atmospheric composition, as well as to modern-day processes such as global biogeochemical cycles and energy fluxes. Understanding soil ecology helps us to comprehend the complexity and evolution of life on Earth.

Algae

The most well-known algae are those living in aquatic environments, such as lakes and oceans. In the soil, we find green algae, yellow-green and diatoms living on the surface and in the top few centimetres. Here we consider the free-living soil algae, although we cannot forget about the terrestrial algae that enter a symbiotic relationship with fungi to form lichens. In the new taxonomic classification of eukaryotes, algae are considered to be 'plant-like protists', because they are closely related to other unicellular eukaryotes. A further discussion of the taxonomic classification of protists is discussed in Section 3.1.

Several hundred species of algae have been isolated from soils around the world. Algal populations typically range from 10^5 to 10^6 cells/g of soil. As photosynthetic organisms, they rely on sunlight as an energy source. These eukaryotic organisms may be unicellular or filamentous. Their cell walls are similar to those of green plants, although diatoms are covered by a siliceous outer layer. However, there are three features of algae that distinguish them from other green plants. First, the algal body is not differentiated into roots, stems and leaves with vascular tissue. Photosynthesis takes place in a thallus and the body is attached to rocks or soil surfaces with hair-like rhizoids. Secondly, there is no embryo to protect the developing zygote (sperm and eggs fuse in water films and the zygote develops outside the plant). Finally, the gametes are produced within a single cell without any sterile cells protecting the gametes.

The most numerous soil algae are the green algae (Chlorophyta), which are divided into three orders: Tetrasporales, Chlorococcales and Chlorosarcinales. There are about 350 species of Chlorophyta that grow on soil and other non-aquatic substrates. Green and yellow-green algae grow particularly well on moist, non-flooded, acidic soils, appearing as a green scum on the surface that turns into a hard, black crust when soils dry out. In temperate regions, algal populations are largest during the wetter spring and autumn months than in the summer. With their hard siliceous coating, diatoms are able to withstand drier soil conditions and large populations are found in well-drained land rich in organic matter.

Perhaps the most famous soil alga is the 10 μm long, unicellular, flagellate green alga *Chlamydomonas reinardtii*, a model system for studying photosynthesis and cell motility. The complete genome (15,000 genes) of this organism was sequenced in 2007. Comparative gene analysis showed that 35% of the protein homologues in the alga were found in plant and human cells, and an additional 10% of the homologues were common between the alga and humans, but not other plants. It was also found that this alga has retained many genes that were lost during the evolution of vascular land plants and has others that are associated with animal cells. The genetic code for a number of critical metabolic processes has been discovered. Some of these genes encode regulatory elements involved in mating and sexual signalling. Others code for transporter proteins that enable the alga to obtain nutrients from the soil solution, and numerous genes are involved in sugar and polysaccharide synthesis as well as in construction of a chloroplast for photosynthesis.

Soil algae are well known for their ability to colonize bare surfaces, but they can be found covering standing logs and even tree leaves. Volcanic soils, desert soils and rock faces are well known to support large populations of soil algae. Together with cyanobacteria, mosses, fungi and other living organisms, the algae form biological soil crusts that are very important for soil stabilization and pedogenesis. Biological soil crusts cover land where vascular plants are absent, and may constitute up to 70% of the soil cover in deserts. The filamentous algae and cyanobacteria swell and contract, depending on moisture conditions, leaving a network of sheaths that bind to the soil surface through rhizoids and extracellular polysaccharides. Crust thickness may range from a few millimetres to 10 cm thick, thus providing some protection against erosive forces and helping to retain soil moisture. In addition, soil algae release carbonic acid, which contributes to soil weathering, nutrient release and soil formation. As pioneer species, these organisms prepare soils for subsequent succession by vascular plants. Although biological soil crusts are well adapted to harsh growing conditions, they are fragile and easily disrupted by human, animal and vehicular traffic. Soil algae are also found in agricultural, grassland and forest ecosystems, but their numbers are rather small due to competition and shading by larger plants.

2.2 Constraints to Microbial Growth

Soil bacteria have the potential to grow very rapidly and complete their life cycle within 2–3 days when soil conditions are favourable and energy sources are abundant. In reality, the average turnover of soil bacterial populations is about 2.5 years, meaning that bacterial activity is constrained most of the time. The same is true for other soil microorganisms. The following are factors that can constrain microbial growth, based on what is known about culturable microorganisms.

Soil water content

Optimal bacterial activity occurs at water potentials between −0.01 and −0.05 MPa (e.g. when soil pores larger than 2–4 μm are filled with water). The soil water content also influences soil aeration and redox reactions (shown also in Chapter 1), the concentration of soluble substances, osmotic pressure and the pH of the soil solution. For these reasons, water management (drainage and irrigation) in agroecosystems and drainage/flooding in natural terrestrial ecosystems is expected to affect soil bacterial activity. The soil algae are more sensitive to dry conditions than the bacteria and actinomycetes, and fungi are the most tolerant of dessication.

Although bacteria and actinomycete activities cease at soil matric potentials less than −1.5 MPa, fungi remain active at matric potentials of −1 to −5 MPa. Plants benefit from symbiotic mycorrhizal fungi, which can extend their hyphae into small micropores that larger plant roots cannot penetrate. The capillary water acquired by the fungi is shared with the host plant, contributing to the survival of both.

Soil temperature

Microbial activity can occur at temperatures below freezing when archaea and bacteria are adapted to very cold climates (e.g. in the Arctic and Antarctic). These archaea and bacteria produce compounds that function like antifreeze and thus protect the integrity of their cellular membranes from freezing. Thermophilic archaea and bacteria produce heat-stable proteins that allow enzymes to function at high temperatures (e.g. in compost heaps, around hot springs and thermal vents). However, soil bacteria are generally most active between 20 and 40°C. Fungal activity is optimal from 20 to 37°C and declines at higher temperatures. Soil actinomycetes are more heat-tolerant and thrive at temperatures up to 45°C.

Soil pH

Some bacteria thrive in extremely acidic or alkaline environments, but most soil bacteria are active within the pH range encountered in terrestrial ecosystems, from pH 4 to 9. Soil bacteria in agroecosystems are most active when the soil pH is near neutral, from pH 6 to 7. Actinomycetes prefer neutral to slightly alkaline soil conditions (pH 6 to 8), while fungi and green algae grow more prolifically in slightly acidic soils (pH 4 to 6).

Salinity

Halophilic archaea are capable of living in extremely saline conditions (including salt springs, salt lakes, sea water and indeed salt pools), but most bacteria do not tolerate highly saline conditions. In general, bacteria can withstand more saline conditions than agricultural plants. Most agricultural crops grow optimally when the soil electrical conductivity ranges from 0 to 4 dS/m, although crops such as cotton and sunflower grow well in moderately saline soils (about 8–12 dS/m) and sunflower can tolerate salinity up to 16 dS/m. Unlike their marine relatives, terrestrial algae do not tolerate highly saline soil conditions and populations decline dramatically when the soil electrical conductivity exceeds 8 dS/m.

Mobility

Soil bacteria are normally adsorbed to soil particles by ion exchange mechanisms, such as ionic bridges involving polyvalent cations. Most have a limited range of movement, but can travel and disperse through the soil in percolating water from rainfall or irrigation. Another means of dispersal is with larger soil biota such as earthworms that consume soil particles containing bacterial cells. Some bacterial cells will pass intact through the gut of the soil animal and get deposited in a new location in the soil. The survival of Gram-negative bacteria is related to their ability to disperse passively into microhabitats such as soil micropores (< 10 μm diameter), where they are protected from predators (Fig. 2.12). Flagellated bacteria and algal cells can move short distances by propelling the cell by means of a whip-like tail on one end of the cell. Actinomycetes and fungi acquire nutrients and water by extending their hyphae in the direction of substrates or in water-filled pores.

Soil bacterial populations may be found most of the time in a resting or inactive state, punctuated by brief periods of activity, growth and reproduction. It is estimated that bacterial cells and colonies can survive for 50 years or longer in vegetative states or as spores. There are several adaptations that permit bacteria to survive during long periods of inactivity:

- Efficient use of substrates: part of the microbial colony may survive by assimilating the cellular contents of lysed cells in the colony.
- Cell maintenance with minimal energy expenditure: an active cell may metabolize $0.4–1.7 \times 10^{-3}$ mg glucose/mg of bacteria/h. The maintenance requirements for the same cell could be as low as $1.6–4.3 \times 10^{-6}$ mg glucose/mg of bacteria/h, nearly 1000 times lower.
- Physical protection: Gram-negative bacteria can survive within microaggregates, where a coating of clay particles and extracellular polysaccharides provides cover and prevents the cells from desiccating, even at low soil moisture contents.
- Production of resistant spores: Gram-positive bacteria produce resistant endospores when the soil moisture content falls below optimum levels.

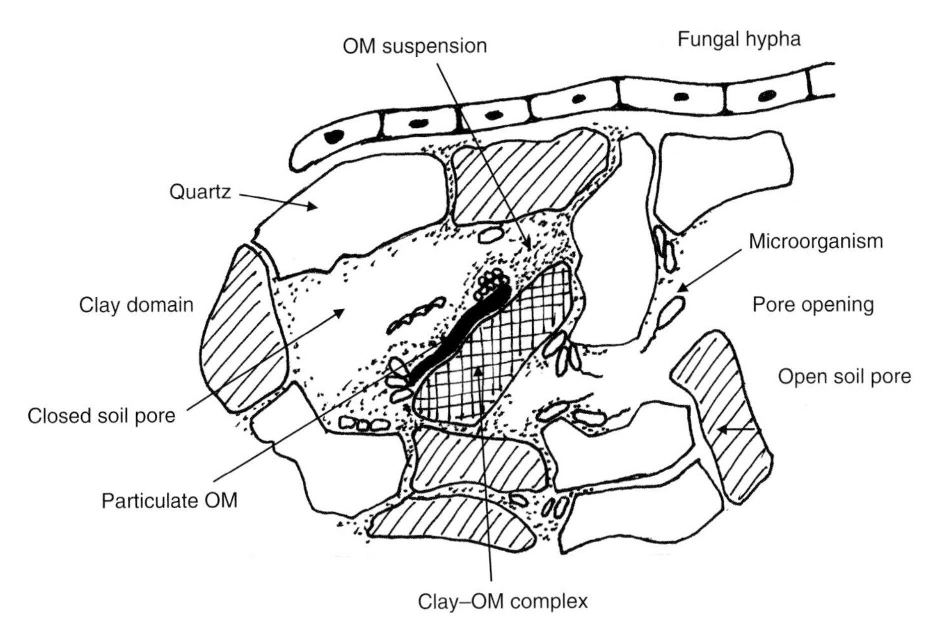

Fig. 2.12. Model of soil aggregate, showing Gram-negative bacteria entering through a micropore and attaching to small clay and clay–organic matter complexes within the aggregate. A fungal hypha would probably be too large to enter the pore.

Inorganic and organic pollutants

The inorganic pollutants in soils are trace metals from natural sources such as volcanic eruptions and forest fires, as well as from anthropogenic activities such as smelting, burning fossil fuels and producing metallic products. Metals can bind on reactive enzyme surfaces and interfere with the three-dimensional movement of the protein that is necessary for substrate hydrolysis. The nitrogenase enzymes responsible for biological N_2 fixation are affected by some trace metals, so biological N_2 fixation is generally lower in metal-contaminated soil than in uncontaminated soils. Algae and cyanobacteria are generally the most sensitive to metal contaminants, heterotrophic bacteria have moderate tolerance, while actinomycetes and saprophytic fungi are least affected. In soils with multiple metal contaminants, saprophytic fungi such as *Penicillium, Aspergillus, Trichoderma, Fusarium, Rhizopus* and *Mucor* dominate the soil microbial communities. Melanized fungi such as *Cladosporium* appear in soils contaminated with copper and mercury.

The toxicity of trace metals to microorganisms depends a good deal on soil physico-chemical conditions, speciation, mobility and chemical reactivity of the substance. Despite the apparent toxicity, many microorganisms survive, grow and even thrive in metal-polluted soils. Several mechanisms allow microorganisms to tolerate such conditions, which affects the cycling of metals between soil biotic and abiotic components. Some general characteristics of these tolerance mechanisms, listed in Table 2.8, are as follows:

- Bacterial plasmids with resistance genes: plasmid-determined resistance is highly species-specific, although most bacterial groups appear to have resistance systems. Resistance mechanisms can involve metal efflux from the cells using an ATPase pump or chemiosmotic pump. Enzyme detoxification is another possibility. For example, Gram-positive and Gram-negative bacteria with mercury resistance genes are able to transport Hg^{2+} to the intracellular detoxifying enzyme mercuric reductase, which reduces Hg^{2+} to elemental Hg°. Intracellular binding of Cu^{2+} in bacteria is controlled by chromosomal genes that also regulate Cu uptake and efflux.
- Fungal vacuoles: regulation of cytosolic metal ion concentrations and detoxification of potentially toxic metal ions occurs when metals are preferentially sequestered by the vacuole via active transport (ATPase pump).

Table 2.8. Microbial tolerance mechanisms with respect to toxic metals and metalloids. (Adapted from Gadd, 2005.)

Tolerance mechanism	Toxic metals/metalloids
Bacterial plasmids with resistance genes leading to metal efflux, enzyme detoxification or intracellular binding	Ag^+, AsO_2^-, AsO_4^{3-}, Cd^{2+}, Co^{2+}, CrO_4^{2-}, Hg^{2+}, Ni^{2+}, Sb^{3+}, TeO_3^{2-}, Tl^+, Zn^{2+}
Fungal vacuoles	Ca^{2+}, Co^{2+}, Cs^+, Fe^{2+}, K^+, Li^+, Mn^{2+}, Ni^{2+}, Sr^{2+}, Zn^{2+}
Fungal cytosol sequestration	Cu^{2+}, Cd^{2+} and possibly Ag^+, Au^{2+}, Hg^{2+}, Ni^{2+}, Pb^{2+}, Sn^{2+}, Zn^{2+}
Modification of soil pH, secretion of chelating agents	Most metals and metalloids
Redox reactions	Cr^{6+}/Cr^{3+}, Fe^{3+}/Fe^{2+}, Hg^{2+}/Hg°, Mn^{4+}/Mn^{2+}, Se^{4+}/Se^{2+}, U^{6+}/U^{4+}
Extracellular binding substances	Most metals and metalloids
Methylation reactions	Metalloids: AsO_2^-, AsO_4^{3-}, SeO_4^{2-},SeO_3^{2-}, TeO_3^{2-}

- Fungal cytosol sequestration: low-molecular weight, cystein-rich proteins (metallothioneins) and peptides derived from glutathione (phytochelatins) bind to the toxic metal, creating a crystallite core containing the toxic metal covered with an outer layer of peptides.
- Modification of soil pH and secretion of chelating agents: acidification of the soil environment by proton efflux increases metal mobility, so the substance is leached away from the microorganism. Organic acids secreted by microbial cells also serve to acidify the soil environment. Organic acids such as citrate and oxalate can form stable complexes with metals, which facilitates their leaching through the soil profile.
- Redox reactions: the reduction of metals such as Fe^{3+} to Fe^{2+} by anaerobic bacteria that use Fe^{3+} as the terminal electron acceptor increases the solubility of iron molecules. In contrast, the reduction of Cr^{6+} to Cr^{3+} reduces the solubility of chromium.
- Extracellular binding substances: metals can be adsorbed and stabilized on the external surface of cell walls. The peptidoglycan carboxyl groups in Gram-positive bacterial cell walls are the main binding site, with phosphate groups contributing significantly in Gram-negative species. Fungal cell wall components including chitin, phenolic polymers and melanins are effective biosorbents. The extracellular polymeric substances (polysaccharides, mucopolysaccharides and proteins) produced by bacteria and fungi are effective at binding metals.
- Methylation reactions: the methylation and subsequent volatilization of methylated selenium derivatives is widespread among soil bacteria and fungi. This may be important in transporting Se between terrestrial and aquatic systems. The most frequently produced volatile substance is dimethyl selenide. Dimethyl telluride and mono-, di- and tri-methylarsine compounds are also produced by soil microorganisms.

Organic pollutants in soils are man-made substances such as pesticide residues, polyaromatic hydrocarbons, hydrocarbons (gas, oil) and chlorinated compounds. All organic pollutants can eventually be degraded by soil microorganisms, but the breakdown of some is very slow. The process of co-metabolism by which actinomycetes degrade recalcitrant substances was discussed in a previous section. Generally organic pollutants are degraded by microbial communities, rather than by single organisms.

2.3 Diversity and Ecological Functions of Soil Microorganisms

The extraordinary contribution of soil microorganisms to ecological integrity in natural and managed ecosystems is possible because of the genetic diversity of the soil microbial community. It was only recently that researchers began to appreciate the extent of genetic diversity in soil systems. In the 1990s, Dr Torsvik and his research team in Norway examined the microbial DNA in a beech forest soil and reported more than 1.5×10^{10} bacteria/g of soil (about 1 teaspoon of soil). They estimated that these bacteria belonged to about 4000 independent bacterial genomes (Torsvik *et al.*, 1990). This work confirmed earlier suspicions that the bacteria we are able to grow in the laboratory represent a tiny fraction (perhaps less than 1%) of all the bacteria that live in the soil. Current advances in molecular biology have led to the discovery of many more soil microorganisms, including archaea that were not previously known to inhabit temperate ecosystems. It is clear that many microbial species remain to be discovered, but how to define a species still remains an open question. Many microorganisms that appear phenotypically and genetically similar have certain adaptations that permit them to perform unique ecological functions. Should these be classified as 'strains' of a particular species or designated as a new species? Estimates of soil microbial diversity in the 1990s suggested there were about 30,000 species for soil bacteria and 1.5 million distinct species of fungi. These numbers may be conservative, depending on how the species concept is applied to genomic data generated by DNA and RNA analysis of microbial communities.

Soil microorganisms are often considered the driving force (engine) of most terrestrial ecosystems because they largely control the turnover and mineralization of organic substrates. Without microorganisms, there would be very little organic matter decomposition and the world would literally be buried in a layer of organic residues. The ecological diversity of soil microorganisms has been recognized for some time, including their ability to:

- produce hydrolytic enzymes capable of decomposing and mineralizing a wide array of organic substrates, including organic nitrogen, phosphorus and sulfur compounds essential for plant growth, especially in natural ecosystems where fertilizers are not normally applied;
- mediate polycondensation reactions that create novel chemical compounds not generated by plants and animals, including polysaccharides that bind soil particles into aggregates;
- release organic acids that accelerate the weathering process and contribute to soil formation; and
- produce antibiotics and specialized chemical compounds that repel predators, protect the cells against freezing or heat stress and aid in binding microbial cells to solid surfaces.

Microorganisms inhabiting the soil have, through gene mutations, genetic transfers and natural selection, adapted to produce populations and communities of organisms with specific forms and physiology that thrive in a particular soil habitat or niche. Let us consider how genetic factors affect the diversity within microbial communities.

Gene mutations are heritable changes in the base sequence of DNA. These may arise from spontaneous mutation, a naturally occurring process that takes place about once in every million replications of DNA. A gene mutation can also be induced by environmental factors, such as exposure to chemical agents or stress during cellular development and growth. When a mutation occurs in the genome of a bacterial cell, any offspring produced by that cell through binary fission will inherit DNA with the same mutation. It is unlikely that microbial cells can control the processes that lead to gene mutation. Natural selection, i.e. the process by which lineages of organisms with favourable, inherited traits become more common in successive generations, can lead to the emergence of microbial genotypes that are adapted to specialize in a certain ecological niche or environment.

Gene transfer occurs when genes are transferred from an external source or donor to a recipient cell. There are several mechanisms that lead to gene transfer in bacteria:

1. Transformation: extracellular DNA is taken up by a bacterium and integrated into its chromosome by displacing a homologous segment of chromosomal DNA.
2. Transduction: genes are transferred between bacterial cells by viruses known as bacteriophages or phages.
3. Conjugation: this requires direct physical contact between bacterial cells. The genetic material transferred may be a plasmid or a portion of a chromosome mobilized by a plasmid.

The most common transfer mechanism for soil bacteria is probably conjugation, since most soil bacteria live in colonies or biofilms on surfaces and thus are in direct contact with neighbouring cells. Bacteria may be able to control gene flow into their cells, and it has been suggested that conjugation occurs when a donor cell receives a signal molecule from the recipient cell.

A complete description of soil microorganisms living in a particular zone within the soil requires information on their activity within the habitat and their genetic diversity. This is illustrated in Fig. 2.13, which shows the complementary nature of culture plates/chemical analysis, DNA-based analysis, microscopy and process studies to understand the diversity, activity and biomass of microorganisms in the laboratory or in nature.

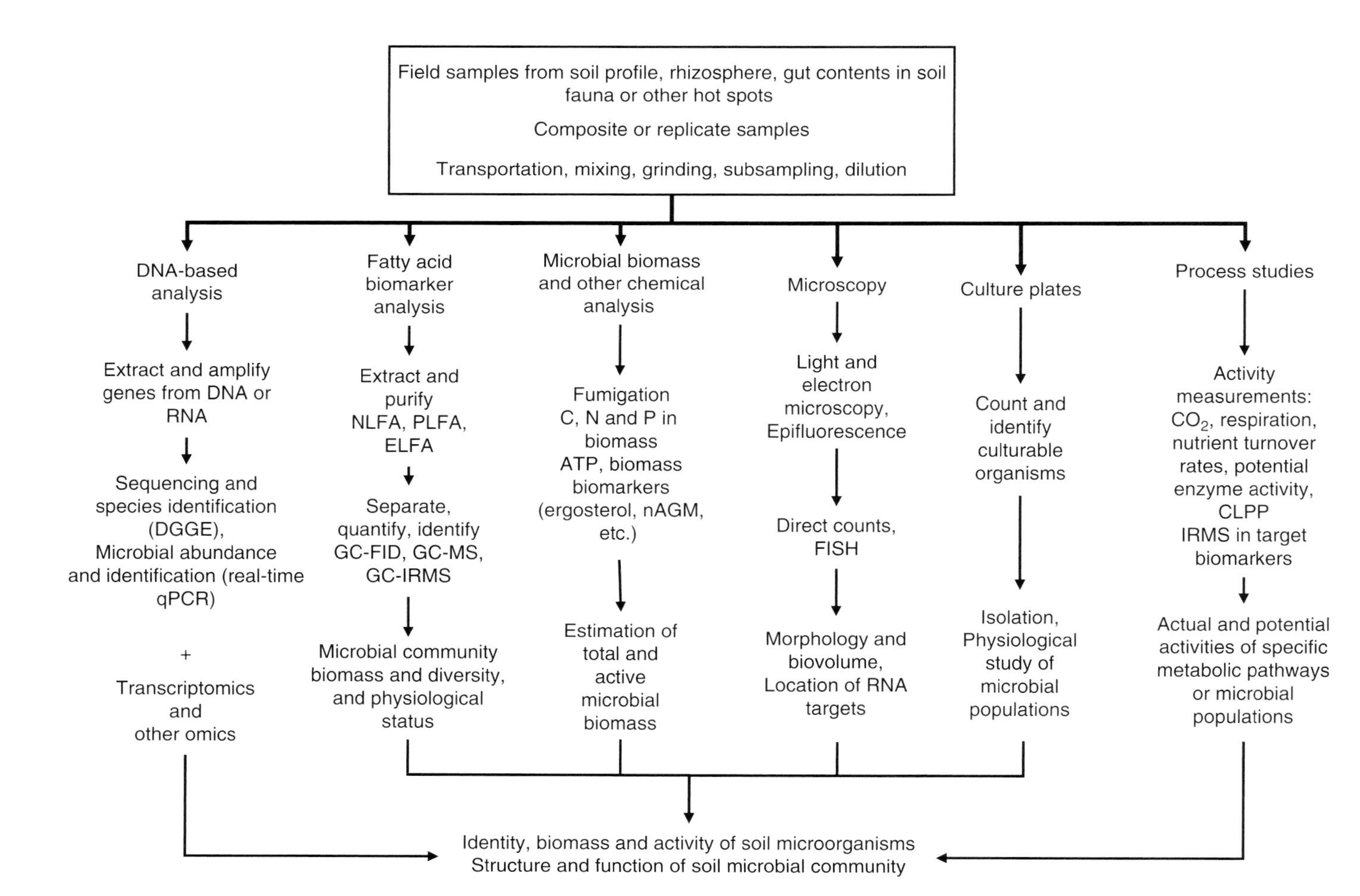

Fig. 2.13. Scheme of currently available approaches for determining microbial diversity, biomass and activity in soil samples.

2.4 Methods for Collecting and Enumerating Soil Microorganisms

The first step in studying soil microorganisms is collecting them from the field. Soil microorganisms are most abundant in the top 2.5–5.0 cm of the soil profile (Fig. 2.14). When the surface soils become warm and dry in the late spring and summer months, microbial cells may become desiccated in the surface layer and thus populations may be larger in the moister layer just below the soil surface. Researchers working in forest soils often collect samples from the litter layer because this respresents a moist, substrate-rich habitat that is highly favourable for bacterial and fungal growth. Samples of the underlying mineral soil can be collected for comparison.

Sampling location and time are also important considerations. Symbiotic microorganisms like the N_2-fixing *Rhizobium* species or arbuscular mycorrhizal fungi such as *Glomus* species are associated with living roots, so plant roots need to be collected during the growing season. Free-living soil microorganisms are abundant around living roots and on decaying plant residues, so sampling is often done during the growing season or soon after senescence.

Soil microorganisms can be collected in composite samples or replicate samples (Fig. 2.13). A composite sample is made by taking multiple soil cores from an area and mixing them together. This provides an 'average' sample from a field or from a homogeneous area within a field. A replicate sample is taken from one experimental unit in a replicated experiment.

It is desirable to minimize handling and processing, to avoid unnecessary disturbance of the soil microorganisms. Soils can be mixed and coarsely sieved through a 2–6 mm mesh sieve in the field to remove large stones and roots. Generally, soils for microbial analysis are transported in an insulated cooler and stored in a refrigerator at 4–6°C to preserve the integrity of the sample. Further mixing and grinding in the laboratory could be necessary prior to analysis. Subsampling generates analytical replicates, so the repeatability of a particular analysis can be verified in the laboratory. When microorganisms are cultured on selective media, it is necessary to prepare serial dilutions so that the number of individuals can be counted – when too many microorganisms grow on a culture plate, it becomes difficult to identify individual colonies

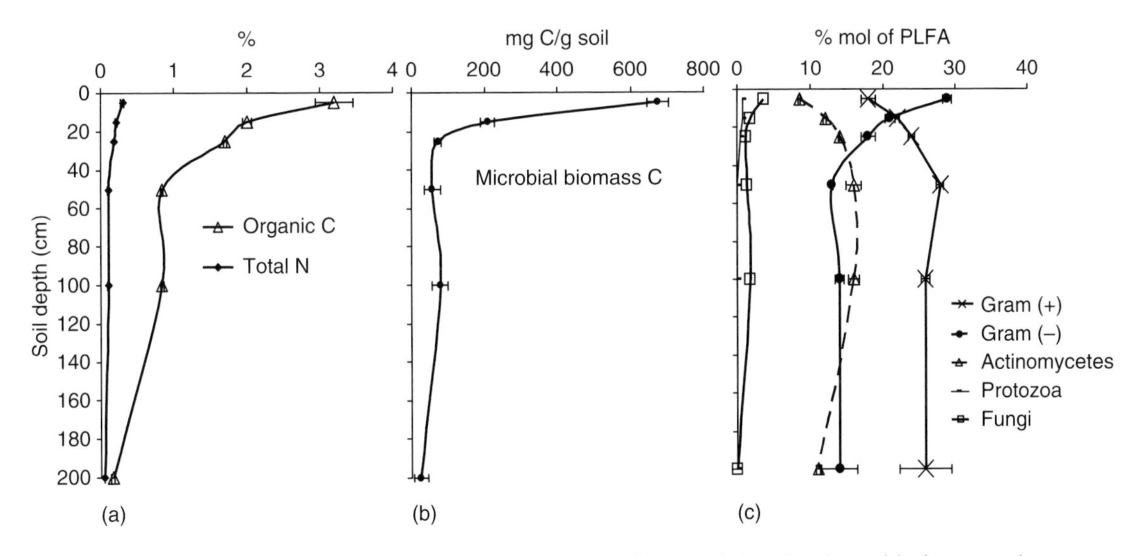

Fig. 2.14. Changes in soil organic matter (a), microbial biomass (b) and relative abundance (c) of some major groups of the microbial community within the soil profile in a Mollisol near Santa Barbara (California, USA). Most of the microbial biomass was found in the first 25 cm. Microbial diversity also declined with depth. Microbes in deeper soil profiles appeared to be more carbon limited than surface microbiota as revealed by PLFA physiological indexes. Abundances of Gram-negative bacteria, fungi and protozoa were the highest at the soil surface, while Gram-positive bacteria and actinomycetes showed increasing relative abundances with soil depth. Researchers suggested that vertical distribution of microbial groups may be linked to carbon availability. (Data from Fierer *et al.*, 2003.)

and estimate the population size. Samples destined for fatty acid analysis should be frozen and then dried while still frozen (freeze-dried) in a lyphilizer to minimize the disturbance of microbial cells. Prior to DNA and RNA analysis, samples can be frozen or stored in ethanol.

There are a number of methods and techniques that can provide insight into the diversity, biology and ecology of the soil microbial community. Some common techniques are shown in Fig. 2.13 and described in more detail below.

Cultural and chemical analyses

Culture plate analyses require the selection of a media containing a carbon source and nutrients known to promote microbial growth that is prepared and poured into Petri dishes or into test tubes. The media may contain agar and solidify upon cooling, or be an energy- and nutrient-rich broth for microbial growth under aerobic or anaerobic conditions. Next, about 1 ml or less of a soil-water mixture (1 g of soil diluted with sterile water to generate serial dilutions of 10^{-3} to 10^{-6} or more dilute concentrations) is added, spread evenly on the surface of agar-based plates or mixed with the broth in test tubes. The Petri dishes or test tubes are placed in an incubator at a favourable temperature for microbial growth (often 37°C) for several days until colonies are detected. The number of colonies on a plate can be counted directly. Bacteria or fungi belonging to target groups can be identified by using selective media or by adding antibiotics to the media. This approach is quite useful for researchers who are looking for specific microorganisms that grow well under laboratory conditions.

The chemical analysis methods aim to measure the mass (biomass) of soil microorganisms. The classic method, involving measuring soil microbial biomass through chloroform fumigation, is called the fumigation-extraction method. Beakers of soil are placed in a vacuum desiccator and exposed to chloroform vapour for several days. The chloroform lyses microbial cells. The nutrients contained in soil plus those released from lysed microbial cells are extracted with a chemical solution and analysed. The microbial biomass carbon, nitrogen and phosphorus concentrations are then estimated by the difference between nutrients in control soils and chloroform-fumigated soils. Another technique involves extracting adenosine triphosphate (ATP), a form of stored energy, from microbial cells. By extracting the microbially bound nutrients or ATP from a soil sample, we get an idea of how many microorganisms may be present in the sample. The classical techniques are inexpensive and give the researcher a general idea of how many microorganisms are present in soil coming from a natural environment; however, they cannot be used to identify species or determine the microbial diversity in the soil environment.

Advanced chemical analysis involves extracting compounds that are 'biomarkers', in order to detect the presence and quantity of certain microorganisms. In fungal cells, ergosterol represents 40–100% of the sterols present and has been used to estimate fungal biomass in a variety of agricultural, grassland and forest soils. The analysis involves extracting the ergosterol with solvents (methanol:ethanol and hexane:propan-2-ol) followed by high-performance liquid chromatography. Phospholipid fatty acids in biological membranes can be extracted with solvents and analysed by gas chromatography to determine the microbial biomass and identify microbial groups, as described in Focus Box 2.5.

DNA-based analysis

The advent of DNA technologies provides insight into the diversity and functional capacity of soil microbial communities, particularly for organisms that cannot be cultured using selective media. For instance, the discovery that the Crenarchaeota are widely distributed in agricultural soils and function as ammonia oxidizers was only possible through molecular methods.

Examples of methods used directly to analyse microbial DNA are DNA:DNA reassociation kinetics, nucleic acid hybridization, microarrays and metagenome sequence analysis. The polymerase chain reaction (PCR) has been used to amplify rRNA genes for microbial identification using denaturing gradient gel electrophoresis (DGGE), terminal restriction fragment length polymorphism (T-RFLP), single strand conformational polymorphism, ribosomal intergenic spacer analysis and sequence analysis of 16S rRNA gene clone libraries. A few of the most common methods are shown in Fig. 2.13 and will be discussed below.

The DGGE protocol consists of six steps: sample collection; nucleic acid extraction and purification; PCR amplification of the target gene; separation of

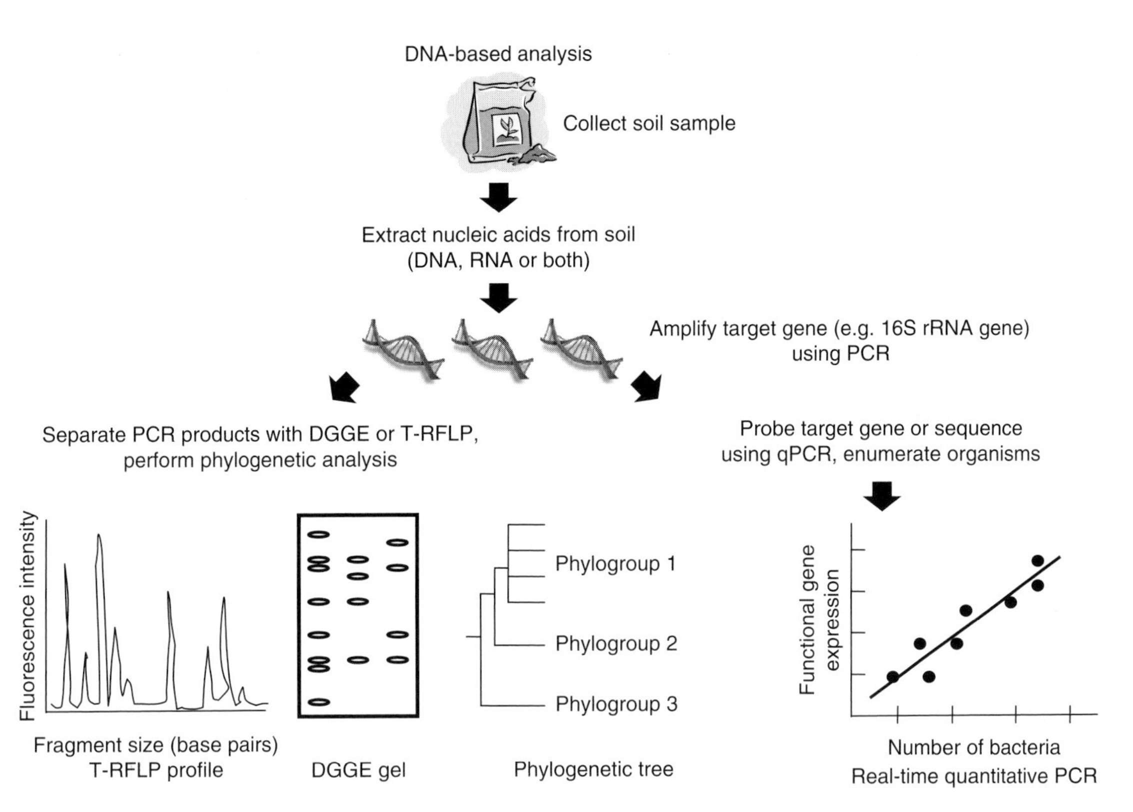

Fig. 2.15. Flow diagram showing the steps for DNA-based analysis of soil microbial communities using the polymerase chain reaction (PCR). An overview of the methods involved in terminal restriction fragment length polymorphism (T-RFLP), denaturing gradient gel electrophoresis (DGGE), phylogenetic trees and real-time quantitative PCR (qPCR) analyses is described in the text. Additional details of these methods are provided by Drenovsky *et al.* (2008).

PCR amplicons by DGGE; visualization of gels; and data collection and analysis (Fig. 2.15). For community analysis, either DNA or RNA can be extracted using primers that target microorganisms from phylogenetic groups, functional groups or individual species (Table 2.9). Since cellular growth is more closely linked to an increase in the total RNA (ribosomes), researchers often look for shifts in microbial activity with rRNA-based DGGE profiles, while rDNA-based DGGE profiles represent the overall microbial community structure.

The T-RFLP analysis is a PCR-based method that offers highly reproducible results and yields a greater number of operational taxonomic units than many other techniques. The advantage of this method is that it provides a comprehensive 'fingerprint' of the microbial community and permits researchers readily to distinguish the microbial origin of extracted DNA or RNA. While it is a rapid, user-friendly technique for distinguishing microbial communities in experimental treatments or from different soils, it is difficult to identify species from complex soil communities with existing

Table 2.9. Examples of primers used to identify soil microbial communities using the PCR-DGGE method.

Target organisms	Nucleic acid	Primers
Eukaryota	18S rRNA	NS1-FR1-GC
Arbuscular mycorrhizal fungi	18S rRNA	NS1-ITS4
Archaea	16S rDNA	PRA46F
Bacteria	16S rRNA	517 r
Actinomycetes*	16S rDNA	F243-R1378 and F984GC-R1378

*Nested PCR requires two sets of primers to complete the amplification

databases with long gene fragments generated from the T-RFLP method.

While DGGE and T-RFLP methods provide a qualitative assessment of soil microbial communities, real-time quantitative PCR (RT-qPCR) offers information about the abundance of specific microbial groups, species and functional genes. The technique is based on a fluorphore probe released from the RNA/DNA template during the PCR reaction (TaqMan assay) or incorporation of a DNA-binding fluorescent dye (SYBRGreenI assay). This technique provides insight into the number of functional gene copies originating from soil microorganisms. For example, genes that produce enzymes such as nitrogenase reductase (*nifH*), nitric oxide reductase (*cnorB*, *cnorBp*), ammonia monoxygenase (*amoA*) and enzymes involved in denitrification (*nirS*, *nirK* and *nosZ* genes) provide insight into soil nitrogen cycling.

Genomics of soil microbial communities is still in its infancy, since the completed whole genome sequences for soil microorganisms are known for hundreds to thousands of bacteria, archaea and fungi (Genomes Online Database, http://www.genomesonline.org). To determine the genome of an organism requires considerable effort, since the DNA must be extracted from a pure culture, fragmented, inserted into plasmid vectors, cloned and sequenced. The resulting cDNA library contains a number of overlapping sequences that need to be aligned into a contiguous sequence, noting the position of genes, operons and promoters. Perhaps of more interest to soil microbiologists is metagenomics, which is the analysis of all the microbial genomes from a soil or other environmental sample. It is not necessary to cultivate the microorganisms or even have prior knowledge of their identity. Databases have been constructed to permit phylogenetic comparison of functional gene sequences, rRNA or other marker genes. Metagenomics allows researchers to compare gene diversity among habitats, find genes for known and new functions and identify the functions of some new genes based on the similarity to other organisms.

Functional genomics is an emerging area for the study of soil microbial communities. Using genetic probes such as 'lux' genes and ice nucleation genes, researchers can determine what conditions stimulate the activity of certain types of microorganisms or physiological capabilities within the microbial community. Microarrays also provide fascinating insight into soil microbial functions. Microarrays are microscopic chips that contain hundreds to thousands of individual gene sequences. The microarray is hybridized to the genomic or metagenomic material, and a fluorescent signal is produced when a match occurs. This technique allows researchers simultaneously to monitor the expression levels of thousands of genes from a single organism or group of organisms in response to environmental stimulus. For instance, a microbial cell exposed to salt stress could be compared with a microbial cell without the same stress, to determine which genes were up-regulated or down-regulated as a result of the stress.

Stable isotope probing is another technique that permits researchers to trace the uptake of stable isotopes of carbon, nitrogen, oxygen, hydrogen and sulfur into the bodies of specific microbial groups. Target molecules from the living microbial cells are extracted and isolated from the soil environment through chemical methods (e.g. PLFA) or PCR-based reactions (e.g. using primers complementary to 16S rDNA or rRNA). Stable isotopes within the PLFAs are detected by gas chromatography followed by isotope ratio mass spectrometry (GC-IRMS). The stable isotopes in DNA or RNA can be quantified by buoyant density gradient centrifugation. For example, the stable isotope ^{13}C is heavier than the more common ^{12}C isotope, thus it is found in the heavier gradient fractions.

Microscopy

Direct microscopy is a technique that uses a light microscope or a scanning electron microscope to obtain information about the morphology and size (biovolume) of microbial cells (bacteria, fungi, algae, etc.). The reproductive structures of bacteria and fungi can be examined in some cases. Microscope slides are often made from laboratory cultures, meaning that the microorganisms viewed with this technique are those that can be grown in the laboratory. It is also possible to coat microscope slides with media and bury them in the field for a period of time, then recover the slide and examine the microbial populations growing on the slide. The disadvantage of this technique is that most media are selective, and only a fraction of the soil microbial community will grow on any medium.

Recent developments in microscopy have focused on techniques to examine the natural microbial communities that do not grow on selective media.

Table 2.10. Examples of oligonucleotide probes relevant to the study of soil microbial communities using the FISH method. A more comprehensive list can be found at probeBase (http://www.microbial-ecology.net/probebase/default.asp).

Target organisms	Probe	Oligonucleotide sequence (5' – 3')
Eukaryota	EUK516	ACC AGA CTT GCC CTC C
Archaea	ARCH915	GTG CTC CCC CGC CAA TTC CT
Crenarchaeota	CREN512	CGG GGG CTG ACA CCA G
Bacteria	EUB338-II	GCA GCC ACC CGT AGG TGT
Alphaproteobacteria	ALF968	GGT AAG GTT CTG CGC GTT
Betaproteobacteria	BET42	GCC TTC CCA CTT CGT TT
Gammaproteobactera	GAM42a	GCC TCC CCA CAT GCT TT
Deltaproteobacteria	SRB385	CGG CGT CGC TGC GTC AGG

Fluorescent *in situ* hybridization (FISH) uses dye-labelled oligonucleotide probes that bind to the bacterial ribosome, which can then be visualized and counted with epifluorescence microscopy, confocal laser scanning microscopy or flow cytometry. Group-specific 16S rRNA FISH probes are available for bacteria and archaea identification based on the 16S rRNA database (Table 2.10). Another application of this technique is to couple FISH probes with microautoradiography (FISH-MAR). In this method, radioactive materials are taken up by metabolically active cells and can be seen clearly as silver grains surrounding the cell when viewed under transmission light microscopy. This permits the researcher to answer questions about the proportion of microorganisms that is metabolically active under various soil conditions.

Process studies

There are many techniques for measuring the activity of soil microorganisms, but most are based on the CO_2 emission from the respiratory and fermentative activities of microbes in soils. The release of CO_2 from soils or consumption of O_2 can be measured in the laboratory or in the field. The assimilation of ^{14}C-glucose and release of ^{14}C-CO_2 from soil is another technique. A similar method is substrate-induced respiration (SIR), which supplies glucose or another carbon source to stimulate the soil microbial activity; the CO_2 released by the microbial community just after substrate addition is considered to be proportional to the biomass of active microbes. Sometimes selective inhibitors are added to determine the efficiency of bacterial or fungal populations in using carbon substrates. Community-level physiological profiles (CLPP) evaluate the ability of microorganisms in soil-water extracts to use a variety of carbon sources and other nutrient substrates, focusing both in the intensity of use and in the diversity of capabilities of the microbial community. This analysis is comercially available for soil extracts in ready-to-use, precharged 96-well microplates called BiologEcoplates™. The mineralization of organic nitrogen compounds leading to the release of ammonium and nitrate is a common indicator of soil microbial activity; it is also possible to measure the ability of soil microorganisms to mineralize organic phosphorus and organic sulfur compounds. The actual and potential denitrification in soils provides insight into the presence of denitrifying bacteria that produce N_2O and N_2 gases. Finally, the energy consumption and output of soil microorganisms can be assessed using miniature circuitry and flow microcalorimeters.

Researchers can assess microbial activity also by studying the potential activity of enzymes found inside active microbial cells (e.g. dehydrogenase, as an indicator of actual respiratory activity). The activity of enzymes outside microbial cells is an indicator of potential degradative activity such as extracellular hydrolases involved in cellulose decompostion (e.g. *β*-glucosidase, cellobiohydrolase from fungi) and lignin breakdown (peroxidase, phenoloxidase, laccase). Other extracellular enzymes of interest indicate inorganic nutrient demand of microbial populations, such as urease, phosphatases and sulfatases that are released from microbial cells and become stabilized on soil organic matter or clay particles. Soil is incubated, generally at 37°C, with a substrate that can be hydrolysed by the target enzyme. The substrate is added with a buffer that will adjust the solution pH to an optimal range for enzyme activity. After the designated incubation period (generally a few hours), chemicals are added to stop the reaction and generate a reaction product (chromophore) that can be measured colorimetrically by absorbance spectrometry, or a fluorescent product such as luciferin (fluorochrome) that can be measured by fluorescence emission spectrometry.

Focus Box 2.5. Soil lipids as biomarkers of soil microbial communities.

Lipids are naturally occurring hydrophobic compounds in biological organisms, deposited as fats and oils, waxes, phospholipids, steroids (i.e. cholesterol) and other related compounds. They are relatively small molecules compared with other biopolymers such as polysaccharides, nucleic acids or proteins. One of the most common types of lipids is fatty acids, long carbon chains with a terminal carboxylic acid that can possess saturated and unsaturated bonds as well as branched chains, hydroxyl substitution and cyclopropyl configurations. The nomenclature of fatty acids is illustrated in Fig. 2.16.

Lipid analysis in an environmental matrix usually requires several steps after sampling, including extraction, purification, concentration and chemical analyses leading to separation, identification and quantification of the targeted lipids. The most common analytical procedures include gas chromatographic separation followed by quantification with a flame ionization detector (GC-FID), usually aided by identification with a mass spectrometer (GC-MS). Statistical analysis and interpretation of the chromatographic profiles usually requires multivariate methodologies such as factorial analyses or clustering.

It is possible to use fatty acids to analyse soil microbial communities, because the cellular membrane of all organisms contains phospholipids and other structural lipids. There are distinct types of fatty acids within various microbial cells, which permits us to use fatty acids as 'biomarkers' to distinguish the different microbial groups (bacteria, fungi) from fatty acids that were synthesized in plants, protozoa, nematodes and other soil organisms (Table 2.11; Fig. 2.17).

The cell membranes of archaea are distinct from all others because the glycerol structural unit is a

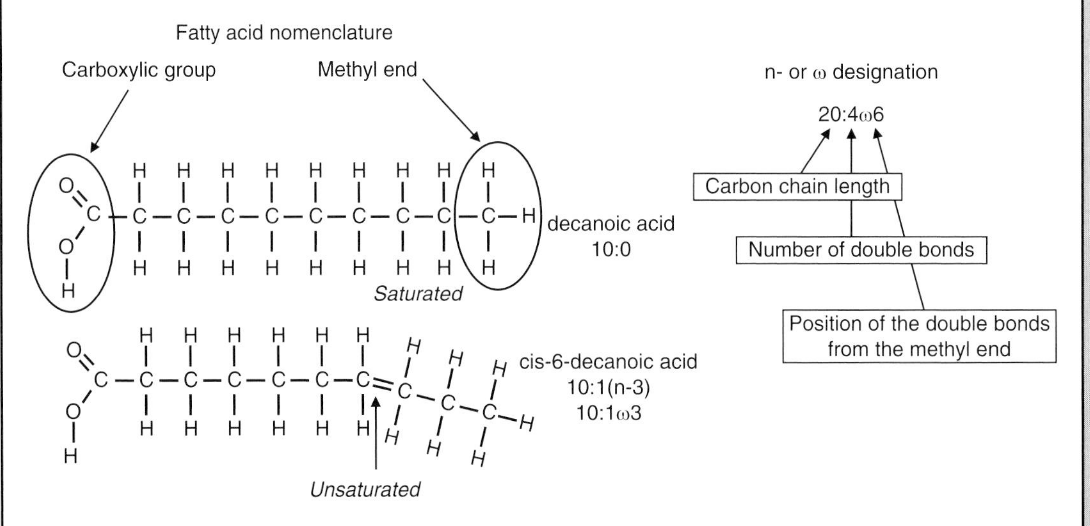

Fig. 2.16. Configuration of fatty acids, showing saturated and unsaturated bonds and the standard nomenclature for naming fatty acids.

Continued

Focus Box 2.5. *Continued*

Table 2.11. Some identified signature fatty acid (FA) biomarkers used to distinguish soil microbial groups and higher organisms.

Group	Fatty acid characteristics	Fatty acid nomenclature
Bacteria		
General bacterial biomarkers	Odd number saturated short-chain FA	15:0, 17:0
Gram-negative	Straight-chain monoenoic	15:1ω4, 16:1ω7, 16:1ω9, 17:1ω9c, 18:1ω7c, 18:1ω9c
Gram-negative	Cyclopropyl FA	cy17:0, cy19:0
Gram-positive	Branched-chain saturated	i14:0, i15:0, a15:0, i16:0, a16:0, i17:0, a17:0
Cyanobacteria	Same as Bacteria plus 18:2ω6	
Actinomycetes	10-Me18:0, 10-Me16:0, 10-Me17:0	
Fungi		
Fungi	18:2ω6 (but also 18:3ω6 and 18:3ω3)	
AM fungi	16:1ω5	
Plants and green algae		
Microalgae	16:3ω3, 16:3ω6, 20:5ω5, 20:5ω3	
Green algae	16:1ω1,3t, 18:3ω3, 18:1ω9	
Higher plants	18:1ω9, 18:1ω11, 18:3ω3, 20:5ω3, 26:0	
Eukaryota (microfauna)		
Protozoa	Long-chain polienoic FA	20:3ω6, 20:4ω6, 20:2ω6,9c, 20:3ω6,9,12c, 20:4ω6,9,12,15c
General Eukaryota	20:2ω6	
In animals and algae but not in higher plants	22:6ω3	

stereoisomer of the molecule found in the cell membranes of bacteria and all eukaryotes, and side branches are formed with a unique ester linkage that lacks a double-bonded oxygen atom (Fig. 2.18). Another key difference is that archaeal membranes are built of isoprene chains (20C atoms long) with side branches. This molecule permits archaea cells to bind together and to form carbon rings on the surface of the cell membrane, which provides protection from extreme environmental conditions such as high temperature or salinity.

Eukaryotic organisms like fungi store energy in neutral lipids such as triacylglycerols, whereas bacteria store energy in compounds such as polyhydroxybutyric acid but not as fatty acids. Then, the specific analysis of the ratios of storage lipids to membrane lipids may provide valuable information about their nutritional status or physiological stress in fungi.

Continued

Focus Box 2.5. *Continued*

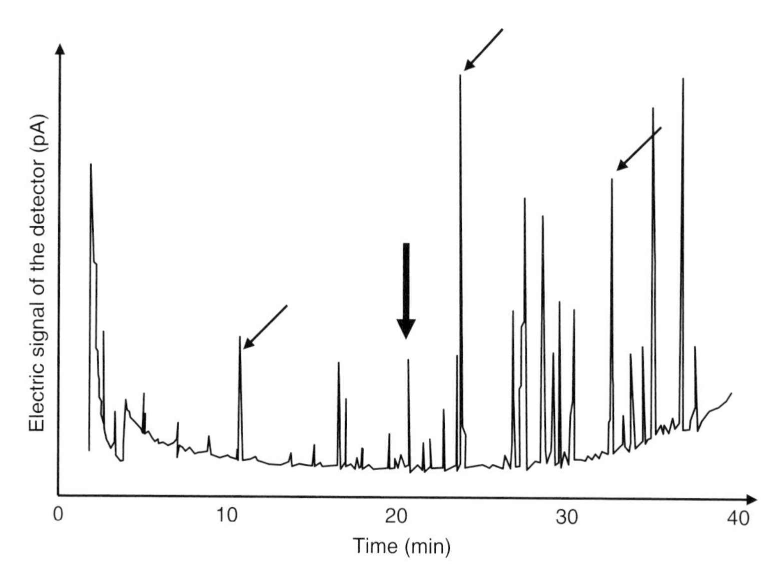

Fig. 2.17. Typical chromatographic profile of fatty acid methyl esters (FAMEs) extracted from a soil microbial community. The whole profile provides a 'fingerprint' of the microbial assemblage. The presence of microbial taxa is confirmed by identifying FA biomarkers (small arrows). The area of these peaks is compared with the peak area of the added internal standard (large arrow) for accurate quantification.

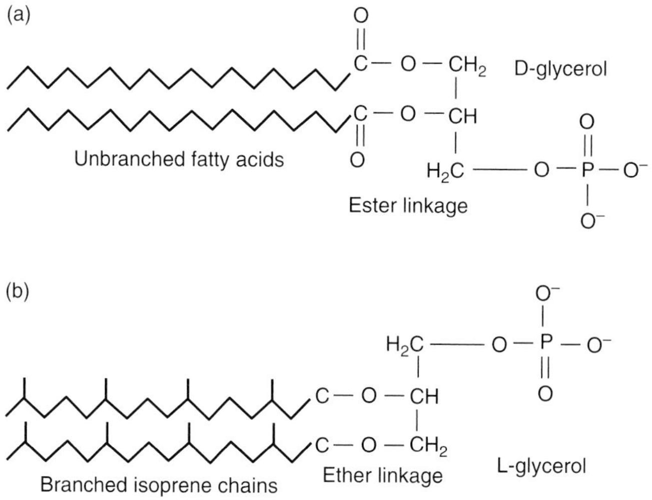

(a)

D-glycerol

Unbranched fatty acids

Ester linkage

(b)

Branched isoprene chains

Ether linkage

L-glycerol

Fig. 2.18. (a) Phospholipids in bacteria and eukaryota have ester linkages. (b) In archaea phospholids join branched isoprene chains with ether linkages.

Continued

Focus Box 2.5. *Continued*

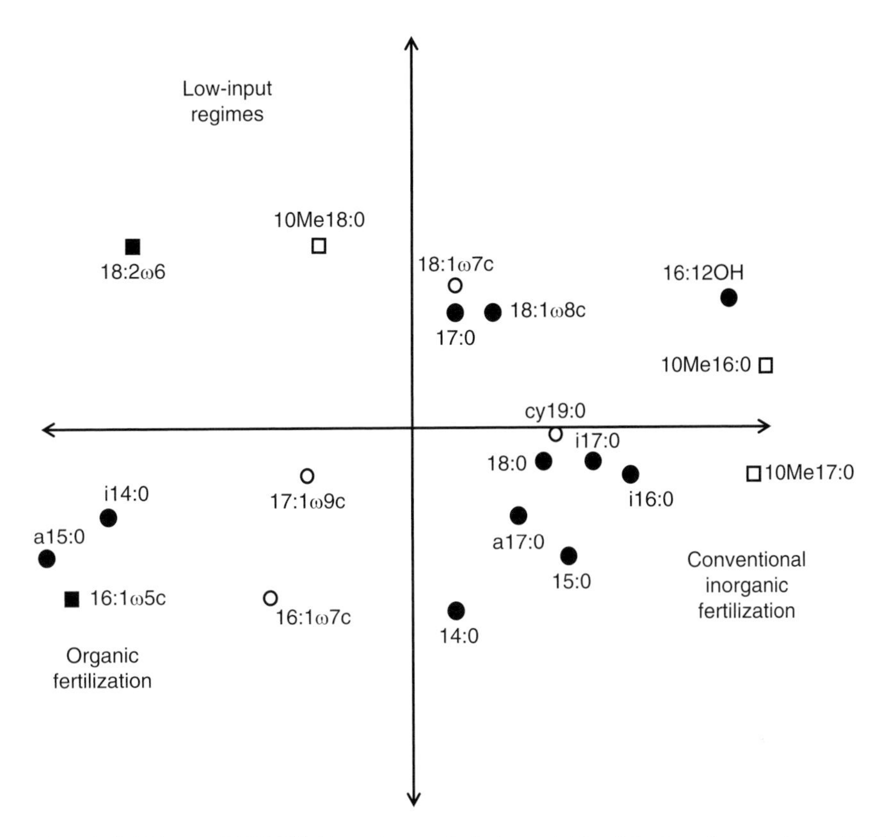

Fig. 2.19. Phospholipid fatty acid (PLFA) biomarker analysis is useful for detecting changes in a microbial community due to changes in land use, land management or soil pollution. Bossio *et al.* (1998) studied the effects of agricultural management, season and soil type on the microbial community PLFA profiles. They found marked changes in the microbial community in organic and conventional agroecosystems. As the ordination plot of multivariate analysis shows, organic soils were enriched with fatty acids from several bacterial and fungal biomarkers, including arbuscular mycorrhizal fungi. They ranked the relative importance of environmental variables in governing the composition of microbial communities in the order: soil type > time > specific farming operation (e.g. cover crop incorporation or side-dressing with mineral fertilizer) > management system > spatial variation in the field. The study of PLFA biomarkers was much more sensitive to changing soil properties and agricultural practices than traditional chemical analysis and process studies, including microbial biomass carbon and nitrogen, substrate-induced respiration, basal respiration and potentially mineralizable nitrogen. Filled dots; Gram-positive biomarkers; open dots, Gram-negative biomarkers; filled squares, fungi; open squares, actinomycetes. (Modified from Bossio *et al.*, 1998.)

3 Soil Microfauna

3.1 Biology of Soil Microfauna

The microfauna are eukaryotic organisms about 2–5500 μm long, much larger than bacteria (Table 3.1). They live in water films and are considered permanent soil dwellers. This group includes the organisms formerly known as protozoa and the nematodes.

Protists (formerly known as Protozoa)

The traditional classification of protozoa was founded on motility groups (flagellates, ameobae and cilates) based on body structures and movement. Phylogenetic description of species based on their ultrastructure, biochemistry and genetics has resulted in a complete reorganization of the taxonomy because species within each motility group come from distinct evolutionary lineages. The great diversity of form, habitat, modes of nutrition and life history exhibited by the protists suggests that they evolved multiple times from simpler prokaryotic ancestors. The Protista are a polyphyletic group whose taxonomy is still being resolved. In recent reviews, the protozoa are considered to be 'animal-like protists' that are closely related to other unicellular eukaryotes, algae and fungi. The common protist groups found in soil, based on the modern taxonomy, are described in Table 3.2.

The 'animal-like' protists are mostly heterotrophic, nonfilamentous, unicellular organisms, although the Euglenida includes species that possess chloroplasts and are capable of photosynthesis. The euglenid cell is covered by a flexible coat (pellicle) that allows the cell to change shape; euglenids can swim using their flagella but can also creep using a peculiar type of 'inching' locomotion known as metaboly. They are more closely related to the algae, which are 'plant-like' protists. The Amoebozoa group includes slime moulds, water moulds and other organisms that are referred to as 'fungus-like' protists because they produce sporangia. The taxonomy and relationships among soil protists are still being resolved with DNA-based methods. A relevant feature of the protists is the occurrence of cell division by meiosis, which is a precursor to true sexual reproduction.

Regardless of their taxonomic classification or trophic group, soil protists are dependent on the presence of liquid water and are therefore only active in those areas in the soil matrix where water-filled pores and water films exist. Even in soils that feel dry to the touch, there may be water in very small capillary pores and a thin film of water covering soil aggregates and organic matter fragments. However, protists are sufficiently large that they are generally found in macropores (greater than 10 μm) and do not enter the smaller pores within microaggregates. Soil pores must be water-filled for protist movement and growth (matric potential between −0.01 MPa and −0.15 MPa).

When soil become too dry or when soil salinity exceeds 25% sodium content, soil protist activity ceases. Populations enter resting states or form cysts that they inhabit until the environmental conditions again become favourable for their activity. Protists may also encyst when food resources are exhausted. The encystment process involves partial dehydration of the cytoplasm and the deposition of cellulose, chitin or protein into a cell wall outside the cell membrane. The cyst can resist very high or low temperatures because the cell and its organelles are protected by the additional cell wall. Testate amoebae gain further protection living inside a test, a hard structure consisting of soil particles glued together by cellular secretions.

Conservative estimates indicate that there are at least 50,000 named species of protists and the global species richness could be much greater. Protists are abundant in soils, with 10^4 to 10^7 active protist

©CAB International 2010. *Soil Ecology and Management* (J.K. Whalen and L. Sampedro)

Table 3.1. Average length and volume of soil protozoa (old terminology) and nematodes, compared with bacteria.

Group	Length (μm)	Volume (μm^3)	Shape
Bacteria	< 1–5	2.5	Spheres, cocci, rods
Protozoa			
Flagellates	2–50	50	Spheres, pear-shaped, banana-shaped, propelled by a whip-like organelle
Amoebae			
Naked	2–600	400	Pseudopodia, movement by cytoplastmic streaming
Testate	45–200	1000	Build oval tests (shells) made of soil
Giant	6000	4×10^9	Enormous naked amoebae
Ciliates	50–1500	3000	Oval, elongated, flat body
Nematodes	250–5500	5000	Long, slender, wormlike

Table 3.2. The taxonomy of soil protists with a description of the general locomotion and trophic functions of each group. (Adapted from Adl *et al.*, 2005.)

Taxonomic group	Example(s)	Locomotion, morphology	Trophic function(s)
Amoebozoa	*Acanthamoebidae*	Amoeboid	Bacterivore
	Flabellinea	Amoeboid	Bacterivore, cytotrophy, detritivore, fungivore, invertebrate consumers
	Tubullinea	Amoeboid, flagellate stages	Bacterivore, cytotrophy, detritivore, fungivore
	Mastigamoebidae	Amoeboid, flagellates	Bacterivore
	Eumycetozoa	Amoeboid, flagellate stages	Bacterivore
Parabasalia	*Cristamonadida*	Flagella	Symbionts
	Spirotrichonymphida	Amoeboid, flagella	Symbionts
	Trichomonadida	Amoeboid, flagella	Symbionts
	Trichonymphida	Amoeboid, flagella	Symbionts
Preaxostyla	*Oxymonadida*	Flagella	Symbionts
Euglenozoa	*Euglenida*	Flagella	Bacterivore, cytotrophy, photosynthetic
	Kinetoplastea	Flagella	Bacterivore, parasites
Heterolobosea	*Acrasidae*	Amoeboid	Bacterivore
	Gruberellidae	Amoeboid	Bacterivore
	Vahlkampfidae	Amoeboid, flagella	Bacterivore, cytotrophy, detritivory, fungivore
Cercozoa	*Cercomonadidae*	Amoeboid with flagella	Bacterivore
	Silicofilosea	Amoeboid	Bacterivore, cytotrophy, detritivore, fungivore
Peronospormycetes		Flagella, thallus	Predators, primary saprotrophs
Cilophora		Flagella	Bacterivore, cytotrophy, detritivory, fungivore
Apicomplexa		Dispersal spores	Parasites

individuals per gram of dry soil or per gram of surface litter in forest soils. In general, the highest species diversity and abundance is found in the litter and organic horizons, decreasing with depth in the mineral soil profile. Protists are also abundant in the rhizosphere, around buried organic residues and in earthworm burrows. Their abundance in these areas is probably related to moisture and the availability of food.

As outlined in Table 3.2, many protists rely on bacteria as their main food source. Bacterivory is generally achieved by phagocytosis, in which the protist engulfs the prey (Fig. 3.1). The amoeboid body form in the Amoebozoa and Heterolobosea groups is particularly favourable for preying on bacteria. The 'plastic' body of an amoeboid protist permits the organism to explore very small cavities and pores in soil aggregates to find food. Some species insert a feeding tube (smaller than fungal hyphae) into cavities and pores smaller than 6 μm in diameter to access food resources. Larger protists can ingest 10^2 to 10^3 bacteria during a feeding session, but the smallest protists can only ingest one bacterium through their cytosome at a time. The protist may not discriminate between prey or it can be highly selective in choosing what bacteria it will consume. When selecting bacteria as prey, the size, cell wall chemistry, nutrition value and toxins secreted by the bacteria seem to be important. For instance, *Pseudomonas* and *Burkholderia* spp. can produce compounds that inhibit the growth of protist species, and there have been reports of bacteria secreting toxins that kill the amoebae.

Fungivores are protists that have the ability to puncture a small hole (1–6 μm diameter) in the fungal hyphae or spores and extend their pseudopodia into the cytoplasm. Although this feeding habit is not particularly widespread, it can be sufficient to reduce the number of spores of plant pathogens such as *Gaeumannomyces graminis* var. *tritici*, which causes take-all disease in wheat, and *Rhizoctonia solani*, the causative agent for 'damping off' of seedlings from many vegetable crops. The protists that engage in this feeding habit are not necessarily obligate fungivores and probably consume bacteria and detritus as well. Their ability to penetrate chitin-rich walls in fungal cells suggests that they may have a role in the decomposition of recalcitrant organic materials, but this remains to be studied.

Detritivory is important for the slime moulds (Amoebozoa: Mycetozoa: Eumyxa) and the water moulds (Peronosporomycetes), which secrete enzymes into the soil solution and absorb soluble nutrients from substrates, in a similar manner to the fungi illustrated in Fig. 2.10. Other detritivores are members of the Amoebozoa that can ingest fragments of microdetritus and digest them inside food vacuoles.

The gut of wood-eating insects such as some termites and cockroaches contains a complex community of bacteria and specialized anaerobic protists that function as symbionts, aiding in the digestion of wood microdetritus. As the debris passess through the insect gut, it is ingested into the food vacuoles of protists of the Parabasalia and Preaxostyla groups. Some digestion occurs in the food vacuole

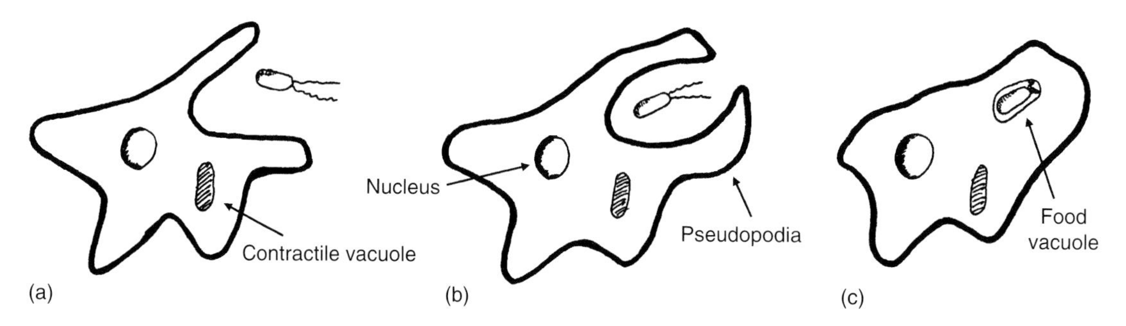

Fig. 3.1. Phagocytosis by amoeboid protists. (a) When the amoeba approaches a bacterium, it forms pseudopodia that surround the prey; (b) the bacterial cell is engulfed and deposited within a food vacuole; (c) digestive enzymes break down the food and the dissolved nutrients diffuse into the cytoplasm, providing energy and nutrients for movement, cellular metabolism and reproduction.

through an association with endosymbiotic bacteria. The partially decomposed cellulosic debris excreted from the protist cell can be further degraded by ectosymbiotic bacteria. Decomposition by-products and turnover of bacteria and protist cells provide energy to the host insect (see Focus Box 5.1.).

Cytotrophic protists prey upon and consume other protists through phagocyotosis or by secreting enzymes that dissolve the prey's cytoplasm. The cellular contents are then absorbed by the predator. Larger individuals in the Ciliophora group have been observed capturing and ingesting smaller amoeboid species. Similarly, many of the testate amoebae will ingest smaller amoebae or insert their pseudopodium into another individual to absorb the cytoplasm. As with higher animals, the predatory protists must expend energy searching for and capturing the prey, ingesting and processing the food resource before they gain energy from this feeding strategy. Therefore, optimal foraging behaviour and population growth can be understood using models that consider prey abundance, the nutritional value of various prey and temperature. Predation on invertebrates, notably nematodes, rotifers and tardigrades, has been reported for amoeboid species from the Amoebozoa and also the Cercozoa: Silicofilosea group. The predaceous Peronospormycetes form sticky cysts that attach to passing prey. The organism then forms a cyst and grows a filamentous thallus that extends into the prey tissues to obtain energy and nutrients.

Nematodes

Nematodes (roundworms) are multicellular eukaryotes belonging to the phylum Nematoda, which has more than 80,000 described species. About 20% of the known nematode species are parasites of plants, animals and humans. Soil nematodes can be free-living or parasitic. Nematodes are among the most numerous organisms found in terrestrial and aquatic ecosystems, from polar regions to tropical regions, high elevations to low elevations and in marine and freshwater environments. They could represent 90% of species of the animal life on the seafloors of the Earth, and are classified into 16–20 orders, depending on the taxonomic authority.

Soil-dwelling nematodes tend to be smaller (150 to 500 μm, weighing 20–60 ng) than those found in fresh water due to the constraints of moving through the soil matrix, but are still very numerous, with 10^7 to 10^8 nematodes/m^2 in temperate soils. The most common soil nematodes are classified in the orders Rhabditida, Tylenchida, Aphelenchida and Dorylaimida. A great deal of attention has focused on the plant-parasitic nematodes due to the extensive damage they can cause to plant roots and the economic losses borne by farmers who encounter this problem. Perhaps the most well-known parasitic nematode is the root-knot nematode (*Meloidogyne* spp.) – about 50 species are found worldwide and they infect the roots of virtually all crop plants, impeding growth and agricultural production. There is considerable interest in developing biological controls for this pest because the traditional chemical control (soil fumigation with methyl bromide) has been phased out in many places due to the fact that methyl bromide is an ozone-depleting substance.

The free-living nematodes inhabit water films and water-filled pore spaces, while plant-parasitic nematodes live in plant roots. Nematodes can move easily through larger soil cracks and pores. When their progress is limited by the narrowness of a soil pore, they try to squeeze through and may succeed after several attempts. Most of the nematodes are found in the top 5–10 cm of soil. As soils dry, nematodes migrate to wetter parts of the soil, but are limited in their ability to disperse more than a few centimetres due to their small size. Nematodes may be eliminated when matric potential is < 6 MPa, but many species can be resuscitated after years of dormancy. Nematodes survive dry soil conditions by anhydrobiosis (body coils and all activity ceases), encystment or protection in galls within plant roots.

The physiology of nematodes is fairly simple, as they lack circulatory, respiratory and endocrine systems. They possess a complete digestive system. Respiration occurs by simple O_2 diffusion through the permeable cuticle. Slime secreted through glands in their skin facilitates their locomotion through soil cracks and crevices. Sense organs consist of papillae and bristles on the head. A resistant outer cuticle provides some protection against environmental stress. Reproduction is amphimictic or parthenogenetic. The generation time for *Rhabditis* may be as rapid as 5 days, while the life cycle of larger species may be 20–50 days when environmental conditions are favourable. Traditional taxonomy is based on the placement of body openings (mouth, excretory and sensory openings), with confirmation of taxa and phylogeny coming from satellite DNA bar coding and other DNA-based methods.

Soil nematodes have a variety of nutritional habits:

- Bacterial feeders pull bacteria from the surface of roots, soil particles, organic matter and the soil suspension. Cuticularized valves within the digestive system crush the particles, aiding in digestion. Many members of the order Rhabditida are bacterivorous.
- Fungal feeders move along until they find a hypha, brace themselves against soil particles and then bite through the hyphal wall. The cell cytoplasm is then pumped through the hollow stylet into the digestive system. Many of the fungivorous nematodes belong to the order Aphelenchida.
- Predators use teeth to ingest their prey, which include protists, rotifers, tardigrades, small worms and other nematodes. They generally have three strong buccal teeth that are used to tear apart the cuticle upon swallowing. The prey's body wall is torn, allowing the predator to suck out the body contents. Soil particles are also ingested, perhaps to help in grinding and digestion of food. The order Mononchida is exclusively predaceous, and so are a few individuals in the order Dorylarmida.
- Saprophytes/omnivores feed primarily on decaying materials. They take up plant and animal residues that have been liquefied and decomposed by bacteria. Some bacteria, fungal and algal cells may also be ingested in the process. It seems likely that most nematodes can adopt an omnivorous feeding habit if their preferred food source is not available. Foraging among decaying materials can make nematodes vulnerable to capture and digestion by animal-trapping fungi like *Arthrobotrys oligospora*, as described in Focus Box 3.1.
- Plant parasites are of great agricultural significance on account of the damage they cause to crops. It appears that they are positively attracted to plants by root secretions. They puncture plant roots with a large stylet and feed on cytoplasm, sucking out the contents of the root cell. Nematodes in the order Tylenchida and a few from the orders Aphelenchida and Dorylarmida are plant parasites.
- Plant-associated nematodes are those that are found in the rhizosphere and consume root exudates and cytoplasm, but without causing any apparent harm to the plant. They may also consume fungal cytoplasm. Included in this group are nematodes in Tylenchidae and Psilenchidae.

Focus Box 3.1. Nematode-trapping fungi.

Plant-parasitic nematodes are a tremendous problem in agriculture, causing billions of dollars of damage to crops worldwide each year. Researchers are searching for biological control agents that can reduce the impact of these nematode pests. Nematophagous soil fungi hold potential in this respect – some are obligate endoparasites that infect adult nematodes with adhesive or non-adhesive spores, while others parasitize nematode eggs.

In addition, there are more than 100 fungi known to use mechanical traps in the form of adhesive networks, adhesive knobs and branches, or constricting rings for capturing and immobilizing soil nematodes (Fig. 3.2). The nematode-trapping fungi are believed to derive the carbon required for growth and reproduction from detritus, but obtain the nitrogen and other nutrients that may not be abundant in the detritus from the decomposition of trapped nematodes. Chemical substances that immobilize nematodes provide another option for reducing nematode populations. The fungus *Pleurotus ostreatus* produces a toxin that immobilizes nematodes, permitting the fungi hyphae to locate, envelop and digest the nematode.

Nematode-trapping fungi are commonly found in decomposing organic matter such as manure piles, leaf litter mounds, tree stumps and decaying roots, with a few also isolated from mineral soil. Since nematode populations are also higher in decomposing organic matter than mineral soil, it has been proposed that organic amendments such as manure or plant residues could enhance the activity of nematode-trapping fungi in agricultural fields and thereby reduce crop damage from plant-parasitic nematodes. Although nematode-trapping fungi are very efficient in capturing and killing nematodes in Petri dishes in the laboratory, it has been difficult to demonstrate this ability in the field and so the existing data are not very conclusive. Further work on nematophagous fungi as possible biological control agents of plant-parasitic nematodes is needed to learn more about their ecology and population biology.

Continued

Focus Box 3.1. *Continued*

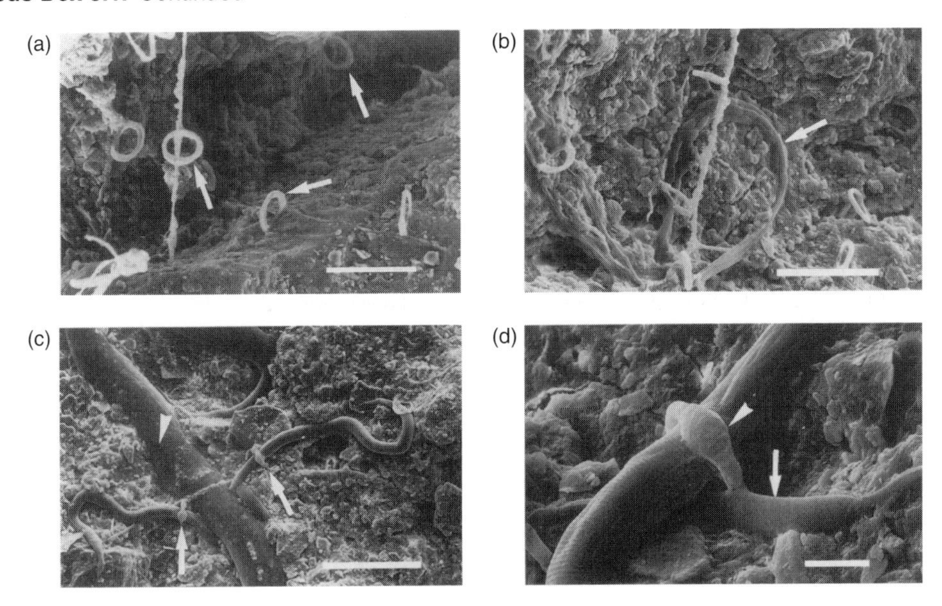

Fig. 3.2. Anatomy and capture activities of the nematophagous fungus *Arthrobotrys dactiloides* growing in soil, visualized by low-temperature scanning electron microscopy. (a) Constricting ring traps (indicated by arrows) formed in the soil; bar = 50 μm. (b) Constricting rings and an infected and digested nematode (arrow); bar = 50 μm. (c) Two nematodes recently captured in constricting ring traps (arrows); the large nematode in the centre (arrowhead) was not captured; bar = 100 μm. (d) Nematode captured in a constricting ring trap; only one cell of the ring has inflated (arrowhead); note the swollen hyphal cell (arrow) closest to the trap; bar = 10 μm. (Jansson *et al.*, 2000. Reprinted with permission from Mycologia. ©The Mycological Society of America.)

3.2 Diversity and Ecological Functions of Soil Microfauna

One of the challenges in assessing the diversity of the soil microfauna relates to the fact that new species are continually discovered. An intensive study of soils in Scotland (UK) during a 3-year period found 365 soil protists, about one-third of the species known to live in the world's soils. Smaller species were more numerous than larger species, suggesting that there is a physical limitation to the niche size for soil protists. One of the key findings of this work was that rare and abundant species at this site were also known to be rare or abundant worldwide, indicating that protists must be ubiquitously dispersed. Dispersal by wind, water and migratory animals seems to be responsible for the cosmopolitan distribution of soil protists (Table 3.3).

Nematode diversity can range from 39 to more than 400 species per study site (Table 3.4). Although there tend to be more species in the non-agricultural sites, populations are generally large and diverse, containing representatives from all trophic groups, regardless of land management practices. In the 71 agricultural sites reported in Table 3.4, the

Table 3.3. Species richness and population size of soil protists in functional groups recorded during a 3-year intensive study in Sourhope, Scotland, UK (Esteban *et al.*, 2006).

Functional group	Number of species	Population size (per g dry soil)
Flagellates	100	46,400
Naked amoebae	53	17,700
Testate amoebae	110	11,000
Ciliates	102	4,310
Total	365	79,410

Table 3.4. Nematode species diversity in soils with natural vegetation and in agroecosystems. (Adapted from Yeates, 2007.)

Vegetation/Location	Total nematode species
Primary tropical forest (Cameroon)	91–431
Rainforest (Cameroon)	153
Temperate forest (Slovakia)	125–182
Forest woodlot (Indiana, USA)	175
Native prairie (Kansas, USA)	228
Chalk grassland (England)	154
Agroecosystems (71 sites)	39–95

nematode populations ranged from 2.3×10^5 to 2.1×10^7 individuals/m^2.

In agroecosystems, the presence and abundance of each feeding group in the nematode assemblage is affected by their occurence in the study region, presence of a suitable food resource, favourable environmental conditions, agricultural management practices and biological interactions (succession, predation, competition). The average values for nematode assemblages from 71 agroecosystems are presented in Table 3.5. Although nematode assemblages appear to be resilient to many agricultural practices, some species are potentially more sensitive than others to tillage.

Soil ecologists are interested in identifying the ecological function of various microfauna and placing them into soil foodwebs so the interactions with other organisms (microorganisms, plants, other soil fauna), the consequences of those interactions on biogeochemical cycles and the sensitivity of external stresses of those interactions can be studied. The major ecological functions of soil microfauna include:

Table 3.5. Proportion of nematodes in each feeding (trophic) group in agroecosystems (values are from 71 sites). (Adapted from Yeates, 2007.)

Feeding group	Proportion of total population (percentage mean)	Proportion of total population (percentage range)
Bacterial feeders	40	13–79
Fungal feeders	10	7–41
Predators	4	0–30
Saprovores/ omnivores	12	0–49
Plant parasites	21	0–60
Plant associated	13	0–38
Total	100	

- Regulation of bacterial and fungal populations: many microfauna feed upon bacterial cells and fungal hyphae, thus regulating the size, the activity and the turnover rates of soil microbial populations, which are ultimately responsible for organic matter decay and nutrient cycling.
- Control of nutrient cycling: protists are voracious consumers of soil bacteria, but they do not absorb all of the carbon and nitrogen contained in bacterial cells. At least one-third to one-half of the nitrogen in bacterial cells may be released into the soil solution when these microbes are consumed by protists. This provides a supply of dissolved nitrogen, phosphorus and other nutrients that is essential for the growth of plants and microorganisms.
- Alteration of litter decomposition rates: predatory microfauna reduce the size of the soil microbial population, as mentioned above. Predators tend to consume larger, older cells of bacteria and fungal hyphae. Removal of the older cells can stimulate the growth of younger bacterial cells and the regrowth of fungal hyphae, which have a higher metabolic activity than older cells. The nitrogen released into the soil solution following predation can be recycled by young, rapidly growing microorganisms. The metabolic activities and functions (enzyme production, substrate degradation, etc.) of these young microorganisms will be faster than older microorganisms, thus stimulating microbial-mediated reactions in the soil. It is still unknown but hypothesized that modifying microfaunal activities would modulate medium-term pedogenetic processes such as formation of soil aggregates and carbon sequestration within microaggregates.
- Biocontrol of disease-causing organisms: the bacterivorous and fungivorous microfauna probably contribute to the natural biological control of pathogenic microorganisms that cause plant diseases, although it is often difficult to know exactly how these organisms are interacting in the natural soil environment. 'Suppressive soils' possess a community of microorganisms

and microfauna that act together to reduce or repress pathogenic organisms (described in more detail in Chapter 11).

The parasitic nematodes that hold promise for controlling insect pests and invasive plants have received more attention. Orchards are susceptible to attack from insects such as the codling moth, which is a vector for the granulovirus that affects fruit crops such as apples, pears and walnuts. Entomopathogenic nematodes in the families Steinernematidae and Heterorhabditidae are vectors for bacteria in the genera *Xenorhabdus* and *Photorhabdus*, respectively. The bacteria are transferred to the leptidopteran when it is infected by the nematode and the insect dies from a bacterial infection (Cross *et al.*, 1999).

The endopathogenic nematode *Ditylenchus drepanocercus* is being used as a biological control agent against the invasive plant *Miconia calvescens* (velvet tree) in Hawaii. Orginally brought to the islands as a decorative plant in the 1960s, the velvet tree was named a Hawaii State Noxious Weed in 1992 because it creates a dense canopy that shades and inhibits the growth of native plant species, and its shallow root system leaves hillside soils more susceptible to erosion (Seixas *et al.*, 2004).

- Cause commercially important diseases of many plants: most plants are susceptible to infection by plant-parasitic nematodes. Maize, potatoes, soybean, sugarbeet, turfgrass, trees, orchards, vineyards, vegetable gardens and flowerbeds are all at risk from nematode attack. Plant-parasitic nematodes include those causing root-knot, cysts and galls on roots, as well as lesions on foliage. Feeding by nematodes causes abnormal plant growth and an infected plant will appear stunted, discoloured or both (for more details on the symptoms and diagnosis of nematode infection in plants, as well as management tips, see Bridge and Starr, 2007 and the 'Nematodes as Agricultural Pests' web site at the University of Nebraska at Lincoln: http://nematode.unl.edu/agripests.htm).

3.3 Methods for Collecting and Enumerating Soil Microfauna

As organisms that live in water films, the number of active protists and nematodes in soil is strongly influenced by rainfall and irrigation events, as well as temperature. The number of protists in a soil can change by 100-fold during a 24 h period following rainfall, warming or cooling. Nematode populations respond more slowly to soil temperature and moisture dynamics due to the longer life cycle, but there are seasonal variations in nematode populations that may be related to weather, the growth of vegetation, litter inputs and fluctuations in bacterial and fungal populations.

There are a number of considerations to be taken when collecting litter and soil samples to examine soil microfauna. The samples should be handled carefully because protists and nematodes are small, fragile animals. Friction from handling or sieving soil samples should be kept to a minimum. Samples can be sieved through a large-mesh sieve (4–6 mm diameter) to remove stones and large root fragments, or the sample can be unsieved. Often the researcher would like to know the size of the active protist and nematode populations in the field. Since protists and nematodes form cysts when soils are dried or they are under stress, soils should be kept at field moisture and measurements made within 1–2 days of collection. Soil can be stored in polyethylene bags in a refrigerator (4–8°C) until analysis.

The recommended method for evaluating protist populations involves direct observation of active cells under a microscope without culturing. A dissection microscope could be used to sort through samples, followed by counting under an inverted microscope with phase contrast objectives capable of achieving 100× to 400× magnification. This method has the following advantages: it requires a minimum of preparation and is not as time-consuming as culture-based methods. The most probable number (MPN) method is not recommended for assessing field populations because it tends to count active and encysted protists, and not all protists will abandon the encysted state at the same rate. Due to the sources of error and longer preparation time involved with the MPN method, it has been abandoned by soil ecologists.

Plant-parasitic nematodes are generally collected from the roots, although eggs and infective juveniles can be extracted from the soil and counted. Free-living nematodes are extracted from soils with passive or active methods before they are counted and sorted into trophic groups or examined for

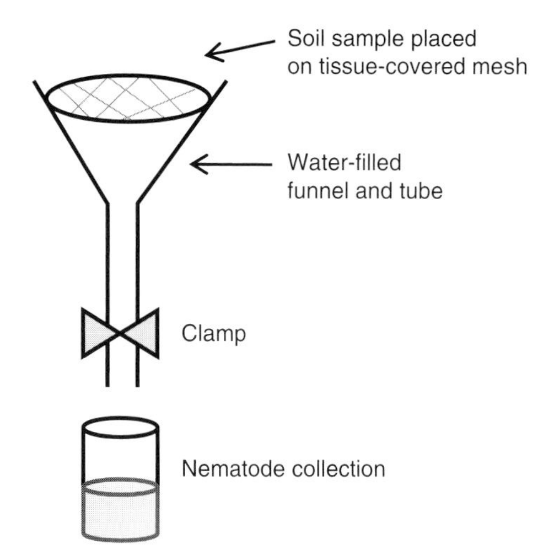

Fig. 3.3. Schematic diagram of a Baermann funnel, used to extract free-living soil nematodes.

taxonomic classification. The passive method of nematode extraction involves suspending soil in a concentrated sucrose solution, then decanting or centrifuging to collect active and encysted nematodes. Active nematodes can be collected with the Baermann funnel method, illustrated in Fig. 3.3. Briefly, soil samples are placed on a fine wire mesh covered with regular tissue paper. The wire mesh is placed in a funnel and the sample is covered with water. A rubber tube and clamp are placed at the bottom of the funnel outlet to keep the soil saturated. Soil particles are retained, but nematodes escape through pores in the tissue paper and move down the funnel stem into free water, which is drained after 1–2 days. After nematodes are collected, they can be examined with a stereomicroscope or an inverted microscope at 10× to 70× magnification.

4 Soil Mesofauna

4.1 Biology of Soil Mesofauna

The mesofauna are eukaryotic multicellular organisms ranging in size from 150 μm to 10 mm long. They are permanent soil dwellers and most are arthropods (insects) with an exoskeleton and jointed legs. These organisms live in the litter layer and in soil cracks and macropores in the top 5–20 cm of the soil profile. The spatially variable microenvironments in soil are the preferred habitat for a myriad of creatures smaller than 1 cm with ancient, complex and, in some cases, unresolved evolutionary origins. The most abundant are the collembola and soil mites, although many other soil organisms are classified as soil mesofauna, including tardigrada, rotifers, proturans, diplurans, pseudoscorpions, symphilids, pauropoda and Enchytraeidae.

Many of the soil-dwelling mesofauna are insects (with the exception of the tardigrades, rotifera and the enchytraeids) and are referred to collectively as soil microarthropods. The most well-studied of these are the collembola and soil mites.

Tardigrada

Tardigrada ('water bears') are easily recognized by the four pairs of stumpy, claw-bearing legs and the conical snout. Newly hatched larvae are 50 μm long and adults can grow to 100–1500 μm. Their habitat is primarily moist litter and soil layers. They consume organic residues and suck the contents from algae and plant cells. Rotifers and nematodes may also be consumed but are not the main food source. Tardigrada are quite intolerant of soil drying and form protective cysts to survive during dry periods. They can suspend metabolic activities and enter a state of cryptobiosis, which allows them to survive extreme temperature, pressure, dehydration and radiation conditions. Even when exposed to the extreme environment of outer space for 10 days, tardigrades that were protected from high-energy UV radiation were capable of producing viable offspring when they returned to Earth, and some even survived full exposure to the sun (Jönsson *et al.*, 2008).

Rotifera

Rotifera are microscopic aquatic animals in the phylum Rotifera that are found in soil water films, in mosses and on lichens. They are not much bigger than the largest unicellular protists (200–500 μm long), but are multicellular with specialized organic systems and a complete digestive tract. The body is cylindrical or spherical and has a ciliated crown or corona (wheel-organ) that is used in locomotion and feeding. The corona is used to move food (fine particulate organic matter and smaller protists) into the mouth. Trophic habits can be determined from the shape of the corona, which may be capable of grinding residues, piercing plant and fungal cells, capturing protists and bacteria and scraping organic matter from soil surfaces. Reproduction is asexual or sexual through parthenogenesis. Rotifera population blooms are common in bacteria-rich environments such as manure heaps, compost piles or cow dung.

Protura

Protura are wingless primitive insects belonging to the family Apterygota (wingless insects in which a true metamorphosis is lacking). They lack eyes and antennae. The fore legs are enlarged and possess many sensilla, and probably function as antennae. Protura are extremely delicate animals, 1–2 mm long, generally yellow, living in wet humous layers of grassland and forest soils. In general, they do not penetrate deeper than 10 cm in the soil. They are probably fungal feeders, as individuals have been observed consuming the hyphae of mycorrhizal fungi and free-living saprophytic fungi.

©CAB International 2010. *Soil Ecology and Management* (J.K. Whalen and L. Sampedro)

Diplurans

Diplurans are primitive blind insects with antennae from 2 to 5 mm long. Many are white or pale yellow. Soil-dwelling species belonging to the Japygidae are characterized by unarticulated, pincer-shaped, brownish pigmented cerci. The antennae have ten or more bead-like segments projecting forward from the head. They are capable of regenerating lost appendages during successive moulting. A single individual may moult as many as 30 times in its life, and the lifespan is typically 1 year. The abdomen has eversible vesicles, which seem to absorb moisture and help maintain the water balance. Diplurans live in ground litter and humous layers, and beneath stones. Many are predators of other insects, such as collembola, soil-dwelling midge larvae and other insect larvae. Besides living prey, dead and injured insects and annelids are also eaten. Some also consume fungi and particulate organic matter.

Pseudoscorpions

Pseudoscorpions ('false scopions') are characteristic and regular inhabitants of ground litter, but usually present in small numbers. The body length is generally 2–8 mm. These arachnids resemble true scorpions with their elongated, pincer-like claws (pedipalps), but they do not possess a stinger at the end of the abdomen as found in true scorpions. The pedipalps contain venom glands and ducts, and are used to inject poison to immobilize their prey. The body colour ranges from yellowish to dark brown, and the claws are sometimes black. Some have two or four eyes, but others are blind. Their food consists of small arthropods, collembola and especially mites such as the Oribatid mites. They have a habit of attaching themselves to the legs and appendages of flies, beetles and other insects, which allows them to 'hitch-hike' to new locations. Pseudoscorpions produce silk to make chambers for overwintering, moulting or brooding. They may produce one or two broods per year, and the lifespan is typically 2 to 3 years. Pseudoscorpions are more resistant to soil dessication than most of the other soil insects.

Symphylids

Symphylids resemble small Chilopods, but they are generally no longer than 1 cm. These white, agile creatures possess 15 tergites and 11 or 12 pairs of legs. They do not have eyes. Symphylids live in the leaf litter and humous layers of soil or beneath stones. They enter cavities in the soil for protection and to lay eggs, but have limited burrowing abilities. Since they are very sensitive to desiccation, populations are not very large in ploughed agricultural soils. Symphylids appear to eat plant detritus exclusively and consume the soft parts of leaves, leaving the harder bits of the leaf skeleton intact.

Pauropoda

Pauropoda are white terrestrial myriapods up to 1 mm in length. These segmented animals possess bifurcated antennae and a cylindrical body with 11 or 12 body segments and 9–11 pairs of legs. Pauropods are blind and have many unusual sensory organs, including five pairs of long sensory hairs on the body. Flagellate-like hairs on the body surface and the globulus at the distal end of the antennae help the animal analyse its surroundings. Almost every species has a distinct anal plate with a characteristic shape (form, size, cuticular structure) that is used for taxonomic identification. Pauropods are dioeceous and progoneate, although parthenogenetic reproduction can occur when environmental conditions are unfavourable. Pauropods live under stones, the plant litter and in the top 5 cm of soil. They cannot burrow, but follow root canals and enter soil crevices, especially to avoid desiccation. They consume mould and detritus or suck the contents from fungal hyphae.

Enchytraeidae

Enchytraeidae ('potworms') are small, segmented white worms from 5 to 15 mm in length. They belong to the Annelida, order Oligochaeta and are classified in the family Enchytraeidae. They have a similar body structure to earthworms. Small bristles (setae) on the segments permit the enchytraeids to grip and move through soils. These worms are hermaphrodites, meaning they possess both male and female reproductive organs. Reproduction can be sexual or asexual. The young develop in a resistant structure called a cocoon. The juvenile enchytraeids hatch from cocoons when environmental conditions are sufficiently moist for their survival. The body is covered with a thin cuticle that permits water diffusion and exchange of O_2 and CO_2 (respiration). Enchytraeids can easily become dehydrated and thus are most abundant in moist substrates (detritus layer, litter layer), where populations can reach 250,000 individuals per square metre (vanVliet *et al.*, 1997). They have a limited ability to burrow, but retreat into the topsoil when surface litter dries out.

Collembola

Collembola ('springtails') are primitive Apterygote insects. They differ from other Apterygota like the Protura and Diplurans because their abdomen contains only six segments. The first abdominal segment carries a ventral tube called a collophore. In some species the collophore supports one or two retractable lobes and enables the animal to adhere to smooth surfaces, but the primary function of the collophore is for excretion of body wastes and water transport. The fourth and sometimes fifth abdominal segments hold a springing organ called the furcula, which permits the animal to jump. A retinaculum on the third abdominal segment holds the furcula secure in the resting position. However, springtails living in deeper soil layers may not have a well-developed furcula and move by creeping or running through the soil, rather than by springing.

Collembola are ubiquitous members of the soil mesofauna, ranging in size from 0.1 mm to 2 cm long and with an individual biomass of 1–20 μg. Populations can reach 10^5 or more individuals/m^2, but vary through a habitat and fluctuate seasonally due to predation and environmental factors that control collembola growth and development. Some collembola can resist dessication and low temperature by entering a state of anhydrobiosis, but most are active when soil matric potential > –5.0 MPa and soil temperatures are between 0 and 25°C.

Twenty-one families and more than 6000 species of collembola have been found in soils. Many species remain unnamed. Species diversity is greater in medium-latitude temperate sites and in the tropics, although a few collembola live in subarctic regions. It is not uncommon to find 60–80 species in a temperate deciduous forest. There are fewer collembola species and populations are smaller in coniferous forests, followed by pasture lands and ploughed agricultural fields.

Collembola exhibit considerable fecundity and reproduction can occur sexually or by parthenogenesis. Reproduction is peculiar. The male deposits spermatophores, which are spherical and have long, hair-like stalks, on the ground. The female comes in contact with these by gliding across the ground in a characteristic manner. Females may lay 100–600 eggs during their lifespan (about 1 year). The development time for embryos to hatch from eggs is about 35–45 days when environmental conditions are favourable (20°C, moist soil conditions). Most insects undergo a series of moults and instars until they reach sexual maturity, but adult collembola can moult up to 50 times during their lifetime, alternating between reproductive and feeding body types.

Most collembola consume decaying vegetation and the microorganisms found on this detritus, but some species are not fastidious in their choice of food. Omnivores in the genus *Hypogastrura* eat decayed plant material, fungal mycelia, fungal spores and soft animal material such as nematodes, dead flies, fly pupae, collembola and dead earthworms. The absence of chewing appendages on their mandibles indicates they consume liquid food, sucking the cellular contents out of the materials mentioned above. The rapid growth rates and large population sizes of collembola could have a significant effect on soil microbial dynamics, nutrient cycling and litter decomposition, but these remain to be quantified.

Many collembola are found around plant roots and thus may be important for rhizosphere processes. Collembola consume plant roots and rhizosphere-associated fungi including mycorrhizae, saprophytic fungi and plant-pathogenic fungi. In a review of food selection experiments, Friberg *et al.* (2005) report that collembola consume plant-pathogenic fungi preferentially when given a choice of various fungi. This may be due to a higher nutritional value of plant-pathogenic fungi, or perhaps because other fungi secrete unpalatable substances or toxins that make them unpalatable to collembola. In their experiment, the collembola *Folsomia hidakana* prevented damping off in cabbage seedlings by reducing the root pathogen *Rhizoctonia solani*, suggesting that collembola could be effective as biological control agents in greenhouses and agricultural fields.

The epiedaphic collembola living in the litter layer are generally larger, move around by springing and have a higher resistance to desiccation than the euedaphic collembola that inhabit soil mineral layers. Litter-dwelling collembola probably feed mainly on plant litter, fungi, bacteria and algae. Some may even consume nematodes when they have the opportunity, although it is not known whether they are active predators in natural environments. Soil-dwelling collembola have limited burrowing abilities and move through macropores in the top 10–15 cm of soil. If soils are compacted and the collembola cannot move into moist soil pores, they can die from desiccation. The soil-dwelling collembola are probably fungivores, although most species are generalist feeders and adapt their diet according to the food resources that are available. Collembola possess cellulose and trehalose enzymes that permit them to digest cellulose, starch and tannins to a limited extent. The degradation of recalcitrant chitinous fungal cells may occur through interactions with bacteria, prior to the ingestion of fungal materials or within the gut content. General morphological, ecological and physiological characteristics of the epiedaphic and euedaphic collembola are shown in Table 4.1.

Soil mites (Acari)

The soil mites (Acari) are chelicerate arthropods related to the spiders and are the most abundant microarthropods in many soils, both in individual numbers and in species diversity. There are 1200 families of mites and about 45,000 species have been described worldwide. They are among the oldest of all terrestrial animals, with fossil specimens dating back to the early Devonian, nearly 400 million years ago. Mites are ubiquitous in every terrestrial, marine and freshwater habitat including polar and alpine regions, deserts and subterranean waters with temperatures as high as 50°C and even deep-sea trenches.

The soil mites are small, ranging from 0.1 to 2.0 mm long and have a biomass of 0.1–120 μg per individual. They are most abundant in the litter layer and to a depth of 15 cm in mineral soil horizons, and are a very diverse group. For instance, in mixed temperate hardwood forest, it is possible to find more than 1 million individuals from 200 species per square metre. Part of the reason for their success is that they have a higher resistance to water and temperature stress than most other microarthropods.

Mites are distinguished from other arachnids because they have no body segmentation. Instead, the mouthparts and associated sensory structures form in a discrete anterior structure known as the gnathosoma. Appendages, the eyeparts (ocelli), the central nervous system, reproductive and digestive systems are all fused into a single, unsegmented body called the opisthosoma. The life cycle of acari begins with sexual reproduction or parthenogenesis. Eggs are deposited in the soil or in the litter layer. Eggs develop into a hexapod prelarva, which then enters a hexapod larval stage. The organism then moults through three octopod nymphal stages before reaching maturity.

Soil ecologists group mites into the following suborders: Cryptostigmata (Oribatei), Prostigmata, Mesostigmata and Astigmata. The distribution and diet of each is summarized in Table 4.2. The preferred diet of Cryptostigmata and Astigmata includes leaf litter, animal manure, bacteria, fungi and algae.

Table 4.1. Morphological, ecological and physiological characteristics of the epiedaphic and euedaphic collembola. (Adapted from Lavelle and Spain, 2001; Ruess *et al.*, 2007.)

Characteristic	Epiedaphic collembola	Euedaphic collembola
Location	Surface litter (L/F layer)	Decomposed litter (H layer) Mineral soil (A horizon)
Size	Large (> 2 μg dry weight)	Small (< 2 μg dry weight)
Morphology	Long antennae and legs Furcae present Pigmented 16 ommatidia	Smaller appendages Furcae absent Unpigmented Ommatidia absent or reduced
Reproduction	Sexual Small progeny Many eggs Seasonal reproduction	Parthenogenetic Large progeny Fewer eggs Continuous reproduction
Foraging activity	Move to acquire food, feed intermittently	Feed in the vicinity, constant foraging
Metabolic activity	High but intermittent	Low but constant
Food	Plant litter and associated decomposer microbiota	Primarily fungal feeders

Table 4.2. Summary of ecological characteristics and population dynamics of the suborders of the Acari. (Data from Coleman and Crossley (2003); Soil Biodiversity Lab at Massey University (http://www.massey.ac.nz/~maminor/mites.html).)

Characteristic	Cryptostigmata	Prostigmata	Mesostigmata	Astigmata
Location Size (length, mm)	Moss, litter and soil 0.2–1.0	Litter and soil 0.2–1.0	Litter and soil 0.3–3.0	Litter and soil 0.2–1.0
Morphology/ development	Hard exoskeleton helps adults to avoid predation Slow-moving	Stigmata located at the base of the mouthparts Fast moving	Body protected by thickly sclerotized plates, laterally placed stigmata Fast moving	Morphology similar to Cryptostigmata. Some instars disperse by phoresy
Reproduction	Parthenogenetic Few large eggs Long life cycle (1–2 years)	Sexual (male guards female during egg-laying) Moderate egg-laying Short life cycle (weeks/months)	Sexual dimorphism is common (males fight for females during courtship, protect females during egg-laying) Moderate egg-laying Short life cycle	Sexual Fecundity: 100–300 eggs/female Short life cycle
Foraging activity	Move to acquire food, feed intermittently	Ranges from ambush to cruise predation behaviour	Predaceous behaviour, inject digestive liquid into prey and ingest dissolved tissues	Move to acquire food, feed intermittently
Metabolic activity	Low but constant	Diverse	High and intermittent	Moderate and constant
Food	Decomposing plant litter, fungi, algae, dead insects	Fungi, small arthropods, living plants, decomposing plant litter	Small arthropods, nematodes and fungi	Manure, fungi, bacteria, N-rich detritus
Organisms of note	Most ancient group of the soil Acari	Spider and gall mites are plant parasites, velvet mites are insect parasites	Phytoseiidae are important for biological control of other mites	Related to 'biting' mites that are vertebrate parasites

Predaceous mites consume smaller mesofauna (microarthropods, nematodes and enchytraeids). Many mites engage in a complex symbiotic association with plants and animals. In tropical rainforests, mites are found feeding on mosses, ferns, leaves, stems, flowers, fruit, lichens, microorganisms, other arthropods and each other. Plant–mite associations can be detrimental to agricultural crops (e.g. spider mites cause crop damage leading to economic losses for farmers) or favourable (e.g. phytoseiid mites are effective biocontrol agents of spider mites).

The ecological roles and functions of the four suborders of soil mites are summarized below:

1. Cryptostigmata (Oribatei) are characterized by their unusually dark pigmentation and strong armour. These are among the commonest inhabitants of the ground litter and the upper soil layers. This is an ancient group, dating back to 420–430 million years ago according to the fossil record. They have diverse forms. Some oribatids have no articulation between the propodosoma (front body) and hysterosoma (hind body), whereas others possess such an articulation. A plate-like extension resembling wings is observed on the hysterosoma of some oribatids. Juvenile polymorphism is common. Juveniles can have a different body shape and consume a diet that is distinct from adults, depending on environmental conditions. Although oribatid populations can be large, the reproduction rate is slow (only one or two generations per year). Females have only a few large eggs and reproduction is peculiar. The male produces spermatophores (spherical bodies on a long stalk) and fixes them on the substratum where the female actively picks them up. Once eggs are hatched, the young pass through six development stages before becoming adults. A few remain active when soils are flooded, and the resistance to desiccation depends on the thickness of their calcareous exoskeleton. Calcium sequestered from fungal cells enhances the thickness of their exoskeleton. Most are detritivorous or fungivorous, as indicated in Table 4.2.

2. Prostigmata are also an ancient group with fossil representatives. They are distinguished by cylindrical, widely spaced coxae of the first pair of legs. Spiracles lie aside the second, third or fourth pair of legs. They are common inhabitants of litter and the top 10 cm of moist soils. The preferred food sources vary, but predaceous forms predominate. Nematodes, enchytraeids and small insect larvae are consumed, as are small worms and animal corpses. Many seem to be facultative carrion eaters, appearing early on carcasses. Some have specific prey. For instance, the 'grasshopper mite' consumes grasshopper eggs and grasshoppers, while the 'velvet mite' consumes termites, 'chiggers' eat collembola and smaller prostigmata specialize on nematodes. Fungal feeders are also found in this group – these types of prostigmata are opportunistic and able to reproduce rapidly after a disturbance or sudden shift in resources.

3. Mesostigmata are not as numerous as the Cryptostigmata or Prostigmata, but are universally present in soils. They have a more rapid life cycle than the other mites, with only two development stages (hexapod larva and quadruped nymph) before becoming adults. Almost all are predators, consuming both living and dead soil animals. They are not especially cryptic and may run along the surface of litter and plants in search of prey (capture predators) or sit quietly and wait for prey to come by their hiding spot (ambush predators). Their elongated and slender first pair of legs are used as feelers. Larger individuals feed on small arthropods or their eggs, while smaller species are nematophagous (consume nematodes).

4. Astigmata are the least widely distributed of the soil mites. They tend to inhabit cool, moist areas with high organic matter and are not especially disturbed by agricultural activities. Astigmatid mites are abundant in agroecosystems following harvest or after the application of animal manure. Their primary food is detritus and the associated microorganisms (bacteria, fungi and algal cells). In one laboratory (Coleman and Crossley, 2003), astigmatid mites in bins of soil from an agricultural site left their bins and entered Tullgren funnels containing soil from a forest site. Thus, they were a source of error in the experiment, since astigmatid mites are not generally found in large numbers in forest soils.

4.2 Diversity and Ecological Functions of Soil Mesofauna

The diversity of soil mesofauna is still under investigation and researchers expect to find more species than have currently been described. The number of species and approximate population sizes in natural ecosystems are given in Table 4.3. Some populations were estimated from the number of individuals in small soil cores (generally 5 cm diameter) taken from a prescribed area and extrapolated to a square metre. Population estimates also come from biodiversity studies, which may be biased because researchers typically collect samples around vegetation, in wet depressions or other favourable habitats to maximize the number of distinct species that may exist in the area. Soil mesofauna are often aggregated in such microsites, possibly because individuals with similar diets and life histories prefer the same type of habitat. It has been hypothesized that some groups exude pheromones that attract other individuals with similar feeding preferences, but this remains to be confirmed.

The ecological functions of soil mesofauna are varied, since this group includes all known trophic groups – saprovores, bacterivores, fungivores, predators and omnivores. Hence, this group has diverse ecological functions, including:

Table 4.3. Mesofauna species and population sizes in terrestrial ecosystems with native or semi-managed vegetation (grasslands, forest). (Data from Wall and Moore, 1999; Barbercheck *et al.*, 2009; Tree of Life Project: http://tolweb.org/tree/)

Taxonomic group	Described species in soil (*n*)	Approximate population size (individuals per m^2)
Tardigrada	700+	2,000–12,000
Rotifera	200	50,000–300,000
Protura	500	9,000
Diplurans	800	100–500
Pseudoscorpions	3,200	100–600
Symphylids	200	300–600
Pauropoda	500+	100–500
Enchytraeidae	600	5,000–150,000
Collembola	6,500	2,000–200,000
Acari	45,000	1,000–1,000,000

- Regulation of bacterial and fungal populations: the bacterivorous and fungivorous mesofauna are predators that consume bacterial cells and fungal hyphae, thus regulating the size and activity of soil microbial populations. Predatory activity on microfauna can also alleviate or/and modify the grazing pressure on soil microbial populations, leading to a trophic cascade.
- Control of nutrient cycling: when soil mesofauna consume soil bacteria and fungi, they absorb 5–10% of the carbon and less than 30% of the nitrogen in microbial cells. The release of nitrogen into the soil solution as proteins, amino acids and NH_4^+ contributes to nitrogen mineralization and ensures the nutrition of plants, microorganisms and soil animals.
- Alteration of litter decomposition rates: by grazing on bacteria and fungi, the mesofauna change the size and activity of soil microbial communities, which directly impacts litter decomposition rates. In addition, the saprophagous mesofauna affect decomposition indirectly through litter comminution. This refers to the fragmentation and redistribution of litter within the soil profile, which occurs when mesofauna select and eat plant residues, mixing them with soil. The grinding action breaks residues into smaller particles. Microorganisms living on the residue or in the soil are ingested; dormant spores are often activated when the mixture passes through the invertebrate gut. The undigested plant–soil–microbial mixture (frass) defecated by the animal usually exhibits more microbial growth and respiration than unprocessed plant litter.
- Contribution to soil pedogenesis: soil mesofauna that mix plant residues with soil particles and activate soil microbial growth provide excellent conditions for the formation of soil aggregates. The soil mesofauna have a modest role in soil pedogenesis compared with the larger soil macrofauna, as will be discussed in Section 5.2.
- Biocontrol agents: as discussed in Section 4.1, fungivorous collembola feed selectively on plant-pathogenic fungi, while predatory acari can control plant-feeding mites. Using a DNA-based approach, a team of researchers in Bill Symondson's group at Cardiff University (Wales, UK) found that the collembolan *Folsomia candida* and the mesostigmatid mite *Stratiolaelaps miles* were effective at reducing the population of the insect-parasitic nematode *Heterorhabditis megidis*. In particular, the collembolan was able to reduce nematode populations by 50% within 10 h (Read *et al.*, 2006). We have barely tapped the potential of mesofauna as biocontrol agents, and this merits further research.

4.3 Methods for Collecting and Enumerating Soil Mesofauna

The soil mesofauna are soft-bodied and easily damaged by rough handling. Thus, samples should not be sieved to avoid damaging the organisms. Intact soil cores can be taken by inserting a metal cylinder or corer into the ground. The bottom edge of the corer should be sharpened to facilitate cutting into the soil. Cylinders or cores with a minimum diameter of 5 cm are recommended to avoid soil compaction, which could inhibit mesofauna from exiting the soil and thus underestimate the species diversity and population size. Since storage conditions can change the mesofauna population due to predation, breeding, moulting or mortality, it is recommended that soils be maintained at field moisture and processed within 1–2 days of collection. Soil can be stored in polyethylene bags in a refrigerator (4–8°C) until analysis.

Tardigrades and rotifers are collected using a similar process as for protists. Microarthropods can be collected by passive flotation with a mixture of oil and water, but it is more common to extract them with a heat gradient apparatus based on the designs of Berlese and Tullgren (Fig. 4.1). Briefly, the soil is

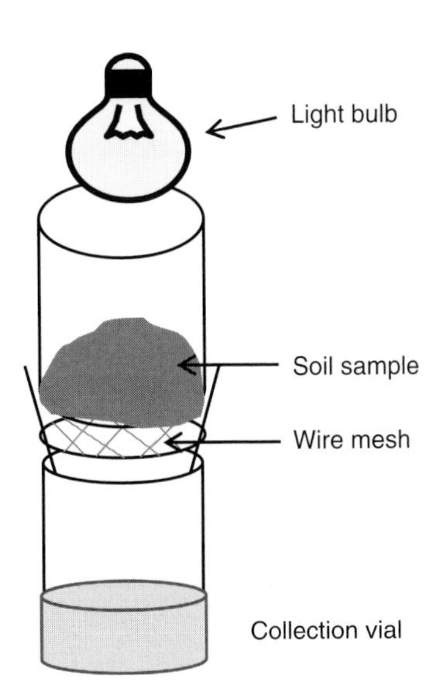

Fig. 4.1. Schematic diagram of a Berlese funnel, used to extract soil microarthropods.

placed on a 1 mm mesh screen, which goes into a funnel. A heat source is suspended over the funnel and the temperature is gradually increased over a 3–5-day period to completely dry the soil sample. The mesofauna move down, away from the heat source, through the mesh and into a collection vial that contains alcohol or a glycerol solution. To facilitate the process, extractors are often set up in a cold room to maintain a hot-dry gradient and stimulate the movement of microarthropods into the collection vial.

Focus Box 4.1. How decomposition of soil organic matter is affected by food resource quality and biotic interactions among the soil fauna.

Decomposition of soil organic matter is linked to the global carbon cycle. Within a region, the soil properties, climate, vegetation and land management are some of the key external factors that control the amount of organic matter that is retained as soil organic matter or lost from terrestrial ecosystems, to the atmosphere (as CO_2 or CH_4) or exported as dissolved organic carbon to rivers and oceans. But organic matter decomposition also depends on intrinsic factors such as diversity and abundance of decomposer community, dynamic properties of the soil foodweb and substrate quality. When we look at various soils, we find that the soil organic matter content is related to the carbon input from plants and other organic residues, as well as the carbon lost through microbial, plant and fauna respiration as carbon dioxide (CO_2). The respiration of soil fauna is believed to be a minor source of carbon lost from soils, so how do they contribute to the decomposition process?

Researchers in Germany (Vetter *et al.*, 2004; Fox *et al.*, 2006) have attempted to answer this question by examining the effect each group of soil fauna, as well as the biotic interactions within the food web, can exert on CO_2 loss from soils. They also studied how these effects depend on the chemical quality of the organic substrates and its structure (i.e. heterogeneity). They devised a large experiment to determine whether soil fauna would stimulate or decrease CO_2 emissions, using an agricultural soil with or without maize litter, and several groups of soil fauna.

The main finding was that the soil fauna were important controllers of soil microbial communities. The microbial biomass was increased by nematodes in bare soil, and by enchytraeids in soil with litter. Some groups increased the degradation of carbon associated with larger particles, but the most labile carbon pool was generally unaffected by soil fauna. The soil CO_2 respiration was increased 1.9-fold by soil fauna in the bare soil, and by only 1.1-fold in the soil with litter (Fig. 4.2).

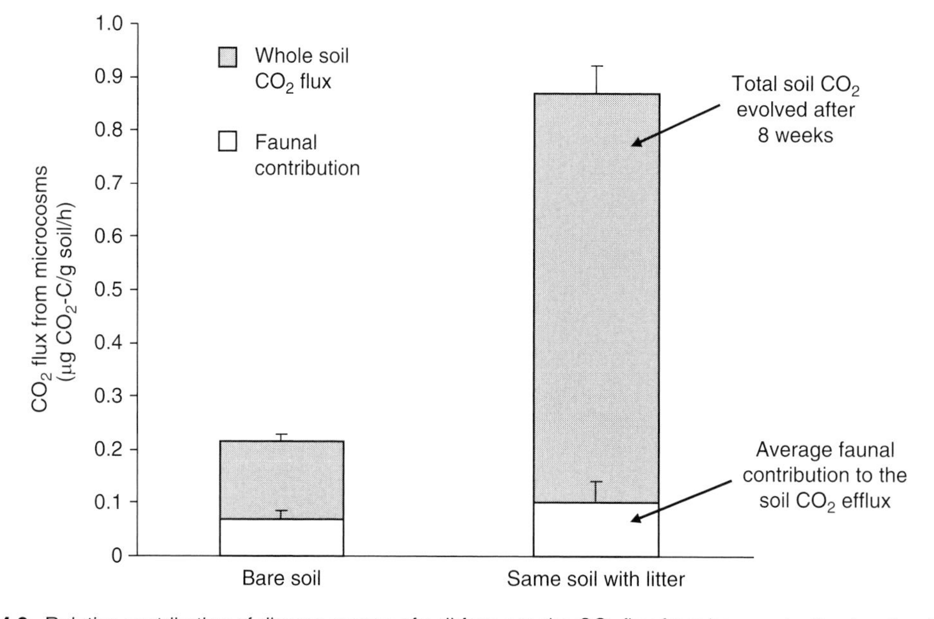

Fig. 4.2. Relative contribution of diverse groups of soil fauna to the CO_2 flux from bare agricultural soil and soil with maize litter on the surface. Microcosms were inoculated with nematodes, enchytraeids, microarthropods and lumbricids at three densities, alone and in combination. (Modified from Vetter *et al.*, 2004.)

Continued

Focus Box 4.1. *Continued*

These authors propose that under conditions of substrate homogeneity, such as in the bare soil, animal effects were stronger, but they were limited by overexploitation of carbon resources and mutual inhibition between groups. In heterogeneous substrates such as litter, animal effects were limited due to incomplete resource exploitation; however, the complementary habitat colonization by different faunal groups in the litter gave rise to positive interaction (Fig. 4.3).

Homogeneous substrates

Uniform colonization pattern
Complete use of the habitat
Relevant overall faunal contribution to CO_2 flux
Strong competition for resources between groups
Greater faunal diversity probably inhibits decomposition

Heterogeneous substrates

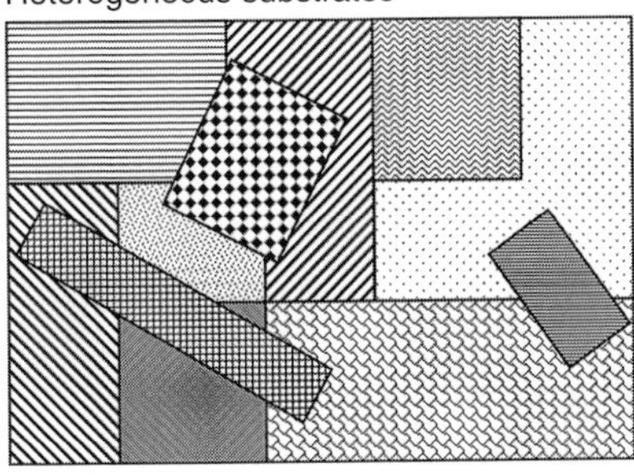

Differential colonization pattern
Partial use of the habitat
Weak overall faunal contribution to CO_2 flux
Weak competition for resources between groups
Greater faunal diversity probably synergyistically increases decomposition

Fig. 4.3. Expected colonization patterns of soil fauna in homogeneous and heterogeneous substrates, and faunal effects on soil functioning as proposed by Vetter *et al.* (2004). Biotic interactions between groups of soil fauna and their joint effect on decomposition are affected by the amount and distribution of organic resources in soils.

5 Soil Macrofauna

5.1 Biology of Soil Macrofauna

Soil macrofauna are eukaryotic multicellular organisms greater than 2 mm long. A number of invertebrates are classified as soil macrofauna, but terrestrial crustaceans (Isopoda) and annelids such as earthworms are also part of this group. A general term for this group is 'cryptozoa', meaning animals that dwell beneath stones, logs, under bark or in other hidden habitats. Generally they emerge at night to forage. Some of the permanent soil dwellers classified as macrofauna are listed below:

Isopods

Isopods ('woodlice') are terrestrial crustaceans involved in the breakdown of litter and wood residues. They are grey, with armoured shells and short antennae. Their respiratory organs may be gills that are kept moist by a special water-conducting system on the underside of the body. They survive desiccation by rolling into a ball or constructing burrows in the soil where they can retreat during dry periods. Isopods are not fastidious in their food choices and feed mainly on well-moistened dead leaves and wood residues, especially those that are partially decomposed by bacteria and fungi. They are important because they consume and fragment dead leaves, thus being responsible for the physical comminution of litter in humid temperate ecosystems. Isopods transform the fresh litter in fecal pellets with much higher specific surface than the original material, which is more suitable for microbial attack.

Diplopoda

Diplopoda ('millipedes') are characterized by double segments, each of which has two pairs of legs. Diplopods have a specialized, two-part mandible that distinguishes them from other insects. Coloration can be white, yellowish or grey. Most feed on decaying plant material. Some species are very active consumers of dead wood, boring between the bark and wood of felled trees and tree stumps. Moist plant residues that have been partially decomposed by fungi are greatly favoured. Diplopods consume small amounts of soil and sand grains during feeding, mixing organic matter and mineral particles and contributing to the formation of soil aggregates. Like Isopods, they are sensitive to desiccation and thus more abundant in moist environments.

Chilopoda

Chilopoda ('centipedes') are regular inhabitants of litter, the top 5 cm of mineral soil and decaying wood. They may be found under bark and stones. All are predators. The first pair of legs is modified into maxillipeds, enlarged claws that contain poison glands. The prey is moistened with digested juices outside the mouth (external digestion) and the liquefied juices are sucked from the prey. They develop slowly and it can take 3 years for a young juvenile to reach reproductive maturity. The lifespan of a centipede can be 6 years.

Scorpions

Scorpions are important as predators of many soil-dwelling insects, although adults generally remain under tree bark and beneath stones, not in the soil. Young scorpions prefer living in the litter layer. They hunt chiefly at night and feed preferentially on beetles. They thrive in hot, dry climates and are perhaps more important as predators in dry and desert ecosystems than temperate forests and grasslands.

©CAB International 2010. *Soil Ecology and Management* (J.K. Whalen and L. Sampedro)

Araneae

Araneae are the true spiders. Being completely predatory, they have practically no role in soil formation and litter decomposition. Instead, they function as top-level predators of soil inhabitants. They often live in dead stalks and tree trunks, where they pursue wood-digesting insects that would settle there. A few spiders construct burrows in the soil and thus make a limited contribution to soil mixing and aeration. The tapestry spiders in the family Ctenizidae construct cylindrical passages in the soil, coat them with a fine web and close them with a trap door that is covered inside with spun silk. Unwary insects and earthworms may fall into the trap and thus be captured. Wolf spiders in the family Theridiidae are large, dark grey spiders inhabiting the litter layer and preying on soil-dwelling insects, mainly beetles.

Opiliones

Opiliones ('harvestmen') are spider-like creatures with eight long, delicate legs. They are related to the spiders, with key differences being that they do not possess silk glands and nor do they produce venom. Harvestmen inhabit the ground litter and humous layer of soils. Some may be found beneath stones and in soil cracks. Their feeding habit may be omnivorous (decaying plant litter and fungi), predatory (feed on small insects) or as scavengers (consume animal residues and dead carcasses). A few predatory opiliones actively search for prey, but most probably use an ambush strategy. Although they possess a pair of eyes, they are unable to form images and use their second pair of legs as antennae to search the environment. The presence of refugia is quite important for the predatory opiliones. Hedgerows around agricultural fields and the crop residue layer in no-till or minimum tillage agroecosystems can provide a suitable refuge for opiliones.

Isoptera

Termites (order Isoptera) are ancient eusocial insects that radiated from subsocial wood-feeding cockroaches, probably during the late Jurassic and early Cretaceous period. Today there are more than 2600 described species found between latitudes of 30 to 51°N and 40 to 45°S (Abe *et al.*, 2000). Termite colonies may contain from 3000 individuals to more than 250,000 individuals. The social structure is organized in a caste system (workers, soldiers, males, winged reproductive queens and queens). Queen termites mate during swarming and can produce 18,000 to 43,000 eggs/day. Juvenile termites (larvae) exhibit polymorphism and neoteny (certain immature organisms can reproduce) and participate in social life. Colonies can exist for up to 50 years, but normally the foraging is local. They are the most abundant soil macrofauna in most tropical ecosystems, where they are keystone species known for their role in soil formation. The importance of termites in soil formation comes from the fact that they thoroughly mix soil minerals and organic residues, add decomposed residues to the soil and liberate essential plant nutrients from organic residues. These effects are mostly seen in the vicinity of the termite mound since most of the foraging is done within a few metres of the nest. Besides decaying wood residues, termites also consume decaying fruits, leaves and carcasses of smaller animals. Before commercial fertilizers were available, many farmers in tropical regions depended on termites to generate nutrient-enriched soils for agriculture.

Termites exhibit a diversified phylogeny (Fig. 5.1), and a broad range of social behaviour and feeding habits. Termites are grouped phylogenetically as 'lower termites' and 'higher termites'. Many termites do consume wood primarily, but there are also soil-feeding termites which can survive on a diet of humified soil organic matter. Regardless of their classification, all termites depend on an obligate symbiosis with gut microflora to digest lignocellulosic materials as a source of energy, nutrients and other essential substances. The symbiotic gut microflora varies from one species to another, and includes a diverse group of methanogenic archaea, spirochaetes and other non-spirochaete bacteria in the higher termites, with an estimated 1000 phylotypes within a single termite species. The lower termites possess archaea and bacteria as well as flagellate protists of the parabasalidsh and oxymonads as symbionts. According to the number of termite species, and to the prokaryote diversity in a single species, which can reach up 100 phylotypes, most of which have no known relatives, termite guts are a relevant reservoir of microbial diversity. Some species are soil-feeders that digest fresh soil organic matter or

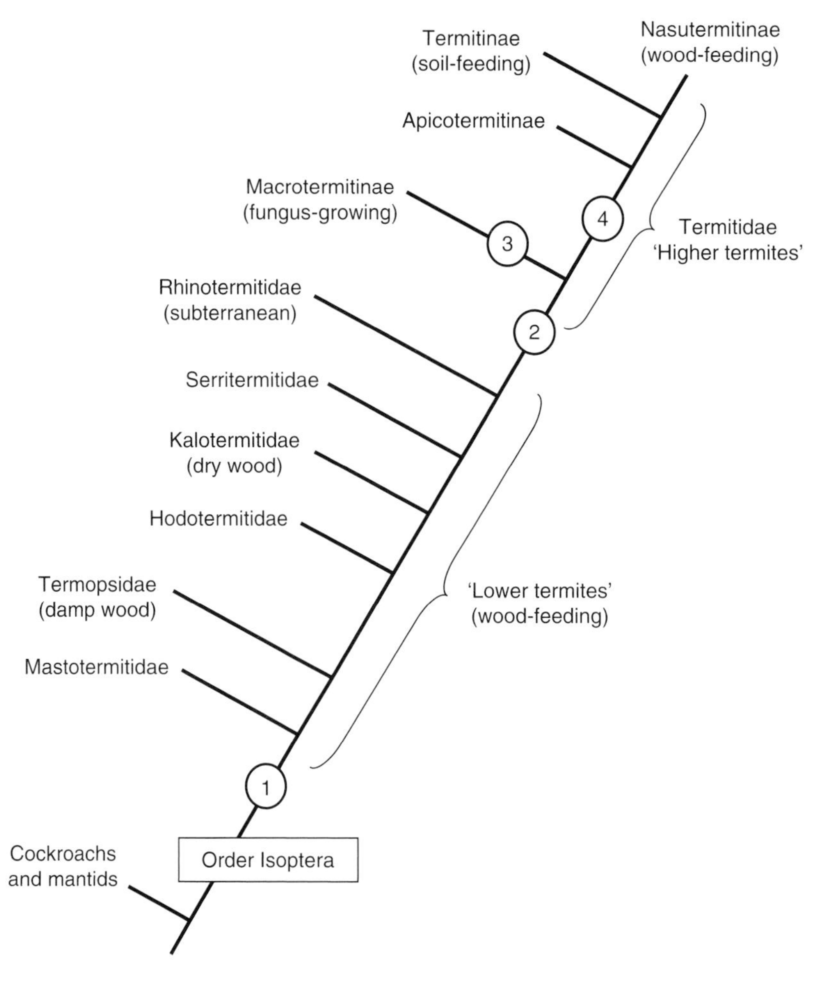

Fig. 5.1. Schematic diagram of the phylogeny of termites and their feeding strategies. The numbers indicate major evolutionary events related to feeding habits and corresponding symbiotic relationships: 1 organisms capable of hindgut cellulose fermentation with the accompanying gut symbionts such as flagellated protists, spirochaetes, methanogenic archaea and other specific groups of bacteria; 2 loss of flagellates; 3 acquisition of *Termitomyces* and fungus-gardening habits; 4 soil-feeding habits. (Modified from Ohkuma *et al.*, 1999; Eggleton, 2001, 2006.)

recalcitrant soil organic matter associated with the humified fraction (humivorous termites) thanks to extremely high pH values in their gut (up to pH 12). The Macrotermitinae have evolved a highly specialized obligate symbiosis with the cellulolitic fungi *Termitomyces*. These termites culture the fungal mycelium with extreme care in specialized garden chambers that are supplied with fresh plant materials, which they cut and place in the gardens just as a farmer might tend his fields. The fungi are harvested and consumed as the main food source in the termite colony.

Throughout the Carboniferous (299–350 million years ago) the terrestrial habitats were rich in lignocellulosic detritus of plant origin. The most efficient decomposers of plant litter then, as today, were bacteria, protozoa and fungi. Cockroach-like insects dominated the terrestrial fauna, many of them phylogenetically related to Isoptera. These fauna facilitated microbial colonization of fresh detritus, first by physical comminution and then by ingesting their own fecal matter that had been partially degraded by microorganisms (coprophagy). The advantage of re-ingesting fecal material is that microorganisms would have partially degraded the cellulose into simpler sugars and would also provide amino acids, vitamins, sterols and other essential compounds for insect nutrition. Thus, the evolution of complex assemblages of gut microbial communities is considered by evolutionary biologists as a process of internalizing the microbial consortia comprising the external rumen (Nalepa *et al.*, 2001). The similarity of the obligate mutualistic bacteria (genus *Blattabacterium*) living in specialized cells of modern-day cockroaches and the termite *Mastotermes darwiniensis* seems to indicate that these organisms have evolved from subsocial, wood-dwelling coackroaches, possibly some time in the late Jurassic or early Cretaceous period (Lo *et al.*, 2003). It has been proposed that this coevolutionary process was also closely related to the evolution of social behaviours. For instance, the transfer of flagellated protists to the progeny must to be done directly by proctodeal trophallaxis (controlled expulsion of undigested hindgut fluids between individuals). As far as we know, the mutualistic gut bacteria are transmitted vertically in the eggs from parents to offspring.

Focus Box 5.1. Symbiotic cellulose digestion in the termite gut.

The digestive tract of a termite has several features that make it a favourable environment for the symbiotic microorganisms and protists. The termite gut can be divided into three regions – foregut, midgut and hindgut – with the hindgut harbouring most of the gut microbiota. Lignocellulosic materials are macerated and partially decomposed by cellulose enzymes secreted in the salivary glands and along the digestive tract before they reach the hindgut (Fig. 5.2). The importance of the hindgut cannot be

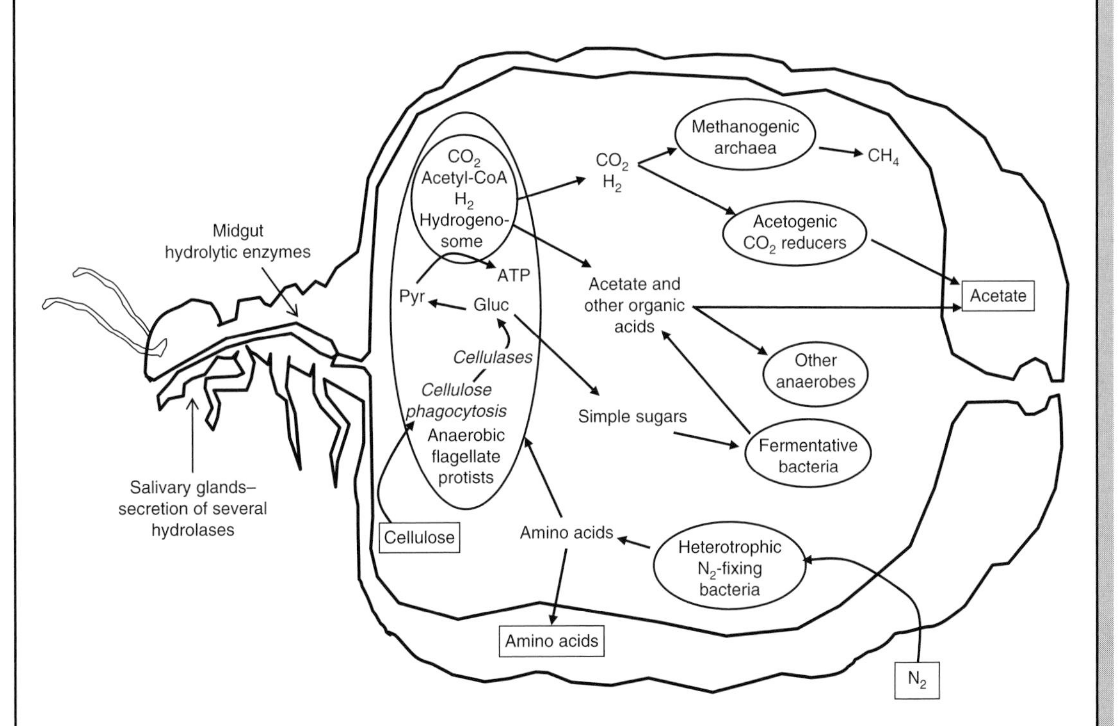

Fig. 5.2. Schematic representation of the metabolic processes occurring in the paunch of a lower termite. Cellulose and lignocellulosic compounds from wood are transformed into acetate and amino acids, which are then assimilated by the termite. Various symbiotic microorganisms may be present and involved in termite metabolism. (Modified from Radek, 1999.)

Continued

Focus Box 5.1. *Continued*

disputed as it constitutes as much as 40% of the body weight of worker termites. It can be considered a continuous-flow chemostatic system where pH, oxygen content and axial flux of material are controlled by the termite. The strict homoeostatic control in the hindgut favours the maintenance of physically differentiated microenvironments where the following complementary biochemical reactions occur (Brune, 1998):

1. Hydrolysis of cellulose and hemicelluloses.
2. Fermentation of the released subproducts to fatty acids that will be resorbed across the gut wall by the host.
3. Dinitrogen fixation.
4. Intestinal recycling of nitrogenated excretion products.

The hydrolysis of cellulose and hemicelluloses is mediated by many microorganisms and particularly the flagellated protists, which possess efficient cellulolytic enzyme assemblages. The breakdown product (glucose) is absorbed by the protists in a specialized organelle called a hydrogenosome, where it is decarboxylated to produce ATP, releasing acetate, CO_2 and H_2. Attached to the protists or living freely in the gut lumen are spirochaetes, bacteria with cells ranging in size from 3 to 100 μm. The spirochaetes are acetogens, producing acetic acid from H_2 plus CO_2. The methanogenic archaea found in the hindgut may also be closely associated with protists or living in the gut wall. These organisms generate methane from CO_2 plus H_2. Another feature of the hindgut microbiota is the presence of prokaryotes possessing the nitrogen fixation gene *nifH*, which encodes dinitrogenase reductase. These organisms are capable of converting N_2 into NH_4^+, which is especially important for termites that feed on carbon-rich residues with a low nitrogen content. Sulfate-reducing bacteria are also common in the hindgut; due to their versatile metabolic capabilities, they are probably involved in several biochemical reactions in the symbiotic digestion process.

The combined metabolic activities of the anaerobic-fermenter flagellates with the acetogenic activity from bacteria, consuming the excess of H_2 from the protists, and other fermentative activities from prokaryotes, release lactate and acetate that are rapidly absorbed by the living termites. The trophic interactions among the various prokaryotes and eukaryotes inhabiting the termite gut are still not fully understood, but remain a fascinating example of a symbiotic relationship that has been retained through millions of years of evolution.

Hymenoptera

Hymenoptera ('ants') were the first predaceous, eusocial insects to live and forage in the litter layer and in soils, according to fossils dating back to the Cretaceous period. The oldest known ants lived in moist tropical forest dominated by gymnosperm trees and a diverse angiosperm flora (flowering plants), which provided a multitude of habitats for foraging ants (Perrichot *et al.*, 2008). At present, ants are classified into 12,000 species distributed among 350 genera, distributed from the Arctic tundra to the tropics.

A mature ant colony is typically 10,000 to 12,000 workers. The castes found in ant colonies include workers, males and queens. Some primitive species have ergatogyne (non-reproductive/reproductive ants), while others have a fertile worker caste in addition to the workers, males and queens. Colonies are established when a virgin queen moves from the parent nest (migration, swarming) and is inseminated. The lifespan of a colony can be 15–20 years. Workers are responsible for foraging, maintaining nest security and caring for the brood. The territoriality can be absolute, with permanent defenders or flexible (no permanent) borders established, but notable inter-colony aggressiveness. Most ants are generalized predators and scavengers, but a wide range of feeding habits occurs within this group. Certain ants are herbivores, like the leaf-cutting ants. Others appear to have a mutualistic relationship with plants, such as species that pollinate plants and disperse seeds. Due to their large numbers, ants are considered keystone organisms in the soil ecosystem. Those that construct nests within the soil are responsible for mixing a significant quantity of organic residues with soil mineral particles. This process contributes to soil aggregation and creates soil pores. Soil around ant mounds is generally enriched in organic matter and has a higher pH than soil undisturbed by ants.

Focus Box 5.2. Evolution in action: new successful social behaviour of invasive ants in southern Europe.

Social organization of ants' nests is a key attribute for understanding their success and ecological prevalence in terrestrial ecosystems. The regular social system is called multicoloniality, where individuals from a nest exhibit systematic aggression to workers of neighbouring nests, fighting until the invaders are completely expelled from their territory. But some species of ants show an even more sophisticated social organization called unicoloniality, whereby individuals can enter, work and mix freely among physically separated nests with no repulsion of individuals from neighbouring colonies. This characteristic is surprising and indeed an evolutionary paradox, because these workers help to raise the unrelated brood.

The human-driven introduction of the Argentine ant (*Linepithema humile*) in Europe at the beginning of 20th century provides a surprising case study of evolution of social strategies in ants. Similar to other invasive species, the colonization of new habitats by the Argentine ant destroyed crops and forests by disrupting the ecological equilibrium of native insect fauna. The native populations of the Argentine ant in South America function under the 'laws' of the regular multicoloniality. However, as Giraud *et al.* (2002) pointed out, the successful colonization of southern Europe by this species was accompanied by a complete breakdown of nestmate recognition ability, leading to a dramatic loss of inter-nest aggression (Fig. 5.3) and to the formation of two immense 'supercolonies', which effectively are two unicolonial populations. The size of the main supercolony extends up to 6000 km (from Italy along the Mediterranean to upper Portugal), comprising millions of nests and billions of

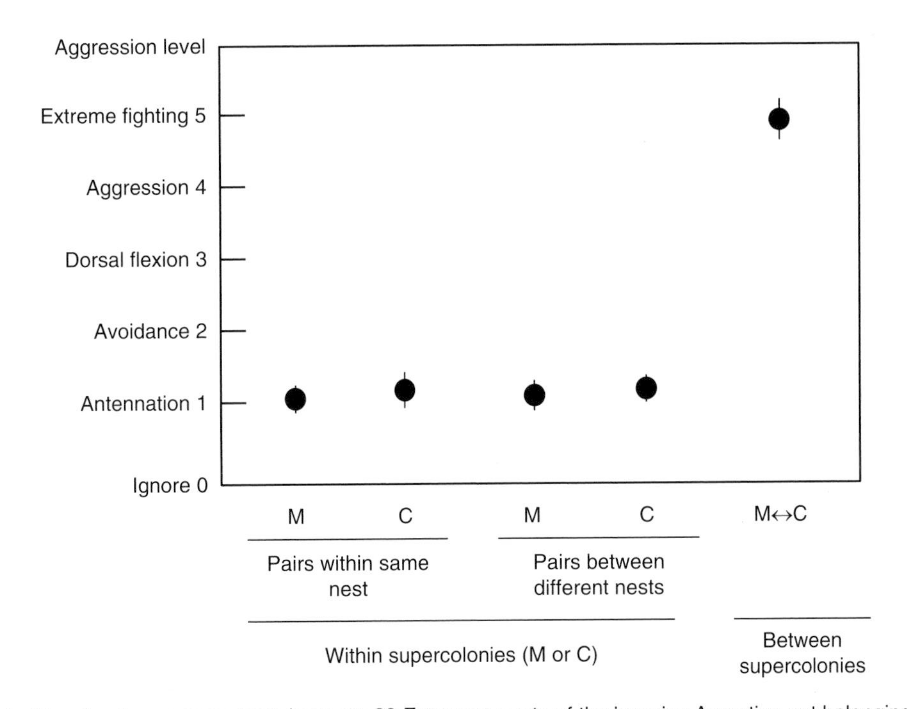

Fig. 5.3. Results of aggression tests between 33 European nests of the invasive Argentine ant belonging to the two 'supercolonies' called 'M' and 'C'. The figure shows the mean maximum aggression level exhibited by combinations of pairs of workers from same nests within each supercolony, between nests within each supercolony and between the two supercolonies. The study showed evidence of two huge and distinct supercolonies in southern Europe (M and C), with no aggression between nests of the same supercolonies despite coming from very distant nests. However, strong fighting was always the response between individuals from the two supercolonies. Behavioural studies in an experimental arena were confirmed by iterative computer simulation. (Data from Giraud *et al.*, 2002.)

Continued

Focus Box 5.2. *Continued*

collaborative workers. Apparently, by reducing the costs associated with territoriality, unicoloniality allows high worker densities and interspecific dominance in invaded habitats.

Initially, it was hypothesized that the genetic bottleneck when the introduced ants arrived in Europe caused loss of all genetic variability at the recognition locus, leading to the formation of a single supercolony. However, Giraud *et al.* (2002) showed that the reduction in genetic variability in the European populations was only 28%, compared with the American ones, based on the study of alleles at 17 microsatellite loci. They proposed that the shift in social organization was more likely the result of a selective cleansing of genetic diversity at the recognition locus (or loci) leading to the formation of two supercolonies.

Colonization of new habitats by the alien species in an environment that was free of native parasites, competitors and predators led to extremely high nest density. Giraud *et al.* (2002) suggest that in this situation, natural selection could favour unicoloniality, reducing genetic diversity at recognition loci. More frequent encounters between close neighbour workers might dramatically increase the costs of defending the territory (fighting is energy and time demanding), possibly outweighing the benefits to workers of defending their territory.

Earthworms

Earthworms are perhaps the most important of the soil macrofauna in temperate regions (forests, grasslands and agricultural fields). These are semi-aquatic invertebrates of the Annelida (family Oligochaeta) that started colonizing terrestrial environments about 600 million years ago. There are an estimated 6000 earthworm species worldwide. The Lumbricidae are a group with peregrine species that have dispersed from Europe to other parts of the world (North America, Australia and New Zealand) as these regions were colonized by European settlers during the past few hundred years.

Earthworms are segmented worms with setae that permit them to grip surfaces. Their movement through the soil is peristaltic and aided by mucus secreted on the body surface. Although they lack eyes, they have sensory organs on their body surface, a complete nervous system, a circulatory system with seven hearts and a well-developed digestive system. Earthworms are true hermaphrodites with male and female sex organs. They are simultaneous hermaphrodites capable of self-fertilization or cross-fertilization. Some species are parthenogenetic. Eggs develop in a cocoon with a thick external chitinous shell that protects them from drought and infection. Earthworms are similar to the enchytraeids and require moist conditions to maintain humidity and facilitate gas exchange through their cuticle. When soil matric potential < -0.1 MPa, some earthworms enter a quiescent state (aestivation) whereas others migrate to depths of 1–2 m to escape dry conditions.

Earthworms are often grouped into ecological categories based on morphological, behavioural and physiological characteristics (Table 5.1; Fig. 5.4). Regardless of their physical location in the soil profile, earthworms consume soil and decomposing plant and animal residues, bacteria, fungi and algae. The quantities consumed can be substantial. An estimated 4 to 10% of the organic matter in surface residue and top 15 cm of soil are consumed by earthworms each year, which suggests that the entire topsoil layer may be processed by earthworms within 10–15 years. Since earthworms literally eat their way through the soil, they have an important role in the mechanical disintegration of particles and the creation of nascent soil aggregates. In addition, chemical transformations occur in the earthworm gut that create clay-humus complexes, stabilizing soil organic matter. Through their interactions with soil microorganisms, earthworms accelerate the decomposition of an important fraction of plant residues and speed up the nutrient recycling from this material. The earthworm body is also rich in nitrogen. A small quantity is released in the mucus that earthworms secrete through their body wall. When earthworms die, the nitrogen contained in their bodies cycles through the microbial biomass and is absorbed by plant roots within a few days.

Table 5.1. Ecological groupings of earthworm species and main characteristics. (Adapted from Lee, 1985; Lowe and Butt, 2005.)

Characteristic	Epigeic species	Endogeic species	Anecic species
Location	Litter layer	Topsoil (0–15 cm depth)	Subsoil (burrows to more than 1 m depth)
Size (cm)	1–5	5–10	> 10
Biomass (g)	0.1–0.5	0.5–1.0	< 5
Pigmentation	Dark red, purple	Pale pink or grey	Dark purple head, pale body and tail
Reproduction	Asexual or sexual	Asexual or sexual	Asexual or sexual
Reproduction rate (cocoons per year)	45–55	10–27	18
Life cycle (days) (field conditions)	40–60 days	80–120 days	About 1 year
Burrowing	Limited	Mostly horizontal, semi-permanent	Mostly vertical, permanent
Casts	Casts deposited in litter layer, rich in organic matter	Casts deposited at soil surface and also in burrows	Casts deposited at soil surface
Food	Organic residues and associated microbiota (bacteria, fungi, protists)	Soil containing partially decomposed organic matter and associated microbiota	Partially decomposed organic matter from the soil surface and associated microbiota

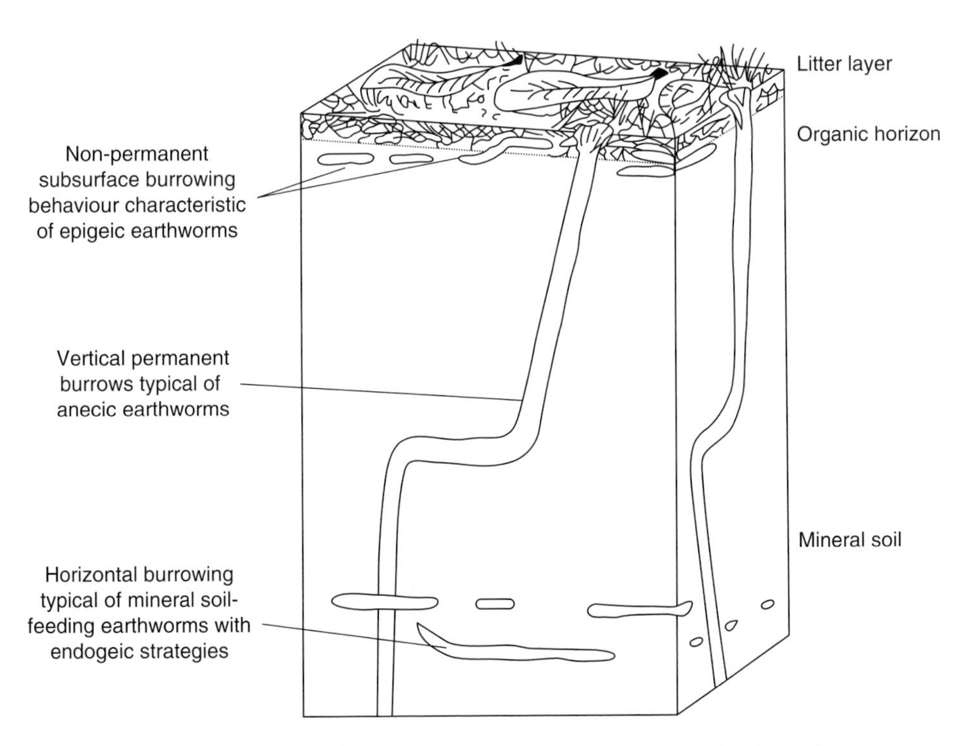

Fig. 5.4. Conceptual representation of the burrowing habits within the soil profile of the different ecological groups of earthworms summarized in Table 5.1.

Burrowing is another earthworm activity of great importance to plants. Earthworm burrows function as macropores, essential for the circulation of water and air in the soil. The lining of the earthworm burrow is a good habitat for bacteria and other soil organisms, creating a nutrient-rich zone for plants. Often, plant roots grow along earthworm burrows, providing better plant anchorage and enhancing root access to essential nutrients and water. Earthworm burrows also decompact the soil, making it fluffier than it would be otherwise and facilitating gas and water exchange.

Focus Box 5.3. Adaptive strategies in hermaphrodite earthworm reproduction.

The principle of reducing costs and maximizing benefits is a general rule in all biological systems, even under the dark cover of the soil surface or in the small pores between soil aggregates. Organisms should allocate available energy to growth, reproduction and maintenance, but energy is limited and trade-offs among those sinks are necessary. Soil fauna, regardless of their body size, feeding strategy or taxonomic group, should allocate energy and other resources in a manner that maximizes their fitness to ensure survival. This focus box looks at a study by Alberto Velando and co-workers, which demonstrated that hermaphrodite earthworms carefully evaluate their mating partners before copulation, adjusting their breeding effort depending on the perceived quality of the partners. They also demonstrated trade-offs made by earthworms when energy and resources are limited – earthworms are expected to prioritize and maintain their actual growth rate at the expense of reducing investment in the current reproduction. This provides evidence that future reproduction might have more impact on their fitness than current reproduction (Aira *et al.*, 2007).

Hermaphrodites are individuals that possess both male and female sexual organs. This trait seems to be retained in sessile animals that may have difficulty encountering a member of the opposite sex. It provides a number of possibilities for the organism when it comes to reproduction – an individual could reproduce asexually or could engage in sexual reproduction. Even when sexual reproduction occurs, it is not guaranteed that an individual will accept the sperm from its partner, so self-fertilization remains an option during sexual reproduction in hermaphrodites.

The earthworm is an excellent species to answer questions about how hermaphrodites reproduce, because they exhibit a variety of reproduction strategies. A number of earthworm species can reproduce asexually by parthenogenesis, a process in which eggs develop without being fertilized and haploid offspring are produced. These include the epigeic species *Dendrobaena octaedra, Dendrobaena rubidus* and *Eisenia tetraedra* and the endogeic species *Aporrectodea rosea, Aporrectodea trapezoides, Octolasion cyaneum* and *Octolasion tyrtaeum* (Lowe and Butt, 2007). Yet, sexual reproduction is common and exclusively practised by earthworms such as *Eisenia andrei* and *Lumbricus terrestris.*

There are a number of advantages to sexual reproduction. Offspring from closely related parents commonly show reduced fitness, particularly under stressful conditions. On the other hand, extreme outcrossing may also decrease offspring fitness because it mixes genomes that have adapted to different environments. Thus, the offspring may not possess genes that will permit them to flourish in either habitat. With outcrossing, there is also a risk of physical or physiological incompatibilities of partners from different populations. Velando *et al.* (2006) demonstrated experimentally that inbreeding and outbreeding both affect the number of potential offspring (number of cocoons) produced by the earthworm *E. andrei* (Fig. 5.5).

In hermaphrodites, the fecundity of the female function is normally limited by the amount of energy available for egg production, and the fecundity of the male function is normally limited by the number of eggs available. Therefore simultaneous hermaphrodites are expected to mate not to get their own eggs fertilized, but rather for the opportunity to fertilize the eggs of their partners. These researchers found that earthworms donated more sperm to non-virgin mates than to virgin mates partners (Fig. 5.6), evidence that earthworms can control the amount of sperm transferred to their sexual partners (Velando *et al.*, 2008). Moreover, as female fecundity was positively correlated with body size, such increases in donated sperm were greater when the earthworms were mated with larger partners. This indicates that hermaphrodites carefully evaluate their partners and subsequently adjust the quantities of sperm that are worthy to invest.

Continued

Focus Box 5.3. *Continued*

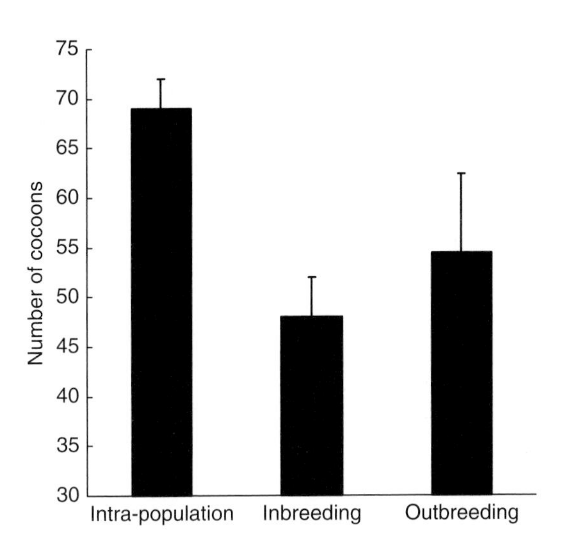

Fig. 5.5. Earthworms adjust their reproductive effort based on their genetic relatedness to potential partners. The figure shows the number of cocoons produced by mating pairs of the earthworm *Eisenia andrei* when mated with sibs and non-sibs from the same population, and with non-sibs from a geographically isolated population. Inbreeding and outbreeding matings caused a significant reduction in cocoon production, especially in genetic lines with high reproductive rates. (Modified from Velando *et al.*, 2006.)

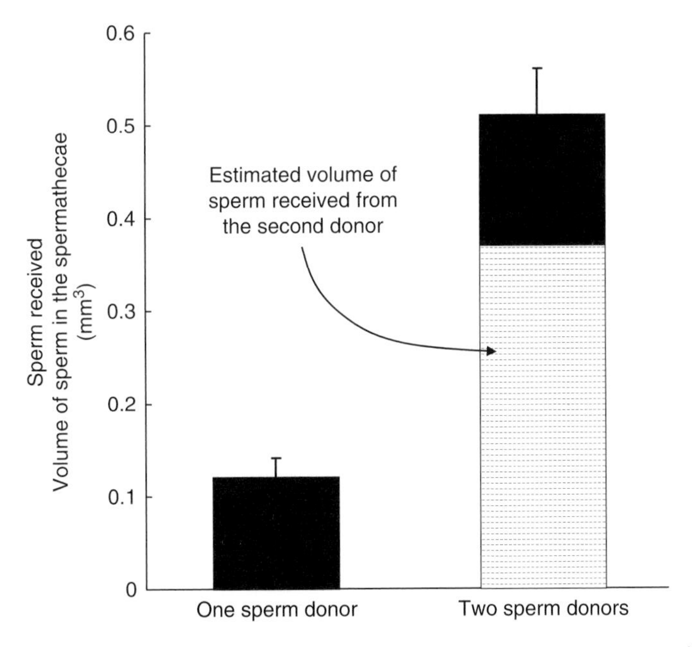

Fig. 5.6. Volume of sperm received by hermaphrodite earthworms after copulation with one (virgin partners) or two donors. Earthworm sperm donation tripled when mating with a non-virgin mate. Donors were selective about donating sperm to mates, suggesting that partner evaluation occurs in this hermaphrodite animal. (Modified from Velando *et al.*, 2006.)

5.2 Diversity and Ecological Functions of Macrofauna

Due to their large size, the macrofauna are easier to collect and enumerate than the soil microorganisms and the smaller soil fauna. However, biodiversity surveys continue to reveal species that have not been described previously. The number of described species and approximate population sizes of the soil macrofauna are given in Table 5.2. The application of DNA-based technologies to confirm species identification is expected to resolve deficiencies in the taxonomy and classification of these organisms in the coming years. Molecular biology also provides information vital to developing phylogenetic trees, deepens our understanding of the relationships among species and provides clues to the ancestory and evolution of our modern soil macrofauna.

The major ecological functions of soil macrofauna are as detritivores and top-level predators. The macrofauna are not numerically dominant, but they have a larger biomass and size than other soil biota and thus have a greater impact on soil functions and interactions in soil foodwebs. Termites, ants and earthworms are perhaps the best known macrofauna due to their immense contribution to decomposition and soil pedogenesis. They are continually moving and mixing large quantities of soil and organic residues, which changes the food resources available to smaller soil fauna and creates a more intimate association between organic and mineral particles, the basis of soil aggregation. They also create habitat for their comfort and survival (e.g. termite and ant mounds, earthworm burrows) that have consequences for soil porosity as well as gas and water exchange. The interaction between soil microorganisms and macrofauna stimulates respiration, litter decomposition and nutrient cycling rates. These groups of soil macrofauna are referred to as 'ecosystem engineers'. They can influence soil properties and ecosystem function directly and indirectly at a scale that transcends, spatially and temporally, the effect of single individuals within populations. They can completely change the soil structure, chemical properties and biological activity in a short period of time. Sometimes they are referred to as 'keystone species', although this term is generally used to refer to top-level predators that control the population dynamics of prey organisms in aquatic ecosystems. The ecological functions of the soil biota ranging from the microorganisms to the macrofauna are summarized in Table 5.3.

5.3 Methods for Collecting and Enumerating Soil Macrofauna

The soil macrofauna can be collected by trapping (e.g. in pitfall traps) or by hand from a prescribed sample area. Organisms can be found under rocks, in the litter layer of the forest floor, under the bark of fallen logs and in the soil. Termite and ant colonies can be excavated from trees, wooden structures or the ground. Take appropriate safety precautions when collecting chilopoda, scorpions and Araneae that possess venom harmful to

Table 5.2. Macrofauna species and population sizes in terrestrial ecosystems with native or semi-managed vegetation (grasslands, forest). (Data from: Gongalsky *et al.*, 2004; Tree of Life Project (http://tolweb.org/tree/); Biology Catalog (http://entowww.tamu.edu/research/collection/hallan/).

Taxonomic group	Described species in soil	Approximate population size (individuals per m^2)
Isopods	5,000	0–200
Diplopoda	10,000	100–800
Chilopoda	3,000	0–100
Scorpions	1,400	0–10
Araneae	35,000	0–100
Opiliones	6,000	0–100
Isoptera (per colony)	2,400	10,000–1,000,000
Hymenoptera (per colony)	12,000	< 1,000–1,000,000
Earthworms	6,000	0–1,000

Table 5.3. Summary of the influence of soil biota on major soil processes, and subsequent applications for pest management.

Group	Decomposition and nutrient cycling	Soil structure	Biological control
Microorganisms	Biochemical breakdown of organic residues, mineralize and immobilize nutrients	Extracellular polysaccharides (chemical binding agent) and hyphae (physical binding) involved in aggregate formation	Group includes fungal pathogens and organisms antagonistic to pathogens Antibiotic production
Microfauna	Regulate and stimulate the turnover of bacteria and fungal populations Alter nutrient cycling rates	Involved through interactions with microorganisms	Group includes nematode parasites and predators of undesirable bacteria, fungi and nematodes
Mesofauna	Regulate fungal and microfaunal populations Alter nutrient cycling rates Fragment plant residues	Produce fecal pellets Create biopores Promote humification Aggregate formation	Collembola and acari could be effective against fungal pathogens, nematode parasites and acari pests
Macrofauna	Fragment and mix plant residues Stimulate microbial activity	Mix organic residues and mineral particles Litter comminution Create biopores Promote humification Macroaggregate formation and disruption Water infiltration	Extracellular polysaccharides from earthworms used in traditional Chinese medicine (antibacterial and antifungal properties)

humans. Although some do possess a hard outer cuticle, the macrofauna tend to be soft-bodied and can be damaged by rough handling such as sieving. It is recommended to sort gently through the litter or soil samples by hand to collect the organisms of interest. For earthworms, soil is typically excavated from a pit 30 cm wide × 30 cm long × 15 cm deep. Deeper-dwelling earthworms may be captured by pouring an expelling solution such as hot mustard into the hole, which irritates the skin and drives the earthworm to the soil surface. Collected specimens can be preserved in 5% formaldehyde solution or 70% ethanol solution until identification is made in the laboratory with the aid of a magnifying glass or dissecting microscope (10× magnification).

6 The Soil Foodweb

The previous chapters have provided an introduction to many soil microorganisms and fauna that inhabit terrestrial ecosystems. The soil foodweb is the assemblage of living organisms and the food resources, including plants, that can be studied to understand energy and nutrient flows in the soil system. Soil foodwebs are often referred to as detrital food webs because they are mainly based upon detrital organic matter, and the major function of soil organisms is to decompose and recycle the nutrients contained in dead residues of plant, animal and microbial origin. The soil foodweb could also be called the root exudate-based foodweb if we considered how carbon and other substances flowing from living plant roots to the root-associated soil serve as food for a host of symbiotic and free-living microorganisms, which in turn are consumed by protists, nematodes and higher trophic levels in the soil foodweb. In either scenario, plants (alive or dead) are at the base of the soil foodweb.

This chapter focuses on the interactions of organisms and plants in the soil foodweb. One might wonder why an entire chapter has been devoted to this topic. Certainly it is interesting to identify the diet of various creatures. However, studying the soil foodweb can help us to gain insight into the importance and relative contribution of each group of organisms in major ecological services such as energy flows, which is essentially carbon cycling, or nutrient cycling. If we want to understand how soils can act as a buffer for atmospheric CO_2, it is absolutely critical to understand how much energy (carbon) is transferred from one trophic level to another before it is finally sequestered into a stable soil carbon pool, and how the biotic and abiotic parameters modulate this process. Similarly, the mineralization of organic nitrogen compounds into the plant-available inorganic NH_4^+ and NO_3^- forms by soil microorganisms and higher trophic groups is of immense importance in sustainable agriculture, where the goal is to optimize nutrient recycling from natural processes and reduce exogenous fertilizer inputs. Understanding the relationships in the soil foodweb can also provide options for biological control, since plant pathogens and parasites can be controlled naturally by other soil organisms, but how are these interactions affected by disturbances and land management? It is clear that soil foodwebs are very complex and much remains to be learned about the trophic structure and relationships among the organisms involved.

6.1 Trophic Groups and Biotic Interactions in the Soil Foodweb

Models of the soil foodweb such as the one shown in Fig. 6.1 provide a picture of the coexistence and interactions among soil organisms, how energy flows from one trophic group to another, and how nutrients are released by the feeding activities and soil disturbance caused by various groups of soil organisms. A typical soil foodweb includes the following trophic groups:

- Primary producers: these are living plants (including algae), which absorb nitrogen and other nutrients from the soil solution and release exudates through their roots that stimulate the activity of soil foodweb organisms.
- Symbiotic soil organisms, fungal pathogens and parasitic nematodes require living plants for their survival. Herbivorous insects and animals consume plants, leaving fecal material and other animal residues behind. Plant debris such as leaves, branches and dead roots is a major food source for many soil organisms. Organic substrates of plant and animal origin are the resource base of the soil foodweb.
- Primary decomposers: the bacteria and saprophytic fungi occupy this trophic level.

©CAB International 2010. *Soil Ecology and Management* (J.K. Whalen and L. Sampedro)

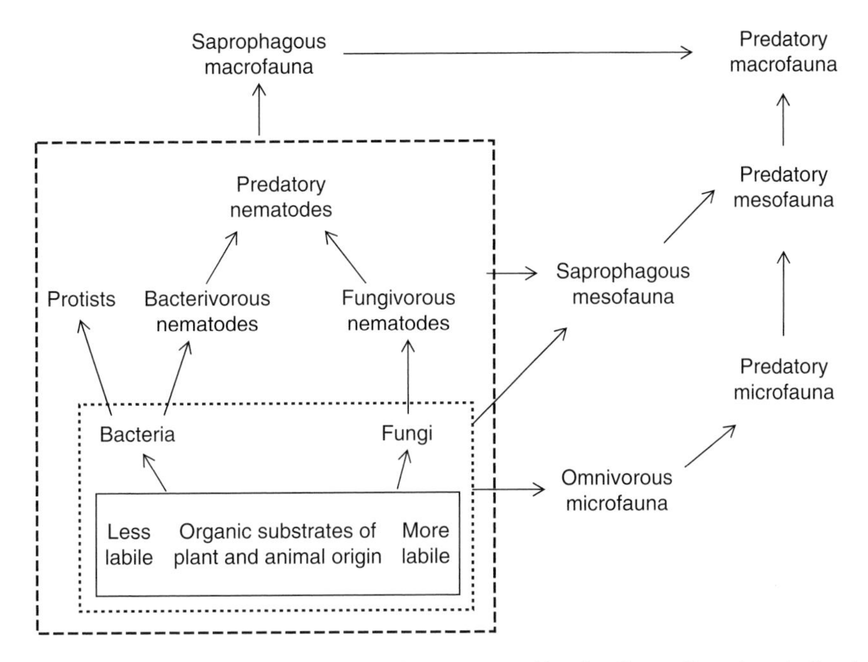

Fig. 6.1. Simplified soil foodweb showing selected trophic groups and feeding fluxes. Organisms in the dotted and dashed boxes may be fed upon selectively or indiscriminately by other trophic groups. (Adapted from Moore *et al.*, 1988.)

- Consumers of bacteria and fungi: protists, bacterivorous and fungivorous microfauna (nematodes) and mesofauna (some collembola and mite species) selectively consume bacterial cells or fungal hyphae.
- Saprophagous mesofauna and macrofauna: many of the mesofauna and macrofauna feed exclusively on decaying organic residues (collembola, mites and earthworms). These organisms often consume bacteria and fungi living on the residue as well as microorganisms adhering to soil particles.
- Predators: at every level of the foodweb are predators that feed on other living creatures, which partially controls the population size of the prey.

It would be very difficult to include all species of soil-inhabiting organisms in each compartment of the foodwebs, due to the number of taxa involved. A soil foodweb model is, by necessity, a simplification of the real world.

The diagrammatic representation of a soil foodweb (Fig. 6.1) provides insight into the interactions among trophic groups due to dietary requirements and feeding habits, yet we cannot forget that there are other biotic interactions among the creatures that live in terrestrial ecosystems. As seen from Table 6.1, some interactions are beneficial (+) for both populations, whereas others are beneficial for one population and have no effect (0) on a second population. Finally, some interactions are detrimental (–) for both populations.

Symbiosis is an obligate or facultative relationship that is beneficial to both partners. In the

Table 6.1. Biotic interactions occurring among soil microorganisms, with plants and soil fauna.

	Effect of interaction on	
Interaction	Population 1	Population 2
Symbiosis	+	+
Commensalism	+	0
Mutualism (Synergism)	+	+ or 0
Amensalism	–	+ or 0
Predation (Parasitism)	–	+
Competition (Antagonism)	–	–

+ positive effect; – negative effect; 0 no effect

absence of one partner, the other remains inactive or functions at a very low level. For example, the interaction between the bacterium *Rhizobium* and leguminous plants is symbiotic. Upon sensing the presence of a host plant, these bacteria secrete substances that break down a small section of the root cell membrane and permit it to enter the plant root. This process leads to 'infection' of the plant roots by the bacteria. Once established within the root, the bacteria grow in a protected structure, the root nodule (Fig. 6.2; Plate 1). The energy source for bacterial growth is carbon fixed by the plant via photosynthesis. The N_2 fixed from the atmosphere by *Rhizobium* is converted to ammonium and uridide compounds, which are then transported to the plant. Thus, the bacteria obtain energy and have a protected environment for growth and reproduction within the root system, while the plant gets essential nitrogen for its growth from the bacteria.

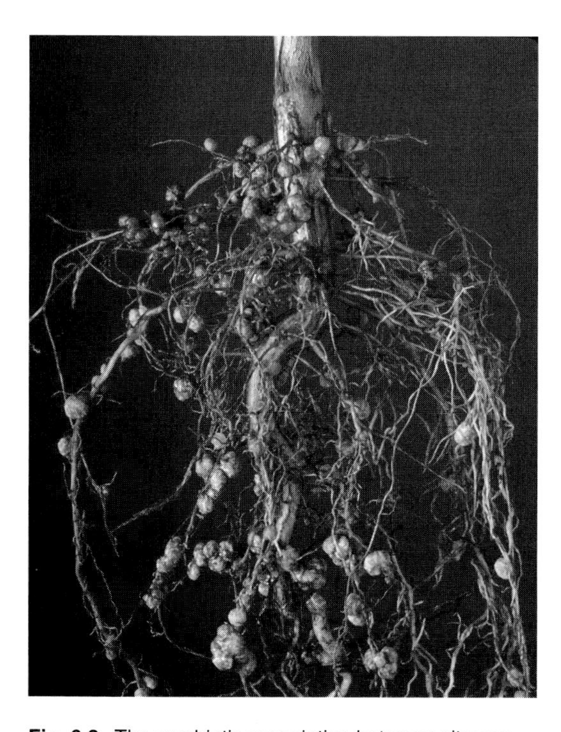

Fig. 6.2. The symbiotic association between nitrogen-fixing bacteria and leguminous plants. The nodules on the root of the soybean plant illustrated contain *Rhizobium* bacteria. Within the nodule, the bacteria fix atmospheric nitrogen, which they share with the plant. In exchange, the plant supplies the bacteria with a source of carbon and energy for growth. (Photo courtesy of David M. Dennis.)

The interaction between *Nitrosomonas* and *Nitrobacter* is an example of **commensalism**. These bacteria are key regulators of nitrification, the conversion of ammonium to nitrate. The reactions involved in nitrification are:

$$NH_4^+ \rightarrow NO_2^- \text{ (catalysed by } \textit{Nitrosomonas}\text{)}$$

$$NO_2^- \rightarrow NO_3^- \text{ (catalysed by } \textit{Nitrobacter}\text{)}$$

It is clear that *Nitrobacter* depends on the activity of *Nitrosomonas* to generate NO_2^- for its reaction, so this interaction is positive (+) for *Nitrobacter*. Yet, the activity of *Nitrobacter* may not have any effect (0) on *Nitrosomonas* since it uses ammonium for its reaction.

The interaction between free-living soil microorganisms and earthworms is an example of mutualism (synergism) in the soil foodweb (Fig. 6.3).

The free-living soil microbes may be inactive or functioning at a low metabolic level when they come in contact with earthworms. Many researchers have observed earthworms secreting mucus, a low-molecular weight mucopolysaccharide, on soil and plant residues as they are ingested, which immediately activates microbial growth. Sometimes the earthworm gathers and covers plant residues with its fecal material (casts) and soil for a period of time before it is consumed. The residues and soil come in contact with mucus from the earthworm body and in the casts, which stimulates microbial activities. When the residues are sufficiently decomposed, they are eaten by the earthworm, along with the microbial cells.

Earthworms also secrete mucus into their intestinal tract to further stimulate microbial activity as the soil-residue mixture passes through their digestive system. The earthworm gut is a favourable microenvironment for microbial activity, particularly for facultative anaerobic bacteria such as the denitrifiers. It is certain that they are activated by passage through the earthworm gut since denitrification activity is much higher in fresh earthworm casts than in soil that was not consumed by earthworms. In general, earthworm casts possess more microbial biomass and higher microbial respiration than the surrounding soil, which indicates that earthworms have a positive effect on soil microbial populations eaten by earthworms. While

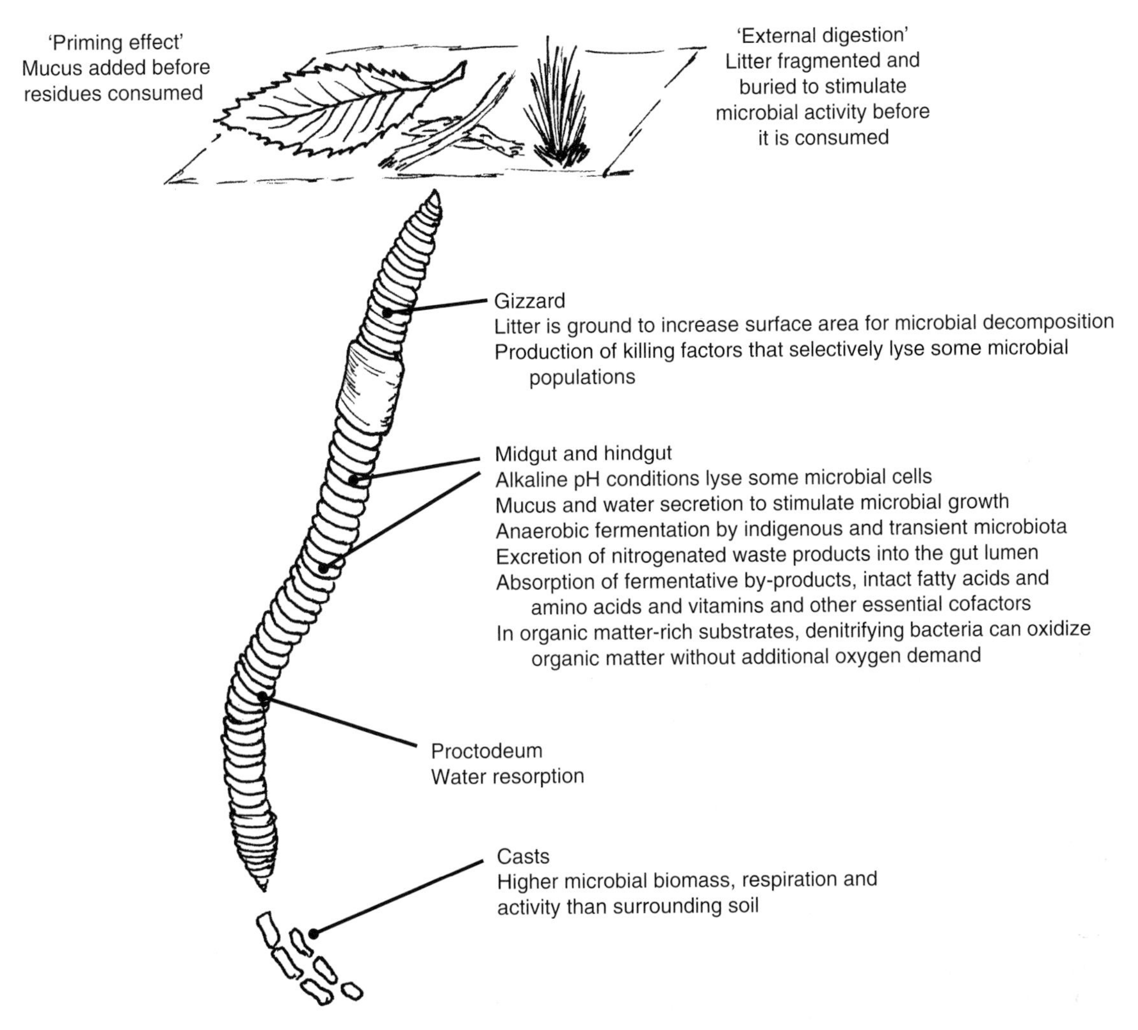

Fig. 6.3. Mutualistic interaction between earthworms and soil microorganisms, leading to the breakdown of organic residues that provides energy for both organisms. The earthworm gut can be considered an absorptive-fermentative tube. At the beginning of the gut tube, the initial grinding of the ingested soil, organic matter and accompanying microbes leads to biochemical breakdown. Afterwards, the midgut and hindgut provide a unique homoeostatic microenvironment for microbial growth based on ingested organic matter, with simultaneous assimilation of newly formed microbial biomass by the earthworm.

some microbial cells can be broken down and digested during this process, many bacteria and fungal spores pass through intact. Some free-living microorganisms secrete hydrolytic enzymes that accelerate the breakdown of organic residues and provide energy and nutrients to the earthworm. There is growing evidence that earthworms depend on an indigenous gut microflora as well as transient microorganisms for survival, as they do not have the capacity to enzymatically degrade their preferred food sources alone nor can they synthesize a number of fatty acids and cofactors essential for their life.

Studies based on fatty acid profiles (Sampedro *et al.*, 2006; Sampedro and Whalen, 2007) showed that microbial biomass was greater, and diversity different, in earthworm gut than in surrounding soil (Fig. 6.4). There were distinct microbial

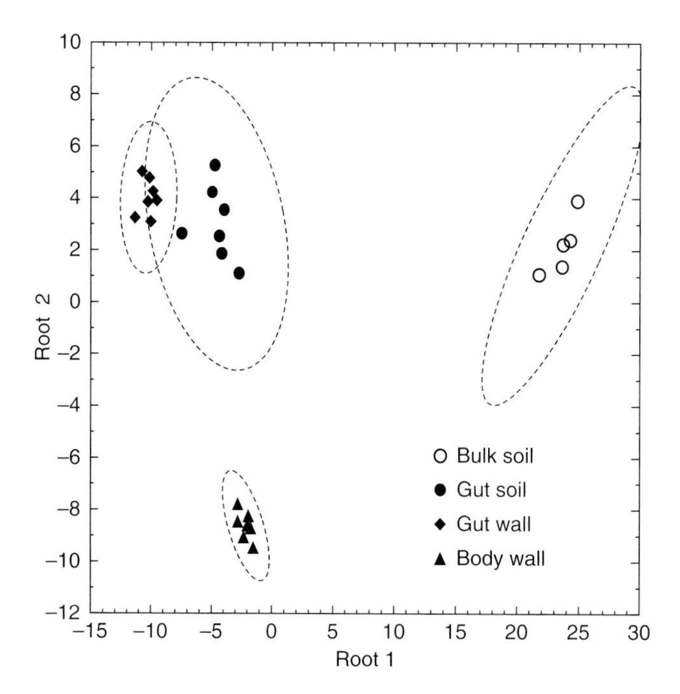

Fig. 6.4. Results of multivariate analysis (discriminant function analysis, DFA) performed on identified fatty acids (FAs) of bacterial and fungal origin from samples of bulk soil, gut soil, gut wall and body wall of the earthworm *Lumbricus terrestris*. The dotted lines indicate the confidence ellipses at $\alpha = 0.05$. The fatty acid profile of the earthworm tissues was more similar to the indigenous gut microorganisms than to those in the ingested soil. Earthworms assimilated intact FAs from microbial origin. This suggests that the homoeostatic gut environment supported the growth of microbes important for earthworm nutrition. (Modified from Sampedro *et al.*, 2006.)

assemblages in each earthworm gut section (midgut, hindgut and proctodeum), suggesting that the microbial community in the earthworm gut is not a casual combination of microorganisms already present in the soil. The transfer of fatty acids from soil microbes to earthworm tissues was observed, and these came from a microbial community assemblage in the gut wall that resembled the gut microorganisms more than microbes from the bulk soil. This indicates that earthworms may derive essential fatty acids and energy from gut microbiota rather than from ingested soil microbes.

Not all interactions in the soil foodweb are positive or have no effect on microbial populations. For instance, amensalistic relationships occur when one population benefits to the detriment of a second population. Antibiotic production by actinomycetes under environmental stress does not necessarily improve their growth, but reduces competition from microorganisms that are more tolerant of the stress. In contrast, predation is nearly always positive for one population and detrimental to the second. The consumption of bacteria by protists (protozoa) is an example of predation in the soil foodweb. Other predators include bacterial-feeding nematodes, fungal-feeding nematodes, soil mites that consume bacteria, fungi and nematodes, and collembola that consume fungi.

Competition occurs when two organisms require the same substrate for physiological processes. Plants and microorganisms compete for ammonium, nitrate, phosphate and other nutrients that are essential for their growth. Heterotrophic microorganisms like saprophytic fungi require ammonium for growth and compete with the bacteria *Nitrosomonas*, which uses ammonium to produce nitrite. Competition is a natural ecological phenomenon that regulates the size of populations in the soil foodweb and indeed in all ecosystems.

Focus Box 6.1. Food chains and foodwebs.

Early ecologists working on the trophic relationships in lakes were the first to introduce the concept of 'food chains'. The term was quickly incorporated into the common language of non-scientists, perhaps more easily than any other ecological term. A food chain is a simple, linear model that shows the transfer of energy and nutrients between groups of organisms that are classified into trophic levels. A typical food chain consists of a primary producer (plant) that is consumed by a herbivore. In turn, the herbivore is eaten by a predator. The energy and nutrients transferred from one trophic level to the next are sufficient to support the populations and communities that inhabit the ecosystem.

Further study of trophic relationships in aquatic and terrestrial ecosystems demonstrated that this simple situation is quite uncommon. It is more realistic to use the term 'foodweb', which represents the interactions among networks of biological organisms – these interactions may be related to diet and feeding habits, but there are other positive and negative interactions among the biota that can be considered in a foodweb.

In soils, the foodweb system is quite complex. Primary producers are a source of energy and nutrients while they are alive and after their death (e.g. dead plant residues). The diet of soil biota can consist of above-ground plant parts (leaves, stems, branches, bark) or below-ground plant parts (roots, root exudates). Interdependency among the soil biota for nutrient cycling is often observed. For example, soil microbial growth tends to be limited by the availability of inorganic nitrogen (NH_4^+, NO_3^-), but grazing by soil fauna recycles some of the inorganic nitrogen that was immobilized in microbial biomass, releasing it into the soil solution and positively stimulating microbial growth. In soil foodwebs, assigning organisms to a specific trophic level is challenging because omnivory is widespread. The trophic level of a species could change due to seasonal fluctuations in soil conditions, litter quality and ontogeny.

Another complexity in the soil foodweb is that we could find two main pathways for energy flow from detritus to the highest trophic positions, depending on whether the detritus is degraded by bacteria or by fungi. Some researchers could consider these pathways to represent simple food chains within the foodweb. The bacterial food chain describes a series of trophic relationships that begins with the breakdown of simple sugars and cellulose by bacteria and other prokaryotes. Bacterial populations are grazed upon by protists and bacterivorous nematodes, which in turn are consumed by higher-level predators (predatory nematodes and mites). The fungal food chain revolves around the brown-rot and white-rot fungi that decompose the more recalcitrant plant components such as hemicellulose and lignin. Energy and nutrients from the fungi are transferred through trophic positions that include fungivorous nematodes, predatory nematodes, collembola and mites.

6.2 Energetic Relations in the Soil Foodweb

Soil foodwebs may have many trophic levels, and the question has arisen as to whether there is sufficient energy to support so many organisms. For populations to survive, individuals must obtain sufficient energy to grow and reproduce. The accumulation of biomass in a heterotrophic population during a period of time is called secondary production (P). This measurement helps researchers to understand the diet, metabolic processes and life history of soil organisms under field conditions.

This is achieved by combining the P value with other data to calculate ecological efficiencies such as:

- P:B ratio is the annual secondary production divided by the mean biomass during 1 year. This ratio tells us how many generations were produced in the field during a year, confirming the development time and reproduction efficiency of individuals in natural field populations.
- P:C ratio is the production to consumption ratio and indicates the proportion of food resources consumed by the animal population during a period of time.
- P:A ratio is the production to assimilation ratio and indicates the proportion of food consumed that was actually assimilated into the animal tissues (corrected for energy or carbon loss through respiration, mucus secretion and other losses).

We could calculate the secondary production for every population in the soil foodweb, but would quickly find that this requires an impossibly large and complicated sampling programme. To overcome this problem, researchers have developed an energetic approach that considers the population sizes, food preferences and consumption rates, mortality rates and energy conversion parameters

Table 6.2. The specific death rates, biomass, assimilation efficiency and production efficiency of trophic goups in a soil foodweb from a North American shortgrass prairie. (Adapted from Hunt *et al.*, 1987.)

Trophic group	Specific death rate (D per year)	Assimilation efficiency (e_{ass}, %)	Production efficiency (e_{prod}, %)	Biomass (mg C/m²/year)
Predatory mites	1.84	60	35	0.16
Nematophagous mites	1.84	90	35	0.16
Predatory nematodes	1.60	50	37	1.08
Omnivorous nematodes	4.36	60	37	0.65
Fungivorous nematodes	1.92	38	37	0.41
Bacterivorous nematodes	2.68	60	37	5.80
Collembola	1.84	50	37	0.46
Mycophagous prostigmata	1.84	50	35	1.36
Crytostigmata	1.20	50	35	1.68
Amoebae	6.00	95	35	3.78
Flagellates	6.00	95	40	0.16
Phytophagous nematodes	1.08	25	40	2.90
Mycorrhizal fungi	1.20	100	37	7.00
Saprophytic fungi	2.00	100	30	63.0
Bacteria	1.20	100	30	304
Detritus	0.00	100	100	3000
Roots	1.00	100	100	300

of each trophic level. Some of the key data for soil organisms living in a shortgrass prairie are given in Table 6.2.

A great deal about soil foodwebs has been revealed from the energetics approach. The assimilation and production efficiencies of soil biota have been used to calculate the amount of energy that is transferred from the resource base to bacteria and fungi and further up the foodweb. The detrital foodweb in a North American shortgrass prairie (Table 6.2) had a sufficient resource base and production efficiencies to support eight trophic levels. In contrast, most aquatic foodwebs and above-ground terrestrial foodwebs possess only four to five trophic levels.

One of the key findings from the study of soil foodwebs was that omnivory under field conditions was much more widespread that previously thought or detected in laboratory feeding trials. Secondly, the flagellates and amoebae have much greater secondary production and faster generation times than nematodes, processing about three times more carbon each year than the omnivorous and predaceous nematodes. Predation on microorganisms releases the nutrients that were immobilized in microbial cells. The protists, especially the amoebae, are responsible for 20–40% of the net nitrogen mineralization under field conditions. The excretion of NH_4^+ (protists and nematodes) and NO_3^- (collembola) supplies inorganic nitrogen in a form that is available to plants and could be important for primary production in managed and natural ecosystems.

Focus Box 6.2. The detrital foodweb of a shortgrass prairie.

Soil detrital foodwebs are complex and show that trophic relationships are often not linear. Another special feature is that they feature 'bottom-up' control, where the populations of microorganisms and soil fauna are controlled by the amount of plant litter and roots entering the system. Moreover, the detritus is spatially heterogeneous in its abundance, availability and quality, which means that the soil biota are constrained by their ability to locate and forage on suitable food resources. In contrast, a foodweb exhibiting 'top-down' control would be one in which the feeding activity of the top predator controls the populations of lower trophic groups.

In soil foodwebs, the food resource is degraded biochemically by bacteria and fungi. The soil fauna are grouped into trophic groups, such as bacterivores and fungivores. Microdetritivores such as enchytraeids and

Continued

Focus Box 6.2. *Continued*

earthworms can use both detritus and microbial biomass for growth. These groups are consumed by specific or generalist predators. Also sharing the soil foodweb are root-feeding organisms belong to the herbivorous group. Despite the clarity of the definitions, it is quite difficult to place species into a single trophic level in soil foodwebs because they may derive their energy from multiple sources. For example, a nematophagous mite may prey on phytophagous, bacterivorous, fungivorous or predatory nematodes, which are from different trophic levels. Thus, soil foodweb organisms are more often grouped into trophic positions, which are less stringently defined than trophic levels.

The soil foodweb of the native North American Shortgrass Steppe at the Central Plains Experimental Range in north-eastern Colorado served as a model for Hunt *et al.* (1987). An exhaustive taxomic study was undertaken to identify the species present, their functional role, population numbers and biomass estimates. A simplified foodweb was devised by placing species into functional groups that use the same food resources and foraging activities, life cycle and population dynamics (Table 6.2). No earthworms or enchytraeids were found at this site, so they were not included.

The trophic relationships between functional groups are presented in Fig. 6.5. This detrital foodweb can be divided into bacteria- and fungus-based food chains, although these two branches are united at the level of predaceous nematodes and mites. Similarly, the decomposer and the primary producer subsystems were separated at the middle trophic position, but were linked by top predators, death and decomposition.

The energy flow and nitrogen cycling through this foodweb was also studied. Most of the nitrogen mineralization was due to bacterial activities (4.5 g N/m^2/year), followed by soil fauna (2.9 g N/m^2/year) and fungi (0.3 g N/m^2/year). Bacterial-feeding amoebae and nematodes together accounted for more than 83% of N mineralized by the soil fauna. This was remarkable, given that bacteria and fungi comprised the majority of total decomposer biomass (Table 6.2).

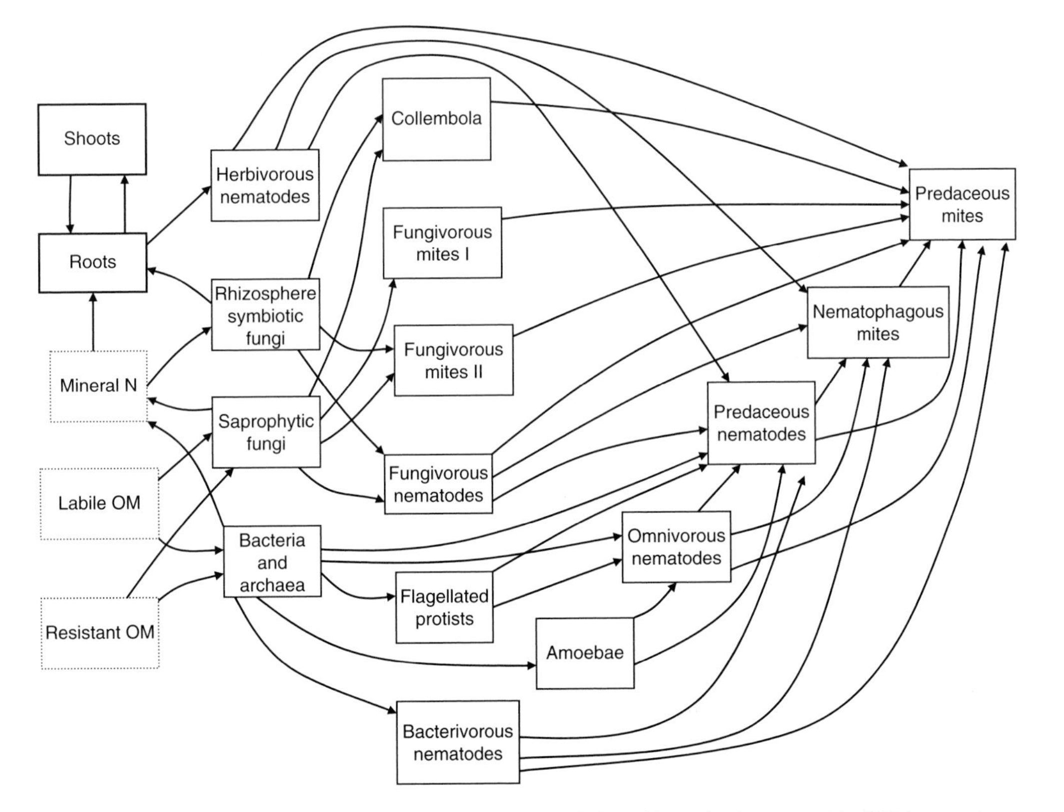

Fig. 6.5. Schematic representation of the detrital soil foodweb found in a shortgrass prairie. OM is organic matter. (Redrawn from Hunt *et al.*, 1987; Moore and de Ruiter, 1991.)

6.3 Spatial Heterogeneity and Temporal Dynamics of the Soil Foodweb

Soil is a complex habitat even over small distances. Even at the microbial scale (less than 1 mm), a soil may contain habitats that are acidic to basic, wet to dry, aerobic to anaerobic, reduced to oxidized and nutrient-poor to nutrient-rich. In some habitats, microorganisms are in intimate contact with plant roots and thus benefit from the efflux of carbon-rich substrates that are secreted from roots. Bacteria found in macropores and in water films are exposed to larger predaceous soil fauna, whereas those living within micropores or within microaggregates may be relatively more protected from predation. Consequently, soil organisms are not distributed evenly across fields or within forests, instead displaying an aggregated distribution in certain zones, such as in pores containing water, shaded areas under plant cover and near freshly deposited organic matter. These zones are considered 'hot spots' of biological activity. It is estimated that the following hot spots represent less than 10% of the total soil volume and support more than 90% of the soil biological activity:

- Rhizosphere: plants release carbon-rich compounds into the soil surrounding their roots, which supports the growth of bacteria and fungi responsible for decomposition and nutrient mineralization. Large microbial populations attract predators as well as other decomposer organisms that feed on plant substrates and the microbial biomass.
- Porosphere: water infiltrates through cracks in the soil, carrying dissolved nutrients and organic substrates that are required for the growth of soil foodweb organisms. Protists and nematodes require water to move in search of microbial prey, while large, soft-bodied enchytraeids and earthworms dehydrate rapidly without sufficient water. When they are not saturated, soil pores also serve as a conduit for gas transfer and exchange between the soil and atmosphere.
- Drilosphere: earthworm burrows are a specialized type of pore that is a highly favourable habitat for smaller foodweb organisms. Mucus secreted through the external body surface helps to stabilize the burrow wall and also serves as a substrate for microbial growth. For instance, the nitrification potential is 20 times greater and the denitrification potential is more than 100 times greater in the drilosphere than in the surrounding soil.
- Detritusphere: undecomposed or partially decomposed residues originating from plants and animals, such as leaf litter, the duff layer in a lawn or dung patches in a pasture. The initial colonizers of this material are generally fungi and bacteria, followed in a few days by bacterivores and fungivores that consume the soil microorganisms, the saprovores and omnivores that consume dissolved and particulate matter released from the residues and lysed microbial cells. Larger soil fauna consume the partially decomposed mass, breaking apart the fibres and mixing the residues with the soil.
- Aggregatosphere: the spaces between aggregates are home to organisms that live on soil surfaces and move through pore spaces. Inside soil aggregates, we can find bacterial colonies, fungal hyphae and microfauna that live in water films; however, the microfauna are generally excluded from micropores less than 10 μm in diameter.
- Termitosphere: the mounds created by the fungus-growing termites (Macrotermitinae) retain moisture and are a rich source of fresh, carbon-rich material, making them a hot spot of fungal activity compared with the surrounding termite-free soil.

The spatial hot spots are produced by a combination of abiotic factors such as (micro)climate, vegetation, land management and soil characteristics, as well as biotic interactions among populations. We can think of the soil system as a series of microsites that supports soil activity to varying degrees. These microsites range in size from a few micrometres to metres, depending on the size of the organism under study. Spatial heterogeneity makes it challenging to model interactions within the soil foodweb, but is valuable in the context of maintaining a diverse assemblage of soil biota. Heterogeneity in the soil habitat and food resources promotes species coexistence in the soil foodweb through resource partitioning.

Not only are the soil ecosystems spatially heterogeneous, they are also temporally dynamic. Seasonal variation in the population size and activity of soil foodweb organisms is well documented. In temperate regions, microbial populations tend to be larger in biomass and more active in spring and autumn than in the dry summer months or when soils are frozen in the winter. Rainfall and irrigation events are important triggers that lead to a 'burst' of microbial activity, soil respiration and nutrient mineralization. Soil microbial communities are very responsive to changes in soil conditions. Even when soils are still partially frozen, denitrifying bacteria are active as seen by N_2O emission from saturated soils during snowmelt, with further N_2O production occurring during the growing season after heavy rainfall events that saturate the soil micropores and provide favourable conditions for denitrifying bacteria (Fig. 6.6).

Populations of soil organisms exhibit temporal dynamics due to genetic and environmental factors. The life history of soil organisms – how quickly individuals grow to sexual maturity, how many offspring they have and their lifespan – is controlled at the genetic and physiological levels, in response to environmental triggers such as soil temperature, moisture, pH and so on. Soil organisms (like any other biota) can modify the expression of biochemical and physiological capabilities in response to changing environmental conditions. This ability is termed 'phenotypic plasticity', and permits soil organisms to survive when they encounter abiotic stresses, such as flooding or drought, and biotic stresses, such as predation. The temporal response of a single population provides insight into how a species responds to environmental change during a period of time. It is unlikely that population dynamics for soil organisms within functional trophic groups will be synchronized due to differences in their life histories. Non-linear responses at these scales are multipled at the soil foodweb and process-level scales, again highlighting the complexity of the soil ecosystem (Fig. 6.7).

The earthworm *Lumbricus rubellus*, which occurs naturally in Europe, can reach sexual maturity in 36 to 112 days. This wide range indicates considerable reproductive plasticity that permits the earthworm to survive in stressful conditions. Polder soils in the Netherlands are grasslands that can be flooded at certain times of the year. In the most frequently flooded soils, *L. rubellus* reached sexual maturity when individuals weighed about 0.4–0.5 g, while in the polders that were seldom flooded, sexually mature individuals weighed 0.9 g or more (Table 6.3).

In detritus-based foodwebs, one major determinant of temporal changes in the structure of the foodweb is the variation in the quality of the resources available to the decomposer community through time. Size and diversity of primary producers, and subsequently the type and amount of dead leaves, dead fine roots, root exudates, litterfall, fallen

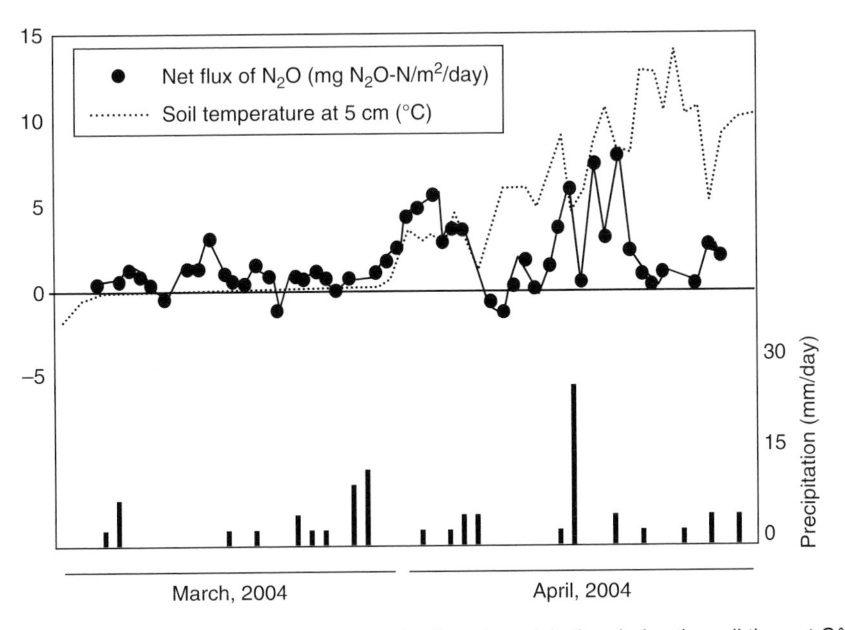

Fig. 6.6. Daily N_2O emissions, soil temperature at 5 cm depth and precipitation during the soil thaw at Côteau-du-Lac, Québec, Canada. The N_2O flux during spring thaw accounted for 15% of the annual measured emissions (5.6 kg N_2O-N/ha) from an agroecosystem under field pea production. (Modified from Pattey *et al.*, 2008.)

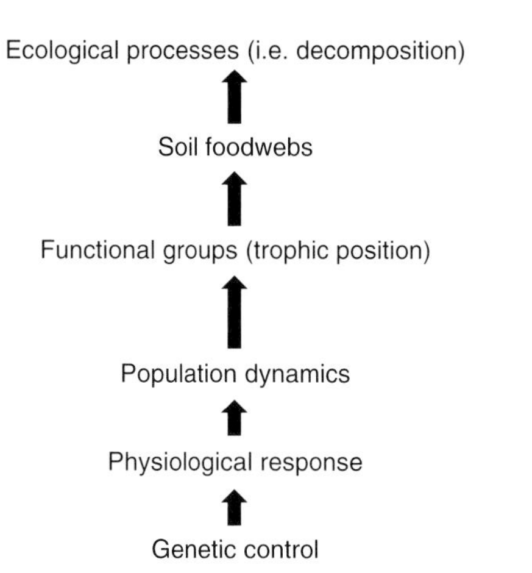

Fig. 6.7. Temporal fluctuations in soil conditions trigger genetic and physiological responses within individuals, which controls population dynamics and influences trophic interactions in soil foodwebs and ultimately ecological processes such as decomposition and nutrient cycling.

trees or ruminant dung vary with time, according to the seasonality of a given site. In addition, the assemblages of soil organisms at a given site change as the quality (e.g. C:N ratio, lignin content, polyphenol content, etc.) of their food source, namely litter and dead materials, changes through time due to organic matter breakdown. This succession of populations is called 'degradative biological succession'.

Table 6.3. Mean weight of *Lumbricus rubellus* adults (sexually mature individuals) collected from sites in a natural flood plain along the Rhine river. (Adapted from Klok *et al.*, 2006.)

Site	Adult weight (g fresh weight)
Frequently flooded	0.40–0.50
Moderately flooded	0.71–0.72
Seldom flooded	0.87–0.97

Dead organic matter is a resource that provides the energy and nutrients needed by the soil decomposer community. Organic matter decomposition, although determined by chemical and physical factors, is the consequence of the growth of different decomposer populations. Exploitation of a given resource by a group of soil organisms causes a depletion of that resource; below certain concentrations, costs of exploitation of that resource overcome the benefits, and that resource is no longer available or useful for that group (Table 6.4). Another group of organisms could have better

Table 6.4. Succession of fungal species on decomposing leaves in a Canadian aspen (*Populus tremuloides*) forest (Visser and Parkinson, 1975).

Fungal species	Living leaves	Net-caught leaves	Litter	F-layer[a] (1–4 cm)	Humus (4–6 cm)
Pleurophomella spermatiospora	✓				
Aureobasidium pullulans	✓	✓			
Cladosporium sp.	✓	✓	✓		
Penicillium janthinellum		✓			
Paecilomyces ochraceus		✓			
Penicillium sp.		✓		✓	✓
Beauveria bassiana			✓		
Phialophora sp.			✓		
Discula sp.			✓	✓	
Mortierella sp.			✓		✓
Mucor sp.			✓		
Penicillium syriacum				✓	
Trichoderma sp.				✓	✓
Phoma sp.				✓	✓
Absidia sp.				✓	✓
Acremonium sp.				✓	✓
Penicillium farinosus				✓	✓
Clindrocarpon sp.					✓
Vollutella ciliate					✓

[a]The F-layer is partially decomposed detritus below newly fallen leaves.

Table 6.5. Types of fungi and soil animals involved in decomposition of plant residues. Although fungi and bacteria are important decomposers, the effect of soil fauna in fragmenting and mixing the litter cannot be overlooked. Depending on the type of plant litter, enchytraeids, earthworms, millipedes, mites or collembola could each have an important role in the decomposition process. (Adapted from Coleman and Crossley, 2003.)

Litter	C/N ratio	Lignin (%)	Fungi	Animals
Lower plants	13–57	7–20	Sugar fungi, ascomycetes	Enchytraeids
Angiosperm	21–71	9–42	Yeasts, fungi imperfecti, ascomycetes	Enchytraeids, earthworms, millipedes
Conifer	63–327	20–58	Ascomycetes, fungi imperfecti, basidiomycetes	Earthworms, mites, collembola
Wood	294–327	17–35	Ascomycetes, fungi imperfecti, basidiomycetes	Mites, collembola

strategies to access the resource, or lower exploitation costs. Besides, exploitation of an organic resource releases CO_2 and partially degraded materials, which could be appropriate resources for another group of organisms, and so on. This natural process leads to a succession along time of peaks of relative abundance of different organic matter qualities within the litter, accompanied by changes in the relative abundances of different soil organisms, depending on the degree of specialization of each organism and how narrow their range of usable resources (Table 6.5).

Thus, in a given site, the relative abundance of different groups in the soil foodweb is higly dynamic due to the overlapping of: (i) the sucession of populations according organic matter decay; and (ii) seasonal changes in the amount and type of organic resources that primary producers 'donate' to the soil foodweb.

The relevance of degrative biological succession for organic matter decomposition and nutrient mineralization has been known for a long time. Nutrients cycle through terrestrial ecosystems from organic to mineral forms, becoming available again for plants as they are released from bacterial and fungal cells. Carbon mineralization and subsequent mineral nutrient regeneration results from the combination of the independent activity of each soil biotic element plus the net synergy between them. But we must not forget that microbes and soil fauna operate under biological laws, subjected to evolutionary forces. Nitrifiers do not care at all about the lack of nitrate in maize fields, nor are white-rot fungi aware of extra-lignin in experimentally fertilized spruce plots. They do not care about the romantic concept of the 'decay succession' – their populations grow and die depending on resource availability, species-specific costs of exploitation, and biotic interactions within the foodweb.

6.4 Importance of Biodiversity in the Soil Foodweb

There is a great deal of 'overlap' in detrital foodwebs due to omnivory – many organisms consume a variety of food resources. Soil organisms also exhibit 'plasticity' in their diets, which allows them to take advantage of a temporary abundance of certain food resources (e.g. leaf litter falling in temperate forests in the autumn, root exudates secreted during the early vegetative growth of grasslands and pastures). The spatial heterogeneity and temporal fluctuations in the soil environment lead to resource partitioning that allows many species that have similar functions to coexist. Due to this overlap, many researchers have asked 'what is the minimum number of species needed to maintain soil ecological functions?' (Moore *et al.*, 2004).

Although we still do not have a firm answer to this question, it is thought that biological diversity in the soil foodweb reduces fluctuations in population dynamics that could lead to excessive growth of certain species or the extinction of others. In addition, biodiversity should provide more channels in the soil foodweb that stabilize the flow of energy and nutrients through the foodweb. The connectedness between bacterial and fungal predators may be important – this is controlled by common high-level predators such as soil mites that consume smaller mites and nematodes. Other key species include earthworms, due to their major role as ecological engineers, building biogenic structures such as burrows and casts that affect soil hydrology, soil organic matter distribution and nutrient cycling. However, it is not necessary to have earthworms to achieve a stable soil foodweb, as was seen in Focus Box 6.2. It appears that the system will be stable as long as there are enough resources to support the energy flow through the biological communities in the soil foodweb. The 'bottom-up' nature of the foodweb permits a shift from bacterial- to fungal-based decomposition, depending on soil management. This seems to suggest ecological processes such as decomposition and nutrient cycling will continue, albeit at a slower rate, if top predators or large detritivores are absent or deliberately removed from the soil foodweb. However, this still remains to be confirmed due to the lack of knowledge of the actual biodiversity, including microbial diversity, in most soil foodwebs.

Further Reading and Web Sites

Acarology Home Page. © Z-Q Zhang, 1996-2008. Hosted at the UK Natural History Museum
http://www.nhm.ac.uk/hosted_sites/acarology/

Biology Catalogue
http://entowww.tamu.edu/research/collection/hallan/

'Checklist of the Collembola of the World'. By Bellinger, P.F., Christiansen, K.A. and Janssens, F. (1996–2009).
http://www.collembola.org

Dirtland
http://commtechlab.msu.edu/sites/dlc-me/zoo/zdmain.html

Earthworm information
http://www.sarep.ucdavis.edu/worms/

Konig, H. and Varma, A. (eds) (2006) *Intestinal Microorganisms of Termites and Other Invertebrates.* Soil Biology Series vol. 6. Springer-Verlag, Berlin, 483 pp.

Nematodes
http://nematode.unl.edu/wormgen.htm

Nematodes as Agricultural Pests
http://nematode.unl.edu/agripests.htm

Online Textbook of Bacteriology
http://textbookofbacteriology.net/index.html

Soil Biology Movies
http://www.agron.iastate.edu/~loynachan/mov/

Soil Biology Primer
http://soils.usda.gov/sqi/concepts/soil_biology/soil_food_web.html

Soil Mites and Other Animals. By Minor M., The Ecology Group, Massey University
http://www.massey.ac.nz/~maminor/mites.html

The Microbial World, University of Wisconsin–Madison
http://bioinfo.bact.wisc.edu/themicrobialworld/homepage.html

Tree of Life Project
http://tolweb.org/tree/

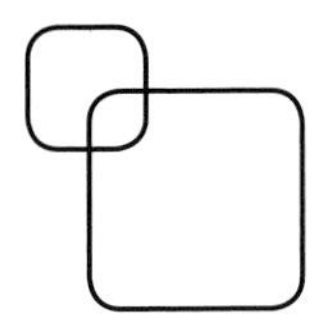

PART III

Ecological and Pedological Functions of the Soil Foodweb

7 Primary Production

Earth is sometimes referred to as the 'green planet' because of the diverse plant life capable of photosynthesis, the process by which solar energy and carbon dioxide from the atmosphere are converted to chemical energy that is stored in molecules such as glucose and transformed into plant biomass. The oxygen released during photosynthesis is essential for life on Earth, and the solar energy captured by plants sustains the trophic relationships of foodwebs in marine, freshwater and terrestrial ecosystems, as discussed in Chapter 6.

In soil foodwebs, plants are the primary producers that occupy the first trophic position and exert a 'bottom-up' control on the animal populations that depend on living plants (root exudate-based foodweb) and dead plant residues (detrital foodweb) for their survival. The energy flow through an ecosystem is a linear, unidirectional process that begins with solar energy entering plants and ends with heat energy lost from biological organisms in higher trophic positions. Although plants are very efficient at capturing solar energy, their survival depends upon soil biota that can recycle matter (e.g. essential plant nutrients) in the abiotic environment. This chapter discusses symbiotic and mutualistic interactions between plants and soil microorganisms that support primary production in terrestrial ecosystems.

7.1 Gross and Net Primary Production

Primary production refers to the accumulation of biomass in plants, similar to the concept of secondary production that was used to describe biomass accumulation in animals (Chapter 6). We begin by reviewing the mechanisms for solar energy capture in plants, based on CO_2 fixation in photosynthesis and CO_2 lost via respiration, according to the equations:

$$\text{Photosynthesis: } 6\ CO_2 + 12\ H_2O \xrightarrow{\text{Light}} C_6H_{12}O_6 + 6\ O_2 + 6\ H_2O \quad [7.1]$$

$$\text{Respiration: } C_6H_{12}O_6 + 6\ O_2 + 6\ H_2O \rightarrow 6\ CO_2 + 12\ H_2O + \text{energy (ATP)} \quad [7.2]$$

Oxygenic photosynthesis by plants is a complex process that occurs in two stages. The light-dependent reactions take place only in the presence of light and occur in the thylakoids of the chloroplast. Chlorophyll absorbs light energy, which triggers a flow of energized, excited electrons from the chlorophyll molecules. Some of the electrons are transformed into chemical energy and used to make adenosine triphosphate (ATP), while other electrons are used to split water molecules to release hydrogen and molecular oxygen (O_2). Electrons and hydrogen protons combine with nicotinamide adenine nucleotide phosphate ($NADP^+$) to make NADPH, a temporary energy storage molecule. The light-independent reactions of photosynthesis, also known as the Calvin cycle, occur in the stroma of the chloroplast. Carbon dioxide molecules are combined with ribulose bisphosphate (RuBP) by the enzyme ribulose-1,5-bisphosphate carboxylase/oxygenase (commonly known as RuBisCO), but this unstable six-carbon molecule is immediately broken into two three-carbon molecules called phosphoglycerate (PGA). The PGA molecules are then converted to phosphoglyceraldehyde (PGAL) using NADPH and ATP generated from the light-dependent reactions. Through a series of reactions, PGAL is rearranged into new RuBP molecules for further capture of CO_2 or exits the Calvin cycle to be used for glucose and carbohydrate synthesis. The sugar compound ($C_6H_{12}O_6$) contains the stored solar energy and serves as the raw material from which structural compounds are made, or as an energy reserve for cellular respiration. Photosynthesis can be achieved through three metabolic pathways: C3, C4 and CAM.

The C3 pathway gets its name from the first step in the reaction, when CO_2 is incorporated into a three-carbon PGA compound. The C3 pathway is the most common in the plant kingdom and the most efficient under cool, moist conditions and normal light because a minimum of enzymatic reactions are involved and no special physiological adaptations are needed. However, if a C3 plant is exposed to light or heat stress, resulting in low CO_2 partial pressure inside the cell, the RuBisCO enzyme binds O_2 to RuBP instead of CO_2. When this happens, some of the intermediate molecules in the Calvin cycle are degraded to CO_2 and H_2O instead of being synthesized into carbohydrates. This process is called photorespiration and reduces the potential biomass accumulation of these plants.

The C4 pathway is common in plants of tropical origin such as maize and sugarcane and gets its name from the fact that the first molecule formed by CO_2 fixation is a four-carbon compound in mesophyll cells using the enzyme PEP carboxylase. The four-carbon compound is transported to bundle sheath cells that surround the leaf veins, where it is broken down and CO_2 is converted to sugar by the regular C3 pathway. Photosynthesis can occur efficiently even under high light intensity and high temperatures – the segregation of the RuBisCO enzyme in specialized cells means that photorespiration does not occur. Also, PEP carboxylase is highly efficient at capturing CO_2 and stomata can be closed during the day, which reduces water loss from the plant.

The crassulacean acid metabolism (CAM) was named for the Crassulacaeae plants, although there are more than 25 plant families capable of CAM photosynthesis. The stomata are open at night and usually closed during the day. The plant fixes CO_2 and combines it with a three-carbon acid to form a four-carbon acid with PEP carboxylase. This compound is stored temporarily in the vacuoles of leaf cells. During the day when the stomata are closed, CO_2 is removed from the four-carbon compound and fixed into sugar by RuBisCO in the C3 pathway. When conditions are extremely arid, the plant can keep the stomata closed night and day. In this CAM-idle mode, the oxygen released from photosynthesis is used for respiration and CO_2 from respiration is recycled by the photosynthesis reactions. Although no growth occurs, the plant can adopt this strategy to survive dry spells without going dormant, so recovery is very quick when water is available again.

Inorganic carbon fixation (calcification) is another pathway that contributes to carbon storage in living cells of single-celled bacteria and algae, as well as some macrophytes. Inorganic carbon (HCO_3^-/CO_3^{2-}) can be transferred across cell membranes, concentrated and transformed into CO_2, which then is fixed into sugar by RuBisCO. Dissolved inorganic carbon is abundant in marine environments and represents up to 10% of the carbon fixation by coccolithophores, foraminifera and pteropods in the upper ocean (Poulton *et al.*, 2007).

Photosynthesis converts solar radiation into stored biomass that is an energy source for other foodweb organisms, but there the conversion process is limited by the fact that only light within 400 to 700 nm in the spectrum (photosynthetically active radiation, PAR) can be used by plants. Some of this light is reflected or transmitted when it travels through the atmosphere. Furthermore, the fixation of one CO_2 molecule during photosynthesis necessitates a quantum requirement of ten (or more) molecules. Based on these limitations, the theoretical maximum efficiency of solar energy conversion by plants is approximately 11%. In practice, however, the photosynthetic efficiency observed in the field is further decreased by factors such as poor absorption of sunlight due to its reflection, suboptimal solar radiation levels on cloudy or hazy days, and energy lost during photosynthesis. The net result is a photosynthetic efficiency of between 3 and 6% of total solar radiation. Plants with the C4 pathway do not lose energy through photorespiration and thus have a higher energy conversion than C3 plants (Fig. 7.1).

At the ecosystem level, we can express the energy capture by plants as gross primary production (GPP), which is the total amount of CO_2 (or total energy capture) that is fixed by the plant through photosynthesis during a period of time. Some researchers express GPP as the sum of the carbon found within plant organs (roots, stems, leaves, flowers and seeds) plus all carbon respired through the leaves and from plant roots during a period of time. The CO_2 lost through respiration is due to energy needed for cellular metabolism (including growth) and maintenance of tissues. In an ecosystem, the values are often expressed as the mass of

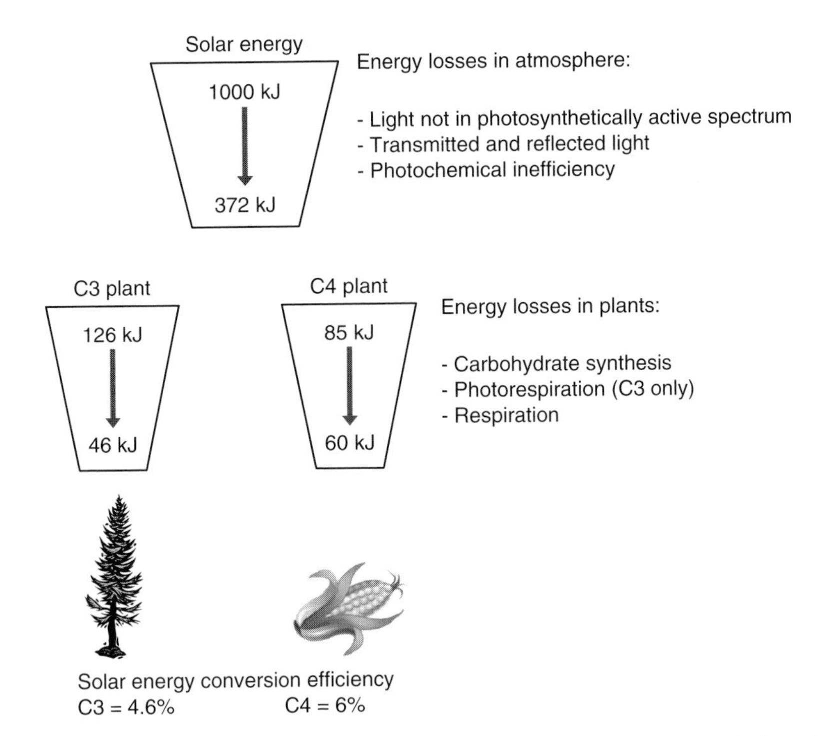

Fig. 7.1. Energy losses calculated for 1000 kJ of incident solar radiation during photosynthesis, from the interception of radiation to the formation of stored chemical energy in plant biomass, considering C3 and C4 photosynthesis pathways. Data are based on a leaf temperature of 30°C and an atmospheric CO_2 concentration of 380 ppm. (Adapted from Zhu *et al.*, 2008.)

carbon per unit area per year. There is growing interest in monitoring the GPP of crops due to their carbon sequestration potential. With satellite-based remote sensing technologies, it is now possible to get an accurate picture of the GPP in agricultural crops at regular intervals during a growing season. Researchers at the University of Nebraska-Lincoln and NASA demonstrated the technology in large maize production fields (49–65 ha) and found that the GPP of maize was quite variable, between 1.88 and 23.1 g C/m^2/day depending on weather conditions and the growth stage of the crop (Gitelson *et al.*, 2008).

In crop production systems, grasslands and short-rotation forests, we are generally more interested in the amount of carbon that can be harvested as the food and fibre. The net primary production (NPP) refers to the carbon that remains in the plant biomass at the end of the growing period. The NPP values for most agricultural crops are fairly well known because economic profits are based upon how much food or fibre can be harvested and sold (Table 7.1). The NPP is also relevant to detrital foodwebs because detritivores and their predators obtain energy and recycle nutrients from residues that are left after crop senescence.

Compared with other ecosystems, agricultural land is moderately productive in terms of its NPP per unit area. The most productive systems are estuaries, swamps and marshes, tropical rainforests and deciduous temperate forests (Fig. 7.2.a). These systems have favourable climate (water, temperature) and sufficient nutrients to support plant communities that can thrive and quickly accumulate biomass in these habitats. Terrestrial regions that are warm and wet are generally more productive, and thus there is a gradient of increasing NPP from polar regions and deserts to the wet tropical rainforest. Cold winters reduce the productivity of temperate grasslands, compared with savannahs, and in boreal forests compared with temperate deciduous forests and tropical rainforests. Differences in moisture reduce the primary production in woodland and shrubland compared with other forests.

Table 7.1. Crop production in Richland County, North Dakota (1992), showing the harvest index, root:shoot ratio, biomass of harvested yield and total biomass yield (NPP) for typical agricultural crops. The biomass values were calculated on a dry matter (DM) basis. (Based on data reported by Prince *et al.*, 2001.)

Commodity	Harvest index[1]	Biomass root: shoot ratio	Biomass of harvested yield (mg DM/ha)	Total biomass yield (NPP) (mg DM/ha)
Maize grain	0.53	0.18	4.27	9.51
Maize silage	1.00	0.18	4.37	5.16
Soybean	0.42	0.15	1.69	4.64
Oats	0.52	0.40	2.77	7.45
Barley	0.50	0.50	3.65	10.94
Wheat	0.39	0.20	3.22	9.91
Sunflower	0.27	0.06	1.77	6.96
Hay	1.00	0.87	3.80	7.11

[1] Proportion of above-ground biomass that is removed from the field when the crop is harvested.

If we wish to know the total NPP in the world, it is necessary to multiply the average NPP per unit area (Fig. 7.2.a) by the area covered by each ecosystem. This calculation reveals that most of the NPP in the world occurs in the open ocean, tropical rainforests, savannahs and tropical seasonal forests (Fig. 7.2.b).

Ever since people began to cultivate plants as a food source for humans and livestock and as a source of fibre for buildings, clothing and cooking/heating, they have been interested in devising ways to achieve maximum yields, which is essentially the maximum NPP for a particular plant in a given site. In the agricultural sector, there have been huge gains in maximizing NPP due to agricultural mechanization, large-scale irrigation, plant breeding programmes, the availability of soluble nutrient sources (fertilizers) and agrochemicals used for pest control. However, these new developments have also been linked to environmental damage due to soil compaction, erosion, salinization, nutrient loading and agrochemical pollution in groundwater and surface waters. The current focus in agriculture and other managed ecosystems is to optimize NPP, that is to achieve high primary production but not at the expense of the environment. There is a sense that we should attempt to maintain or increase NPP with less energy and fewer synthetic inputs than were used in the past. This could mean a greater reliance on beneficial biological interactions such as symbiosis and mutualism, which occur between plant roots and soil organisms.

7.2 Plant Roots

The importance of plant roots is not appreciated by many people because roots grow underground and are not visible. Yet, the rooting system of a plant is often more extensive than its above-ground organs. For example, the roots of an annual crop such as maize grow laterally 30 to 45 cm from the stalk, extending as far as 1.2 m from the stem. About 90% of the roots will be found in the top 1 m of soil, although some can be found to a depth of 2.5 m in deep soils without rocks or other barriers to root growth. This rooting system allows the plant to obtain the water and nutrients needed for its growth. Over the course of a growing season, about 40% of the water used by the crop will come from the first 35 cm of the soil profile, with 30% from the next 35 cm depth and 30% from depths greater than 70 cm in the soil profile.

The rooting system of annual crops is rather small when compared with trees, which can possess two to three times more biomass in roots than in the above-ground components (trunk, branches, leaves). Tamarisk (*Tamarix aphylla*) trees in the desert have long taproots that grow to a depth of 50 m, allowing them to reach underground water supplies. This adaptation allows them to survive in arid, saline environments in Eurasia and Africa. First introduced to the western USA as an ornamental plant in the 1800s, tamarisk (salt cedar) has proved to be an undesirable invasive plant and efforts are underway to eradicate this tree. Due to its efficacy at intercepting water, tamarisk disrupts the structure and stability of

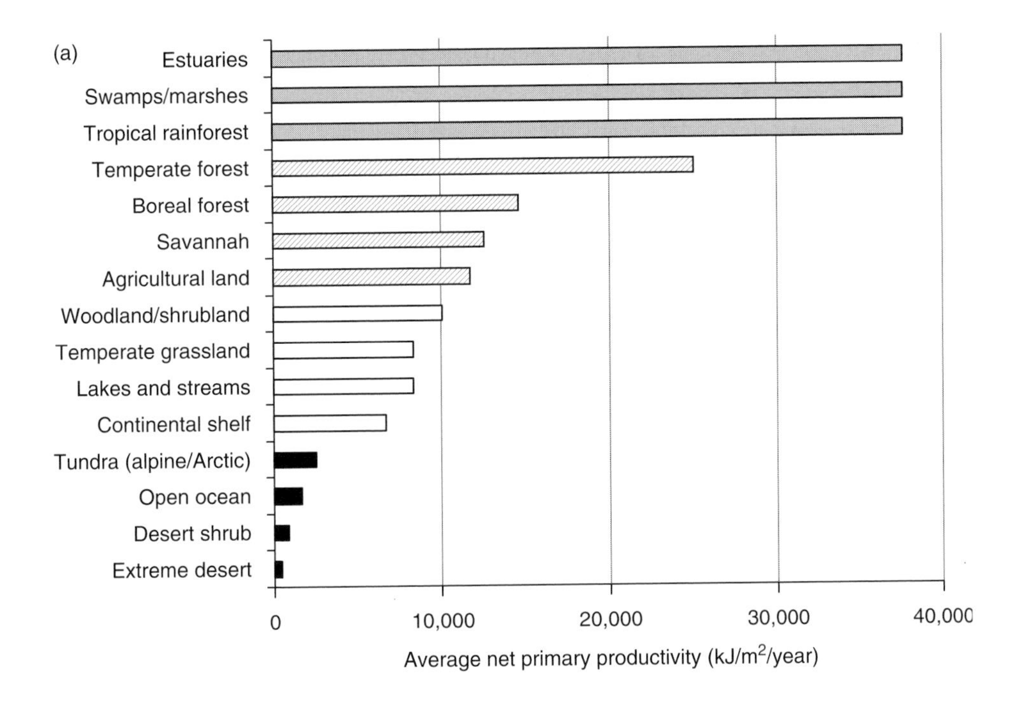

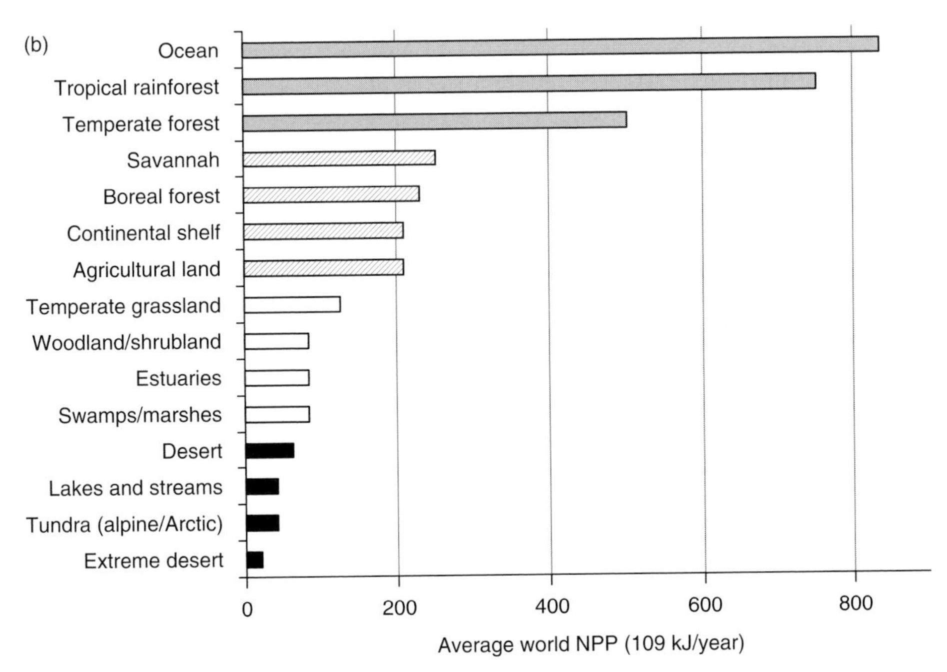

Fig. 7.2. (a) Net primary production (NPP) per unit area of the world's common ecosystems and (b) the average world NPP of various ecosystems. (Adapted from: http://www.globalchange.umich.edu. Accessed on 28 December 2008.)

semi-arid ecosystems by outcompeting and replacing native plant species. Although they provide some shelter, the foliage and flowers of this plant are of low food value for native wildlife, compared with the native plants.

Tree roots may extend vertically through a taproot system or horizontally through shallow lateral roots. The eucalyptus (*Eucalyptus* sp.) tree is an important source of timber and fuel, gives shade, acts as a windbreak and furnishes gum, resin, oil and nectar in its native Australia and as an introduced tree in Africa, Eurasia, Europe, North and South America. One of the reasons that eucalyptus has been so successful in such a range of habitats is due to its root system. Eucalyptus trees have a strong taproot that extends about 2 m to secure and support the above-ground components, and produces a dense network of lateral roots that can extend for more than 10 m around the tree to search for water. This can be a problem when the tree is grown near buildings and other facilities since its roots and rootlets can disrupt ditches, crack cisterns, clog water pipes and damage septic tanks in a search for water. To avoid problems, it is recommended that eucalyptus and other trees that produce extensive lateral roots be planted at some distance from sensitive areas or pruned to constrain their size and therefore the extent of the root growth.

Since we cannot easily see or excavate plant roots, we often estimate the root biomass using a root to shoot ratio. The root:shoot ratio describes the proportion of the plant biomass allocated to roots. In annual crops, the root biomass ranges from 6% (sunflower) to 50% (barley) of the shoot biomass (Table 7.1). A hay crop generally consists of a mixture of annual and perennial plants and the root biomass represents 87% of the shoot biomass. Perennial plants, particularly grasses, allocate more energy to the development of a strong root system than annuals. As shown in Table 7.2, roots can contribute to more than 50% of the net primary production in all grasslands (savannah, tropical and temperate). Fine roots in deciduous forests grow and die as do leaves, contributing to the annual below-ground dead organic matter input that is approximately equivalent to leaf fall.

Table 7.2. Root:shoot ratios for vegetation in major terrestrial biomes. (Adapted from Mokany *et al.*, 2006.)

Vegetation category	Shoot biomass (mg/ha)	Root:shoot ratio
Tropical rainforest	> 125	0.24
	< 125	0.21
Temperate coniferous forest	> 150	0.20
	50–150	0.29
	< 50	0.40
Temperate deciduous forest	> 150	0.24
	75–150	0.23
	< 75	0.39
Boreal forest	> 75	0.39
	< 75	0.24
Woodland/shrubland		0.32–1.84
Savannah		0.64
Tropical grassland		1.89
Temperate grassland		4.22–4.50
Tundra		4.80
Desert		1.06
Extreme desert		4.09
Tidal marsh		1.10

Root morphology and functions

Plant roots branch extensively through the soil, forming a network that anchors the plant firmly in place and absorbs water and dissolved nutrients contained in the soil solution. One of the primary functions of the roots is to keep the plant anchored so it can resist high winds, heavy rains and soil movement during freezing/thawing cycles. A strong root system is absolutely essential for the survival of the plant, especially if it is growing in an area that is exposed to frequent storms.

Another important function of the root system is to absorb water and dissolved nutrients that are required for growth. The majority of the water needed for plant growth comes from the soil, absorbed through the roots. There are two types of root systems that are adapted to obtain water: taproot and fibrous root systems.

A taproot system consists of a main root with many smaller lateral branches extending from it. Dandelions have a well-developed taproot, as anyone who has ever tried to pull one from the ground can attest. Most dicots and gymnosperms have taproot systems, which allow the plant to extend deep into the soil to obtain water from underground sources. In hickory trees, the taproot is retained as the tree ages, but most mature trees lose the taproot and have a root system consisting of large, shallow lateral roots from which other roots branch off and grow downward.

In a fibrous root system, we find many roots of the same size developing at the end of the stem. These in turn have many smaller lateral roots branching off. Most monocots have fibrous rooting systems, including grasses, cereals and onions. A fibrous root system has many roots concentrated near the soil surface, thus most of the water is obtained from soil macropores after rainfall or irrigation events, as well as from the capillary water reserves in the soil that are accessible to the smallest roots of these plants.

Certain plants have a modified root system with features of both taproot and fibrous rooting systems. The root system of a grapevine consists of a main taproot that grows vertically to perhaps 2 m or more in the soil profile. At a depth of 30–50 cm, many fine, fibrous lateral roots branch off from the taproot and extend through the soil to acquire nutrients. At greater depths in the soil profile, the root may divide into three or four taproots, each of which exhibits extensive fibrous roots at depths below 120–150 cm. The purpose of these deep lateral roots is to absorb water.

The function of plant roots can also be inferred by their appearance. Coarse roots (> 2 mm) contain a large amount of suberin, a waterproof fatty material deposited in the Casparian strip of plant cells, and lignin, a complex chemical substance that confers structural stability to the root. Coarse roots are strong and resist attack from insects and diseases. They provide the anchorage needed to support perennial plants for many years, and serve as a conduit for water and material transport in the plant. Thus, water and dissolved nutrients travel through the xylem of coarse roots to the above-ground organs, and carbon fixed through photosynthesis in the leaves can be translocated through the phloem of stems and coarse roots to support metabolic activities and growth in the root system.

Plants also possess fine roots (< 2 mm) that contain little suberin or lignin. These are generally pale coloured and covered with small root hairs, which increases the absorptive surface area. Virtually all of the water and dissolved nutrients needed for plant metabolism are absorbed through the fine roots. Just as leaves sprout, grow and finally senesce and fall from stems and branches, so do fine roots. The lifespan of fine roots ranges from days to months, meaning that they are continually replaced as the plant grows. Since they are not suberized or lignified, fine roots are the site of colonization by symbiotic microorganisms such as nitrogen-fixing bacteria and mycorrhizal fungi. Fine roots are an important component of the diet of insects such as beetle larvae, which graze upon live roots. Most of the saprophytic organisms will consume senescent and dead roots. The fine roots are also the site of entry of many root pathogens.

The majority of plant roots are found in the top 15–20 cm of soil, especially in annual plants. Root systems develop seemingly at random, but roots tend to maintain a minimum distance from each other to avoid crossing and to exploit the extraction of water and nutrients from the soil.

The distance between lateral roots typically ranges from 1.5 cm to 3.0 cm; however, the extent of lateral root spread from the plant stem tends to be greater in arid than humid regions, probably to maximize water interception (Casper *et al.*, 2003). Lateral root growth is also affected by soil texture, with more extensive root growth occurring in coarse-textured and uncompacted soils because they offer less impedance to root growth, enabling plants to explore a larger soil volume. Root density is also affected by the age or developmental stage of a plant (Table 7.3).

Other factors that affect the quantity and length of roots produced by a plant are the plant genotype and lifespan, soil fertility, soil water availability and herbivory. As mentioned above, dicots tend to possess taproot systems whereas monocots generally have fibrous rooting systems, while the root:shoot ratios differ among annual and perennial herbaceous plants, woody shrubs and trees as shown above in Table 7.2. Although younger plants have fewer roots per unit area, the growth rate and activity of roots is expected to be greater in a young plant than in an older plant. Fine roots demonstrate chimiotropism, in that they tend to proliferate in areas of the soil where there are high concentrations of dissolved nutrients, facilitating the absorption of essential plant nutrients. Soil water conditions also exert an important effect on root growth. As discussed in Chapter 1, roots do not normally grow in waterlogged soils unless the plant possesses special adaptations that permit it to acquire sufficient oxygen for root metabolic functions. In arid soils, plants allocate more resources to produce roots to maximize water uptake. Thus, the distribution and rooting depth of many plants is constrained by site-specific soil conditions such

Table 7.3. Maize root density as influenced by maize developmental stages (Liedgens and Richner, 2001).

Developmental stage	Growth stage	Number of roots/cm²
3 expanded leaves	V3	0.33–1.07
6 expanded leaves	V6	0.91–1.15
9 expanded leaves	V9	1.93–2.30
12 expanded leaves	V12	2.30–2.63
Tasselling	VT	2.26–2.84

as texture, hydrology and drainage. A low level of root herbivory can stimulate root growth as the plant allocates more photosynthates below ground to compensate for loss of root biomass; none the less, high herbivory would damage roots and impact plant growth negatively. Root-feeding Coleoptera larvae such as Western corn rootworm (*Diabrotica virgifera virgifera* Leconte) can cause slight injury (feeding scars and tip injury on a root system) or very severe injury (elimination of root nodes). When herbivory is high, the plant is weakened to the point that stalks break and yield losses occur. Chemical insecticides are not very effective in controlling below-ground herbivores, so the focus has been on rotating with non-host crops and breeding resistant hybrids, including genetically modified Bt maize. The Bt maize hybrid contains genetic material from *Bacillus thuringiensis* and expresses the Cry3Bb1 toxin that is lethal to rootworms.

Water and nutrient acquisition by roots

The movement of dissolved nutrients and water from the soil into root cells is a passive process. The transport of water through the plant is also governed by physical forces that do not require energy expenditure from the plant. Once in the root cell, water passes through various tissues until it reaches the tracheids and vessel elements of the root xylem, and then is transported through the xylem cells to stems and leaves. According to the tension-cohesion theory, water is pulled up the plant as a result of tension generated from evapotranspiration from the above-ground plant organs. The upward movement of water is possible only when there is an unbroken column of water in the xylem throughout the plant. Another mechanism for water movement in the plant is known as root pressure, in which the accumulation of water in root tissues produces a pressure that forces the water up the xylem.

The absorption and transport of nutrients in plant tissues is more complicated. First of all, nutrients must be present as ions or as dissolved compounds with a low molecular weight that can be carried in the water moving through the soil to the plant root. For example, nitrogen is transported primarily as ammonium (NH_4^+) and nitrate (NO_3^-) ions, although small quantities of urea ($(NH_2)CO$), N-rich amino acids and small proteins may also be transported in the soil water. The first step in nutrient absorption is passive. Dissolved nutrients move through the root cell wall, from the epidermis to the endodermis. For the nutrients to reach the xylem, they must be transported through the Casparian strip, which functions as an impermeable, suberized barrier in the endodermis. This is an active absorption process that requires energy and is selective because carrier proteins must bind and transport the nutrients through the Casparian strip before they can enter the xylem. The essential elements for plant growth are listed in Table 7.4.

Nutrient absorption by roots is strongly dependent on the nutrient concentration in the soil solution.

Table 7.4. Average concentrations of essential elements in plants.

Plant nutrient	Element	Average concentration[a]
Hydrogen (%)	H	6.0
Oxygen (%)	O	45
Carbon (%)	C	45
Nitrogen (%)	N	1.5
Potassium (%)	K	1.0
Calcium (%)	Ca	0.5
Magnesium (%)	Mg	0.2
Phosphorus (%)	P	0.2
Sulfur (%)	S	0.1
Chloride (ppm)	Cl	100
Iron (ppm)	Fe	100
Boron (ppm)	B	20
Manganese (ppm)	Mg	50
Zinc (ppm)	Zn	20
Copper (ppm)	Cu	6
Molybdenum (ppm)	Mo	0.1
Nickel (ppm)	Ni	0.1

[a]Concentration expressed by weight on a dry matter basis

In many forest and grassland ecosystems, this is related to the inherent soil fertility, namely how readily the nutrients are released from rocks, the soil clay fraction and soil organic matter, into the soil solution by dissolution, desorption and mineralization processes. In agricultural soils, the dissolution, desorption and mineralization processes remain important but fertilizers are generally applied to ensure that there will be sufficient nutrients in the soil solution for plant growth.

The relative mobility of nutrients determines how quickly they move through the soil to the root hair where they are absorbed. This is governed by the ionic form and size of the nutrient in the soil solution and transport processes. For example, nitrate (NO_3^-) is a small, negatively charged ion that adsorbs poorly to soil surfaces. Consequently, it moves readily in the soil water by mass flow. In contrast, phosphates like $H_2PO_4^-$ tend to become adsorbed on soil surfaces and precipitate with calcium, iron and aluminium ions. Phosphate ions move by diffusion, a slow and tortuous process. Plants absorb phosphate ions from a very small volume of soil immediately surrounding the roots. Plants can obtain some cations like potassium (K^+) through root interception, an exchange reaction that causes desorption of a hydrogen ion (H^+) from the root surface and the absorption of K^+ from a soil particle. Root interception provides some nutrients, but most of the ions absorbed by plants move from the soil solution to the roots through mass flow and diffusion (Fig. 7.3).

Even if plants have a very extensive root system, they can only absorb nutrients through fine roots. In trees, the fine roots constitute 20–50% of the total root biomass (Bartelink, 1998) and the remainder is dedicated to structural roots that provide support for the canopy. In temperate grasslands, the peak production in fine roots occurs within 15 days of maximum leaf production, indicating synchrony between the allocation of photosynthates from leaves to roots and transport of nutrients from roots to leaves. There is a greater lag period (about 45 days) between maximum leaf and

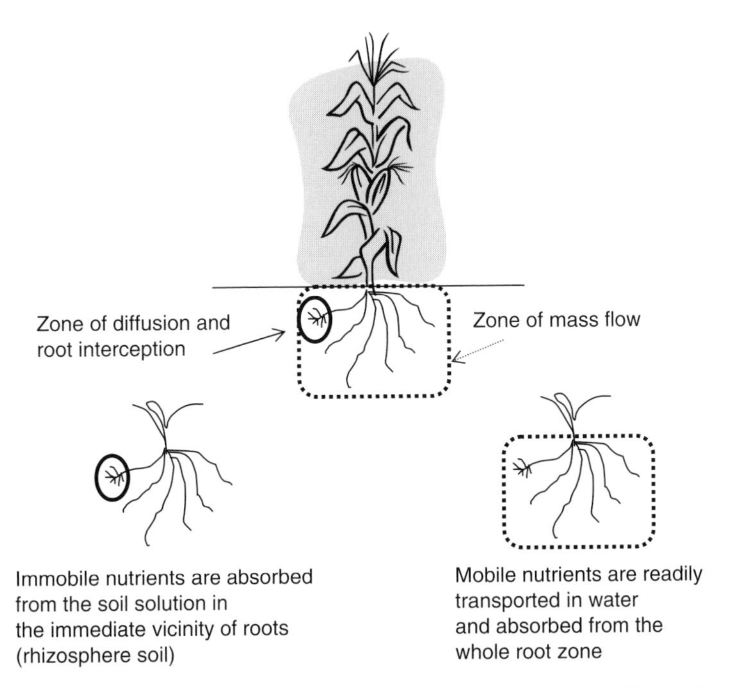

Fig. 7.3. Soluble ions are absorbed by plant roots by mass flow, root interception and diffusion. Immobile nutrients such as phosphate ($H_2PO_4^-$) are often bound to soil surfaces and exchange sites by adsorption and fixation reactions, and are absorbed from the soil solution by diffusion and root interception processes. Mobile nutrients such as nitrate (NO_3^-) do not bind tightly to soil surfaces and are easily transported in soil water throughout the root zone, where they can be absorbed by roots through mass flow. (Based on Benton Jones, 2003.)

root production in temperate forests, probably because it takes more time for photosynthates to be allocated progressively from shallow to deeper roots. Consequently, nutrient absorption by trees can continue for some time after deciduous leaves have fallen, even into the winter (Steinaker and Wilson, 2008).

Annual plants do not exhibit a delay in absorbing nutrients because their root system is composed mostly of fine roots, especially during the early stages of development. A 45-day-old maize plant with a root biomass of 500 g has about ten times more absorption capacity than needed (about 30–50 g of fine roots will be able to acquire the water and nutrients required for its growth). Therefore, plant growth is generally limited more by the nutrient concentration and mobility in the soil solution than by the capacity of the root system to absorb nutrients.

7.3 Plant Root Interactions with Symbiotic Organisms

Plants need a well-developed root system to acquire essential nutrients. In unfertilized soils, the low nutrient concentration and low mobility of required ions are a major impediment to plant growth. Even when soils are fertilized, the placement of fertilizer must be considered carefully to ensure that small seedlings can get the nutrients needed for rapid growth and high productivity. Symbiotic interactions with soil organisms permit natural plant communities and crop plants to thrive even in very low-fertility soils. Among the most well-known symbiotic organisms are the nitrogen-fixing bacteria (e.g. *Rhizobium* and *Bradyrhizobium*) and nitrogen-fixing actinomycetes (*Frankia*). The roots of most plants are colonized with fungal symbionts called mycorrhizae that aid the plants in absorbing nutrients and water. Symbiotic plant–microbial interactions are found in all biomes, indicating that life together is clearly beneficial to both partners.

Nitrogen-fixing bacteria

Biological nitrogen fixation is widespread in agricultural and natural systems, through symbiotic plant–microbial interactions or with free-living bacteria (Fig. 7.4). The widespread occurrence of biological nitrogen fixation suggests that it is a very important process, which has been estimated to capture 122 Tg N/year at a global scale. About half the N_2 fixation occurs in cultivated agroecosystems, which fix about 50–70 Tg N/year (Table 7.5).

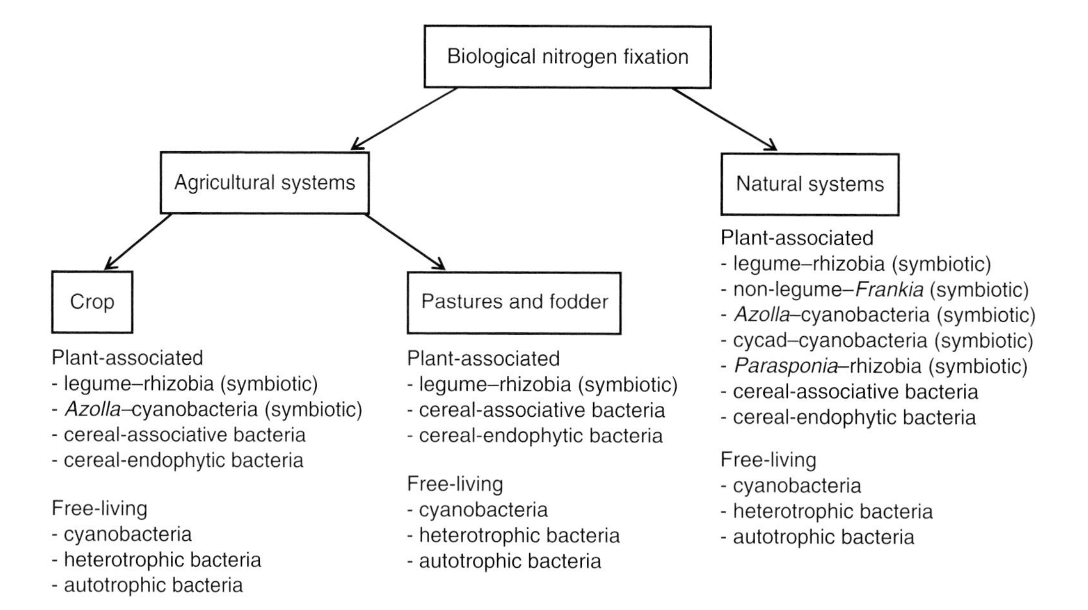

Fig. 7.4. Biological nitrogen-fixing organisms in agricultural and natural systems.

Table 7.5. Estimates of global N_2 fixation occurring annually in agricultural land. (Adapted from Herridge *et al.*, 2008.)

Land use	N_2-fixing agent	Annual N_2 fixation input (Tg)
Pasture and fodder legumes	Legume–rhizobia	12–25
Rice	Azolla–cyanobacteria, cyanobacteria	5
Sugarcane	Endophytic, associative and free-living bacteria	0.5
Non-legume crops	Endophytic, associative and free-living bacteria	< 4
Extensive savannahs	Endophytic, associative and free-living bacteria	< 14

The plant-associated symbioses can be grouped into four categories: (i) legumes and rhizobia; (ii) actinorhizal plants and *Frankia*; (iii) the tropical tree *Parasponia* sp. and rhizobia; and (iv) cycads and cyanobacteria. Although there is similarity in the genetics, development and function of symbiotic associations, there remains a great diversity in many aspects of the root-based N_2-fixing symbioses. As discussed by Vessey *et al.* (2005), there are a number of physiological conditions that must be met for a successful symbiosis between the host plant and the N_2-fixing organism. These include:

- the ability of the microsymbiont to infect and colonize host plant organs;
- the ability of the host plant to supply energy and nutrients to the microsymbiont;
- the ability of the host plant and the microsymbiont to regulate O_2 flux; and
- the ability to transfer the fixed N from the microsymbiont to the host plant.

The legume–rhizobia symbiosis is the best studied example of how the N_2-fixing symbiosis occurs. There are more than 650 genera and 18,000 species of legumes, belonging to the subfamilies Mimosoideae, Caesalpinoideae and Papilionoideae. The production of protein-rich grain and feed from legumes is extremely important for human and livestock nutrition. Legumes form a symbiosis with α-proteobacteria of the order Rhizobiales, including species of *Rhizobium, Bradyrhizobium, Sinorhizobium, Azorhizobium* and *Mesorhizobium*. There is evidence that β-proteobacteria may also participate in these kinds of relationships. Due to the agricultural importance and widespread production of legumes, much research has focused on understanding and optimizing the legume–rhizobia symbiosis.

The first step in the legume–rhizobia symbiosis is the invasion of the plant by the microsymbiont. Bacterial chemotaxis toward plant root exudates is a crucial event. Various root exudates serve as chemoattractants for the bacteria, including sugars, amino acids, succinate, malate, fumarate and aromatic compounds. Next, the rhizobia must bind to the plant surface. The binding site is probably a lectin protein receptor on the root surface. First, bacterial polysaccharides are weakly bound to the protein receptor or adhere to polysaccharides on the root surface. Secondly, the bacteria synthesize cellulose that causes a tight, irreversible binding and formation of bacterial aggregates on the host surface.

Nitrogen fixation can only occur after rhizobia invade the root or stem cortex. This requires a 'molecular dialogue' of signal generation and perception that leads to a gradual and coordinated differentiation and adjustment of the physiology and metabolism of both partners. The early signals from the legume are micromolar or nanomolar concentrations of flavonoids or isoflavonoids (e.g. genistein, naringenin, luteolin). Flavonoids act as primary signals, released in great concentrations near the emerging root hair zone, which is the most favourable site for rhizobium infection. Upon receiving the flavonoid signal, the microsymbiont expresses a *nodD* gene product that forms a protein–phenolic complex and regulates the expression of the structural nodulation (*nodABC*) genes.

Rhizobia use the *nod* genes to produce highly specific signal molecules called lipo-chito-oligosaccharide nod factors. The nod factors are usually four or five β-1-4-linked N-acetyl glucosamine residues with a long acyl chain that is attached to a terminal glucosamine. The substitution of functional groups (sulfuryl, methyl, acetyl, etc.) on to the backbone determines the specificity of the signal molecule for the desired host. When nod factors are detected by the legume host, a major developmental change occurs in the plant, which is necessary for rhizobial entry into the host. Even when nanomolar or femtomolar concentrations of purified rhizobial nod factors are applied to the roots of an appropriate legume host, a response is observed. The tip of the root hair, to which the rhizobia are bound, curls back on itself, trapping the bacteria. Rhizobia degrade the plant cell walls and are moved into the plant through an intracellular infection thread.

Nod factors produced by the rhizobia induce cell division and gene expression in the root cortex and pericycle, which is necessary for the development of the nodule. Division of the cortical cells forms the nodule primordium, and the infection threads penetrate and ramify into the primordium, initiating a process that leads to cell enlargement. The primordia of indeterminate nodules are initiated within the inner-root cortex, while determinate nodules are initiated in the outer-root cortex. Indeterminate nodules maintain a persistent apical meristem and continue to grow throughout the lifespan of the nodule, whereas determinate nodules do not have an active meristem and have a defined lifespan. On lucerne plants, the indeterminate nodules are long and elongated (about 1 cm diameter), with 10–50 nodules clustered on the primary taproot. When crushed, the nodule has a pink or red centre due to the presence of leghaemoglobin, which indicates that the rhizobia are fixing N_2. Numerous small determinate nodules (< 2 mm diameter) may be scattered over the entire root system. When they are no longer active, the centre turns white or pale green before the nodule is discarded by the plant.

In both types of nodules, the rhizobia multiply within the infection thread, but remain confined by the plant cell wall. As the primordia develop into a nodule, the bacteria are released from the tip of the infection thread by endocytosis and differentiate into bacteriods. The bacteriod cells have special physiological characteristics, notably the ability to produce nitrogenase. Another special feature of nodules is that the bacteriod's outer membrane is clearly delineated from the plant by a symbiosome membrane. The symbiosome membrane serves both as a physical interface and a mediator of metabolite exchange between the rhizobia symbiont and the host plant. The N_2 fixation by the microsymbiont within nodules occurs as follows:

$$N_2 + 3\,H_2 + \text{energy} \xrightarrow{\text{Nitrogenase}} 2NH_3 + H_2O \rightarrow 2\,NH_4^+ \rightarrow \text{protein} \qquad [7.3]$$

The first step in the reaction requires energy, which comes from photosynthesized carbon (sucrose) transported through the phloem. In soybean, about 20–30% of photosynthates are allocated to support rhizobia in nodules. After sucrose is broken down via glycolysis, the three dicarboxylic organic acids are transported across the peribacteroid membrane and serve as the energy source for the bacteria. The bacteria produce an enzyme, nitrogenase, which converts nitrogen and hydrogen gases into gaseous ammonia. In legume nodules, the NH_3 quickly diffuses out of the bacteroid's protoplasm into the symbiosome's space, where it is protonated to NH_4^+. An ammonium-transporting system present in free-living rhizobia is suppressed in bacteroids, thereby stopping the potential for channelling the NH_4^+ back into the bacteroid. An ion channel facilitates the transport of NH_4^+ into the plant cytosol where it is transformed into simple protein compounds (amides or ureides) and eventually converted to protein by the plant. Nitrogenase is sensitive to inhibition by O_2, but the enzyme is protected in the nodule because an O_2 diffusion barrier exists in a region of densely packed cells in the inner cortex of legume nodules.

There is considerable interest in promoting more biological N_2 fixation to reduce our reliance on fertilizer N for the production of food and forages. Fertilizer manufacturers produce approximately 8×10^{10} kg NH_3/year through industrial N_2 fixation (Haber-Bosch process), which is more energetically costly than biological N_2 fixation. In addition, the low efficiency of fertilizer N use leads to gaseous N emissions and N losses to the environment via leaching and erosion. The estimated N_2 fixation capacity of various agricultural crops is provided in Table 7.6.

The N_2 fixation by legumes accounts for about 40–75% of the nitrogen requirement of the crop. The rest of the nitrogen required by the crop is absorbed from the soil solution (NH_4^+ and NO_3^-). The variation in N_2 fixation values is related to growing conditions (climate, soils), cultivars and the biotic interaction between the host and the microsymbiont. Higher values are possible when producers inoculate the seed with strain(s) of rhizobia that are highly efficient at fixing N_2 with that plant. Most commercial inoculants are relatively inexpensive and provide a significant boost in crop production, so inoculation is an economical option for producers. Other considerations for promoting N_2 fixation include:

Table 7.6. Estimates of N_2 fixation per unit area of grain and forage legumes. (Adapted from: Carlsson and Huss-Danell, 2003; Herridge *et al.*, 2008.)

Legume crop	% plant N derived from N_2 fixation	N_2 fixed (kg N/ha/year)
Grain legume		
Common bean (*Phaseolus* sp.)	36–40	30–50
Chickpea (*Cicer arietinum* L.)	63–65	40–60
Pea (*Pisum sativum* L.)	63–65	30–85
Lentil (*Lens culinaris* L.)	63–65	30–50
Fababean (*Vicia faba* L.)	68–75	80–120
Groundnut (*Arachris hypogaea* L.)	58–68	60–100
Soybean (*Glycine max* (L.) Merr.)	58–68	60–170
Forage legume		
Lucerne (*Medicago sativa* L.)	49–88	50–300
White clover (*Trifolium repens* L.)	40–93	30–150
Red clover (*Trifolium pratense* L.)	40–93	70–160
Vetch (*Vicia* sp.)	50–75	80–140
Trefoil (*Lotus corniculatus* L.)	60–95	30–150
Crimson clover (*Trifolium incarnatum* L.)	70–90	30–180

- select an appropriate cultivar for your region;
- install drainage to prevent root damage from flooding;
- irrigation may be recommended in semi-arid or arid regions;
- add lime to adjust soil pH to the optimal range for the legume;
- check the initial soil fertility level and add nutrients (especially phosphorus, calcium, molybdenum, boron, cobalt, iron and copper) to support legume growth; and
- use good management practices to minimize competition from weeds and avoid over-harvesting, particularly in the first year.

As shown in Fig. 7.4, there are a number of other N_2-fixing interactions that are important for primary production in agricultural and natural systems. In all cases, prokaryotes having the ability to fix N_2 possess the *nif* gene that encodes for nitrogenase, the enzyme required to transform H_2 and N_2 into NH_3. Details of selected N_2-fixing symbioses and plant-associated interactions are provided below.

Azolla–*cyanobacteria*

The aquatic water fern *Azolla* and the cyanobacterium *Anabaena azollae* grow together on the surface of streams, ponds and rice paddies throughout tropical and temperate regions. This cyanobacterium is universally present in ovoid cavities within the fern's leaves. Short filaments of the cyanobacteria are found on top of the germinating fern megaspore and are probably trapped by the embryonic *Azolla* plant during differentiation of the shoot apex and the dorsal lobe primordial of the first leaves. A fully formed leaf consists of a thick green or red dorsal (upper) lobe that floats on the surface of the water and a thinner ventral lobe immersed in the water. In this lobe, the cyanobacteria develop a thick-walled heterocyst, lose their ability to photosynthesize (thylakoid membranes become inactivated) and are devoted to N_2 fixation (see equation 7.3). Ammonia produced by the cyanobacteria diffuses into the *Azolla* tissues and is converted to ammonium, amino acids and proteins.

The daily N_2 fixation rates from this symbiosis are in the range of 0.4–3.6 kg N/ha, which supplies at least 80% of the *Azolla*'s nitrogen requirements. During a growing season, the *Azolla–Anabaena* symbiosis could fix 25–170 kg N/ha. In rice paddies, about 30–40 kg N/ha can be obtained from this 'green manure', reducing the need for commercial fertilizer N. Research is under way to select new varieties of *Azolla* that will flourish under various climatic and seasonal conditions.

Non-legumes and Frankia

Actinomycetes of the genus *Frankia* form nodules and are capable of symbiotic N_2-fixing association with a variety of actinorhizal plants. These include 200 species of woody shrubs and trees, predominantly in temperate regions but also extending into the tropics, especially the Casuarinaceae. The actinorhizal plants typically grow in marginally fertile soils and many are early-successional plants. Hence, actinorhizal plants are very important in the nitrogen cycle of forests and in the re-vegetation of various landscapes. Actinorhizal plants have been used in erosion control, soil reclamation, agroforestry and for stabilizing desert and coastal dunes (e.g. as shelter belts).

The mechanism of *Frankia* infection in its actinorhizal host is not entirely understood. Equivalents of rhizobial nod factors have not been identified for Frankia, although strains can enter the host roots intracellularly via root hairs or intercellularly, depending on host plant species. Intracellular infection takes place in the Betulaceae, Casuarinaceae and Myricaceae. The *Frankia* hypha is trapped by curling root hair and an infection thread develops, although it has different characteristics from rhizobia infection threads. Cellular division in the root cortex is induced and the infection thread grows into dividing cortical cells and fills them with *Frankia* hyphae (pre-nodule), leading to the development of a nodule primordium in the pericycle and finally nodule lobes that extend from the root into the soil. Intercellular infection takes place in the Rhamnaceae, Elaeagnaceae and Rosaceae familes, where the *Frankia* hyphae enter the root between epidermal cells and colonize the root cortex. The plant responds by secreting pectin- and protein-rich material into the intercellular spaces and the formation of a nodule primordium is induced in the root pericycle. Sucrose and possibly other energy sources (sorbitol, hexoses) from the host fuel the N_2 fixation by *Frankia*. Rates of N_2 fixation are in the range of 30–50 g N per tree during a growing season, but the actual rates in the field are often lower due to environmental stresses such as drought or nutrient limitation.

Cycads–cyanobacteria

The cycads are an ancient life form that dominated the Earth's forests from Greenland to Antarctica about 250 to 65 million years ago, before the advent of the angiosperms. These seed-bearing gymnosperms are evergreen and have a palm-like appearance with a thick, columnar stem and rosettes of long, pinnately compound leaves. There are about 240 species in the order Cycadales. It is believed that the cycad–cyanobacteria symbiosis developed to meet plant N requirements during a period when the climate was warmer and wetter and had higher atmospheric CO_2 concentrations than the present day.

Cycads can form a symbiosis with a variety of cyanobacteria, with filamentous heterocystous species of the genus *Nostoc* being the most common microsymbionts. The cyanobacteria invade a particular root type, referred to as coralloid roots due to their 'coral-like' appearance. These specialized roots reach up to 10 cm in diameter and weigh up to 500 g. The growth of precoralloid roots brings the root cap close to the soil surface, where the free-living photoautotrophic cyanobacteria live. It is not clear how the cyanobacteria enter the root – they could enter through apical lenticels, through the papillose sheath, through cracks in the dermal layer or through channels from the root surface. The molecular signals and genes responsible for the infection and root colonization processes are still under investigation. Once inside the root, the N_2-fixing heterocysts begin to differentiate and soon reach numbers never observed in the free-living state, probably due to over-expression of *hetR*, the heterocyst regulatory gene. The host supplies energy, possibly in soluble forms (glucose, fructose) or more complex carbohydrates (mucilage) to fuel N_2 fixation. The microsymbiont transforms NH_3 into NH_4^+ and finally to glutamine or citrulline that is translocated to the cycad. The N_2-fixation rates from this symbiosis are about 8–20 kg N/ha/year under field conditions.

Parasponia*–rhizobia*

The *Parasponia* are flowering plants in the order Rosales and is the only genus outside the legumes known to engage in an N_2-fixing symbiosis with rhizobia. This was discovered in the 1970s in five species of tropical trees that are native to the Indo-Malaysian archipelago. These trees are often pioneer species on very nutrient-poor soils and thus the symbiosis permits the plant to survive in an otherwise inhospitable environment.

The molecular dialogue between the host and rhizobia is achieved through flavonoids and nod factors, but the rhizobia do not infect the plant through root hairs. Instead, the rhizobia erode the root epidermis below the site of bacterial colonization on the root surface or enter roots through cracks. Thick-walled infection threads of plant origin act as a conduit for the rhizobia to nodule primordia developing from the root pericycle. The nitrogenase activity of rhizobia nodules in *Parasponia* is much lower than in the legume–rhizobia symbiosis, probably because the host does not supply sufficient energy to optimize nitrogenase activity. The carbon supply to rhizobia in *Parasponia* and estimates of N_2 fixation rates due to this symbiosis under field conditions still remain to be confirmed.

Plant-associated N_2 fixation (cereal-associated and cereal-endophytic bacteria)

There are a number of N_2-fixing prokaryotes living in the rhizosphere or within the plant tissues of grasses such as sugarcane, rice, wheat, sorghum, maize and others. Among the N_2-fixing bacteria found in the rhizosphere are *Beijerinckia fluminense* as well as *Azospirillum, Azotobacter, Bacillus, Derxia, Enterobacter* and *Erwinia* species. None of these organisms forms nodules, but live on the surface of fine roots, use carbon compounds secreted by the roots as an energy source to catalyse the reaction, and transfer a portion of the NH_4^+ captured from N_2 fixation to the cereal crop. The *Azospirillum* spp. may also be found within the epidermal cells of plants, probably entering through wounds or cracks at lateral root junctions.

The N_2-fixing endophytes such as *Herbaspirillum* spp. and *Gluconacetobacter diazotrophicus* (formerly classified as the genus *Azotobacter*) are obligate or facultative endophytes of grasses that do not survive for long in soil. They may be transferred from one plant to another via seeds, vegetative propagation or dead plant material and possibly by sap-feeding insects. They live in the roots, stems and leaves of grasses, existing in non-specialized plant tissues such as xylem vessels and intercellular spaces.

There is considerable variability in N_2 fixation from the plant-associated N_2-fixing prokaryotes. In sugarcane, these organisms can supply 0–60% of the nitrogen requirement under field conditions. Temperate cropland that is cultivated for the production of cereals and non-leguminous oilseeds may gain 1–25 kg N/ha/year from free-living heterotrophic bacteria. Grazed tropical savannahs probably obtain ~10 kg N/ha/year from plant-associated N_2 fixation (James and Olivares, 1998).

Mycorrhizal fungi

Many plant species evolved in natural ecosystems with low soil fertility. They have been able to thrive despite the low concentrations of essential nutrients through a relationship with soil fungi called mycorrhizae. Some are obligate symbionts that rely on plant photosynthates for energy, whereas others are facultative symbionts that can also mineralize organic carbon from non-living sources. The fungal hyphae proliferate from the host's root and extend into the soil, far beyond the normal range of the plant roots. Mycorrhizae effectively increase the root surface area of the host plant and enhance nutrient and water uptake. Most trees, shrubs, grasses and even cactus have mycorrhizae living in their roots. Virtually all agricultural crops, with the exception of the Brassicaceae family (cauliflower, cabbage, canola and others) and a few others, are able to associate with these fungi. Seven types of mycorrhizae (Table 7.7) have been categorized on the basis of morphology and anatomy:

1. Ectomycorrhizae: some of these fungi are facultative symbionts. The fungal mycelium grows like a sheath (mantle), completely covering the fine root surface. The mycelium does not penetrate beyond the root epidermis. Fungal hyphae grow outward, into the soil, and also in a Hartig net (thick net of hyphae) between the epidermal and cortical cells (Fig. 7.6). Often the hyphal strands extending into the soil weave together to form rhizomorphs, which are specialized for long-distance transport of nutrients and water. The ectomycorrhizal fungi (EMF) are generally Basidiomycetes, with a few actinomycetes in this group. The plant symbionts include gymnosperms and woody angiosperms such as the Pinaceae, Fagaceae and Betulaceae in temperate forests. The hyphae provide a large surface area for the interchange of nutrients between the host and the fungi. They also release extracellular enzymes that mineralize organic matter, releasing inorganic

Table 7.7. Mycorrhizal fungi, their plant hosts and general characteristics.

Mycorrhizal type	Host	Taxonomic group	Characteristic structures	Major ecological functions
Ectomycorrhizae	Mostly gymnosperms, woody angiosperms	Mostly Basidiomycetes	Hartig net, mantle and rhizomorphs	Nutrient uptake, mineralization of organic matter, soil aggregation
Endomycorrhizae (vesicular arbuscular mycorrhizae, VAM)	Bryophytes, pteriodophytes, gymnosperms, most angiosperms	Glomelomycota	Arbuscules, vesicles and auxiliary cells	Nutrient uptake, soil aggregation
Ericoid	Ericales, Monotropaceae	Ascomycetes and Basidiomycetes	Hyphae in cells or mantle and Hartig net	Carbon transfer between plants, nutrient uptake
Arbutoid	*Arbutus* and other genera	Basidiomycetes	Mantle and Hartig net, intracellular penetration with a dolipore septum	Nutrient uptake
Monotropoid	Monotropaceae	Basidiomycetes	Mycelial sheath, Hartig net and fungal peg	Nutrient uptake and carbon transfer to achlorophyllous plants
Ect-endomycorrhizae	Pinaceae	Ascomycetes (*Wilcoxina* sp.)	Hartig net with some cell penetration, thin mantle	Nutrient uptake, mineralization of organic matter
Orchidaceous	Orchidaceae	Basidiomycetes	Hyphal coils	Supply carbon and vitamins to embryo

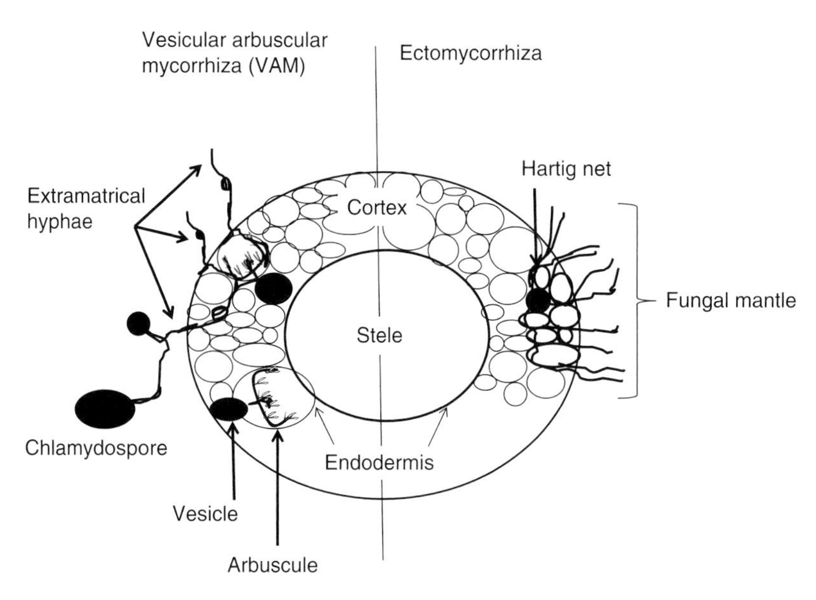

Fig. 7.5. Schematic diagram of a transverse section of a fine root showing a vesicular arbuscular mycorrhiza and an ectomycorrhiza colonizing root tissues. Note the difference in structures produced by these two types of symbiotic fungi.

nutrients into the soil solution for fungal-host uptake and growth.

2. Endomycorrhizae (vesicular arbuscular mycorrhizae, VAM): these fungi are obligate symbionts. Aseptate hyphae of the fungus penetrate the cell wall of fine roots and grow in the root cortex between the epidermis and the endodermis. The fungus produces special intracellular structures (vesicles and arbuscules) during development. As the fungus grows, the host cell membrane invaginates and envelops the fungus, creating a new compartment where fungal arbuscules can be deposited (Fig. 7.5). This avoids direct contact between the plant and fungal cytoplasm, but allows for efficient transfer of nutrients between the symbionts. The vesicles are thin-walled, lipid-filled structures that usually form in intercellular spaces. They may be used for storage or for fungal reproduction in asexual spores, although spores are more often produced at the ends of hyphae in the soil, outside the plant root. The spores can range in diameter from 10 μm for *Glomus tenue* to more than 1000 μm in some *Scutellospora* spp. The fungus also produces hyphae (2–27 μm diameter) that extend into the soil.

The arbuscular mycorrhizal fungi were traditionally classified in the order Glomales in the Zygomycetes, but have been transferred into a new fungal phylum called the Glomelomycota. Endomycorrhizae form with about 70% of plants in nature, including all the grasses (but not sedges, which generally lack mycorrhizae), most herbaceous plants and some trees. The extensive hyphae often make a significant contribution to the phosphorus nutrition of the host plant.

3. Ericoid mycorrhizae: this fungus has similarities to the endomycorrhizae because it grows extensively in the root cortex, but no arbuscules are formed. The fungal symbiont penetrates the cortical cell wall of fine roots and invaginates the plasmalemma. Infected cells appear to be fully packed with fungal hyphae, but the hyphal extension into soil is quite short. While the root surface is covered with hyphae, no mantle is observed.

The ericoid mycorrhizae are mostly ascomycetes of the genus *Hymenoscyphus* that form symbiosis with plants such as *Calluna* (heather), *Rhododendron* (azaleas and rhododendrons) and *Vaccinium* (blueberries). This symbiosis is commonly found in acidic, peatland soils.

4. Arbutoid mycorrhizae: the arbutoid fungi have characteristics of both ecto- and endo-mycorrhizae. Intracellular penetration can occur, a mantle forms and a Hartig net is present in the outer cells of the cortex. The presence of the mantle means that all nutrients must be absorbed by the mycorrhizae before they pass into the plant root. They are distinguished from the ericoid mycorrhizae because the internal hyphae possess a dolipore septum, which may add rigidity to the filament and control nutrient transport through the hyphae. The fungi in this group are Basidiomycetes that associate with trees and shrubs of the genera *Arbutus* (Pacific madrone), *Arctostaphylos* (manzanita), *Arctous alpinus* (mountain bearberry) and several species of the Pyrolaceae.

5. Monotropoid mycorrhizae: these mycorrhizae are similar to the ectomycorrhizae and form a thick, compact mycelial sheath and a Hartig net. They exhibit a distinctive type of intracellular penetration in cortical cells by inserting a fungal peg that becomes extensively invaginated. The final structure resembles a specialized transfer cell that plants form to achieve rapid, short-distance solute transport.

The term ‘monotropoid mycorrhizae’ refers specifically to Basidiomycetes that form mycorrhizae on plants that lack chlorophyll (achlorophyllous) in the family Monotropaceae (e.g. Indian pipe). The fungus forming the monotropoid mycorrhiza also forms an ectomycorrhizal relationship with a tree, thereby creating a link through which carbon can flow from the photosynthetically active tree to the heterotrophic parasitic plant. Both plants probably derive most of their nutrients from the fungus.

6. Ect-endomycorrhizae: these mycorrhizae form a Hartig net in the cortex of the root, but develop little or no sheath. Intracellular penetration of cortical cells takes place and so they are similar to the arbutoid type. The fungi involved in this symbiosis are ascomycetes of the genus *Wilcoxina* and the mycorrhiza is only formed with genera in the Pinaceae.

7. Orchidaceous mycorrhizae: the fungal association is of the endomycorrhizal type, where the fungus penetrates the cell wall, invaginates the cell membrane and forms a hyphal coil within the cell. The hyphae can be spread internally from cell to cell. Within a few days or weeks, the internal hyphae collapse or are digested by the host cell. A number of Basidiomycetes can be involved in this symbiosis and seem to be related to those that are wood-rot fungi or pathogens. Plants of the

Orchidaceae are the host. Orchids typically have very small seeds with little nutrient reserve. The mycorrhizae colonize the plant shortly after germination. The digestion of the internal fungal hyphae supplies carbon and vitamins to the developing embryo. Some achlorophyllous species depend on the fungal partner to supply carbon throughout its life.

In some of the mycorrhizae described above, there are specific fungal–host interactions, but many mycorrhizae form when a fungus encounters a potential host from a broad taxonomic group. Although host specificity is not a requirement for colonization in such a case, it is expected that the degree of colonization of the root system and hyphal growth will differ depending on the host. The 'molecular dialogue' that occurs between a fungal symbiont and a plant host is probably similar to that occurring in the legume–rhizobia symbiosis, although much remains to be learned about how this occurs in nature. The genetic basis of the arbuscular mycorrhizae symbiosis (Fig. 7.6) has

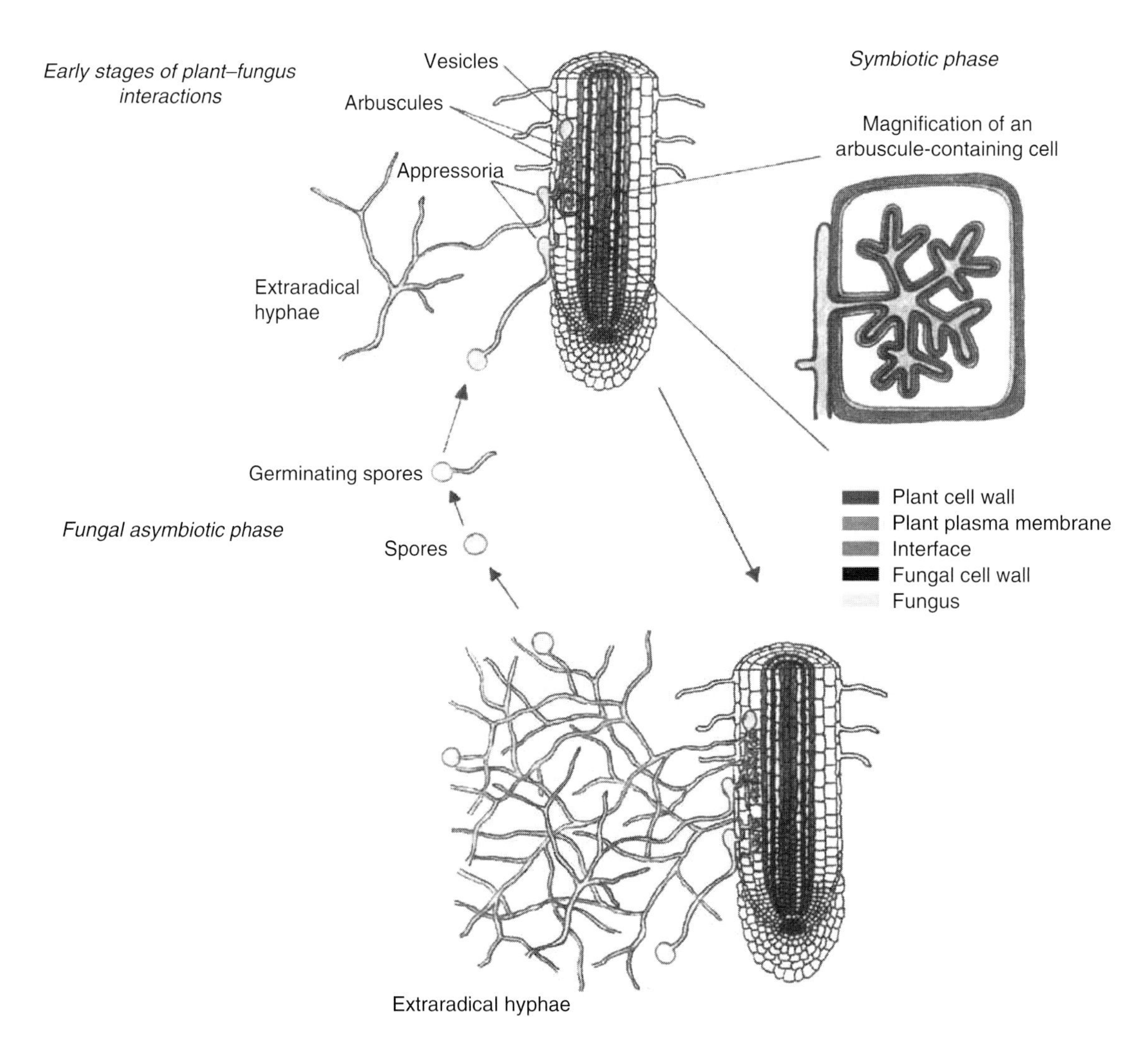

Fig. 7.6. Different stages occurring during root colonization by an arbuscular mycorrhizal fungus. (Reproduced with kind permission from Springer Science+Business Media® from *Mycorrhiza, Fungal and Plant Gene Expression in Arbuscular Mycorrhizal Symbiosis*, vol 16, Balestrini and Lanfranco, 2006, Fig. 1, p. 510.)

been described by Balestrini and Lanfranco (2006) and consists of the following steps:

1. Fungal asymbiotic stage: fungal spores germinate and produce a limited amount of mycelium. Over-expression of the *GmGIN1* gene by the fungus may be a molecular signal to the plant root. It is not known whether the fungus germinates in response to a signal from the plant root, rhizosphere bacteria or both.
2. Early stages of plant–fungus interactions: genes encoding cell wall proteins lead to a reorganization in the host plant's epidermal cells upon adhesion of the fungus to the root surface. Plant signalling molecules (sesquiterpenes) stimulate fungal hyphal branching towards the root. The fungus forms the appressorium, a structure that makes contact with the roots and allows the hyphae to penetrate into the root tissue.
3. Symbiotic phase: the intracellular colonization by the fungus dramatically changes the morphological organization of the host cells and leads to the induction and up-regulation of an array of genes responsible for cell wall functions. Transport systems for sugars, phosphate and nitrate are activated. Enzymes capable of synthesizing glycogen and hydrolysing sucrose are produced. Finally, the extension of extraradial fungal hyphae into the surrounding soil leads to the induction of systems for phosphate transport, lipid utilization and carbon and nitrogen metabolism in the host plant.

Mycorrhizal associations are favourable to plants because hyphae are able to grow and explore areas in the soil profile that cannot be accessed by larger plant roots, which contributes to the uptake of water and nutrients to the plant. While the fungal hyphae in soil account for less than 20% of the total nutrient-absorbing surface mass, they represented nearly 80% of the absorbing surface area of pine seedlings (Rousseau *et al.*, 1994). The large absorptive area is related to the tiny size of mycorrhizal hyphal tips (as small as $2\,\mu m$) that can penetrate even the smallest ultramicropores. In contrast, fine root hairs of most species cannot enter spaces smaller than $30\,\mu m$. Mycorrhizae facilitate the diffusion of immobile nutrients (P, Cu, Zn and Mn) and increase the effective rooting zone from which mobile nutrients (N, K, Ca and Mg) can be accessed. In arid environments, ectomycorrhizae enhance plant water uptake when soil water potentials in the rooting zone are between −1.5 and 2.0 MPa, which is below the permanent wilting point (Allen, 2007). Mycorrhizal symbiosis is vital for plant survival in semi-arid and arid ecosystems.

Not only do mycorrhizae improve the uptake of nutrients and water in the soil solution, but they can increase the concentrations of soluble ions, such as phosphate ($H_2PO_4^-$). Phosphorus nutrition is critical for plant energy relations and flower and grain production. Most of the phosphorus in soils is not available for plant uptake because it is bound to soil minerals, precipitated with metal oxides such as iron and aluminium or immobilized in organic residues and microbial cells. Mycorrhizae can alter the soil environment to increase the soluble phosphate in soil solution by secreting low-molecular weight organic acids such as oxalate that can: (i) replace $H_2PO_4^-$ bound in adsorption sites on soil mineral surfaces; (ii) dissolve iron phosphate and aluminium phosphate compounds, releasing $H_2PO_4^-$ into soil solution; and (iii) form complexes with iron, aluminium and calcium in soil solution, thus preventing the precipitation of $H_2PO_4^-$.

Mycorrhizal fungi, as well as plant roots and other microorganisms, produce extracellular phosphatase enzymes that hydrolyse organic phosphate compounds, releasing soluble $H_2PO_4^-$ into soil solution. In addition, the ericoid and ectomycorrhizal fungi produce hydrolytic enzymes that can degrade organic residues for the purpose of obtaining NH_4^+ from organic nitrogen compounds. In return, the plant supplies carbon and a safe habitat for the mycorrhizal fungi to grow and thrive. Between 10 and 30% of the total carbon assimilated by plants may be transferred to the fungal partner. Generally, the fungal symbiont allows the host to increase photosynthetic activity, which compensates for the energy drain on the plant. Although some ectomycorrhizal fungi could obtain carbon from organic residues, they probably rely on the autotrophic host for carbon. The ectomycorrhizae and ericoid mycorrhizae transform carbohydrates into fungal-specific storage carbohydrates such as mannitol and trehalose, while the arbuscular mycorrhizae store lipids in their vesicles.

Ecologically, mycorrhizae are vital for primary production, are involved in carbon flow and nutrient cycling and contribute to soil aggregation. The hyphal networks of mycorrhizae act as a net that attracts and binds soil particles, contributing to the formation of aggregates. This is significant at the ecosystem level due to the large number of hyphae in soil, with an estimated 1–20 m of these small hyphae per gram of soil. Fungal cells also produce glomalin, an extracellular protein that binds to clay particles and thus serves as a cementing agent that holds aggregates together at the micron scale. Mycorrhizae also interact positively with N_2-fixing bacteria in leguminous plants. Legumes tend to possess a taproot system with short, fine roots, which are insufficient to obtain sufficient phosphorus for optimal production, and thus benefit from arbuscular mycorrhizae. The colonization of fine roots by mycorrhizae enhances the plant defence against soil-borne pathogenic fungi, insect herbivores and nematodes. The main mechanism is thought to be physical – the presence of dense hyphal networks on the root surface and in the intracellular space tends to 'block' other fungi that would occupy the same niche. Insect herbivores and nematodes are expected to possess cellulase enzymes to digest root cells, but may not be able to break down chitin-rich fungal cells covering the root surface. As discussed by Bi *et al.* (2007), arbuscular mycorrhizae stimulate biochemical and molecular responses in host plants that can lead to:

- increased production of many allelochemicals (phenolic compounds, terpenes, alkaloids and essential oils);
- induction of plant defence genes such as phenylalanine ammonia-lyase (PAL), chalcone synthase (CHS) and chitinase-coding genes; and
- activation of the jasmonate-signalling pathway.

Thus, mycorrhizal fungi may provide physical protection and induce plant defences that reduce the incidence and severity of root diseases and root herbivory.

Although the plant–mycorrhizal association occurs naturally, it can be enhanced by careful management. In agricultural systems where the goal is to reduce the reliance on fertilizer inputs and rely more on symbiotic relationships to support crop growth, the following practices could be helpful:

- Select crops or establish crop rotations with various host plants to promote the growth and diversity of mycorrhizae.
- Avoid frequent, deep soil disturbance with tillage implements to avoid damaging fragile mycorrhizal hyphae.
- Apply organic amendments or leave organic residues in the agroecosystem to stimulate mycorrhizal activity.
- Install a good drainage system and avoid compaction, because mycorrhizae grow best in well-aerated soils.

Commercial inoculants of *Glomus* sp., an arbuscular mycorrhiza, can be purchased and introduced into the rhizosphere of the target plant. The practice of inoculating with arbuscular mycorrhizae is not as widespread as the inoculation of legumes with rhizobia at present. Inoculants could be quite helpful in promoting plant growth in soils with low populations of indigenous mycorrhizae. For example, trees and shrubs could benefit from mycorrhizal inoculants in efforts to reclaim disturbed or marginal lands (e.g. mine tailings, polluted soils, saline sites).

7.4 Plant Root Interactions with Free-living Soil Organisms

Interactions between plant roots and free-living soil organisms occur in the rhizosphere. The rhizosphere includes the soil and soil solution that circulates through pores within a few millimetres of plant roots. It is difficult to collect material from the rhizosphere because it usually involves disturbing the root system. To collect soil organisms from the rhizosphere, begin by excavating an intact soil block containing the plant roots. After gently separating the roots and shaking them, the soil that remains adhered to lateral roots and root hairs is considered to represent the rhizosphere soil. The term 'rhizoplane' is also used to describe root–microbial processes at the root surface, such as the interaction between roots and ectomycorrhizae or roots and N_2-fixing cereal endophytes. These are different from the interactions between free-living rhizosphere bacteria and fungi that live within a few millimetres of the plant root and are not attached to the root surface. Due to the methodological challenges in clearly distinguishing rhizosphere soil from the bulk soil, some researchers consider the soil within a few centimetres of a plant root to be root-associated soil that could provide a favourable environment for free-living soil organisms.

The rhizosphere is a favourable habitat for many free-living soil organisms that are involved in nutrient mineralization and release nutrients into the soil solution. These include the free-living bacteria and fungi that produce the enzymes necessary to degrade organic matter, as well as the microfauna that graze upon bacteria (protists and bacterivorous nematodes) and fungi (fungivorous nematodes). Grazing of microbial biomass provides the energy and nutrients that these organisms require for growth, and also releases nutrients such as NH_4^+ and $H_2PO_4^-$ into the soil solution, where they can be acquired by plant roots and mycorrhizal hyphae. Larger soil animals such as collembola and earthworms create pores where roots may grow, and deposit nutrient-rich fecal pellets and casts. The increase in plant-available nutrients due to the activities of the free-living rhizosphere microfauna is beneficial to plant nutrition. Changes in porosity due to the larger soil fauna may allow roots to penetrate more deeply in the soil profile, enhancing anchorage and resource acquisition.

Plant roots release compounds that attract many free-living soil organisms through processes known as root secretion and rhizodeposition. Root secretions are substances that are produced by plants in response to environmental conditions or stresses. Rhizodeposition refers to the process by which soluble and insoluble compounds, together with sloughed cells, are released from the plant root into the soil (Fig. 7.7).

The carbon input to the soil through plant roots is considerable. Carbon fixed by photosynthesis is transported through the phloem to the roots of annual plants to support their growth and metabolic functions. From 1 to 25% of photosynthesized carbon may be transferred to the soil via rhizodeposition and thus constitutes a major input of energy for soil foodweb organisms. On average, annual crops and perennial grasses transfer 20% of the carbon from photosynthates to the roots during the growing season (Fig. 7.8).

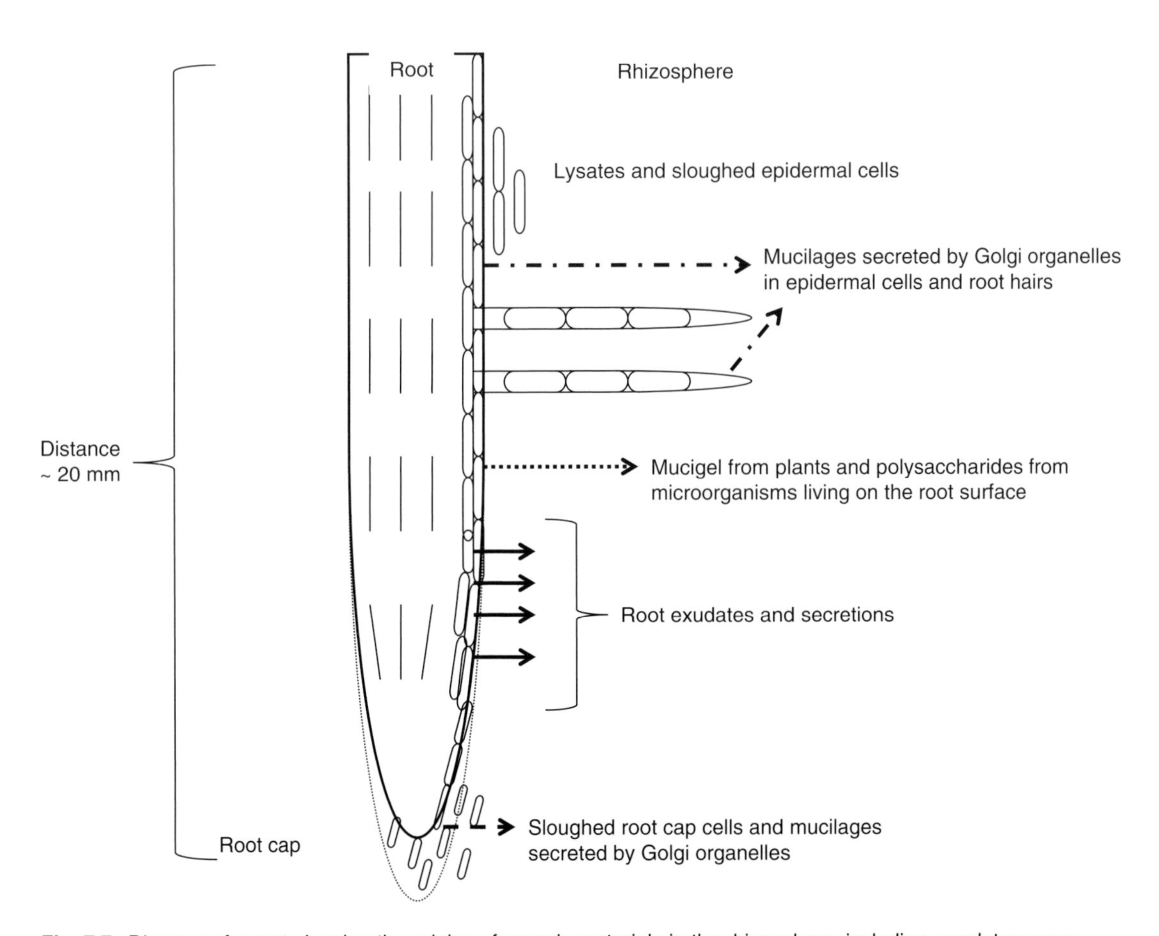

Fig. 7.7. Diagram of a root showing the origins of organic materials in the rhizosphere, including: exudates; secretions; plant mucilages secreted by Golgi organelles in the root cap cells, from epidermal cells and root hairs; mucigel; and lysates and sloughed cells. (Adapted from Lavelle and Spain, 2001.)

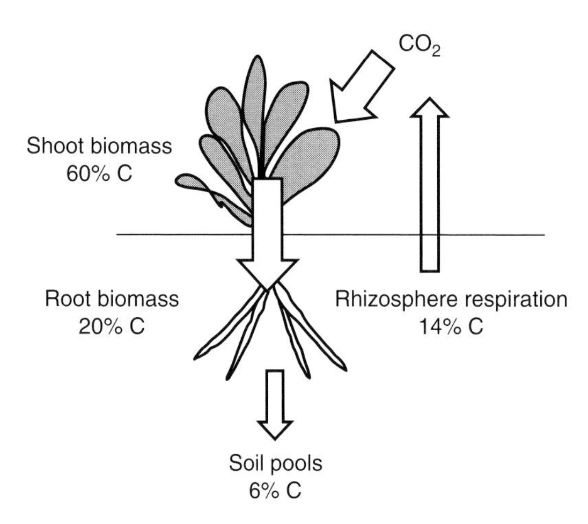

Fig. 7.8. Net fluxes of carbon in annual plants and perennial grassland species. (Adapted from Nguyen, 2003.)

Rhizodeposits are a mixture of carbohydrates (simple sugars and polysaccharides), amino compounds, organic acids, nucleotides, flavones, enzymes and growth factors (Table 7.8). They are often grouped into the following categories:

1. Exudates are water-soluble, low-molecular weight compounds leached from the roots without metabolic control by the plant. These compounds could be easily degraded and absorbed by soil microorganisms.
2. Secretions are low-molecular weight compounds that are released from the plant in response to an environmental stress. These are available for absorption by soil microorganisms.
3. Plant mucilages are insoluble organic compounds that may come from:

(i) mucilage secreted by Golgi organelles in the root cap cells;
(ii) hydrolysates of the polysaccharides of the primary cell wall between the epidermal cells of the primary wall and sloughed root cap cells;
(iii) mucilage secreted by epidermal cells and root hairs;
(iv) mucilage produced by bacterial degradation of dead epidermal cells.

4. Mucigel is the gelatinous material at the surface of roots grown in non-sterile soil. It includes plant mucilages, bacteria and their metabolites, together with colloidal mineral and organic matter from the soil.
5. Lysates and sloughed cells come from the epidermis and cortex of plant roots (dead and alive).

Root exudates and secretions are rich carbon sources that can be readily absorbed by microbial cells and are of major importance for microbial growth. The mucilages, mucigels, lysates and sloughed cells are composed of carbon compounds that are not as

Table 7.8. Compounds released by plant roots in the process of rhizodeposition. (Adapted from Giri *et al.*, 2005.)

Compound(s)	Exudate components
Sugars	Glucose, fructose, sucrose, maltose, galactose, rhamnose, ribose, xylose, arabinose, raffinose, oligosaccharide
Amino compounds	Asparagine, α-alaline, glutamine, aspartic acid, leucine/isolueine, serine, γ-aminobutyric acid, glycine, cystine/cysteine, methionine, phenylalanine, tyrosine, threonine, lysine, proline, trypotophae, β-alanine, arginine, homoserine, cystathionine
Organic acids	Tartaric, oxalic, citric, malic, propionic, butyric, succinic, fumaric, glycolic, valeric, malonic
Fatty acids and sterols	Palmitic, stearic, oleic, linoleic, linolenic acids; cholesterol, campesterol, stigmasterol, sitosterol
Growth factors	Biotin, thiamine, niacin, pantothenate, choline, inositol, pyridoxine, aminobenzoic acid, N-methyl nicotinic acid
Nucleotides, flavinoids and enzymes	Flavonine, adenine, guanine, uridine/cytidine, phosphatase, invertase, amylase, protease, polygalacturonase
Miscellaneous compounds	Auxins, scopoletin, fluorescent substances, hydrocyanic acid, glycosides, saponin (glucosides), organic phosphorus compounds, nematode cyst or egg-hatching factors, nematode attractants, fungal mycelium growth stimulants and inhibitors

easily degraded by soil microorganisms. Rather than serving as an energy source, these compounds are more likely binding agents that hold soil particles together, creating a favourable habitat for the growth of microbial colonies and aiding in the formation of stable soil aggregates.

Sugars and organic acids exuded by roots are an important energy source for free-living microorganisms, although they are not used equally by all groups. As illustrated in Table 7.9, Gram-negative bacteria living in the rhizosphere obtained much more carbon from root exudates than the fungi or actinomycetes. This is consistent with the growth habit and ecology of each group. The Gram-negative bacteria are copiotrophic rhizosphere colonizers that thrive in nutrient-rich environments, whereas the Gram-positive bacteria, actinomycetes and fungi are considered oligotrophic organisms that depend more on the decomposition of complex carbon compounds for their nutrition.

Root exudates mediate other interactions with microorganisms and soil fauna. Proteins may serve as 'recognition signals' that indicate the presence of a growing root. In fact, the actively growing roots may represent less than 10% of the root biomass in perennial plants. The secretion of proteins may stimulate microorganisms to actively disperse towards the growing root tips. This could be achieved by motile, flagellate cells or through hyphal growth (free-living actinomycetes and fungi). Still other microorganisms may move with growing roots via passive mechanisms, either passive transport of microbial cells in the soil water that moves along the root surface due to turgor pressure and transpiration by the plant, or passive transport of microorganisms in front of the root apex during root elongation.

Table 7.9. Plant-derived carbon in microorganisms isolated from the rhizosphere of annual ryegrass (*Lolium perenne*), determined from ^{13}C-phospholipid fatty acid biomarker analysis. (Data from Paterson *et al.*, 2007.)

Microbial group	Plant-derived carbon in microbial biomass (%)
Gram-negative bacteria	27–35
Gram-positive bacteria	4.8–5.2
Fungi	2.3–7.7
Actinomycetes	0.3–0.4

Organic acids in root exudates can chelate metals, alter soil pH in the rhizosphere and aid in the absorption and translocation of nutrients, notably phosphate. In the rhizosphere, organic acids may be produced by roots, by symbiotic mycorrhizae or by free-living rhizosphere bacteria and fungi. Fatty acids stimulate the growth of ectomycorrhizal fungi, whereas growth factors may stimulate growth in the microbial community or selected populations. Flavonoids are signalling molecules required for the development of legume–rhizobia symbiosis. Root exudates influence the abundance and behaviour of nematodes as well. Egg hatching of nematodes can be induced by root exudates. Most nematode larvae are found in the region of cell elongation behind the root tip. They are attracted to this region by attractant compounds, and perhaps also due to the food supply (bacteria and fungi inhabiting the rhizosphere).

Allelopathic compounds released by roots can inhibit the growth of other plants; these and other miscellaneous compounds may stimulate or inhibit the growth of soil biota. It is not known whether mesofauna and macrofauna are attracted by root exudate compounds, but large populations are found in the root-associated soil and also on organic residues, probably because this is where their food supply is found.

The quantities and chemical composition of root exudates vary with the soil and environmental stresses, and the plant species. In an ecological context, root exudation is part of the plant's growth strategy and essential to its survival. Examples of factors controlling root exudates and rhizodeposition are provided below:

- Plant species: certain plants secrete allelopathic compounds through their roots to inhibit the growth of other plants of the same or different species (intra- and interspecific competition). Black walnut (*Juglans nigra* L.) produces juglone (5 hydroxy-1,4 napthoquinone), a respiration inhibitor, in all plant parts. Plants belonging to the Solanaceae family, such as tomato, pepper and aubergine, are very sensitive to this allelotoxin. When exposed, they will exhibit symptoms such as wilting, chlorosis and eventually death. Since juglone is not very soluble, the greatest impact occurs within the tree canopy. Asparagus (*Asparagus officinalis* L.) secretes cinnamic acid and other compounds from its roots to inhibit the roots of neighbouring plants from growing into its root zone.

When an asparagus plantation is removed from a field, producers must carefully select the next crop in the rotation because dead and decaying asparagus roots continue to release allelochemicals that inhibit the growth of subsequent asparagus, tomato, barley and lettuce crops.

- Age of plant roots: young, rapidly growing fine roots release water-soluble substances such as sugars, organic acids and amino acids through their root tips, which serve as a 'priming' substrate for microorganisms in the path of the growing root. Just behind the root tip, the release of soluble, insoluble and volatile root exudates attracts and supports large populations of soil microorganisms, especially Gram-negative bacteria. Along the older root parts, the primary substrates include cellulose, suberin and other recalcitrant cell wall materials from sloughed root cortex cells. These support smaller populations of slower-growing microorganisms that can degrade and use these carbon sources for energy.
- Plant health: plant roots that have been infected by parasitic nematodes may release compounds that attract biocontrol organisms such as *Pasteuria penetrans* and nematrophic or predatory fungi. The importance of root exudates in the relationship between plant host–pathogen/parasite–biocontrol agent is still under investigation.
- Plant nutritional status/soil environmental conditions: some root exudates impede the growth of *Nitrosomonas*, the bacterium responsible for nitrification. By blocking the conversion of ammonium to nitrite, the plant is able to conserve nitrogen in infertile soils.
- Defoliation or below-ground herbivory: grasses that experience light to moderate defoliation exhibit an elevated level of root exudation, which stimulates the microorganisms responsible for N mineralization and nitrification processes, providing a surplus of nutrients useful for plant compensatory growth.
- The presence of symbiotic soil microorganisms: the detection and infection of plant roots by symbiotic N_2-fixing bacteria or mycorrhizal fungi is the result of a molecular dialogue. Complementary signalling compounds released by plant roots and from the symbionts govern the genetic, biochemical and physiological changes in both organisms that are required for a successful symbiosis.
- The presence of free-living soil microorganisms: root exudates are a good source of energy for plant growth-promoting organisms, a group of free-living microorganisms capable of producing compounds that act like plant growth hormones (auxins, gibberellins) and stimulate plant growth.

In conclusion, the rhizosphere is a zone that supports interactions between plants and free-living soil biota. These interactions vary spatially and temporally, and still much remains to be learned about the microbial ecology in the rhizosphere. Root exudates provide energy and nutrition to microbial communities, but also send molecular signals into the soil that induce responses by microorganisms and other fauna. There is great interest in identifying the plant growth-promoting organisms that can contribute to crop nutrition and plant protection against parasites and pathogens. Since these interactions could improve plant health, vigour and production in managed and natural ecosystems, further study on the rhizosphere is warranted.

Focus Box 7.1. Plant growth-promoting rhizobacteria.

Some bacteria inhabiting the rhizosphere are called plant growth-promoting rhizobacteria (PGPR) due to their ability to stimulate plant growth. Bacterial populations that can improve plant nutrition, produce plant growth regulators or prevent attack by pathogenic fungi are all considered PGPR (Table 7.10). This broad category includes bacteria that live within roots, such as N_2-fixing rhizobia, and those that are free-living in the rhizosphere. Some PGPR are commercially available as inoculants and are applied to agricultural land, forest plantations and to degraded lands that are being decontaminated or reclaimed by phytoremediation. There is considerable research effort underway to isolate and identify more PGPR. Metagenomics is a promising approach that could lead to the discovery of novel plant growth-promoting genes and gene products and identify microorganisms that have not yet been cultured in the laboratory.

Continued

Focus Box 7.1. *Continued*

Table 7.10. Beneficial activities of selected plant growth-promoting rhizobacteria (PGPR).

Activity	Representative bacteria
Biological N_2 fixation	*Rhizobium* (symbiotic) *Herbaspirillum* spp. (associated endophyte) *Azospirillum* (free-living)
Production of plant growth hormones (auxins, cytokinin, gibberellins)	*Azospirillum*, *Bacillus*, *Erwinia*
Protection against fungal root pathogens and parasitic nematodes	*Pasteuria penetrans*, *Pseudomonas fluorescens* and *Bacillus* spp.
Enhance nutrient use efficiency	*Azospirillum brasilense* (increased uptake of NO_3^-, K^+ and $H_2PO_4^-$)
Increase availability of inorganic phosphorus compounds to plants	*Bacillus megaterium*, *Burkholderia caryophylli*, *Pseudomonas cichorii*, *Pseudomonas syringae* (phosphorus-solubilizing bacteria)
Increase availability of iron to plants	*Vibrio* sp. (produce iron-chelating siderophores)
Mineralization of organic nitrogen, phosphorus and sulfur to plant-available nutrient forms	Heterotrophic bacteria that produce proteases/ureases, extracellular phosphatase and sulfatase enzymes
Degrade synthetic organic compounds and contaminants	Heterotrophic bacteria (generally in association with actinomycetes and fungi via co-metabolism)

8 Decomposition

Atmospheric concentrations of CO_2 increased from pre-industrial levels of 280 parts per million (ppm) in 1850 to 310ppm in 1950, and increased further to 380ppm in 2005. There is a large body of scientific evidence indicating rising CO_2 concentrations are linked to an increase in global temperature, melting of glaciers and polar ice, rising sea levels and extreme storms. Two important anthropogenic factors contributing to the increase in atmospheric CO_2 concentrations are fossil fuel combustion, with average emissions of 7.2Pg C/year, and land use change (Fig. 8.1). Fossil fuel burning suddenly takes C that has been inert in underground stores for millions of years and adds it to the pools of exchangeable C already present in the atmosphere, terrestrial ecosystems and the ocean. Land use change, such as the conversion of forests or grasslands for agricultural activities, removes carbon-rich vegetation and accelerates decomposition.

How land use change affects the carbon balance is of great interest for soil ecologists. The global carbon cycle is a delicate balance between carbon fixation by green plants, which removes CO_2 from the atmosphere, and respiration, which releases CO_2 back to the atmosphere. Respiration is a normal metabolic process for all living organisms including plants, microorganisms, animals and humans. In soils, CO_2 is released from plant roots and soil biota that are involved in decomposing organic residues and soil organic matter. The detrital soil foodweb organisms depend on decomposition of carbon-rich substrates to acquire the energy needed for their survival, and in turn recycle nutrients contained in the organic residues for the benefit of plants. In this chapter, we will examine the global carbon cycle briefly before discussing the ecological importance of decomposition in terrestrial ecosystems.

8.1 The Global Carbon Cycle

As illustrated in Fig. 8.2, oceans constitute the largest carbon reserve on Earth. In terrestrial ecosystems, the soil organic matter pool contains about four times more carbon (about 2000Pg C; $1 Pg = 10^{15}$ g) than is stored in the plant biomass (500Pg C). Globally, there is more carbon stored in soil organic matter and living biomass than there is carbon in the atmosphere (760Pg C). The fossil carbon stocks contain an additional 6562Pg C. The burning of fossil fuel releases CO_2 into the atmosphere at a rate of 7.2Pg C/year. This has stimulated autotrophic carbon fixation in terrestrial ecosystems and oceans, which partially offsets CO_2 emissions but cannot adsorb all of the CO_2 released from fossil fuels. The net effect is an input of 4.1Pg C/year to the global carbon cycle that perpetuates the increase in the atmospheric CO_2 pool.

The carbon pools and fluxes in a terrestrial ecosystem are shown in Fig. 8.3. Trees assimilate CO_2 from the atmosphere every year via photosynthesis. Some of the carbon remains in the tree, but residues enter the soil every year (dead leaves, branches and twigs, as well as dead roots). As the dead residues are decomposed, nutrients are released into the soil solution to permit further plant growth through the process of mineralization. Some CO_2 is released back to the atmosphere due to the metabolic processes of soil fauna and plant respiration. The rest of the residues remain in the soil, where they are transformed into humus and enter the soil organic matter pool. Every year, some CO_2 is released from the soil organic matter pool, which is susceptible to decomposition, although at a slower rate than fresh organic residues.

8.2 Decomposition of Organic Residues

The action of microorganisms and larger fauna in the soil foodweb leads to the decomposition of all types of organic residues. By definition, decomposition is a process through which soil microorgan-

©CAB International 2010. *Soil Ecology and Management* (J.K. Whalen and L. Sampedro)

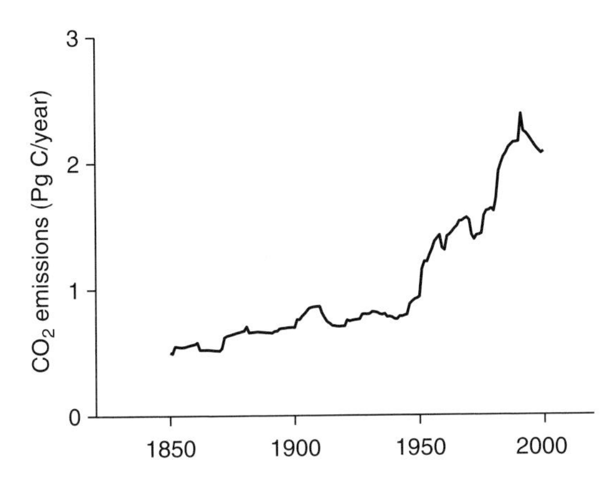

Fig. 8.1. The increase in global CO_2 emissions due to land use change from 1850 to 2005. Values are in petagrams (1 Pg = 10^{15} g). In the 1990s, the annual increase in CO_2 emissions was 1.6 Pg C/year on average (estimates ranged from 0.5 to 2.7 Pg C/year). (Data from: Houghton, 2008; CDIAC (http://cdiac.ornl.gov/trends/landuse/houghton/houghton.html), accessed on 4 January 2009.)

isms transform organic materials from identifiable plant, animal and microbial residues into CO_2, inorganic nutrients and humus. During the decomposition process:

1. Some of the carbon contained in the organic material is used for microbial metabolism and is released to the atmosphere as CO_2 through aerobic respiration and methane (CH_4) in anaerobic respiration. Gaseous nitrogen (NO_x, N_2O, N_2) by-products are released as a consequence of mineralization and denitrification, both of which require an organic carbon source for energy.
2. Some of the nutrients in the organic material are converted from organic forms to inorganic forms through mineralization. For example, organic nitrogen compounds, which cannot be absorbed by plants, are hydrolysed to release plant-available NH_4^+.
3. The remainder of the organic materials is condensed into degradation-resistant organic polymers. These resistant or recalcitrant materials adhere to soil particles and are considered to be a pool of humified soil organic matter, or humus. The humus decomposes very slowly, compared with fresh plant or animal residues.

The relationships between decomposition and important ecosystem processes in the atmosphere, terrestrial ecosystems and aquatic systems are shown in Fig. 8.4.

It is important to remember that not all the organic carbon entering terrestrial ecosystems is

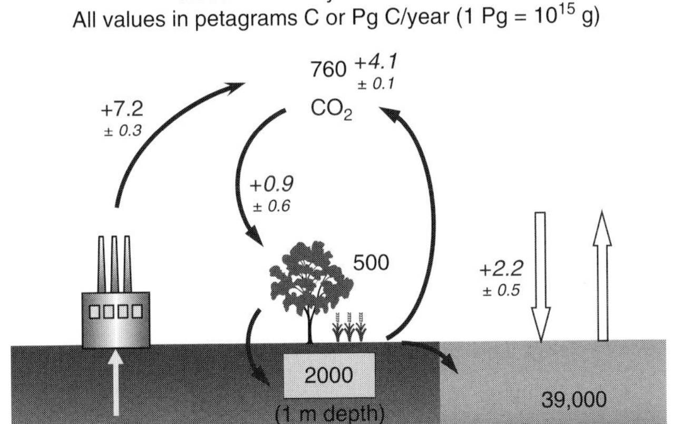

Fig. 8.2. The global carbon cycle for the 1990s. When fossil fuels are burned, they release CO_2 into the atmosphere at a faster rate than the CO_2 can be absorbed by terrestrial ecosystems or in the oceans, which is causing an increase in atmospheric CO_2 levels. Values in italics are estimates. (Source: illustration reproduced with permission from H. Janzen; data from IPCC, 2007.)

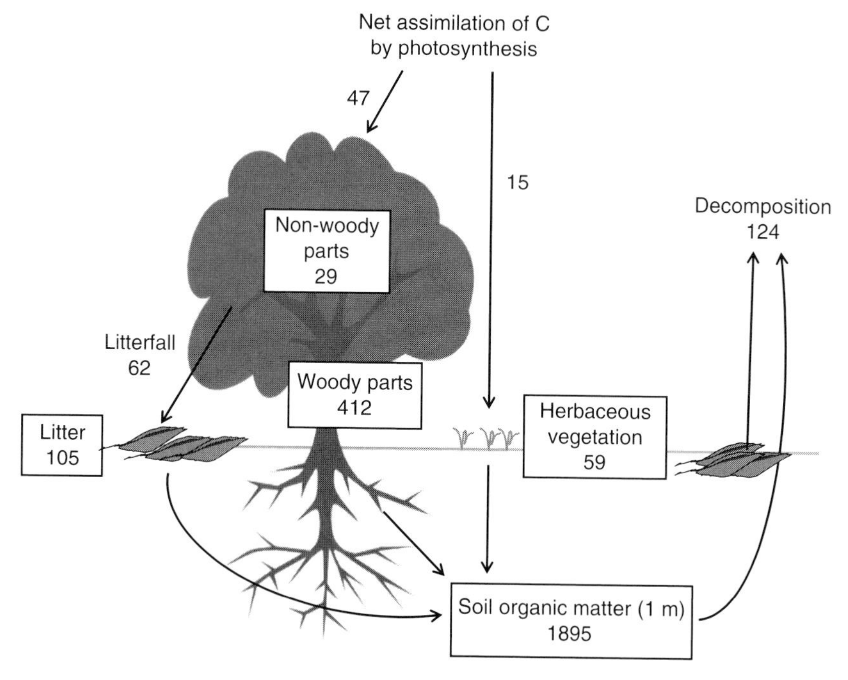

Fig. 8.3. Terrestrial carbon cycle showing fluxes and reservoirs of the various components. Photosynthesis results in net assimilation of atmospheric CO_2-C. The carbon input through photosynthesis and litterfall is balanced by the decomposition of plant residue and litter, and soil organic matter. The compartments and fluxes are indicated in Pg C or Pg C/year (1 Pg = 10^{15} g). (Source: concept from Post *et al.*, 1990, with data from IPCC, 2007.)

completely reduced to CO_2. The rest undergoes humification, which forms soil organic matter. The biochemical reactions that lead to decomposition, humification and mineralization are mainly catalysed enzymatically by soil microorganisms. Soil fauna have a limited ability to enzymatically decompose organic residues. Their major contribution to decomposition comes from grazing upon the soil microbial biomass (especially the microfauna) or by fragmenting and mixing organic residues into the soil, bringing them into more intimate contact with the decomposer microorganisms.

Various organic substrates enter the soil and undergo decomposition by the soil foodweb organisms. Root exudates and secretions such as sugars, amino acids and organic acids are readily absorbed by microbial cells and are of major importance for microbial growth in the rhizosphere. The more complex organic substrates in mucilages, mucigels, lysates and sloughed cells from plant roots can also be decomposed to provide energy and nutrients for rhizosphere microbes (Table 8.1). Microorganisms and microfauna (protists, in particular) are generally short-lived and can themselves be decomposed by succeeding populations. When an organism dies, its cells lyse and gradually disintegrate as the cell membranes break down, releasing carbohydrates, proteins, lipids and other substances that can be used by living microbial cells. Bacterial cell walls are composed of peptidoglycans, which are more easily decomposed than cellulose-rich or chitin-rich fungal cells.

In all terrestrial ecosystems, dead plant residues constitute a major input of organic substrates for the detrital foodweb organisms. In native prairie, the die-back of grasses and legumes at the end of the growing season results in a layer of dead plant debris (leaves, stems) at the soil surface. Considerable below-ground residue is also deposited each year from root death. In forest ecosystems, fine roots die each year and are available for decomposition by saprovorous organisms. Senescent leaves, branches, bark and other plant debris that fall to the soil surface accumulate in a

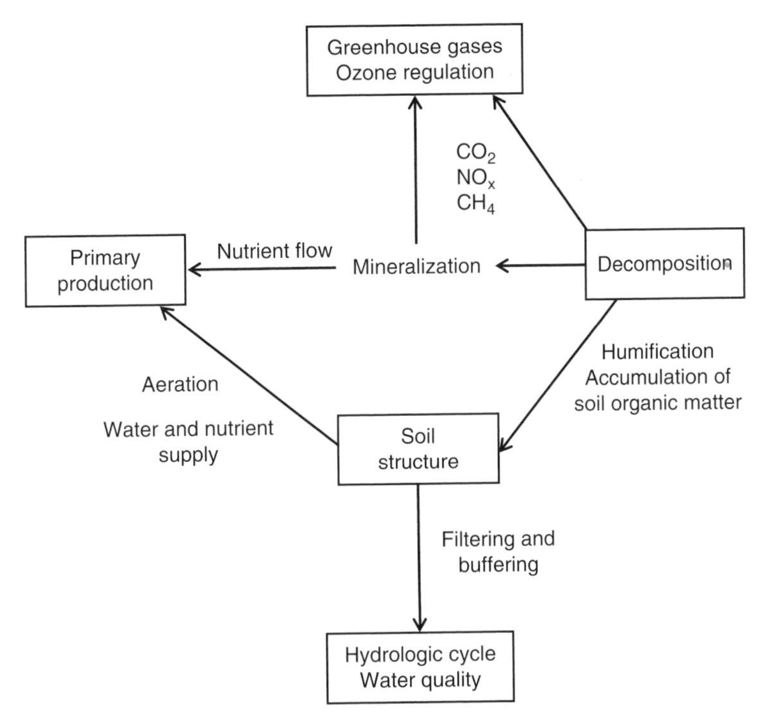

Fig. 8.4. Relationships between decomposition and major processes in terrestrial ecosystems. (Based on the concept of Lavelle and Spain, 2001.)

Table 8.1. Estimated carbon available to decomposers from root biomass and rhizodeposition by winter wheat from seedling emergence to flowering (maximum root biomass). (Data from Rees *et al.*, 2005.)

Plant component	Carbon input (kg C/ha)
Living root biomass	504
Exudates	285
Root cap cells	3
Sloughed root hairs	22

Table 8.2. Residue biomass input from above-ground (leaves, stems, etc.) and root components in selected agroecosystems. (Adapted from Johnson *et al.*, 2006.)

Agricultural crop	Grain yield (thousand kg/ha)	Above-ground residue (thousand kg C/ha)	Root residue (thousand kg C/ha)
Maize grain	8.92–12.2	3.12–4.28	4.01–5.50
Wheat	1.84–6.73	0.90–3.30	0.78–3.59
Soybean	1.55–2.92	0.73–1.37	0.81–1.52
Sorghum	1.69–5.96	1.53–2.68	1.33–3.04

thick layer. Fungal decomposition is very important for the initial breakdown of residues in native prairies and forests due to the fact that most fungi possess the enzymes necessary to degrade the complex, ligno-cellulosic materials. Once the residue has been partially degraded, actinomycetes and bacteria become involved in further decomposition reactions.

In agricultural systems, the unharvested remnants of crops are subject to decomposition (Table 8.2). This includes leaf, stem and root tissues not removed from the field. Depending on the tillage practices in the agroecosystem, the residues may be decomposed rapidly by a bacterial food chain or slowly by a fungal food chain. Conventional tillage involves ploughing and disking the soil to depths of 10–20 cm. This type of tillage stimulates decomposition by the bacteria and bacterivorous organisms in the soil foodweb because it fragments and mixes residues more thoroughly with the soil. Tillage tends to break apart macroaggregates and disrupt macropores, creating a soil matrix with more microaggregates, smaller pores and water availability that favours bacterial growth. At the same time, tillage disrupts fungal hyphae and can temporarily reduce fungal activity. Reduced tillage lightly incorporates crop residues to a depth of 5–10 cm, while no-till operations leave the crop residues on the surface, similar to the litter layers that develop in natural ecosystems. Residue incorporation in a no-till agroecosystem occurs through the action of soil fauna such as earthworms, which can fragment litter and redistribute it in the soil profile.

Organic residues of animal origin are added to some terrestrial ecosystems. Land application of biosolids (sewage sludge) and animal manure occurs primarily in agroecosystems. Not only does this practice permit recycling of nutrients contained in these wastes, but it can also increase the soil organic carbon concentration through two mechanisms: (i) biosolids and animal manure contain about 50% organic carbon, which can be decomposed and humified into the soil organic carbon pool by detrital foodweb organisms; and (ii) these nutrient-rich materials can promote crop production. This increases the amount of grain, oilseed or hay harvested from the agroecosystem, and leaves behind more crop residues after harvest.

The decomposition process

Decomposition involves the chemical breakdown of complex organic compounds into simpler compounds (Fig. 8.5). Plant residues and other complex compounds can be considered as the 'fuel', and soil microorganisms the 'engine' that transforms the fuel into simpler compounds and emits gases (primarily CO_2, although other gases can be generated in anaerobic decomposition). Decomposition occurs quickly in well-aerated soils (aerobic conditions), and more slowly in anaerobic (waterlogged) soils. In most soils, the aerobic pathway is of greatest importance.

When plant residues enter the soil, there is an initial flush of decomposition followed by slower, steady breakdown. The 'burst' of activity occurs as microorganisms compete for simple substrates such as soluble sugars, organic acids, peptide fragments and amino acids. For complex and resistant substrates such as carbohydrate polymers, extracellular enzymes from one or more microbial populations are needed to break the polymer into simpler monomeric compounds. The final step is the assimilation of the decay products into microbial cells, where they are further oxidized with intracellular enymes to obtain energy and carbon for growth, metabolic activities and reproduction. In aerobic metabolism, the respiration by-products are CO_2 and water. Anaerobic metabolism produces CO_2, methane, alcohols, organic acids, ammonia and hydrogen gas (Fig. 8.5). Of these two processes, aerobic metabolism is the more efficient at converting carbon substrates into energy and biomass.

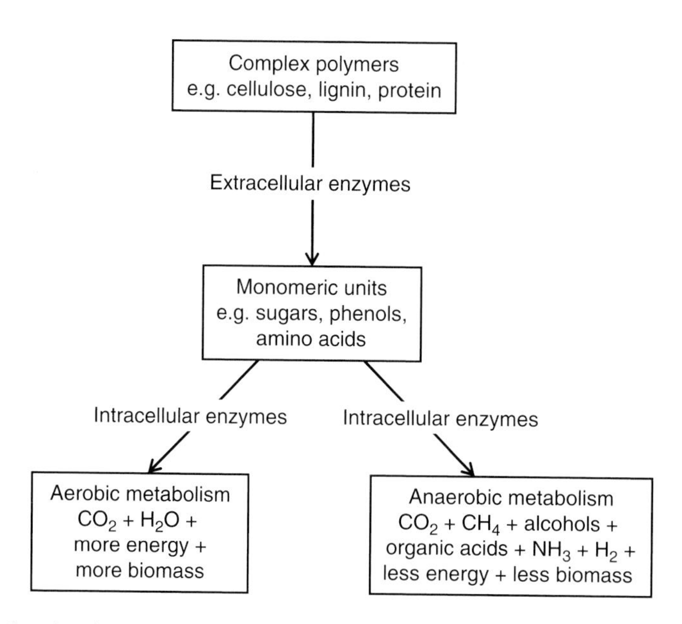

Fig. 8.5. General reactions in microbial decomposition under conditions favouring aerobic and anaerobic processes.

About one-half to two-thirds of plant residues entering the soil are decomposed in 1 year. Decomposition begins even before the residues hit the ground. Studies of fungal succession have emphasized the persistence of endophytic and epiphytic phyllosphere fungi from live leaves to freshly fallen leaves and their frequent occurrence in the early stages of decomposition when litter mass loss and chemical changes take place most rapidly. Being the first colonizers, phyllosphere fungi have the advantage of gaining access to readily available organic compounds in freshly fallen leaves, before the soil-inhabiting microorganisms that colonize after litter fall. After the initial flush or 'burst' of activity from the phyllosphere fungi and early colonizers (fungi, bacteria and actinomycetes), the material is subject to slow, steady decomposition and gradually becomes chemically and physically stabilized, which limits further microbial attack (Fig. 8.6).

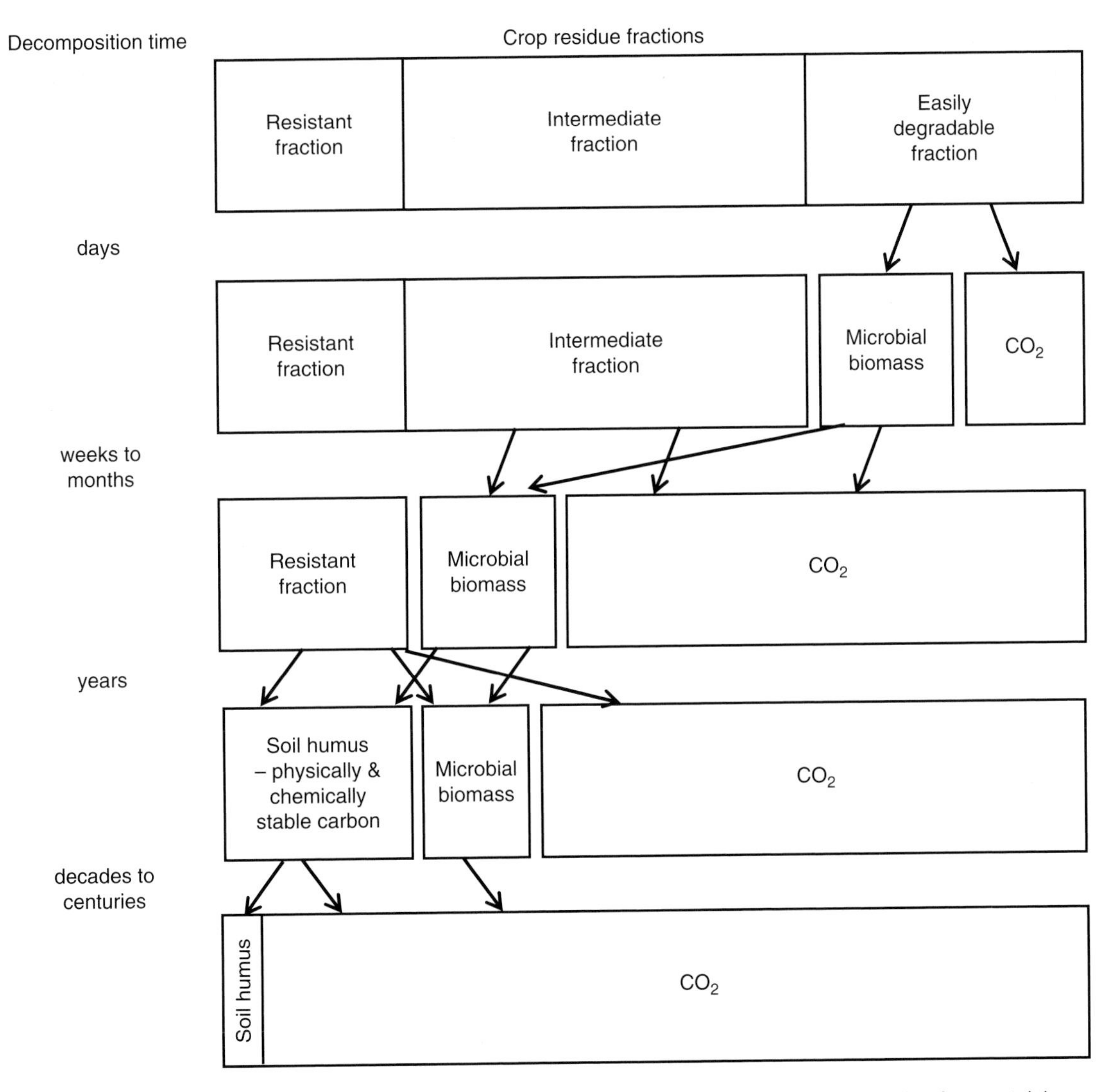

Fig. 8.6. Decomposition of crop residue after incorporation in soil. The crop residue can be thought of as containing easily degradable, intermediate and resistant fractions of carbon, based on its chemical composition. Changing box sizes indicate the relative quantities of carbon in the various fractions as decomposition progresses toward completion, which can take decades or centuries in temperate ecosystems. (Modified from Wagner and Wolf, 1998.)

Chemical stabilization of residues results from the depletion of easily decomposable substances and the accumulation of more resistant materials. Some resistant chemical compounds are also generated during the decomposition process, by-products of microbial metabolism and complex carbon compounds that were formed through polycondensation reactions and other processes. Eventually, the undecomposed residues and other by-products are considered to become soil humus, which can persist in the soil for many decades or even centuries before it is eventually decomposed to CO_2 and released to the atmosphere (Fig. 8.6).

Physical stabilization occurs through bioturbation when organic residues are consumed and mixed with soil by the mesofauna and macrofauna. Lacking teeth, these organisms consume soil mineral particles to aid in the grinding and digestion of organic substrates. As the material passes through the digestive tract, some is hydrolysed to provide energy to the animal, while the undigested organic matter binds to clay particles, creating macroaggregates that are deposited on the soil surface and throughout the soil profile. Organic matter inside the aggregate is not easily accessible and thus physically protected from decomposition by microorganisms. Earthworms also mix organic residues that they do not consume in the soil profile, which is especially in reduced tillage and no-tillage fields. In ploughed fields, mechanical tillage is probably more important than earthworm activity for fragmenting and mixing organic residues with soil particles.

Focus Box 8.1. Detritus as a resource for microbial growth: the importance of extracellular enzymes and their hydrolytic activity.

Organic carbon in detritus, the dead tissues of plant, animal and microbial origin at the soil surface and within the soil profile, is the essential resource for the decomposer community. With the exception of organisms that are associated directly with living plants (e.g. symbiotic bacteria and mycorrhizae), most of the biota in the soil foodweb depend upon detritus as the source of the energy, reduction power, carbon and nutrients.

Three main characteristics determine the quality of the organic matter for decomposers: (i) the type of chemical bonds between the carbon atoms determines the energy liberated when bonds are broken; (ii) the molecular size and three-dimensional complexity of carbon compounds; and (iii) the content of inorganic nutrients associated with the organic matter. The ratio between energy and nutrients is extremely important for microbial growth, because a very energetic substrate is not useful if the necessary nutrients and co-factors for cell growth and division are not available.

The breakdown of molecular bonds in complex carbon compounds is achieved by microbial enzymes. Enzymes are catalytic proteins; in terms of the energy balance in a microbial cell, the cost of their synthesis should be lower than the gain obtained when energy and nutrients liberated from organic matter become available for microbial growth. The maximum efficiency is achieved when microbial cells absorb small organic compounds such as simple sugars, with labile and energetic carbon bonds, that can be transported through the extracellular membranes with minimal energy expenditure and hydrolysed by enzymes within the cell. Such organic compounds constitute an extremely small part of the carbon stored in the detritus of soils and sediments. Most of the energy reserves are in large organic polymers, which must be attacked first with extracellular enzymes secreted from the microbial cell into the soil environment.

Secretion of extracellular enzymes is a 'high-risk investment' for microbes: (i) biosynthesis of these enzymes is energetically expensive, as they are typically large and complex; (ii) the target of those enzymes is usually a large organic polymer with complex tertiary structure, which presents a challenge for substrates to enter the enzyme's active site. Commonly, the joint action of a diverse array of enzymes is required to depolymerize the three-dimensional organic complex, converting sections into linear chains that can then be acted upon by enzymes that break specific bonds; and (iii) it is possible that some proportion of secreted enzymes do not reach their objective – hydrolysis – due to physical interference with the soil matrix, insufficient enzyme or substrate concentration, suboptimal pH or negative chemical interactions with soil particles or other molecules in the soil solution.

Despite these barriers, it has been shown that extracellular enzymes are readily stabilized in soils, binding to clay and organic matter surfaces. As long as the active site is not blocked, these enzymes remain capable of hydrolysing organic substrates in the soil microenvironment some time after they were secreted, even longer than the lifespan of the microbe which produced them.

Continued

Focus Box 8.1. *Continued*

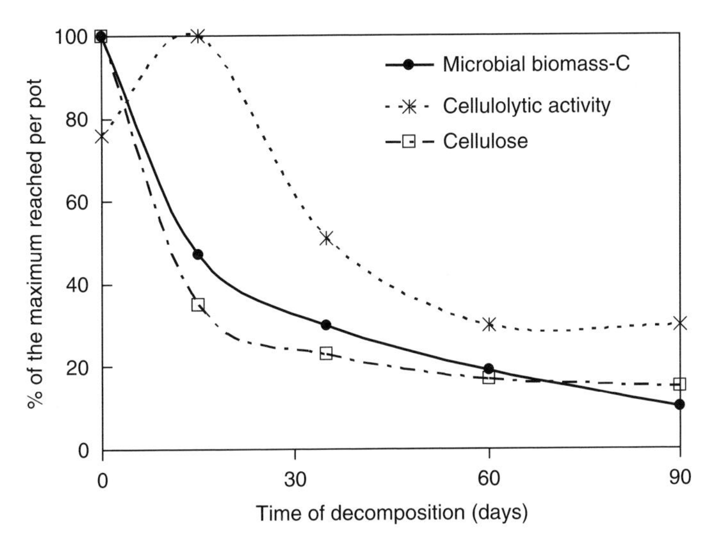

Fig. 8.7. Evolution of microbial biomass carbon, the activity of extracellular cellulolytic enzymes in the substrate and the cellulose concentration during decomposition of lignocellulosic agricultural wastes. Maximum cellulolytic activity was observed about 15–20 days after the microbial populations reached their maximum biomass. (Source: Sampedro, 1999.)

Incompletely decomposed organic residues, microbial exudates, dead fungal hyphae and plant roots attract clay particles and can be transformed into microaggregates, which physically protect the organic matter from further decomposition. The contribution of soil organisms to aggregate formation and soil structure is discussed further in Chapter 10. We should remember that physical stabilization is not permanent, as aggregates will eventually break down due to land management or through natural pedological processes over a period of decades. When this occurs, the organic matter within aggregates is susceptible to decay and CO_2 release to the atmosphere (Fig. 8.6).

As mentioned, there is considerable diversity in the organic substrates available for decomposition: root exudates and sloughed-off plant root cells, lysed microbial biomass, plant residues, animal manure and soil organic matter. The decomposition rate of these substrates is affected by multiple factors, including:

- Climate: temperature and rainfall exert a very important control on soil microbial activity, as discussed in Part II. When the soil temperature and moisture conditions are ideal for microbial activity, decomposition will proceed quickly.
- Soil conditions: the optimal soil pH for bacteria-mediated decomposition is 6.5 to 8.0, whereas fungi function best at soil pH between 5.5 and 6.5 (see Part II). Soil bacteria have a greater tolerance to flooded conditions than fungi, and are probably responsible for most decomposition that occurs in anaerobic soils. Fungi contribute principally to aerobic decomposition. Soil texture affects the habitat available for soil microorganisms and the amount of predation on soil microbial communities by larger foodweb organisms. Besides carbon, soil microorganisms require nitrogen, phosphorus, potassium and other nutrients to sustain their growth. Insufficient quantities of any essential nutrient could limit microbial activities.
- Biotic factors: the size, diversity and activity of the microbial community affect the decomposition rate, as well as interactions between soil microorganisms and larger soil biota. Some of these interactions may accelerate decomposition (i.e. predation of bacteria by protists,

competition for organic substrates among soil microorganisms, bioturbation and fragmentation of organic residues by earthworms).

- Characteristics of the organic substrate: the physical size of organic residues is critical: small, fragmented residues have a greater surface area for microbial colonization and decomposition than large, coarse residues. Thus, fragmented or chopped organic residues will decompose more quickly than large, intact residues. The chemical composition of organic substrates (type of chemical bonds) and their three-dimensional structure are of particular importance and will be discussed in detail below. Simple substrates such as glucose and amino acids can be readily absorbed through cellular membranes and completely metabolized by soil microorganisms within a few days. More complex substrates must be depolymerized and partially degraded by extracellular enzymes before they are sufficiently small to pass through the microbial cell membrane. Lysed microbial cells containing carbohydrate polymers and proteins can be decomposed readily, whereas chitin-rich fungal cell walls are more difficult to break down. In plant residues, the lignin content and C:N ratio are considered good indicators of the relative speed of decomposition.
- Agricultural management: tillage (ploughing, harrowing, disking and roto-tilling) will fragment and mix the residues with the soil, speeding the rate of decomposition. Agricultural producers control the amount of organic residues that enters the agroecosystem because they set combine harvesters to cut the crop at a certain height (roots and straw that are not removed from the field will eventually be decomposed by soil foodweb organisms). The decision to leave straw in the field after harvesting a grain crop or to remove it from the field for other uses (e.g. animal bedding) is a decision that affects the quantity of organic residues entering the agroecosystem. The time of year that the residue is added or incorporated into the soil also affects the decomposition rate. When residues are mixed with the soil in the autumn, some decomposition occurs during the frost-free period and from snowmelt to spring cultivation and planting, whereas spring-applied residues have a much shorter decomposition period before crops are planted. Burning of surface residues should be strictly avoided in temperate regions since this practice leads to nutrient loss from the agroecosystem and removes resources needed to sustain the soil foodweb. Even in tropical regions where slash and burn agriculture is common, some producers have switched to a slash and mulch system that crushes and fragments the vegetation and allows natural decomposition and nutrient recycling to occur. Microbial inoculants can be added to speed up decomposition, but it is generally more practical to use machinery to chop/fragment the organic residues, since the naturally occurring soil foodweb organisms are capable of decomposing organic material as long as it can be colonized by microorganisms or consumed by soil fauna prior to microbial colonization. If decomposition would be limited by a lack of nitrogen or other essential nutrients, producers can add animal manure or nitrogen-rich organic residues such as green manure to support microbial growth as well as plant growth.
- Pasture management: the stocking density of grazing animals affects the amount of organic residues that is left as residue in grasslands. Animal manure deposited by grazing animals and from other sources (biosolids and manure slurry applications) can accelerate the decomposition rate. In some areas, grasslands are irrigated with wastewater from municipalities, food-processing facilities and other industries, which contains nutrients and affects soil water availability. Residue biomass and chemical composition is different in improved grasslands that have been over-seeded with legumes and non-native grasses from native grassland. Fire started by natural causes or by humans affects decomposition: fire removes surface residues and leaves behind ash with a high concentration of some essential nutrients (P, K, Ca and Mg). The regrowth of prairie grasses may lead to a stimulation of rhizodeposition.
- Forest management: the frequency and extent of forest harvesting (e.g. clearcutting versus selective cutting) affects the depth of the litter layer as well as the amount of litterfall and root biomass that is available for decomposition in a forest. These factors are also affected by the diversity of trees and understorey plants that have regenerated naturally or been planted deliberately, as is the case in forest plantations. The chemical composition of residues is affected by these forest management practices,

by the age of the forest at harvest because older woody species tend to have more recalcitrant structural carbon compounds than younger plants, and by the dominant trees present (coniferous trees have more recalcitrant substances in their litter than deciduous trees). Fire started by lightning strikes or by humans is another factor that affects decomposition for the same reasons as above.

8.3 Chemical Composition of Organic Residues

As shown in Fig. 8.6, the chemical characteristics of organic residues are quite important to consider when estimating the decomposition rate. Some chemical components are readily decomposed and others are far more difficult to decompose. The major components of plant residues are shown in Table 8.3.

The carbon content of plant residues establishes the total carbon contained in a residue, which is useful in calculating the carbon balance in an agroecosystem: carbon input from crop residues – carbon output from CO_2 emissions. Nitrogen in plant residues is generally found in proteins, polymers of amino acids linked by peptide bonds (Fig. 8.8). All soil bacteria, fungi and actinomycetes produce extracellular proteolytic enzymes such as proteases or peptidases that can hydrolyse proteins into amino acids. The first enzymatic breakdown must be done outside the microbial cell because the protein molecule is too large to pass through the outer cell membrane of these organisms. Following this, amino acids are actively absorbed through the outer membrane and transported into the cell. The amino acids undergo glycolysis, which releases energy and nitrogen for cellular metabolism.

Ash is the mineral fraction of the plant residue, containing essential nutrients such as potassium, calcium, magnesium and micronutrients, as well as trace metals that are not essential for plant growth. In Table 8.3, the soluble sugars include simple sugars and starch. These compounds are produced during photosynthesis and stored in the amyloplasts as an energy reserve for periods when photosynthesis is not possible (e.g. night) or when photosynthetic rates are low (e.g. on overcast days). Other soluble components in plant residues include free amino acids and organic acids. Bacteria, actinomycetes and fungi can absorb sugars, free amino acids and organic acids through their cellular membrane and begin to use them for metabolic processes immediately. The Zygomycetes ('sugar fungi') and many other microorganisms grow preferentially on sugars. Starch exists as a linear or branched polymer of glucose. The linear polymer is known as amylose and is broken down by the extracellular enzyme amylase into the disaccharide maltose. After it is transported into the microbial

Table 8.3. Chemical composition of leaves collected from maize (*Zea mays* L.), soybean (*Glycine max* (L.) Merr.), lucerne (*Medicago sativa* L.) and switchgrass (*Panicum virgatum* L.). (Modified from Johnson *et al.*, 2007.)

Component[1]	Maize	Soybean	Lucerne	Switchgrass
Carbon	416	439	454	445
Nitrogen	14	16	44	13
Ash[2]	77	2	1	29
Soluble sugars[3]	34	32	72	36
Cellulose	340	207	133	418
Hemicellulose	461	245	177	498
Lignin	132	114	65	96

[1]Values are given as g/kg dry mass.

[2]Substance remaining after ignition of the plant material at 550°C (contains mineral nutrients).

[3]Glucose, fructose, sucrose and starch.

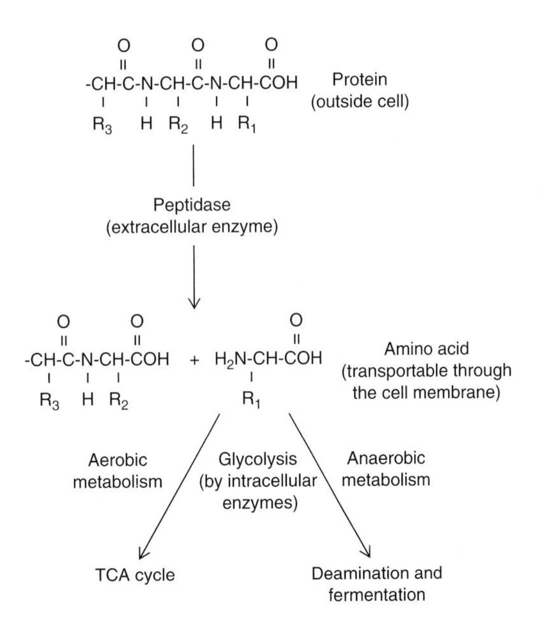

Fig. 8.8. Degradation of protein by hydrolysis of the peptide bond between O=C and N–H.

cell, maltose is hydrolysed by α-1-4 glucosidase (maltase) to yield glucose for glycolysis. Branched starch polymers known as amylopectin can be degraded by β-amylase, glucoamylase and pullulanase, yielding simple sugars such as dextrins, maltose and glucose within a few days to a week after bacteria and fungi colonize the plant residue.

The other plant components are carbon-rich and an important source of energy for soil microorganisms. Cellulose is the most common structural polysaccharide in plants and provides structural rigidity and strength to the plant tissues. It is composed of linear chains of glucose units joined by β-1-4 linkages (up to 10,000 glucose units in a cellulose molecule). The cellulose content in leaf tissue of agricultural crops can range from 13 to 42% of the plant biomass (Table 8.3). Cellulose is also an important component of some fungal and algal cell walls.

Cellulose is insoluble and too large to enter the microbial cell, so it must broken into smaller compounds that can be transported through the cell membrane. There are three steps in cellulose decomposition. The first two steps involve depolymerization of the three-dimensional structure, eventually transforming the three-dimensional structure into a long linear chain. The cellulase enzymes involved in these steps are the extracellular enzymes β-1,4-endoglucanase (attacks internal bonds in the three-dimensional structure) and β-1,4-exoglucanase (attacks external bonds in the three-dimensional structure). Cellulase enzymes function in both aerobic and anaerobic conditions. Only a few soil organisms produce cellulases, including the soil fungi *Trichoderma, Aspergillus, Penicillium* and *Fusarium*, as well as soil actinomycetes and bacteria belonging to *Streptomyces, Pseudomonas* and *Bacillus* species.

The cellobiose molecule liberated by these enzymes can be transported through the outer cell membrane of soil microorganisms. Once inside the cell, this molecule is further hydrolysed to glucose by β-1,4-glucosidase, which is then used as an energy source in microbial metabolism. Virtually all soil microorganisms produce this enzyme and can obtain glucose from cellobiose (Fig. 8.9).

Hemicellulose is the second most common carbohydrate constituent in plant residues, accounting for 18–50% of the biomass in agricultural residues (Table 8.3). The most common hemicelluloses in plant residues are pectin and xylanes. Pectin molecules are deposited in the middle lamella of the primary cell wall, helping to cement and stabilize neighbouring cells, facilitating the transfer of materials from cell to cell. The fibrous hemicellulose xyloglucan coats cellulose microfibrils and binds to other xylanes via hydrogen bonding. This provides structural stability and rigidity to the plant cell.

The hemicellulose polymer contains hexoses (6-C sugars), pentoses (5-C sugars) and uronic acids. A major difference between cellulose and hemicellulose is that COOH is substituted for CH_2OH on the 5-C position of the 6-C sugar. This substitution creates a uronic acid subunit, a branched polymer. Hemicellulose is therefore more complex structurally than the long linear chains of glucose that make up a cellulose molecule. As an example, the decomposition of pectin by extracellular enzymes called pectinases cleaves the polymer into 6-C galacturonic acid subunits that can be transported into the cell to be used for glycolysis.

Many of the same fungi and bacteria involved in cellulose decomposition are important for decomposing hemicelluloses. Fungi are the most important, and 'brown rot' fungi being especially well known for their role in hemicellulose and cellulose decomposition. The action of 'brown rot' fungi results in a decline in the hemicellulose and cellulose content of residues; the resource becomes proportionately richer in lignin and phenolic materials that cannot be decomposed by these organisms. Fungal pathogens, symbiotic N_2-fixing bacteria, associative N_2-fixing prokaryotes and mycorrhizal fungi also produce pectinases, which allows them to break down cell walls and enter the plant root and other plant organs.

The decomposition of hemicellulose in plant residues is generally more rapid than the cellulose decomposition rate. This may be due to the fact that hemicellulose polymers are relatively small (50–200 sugar units) compared with cellulose polymers. The biochemical reactions leading to the breakdown of hemicellulose are similar to those shown for cellulose; the initial depolymerization of the molecule occurs outside the cell. The breakdown products (monomeric sugar and uronic acid) are transported into the cell for subsequent metabolism.

Lignin is the most resistant material to decomposition. This component is deposited in the primary and secondary cell walls of plants containing vessels (e.g. xylem) and provides structural strength by reducing elasticity, and increasing hardness and tensile compaction. Lignin also makes the cell impermeable to water. It is not as abundant in annual plants

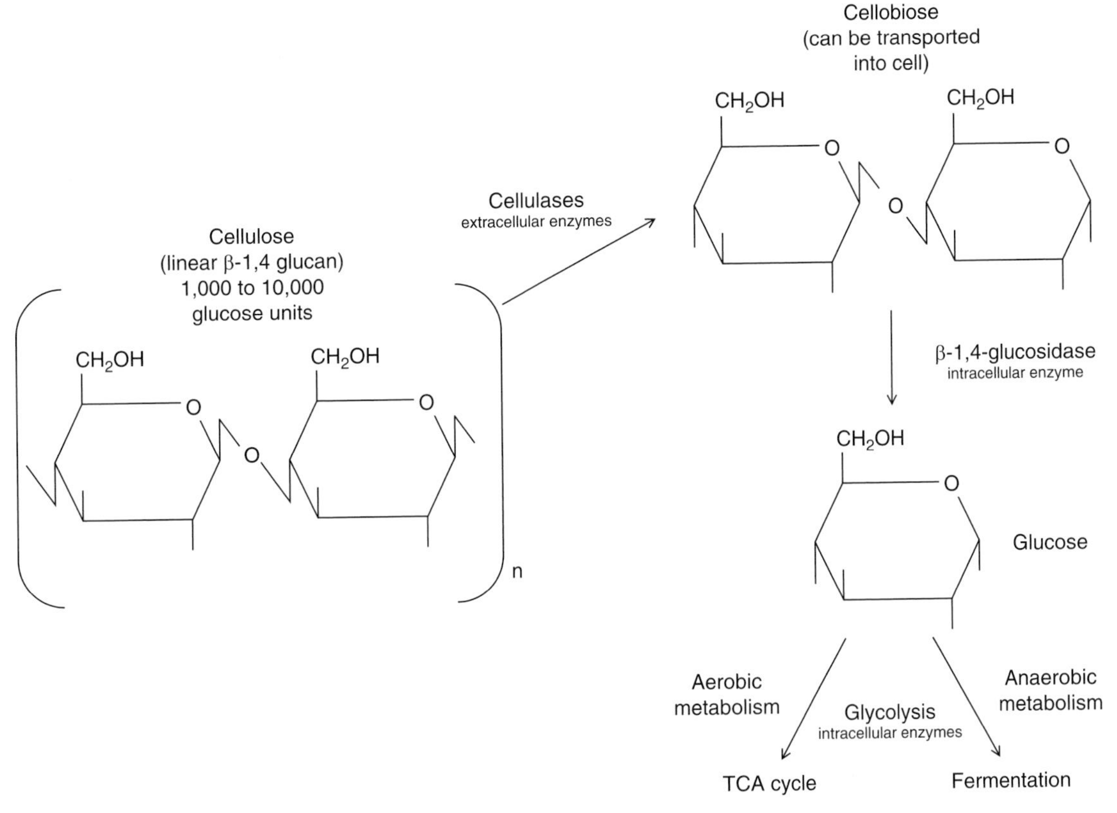

Fig. 8.9. Decomposition of cellulose under aerobic and anaerobic soil conditions. Where no side-bonded group is indicated, the carbon in the glucose ring structure is bonded to a hydroxyl group (–OH).

as cellulose and hemicellulose. The lignin content of plants increases with age, and the chemical composition varies among plant organs and species. During late vegetative growth, maize plants contain about 2–3% lignin in the leaves and stems, and 6–7% lignin in the roots. Maize tissues taken at maturity contain 10–11% lignin in leaves, stems and roots (Johnson *et al.*, 2007). Woody plants (shrubs and trees) may contain more than 35% lignin.

The basic building block of lignin is a phenylpropene unit with a hydroxylated 6-C aromatic benzene ring (phenol) and a 3-C linear side chain. The three most important starting compounds are coumaryl alcohol, coniferyl alcohol and sinapyl alcohol (Fig. 8.10). The lignin of ferns and their relatives (pteridophytes) consists mainly of coniferyl alcohol polymers. In herbaceous dicots, the secondary cell wall contains coniferyl and sinapyl alcohol in roughly equal amounts. However, angiosperms contain a mixture of guaicyl and sinapyl lignin, while coniferous gymnosperms of the Pinaceae have mostly guaicyl lignin and no sinapyl lignin. Coumaryl alcohol is found in trace amounts in plants. This is because the starting materials undergo polymerization to form *p*-hydroxylphenyl and guaiacyl alcohols, which are deposited in the primary cell wall. During secondary wall development, coniferyl and sinapyl alcohol are co-polymerized to form a mixture of guaiacyl and syringyl lignin units in monocots. Other polymerization reactions lead to the formation of vanillyl and cinnamyl phenols.

Lignin generally contains 500–600 phenylpropene units with randomly condensed units. The structure can be very complicated because the randomly condensed units can be linked through C-C bonds or C-O-C bonds. In addition, the binding can occur through ring–ring linkages, as side chain–side chain linkages or as ring–side chain

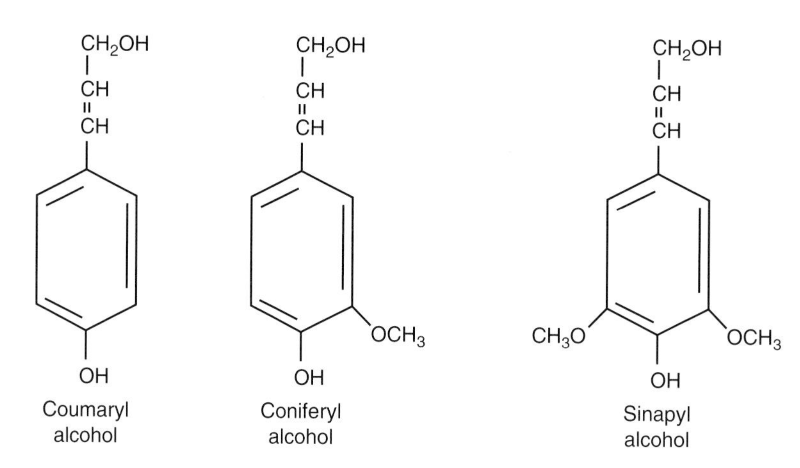

Fig. 8.10. Examples of the phenylpropene units that are the starting material for lignin formation in vessel-containing plants. (Modifed from Wagner and Wolf, 1998.)

linkages. This leads to a great deal of variation in the types of lignin that are deposited in plant cell walls.

Lignin decomposition is complex, but may proceed as follows:

1. Exposed side chains are hydrolysed by phenol-oxidases and removed from the main lignin molecule.
2. Depolymerization results in the liberation of individual phenolic units with side chains.
3. Individual phenolic units can be transported inside microbial cells.
4. Side chains are removed; if hydroxyl groups exist on adjacent C atoms in the phenolic ring, the ring can then be cleaved.
5. Once the ring is cleaved, it is converted to a straight-chain C compound that can be used in the TCA cycle.

The first two steps in this process occur outside the microbial cell. Fungi, especially Basidiomycetes (white-rot fungi), are the major microbial group responsible for lignin degradation. Soil actinomycetes (*Streptomyces*) and bacteria are also involved in some of the hydrolysis steps.

Plant cells produce small quantities of other recalcitrant substances that will be decomposed slowly when residues are incorporated in the soil. Resin is produced especially by trees to block injuries and reduce the risk of herbivory or invasion by pathogens. Suberin is deposited near the vascular tissue of roots to regulate ion and water movement in and out of the plant (e.g. in the Casparian strip of the endodermis) and between neighbouring cells (e.g. in the suberin lamella of primary cell walls). Suberin is also present in the above-ground organs of woody species. Cutins are deposited in the periderm and cuticle of shoots and leaves, making the outer layer impermeable to water. This waxy material serves as a cuticle, guarding the integrity of dead, mature cells that protect and support the living plant tissues. Callose is a β-1,3-glucan that encloses the sieve plates of sieve elements upon wounding. This substance appears to be a barrier that prevents cell loss upon wounding and blocks foreign substances from entering the plant. Callose also develops in the pollen tube and protects the developing pollen grain before it can be coated by sporopollenin, a hydrophobic substance polymerized from carotene molecules. Without the protective sporopollenin, the gametes could be degraded and the reproductive success of the plant will be diminished. Other materials contained in or produced by plant cells include low-molecular weight compounds (dyes, alcohols, terpenes, tannins, etc.), oligosaccharides, proteins (usually glycoproteins) and mucilaginous substances, which vary in their relative susceptibility to decomposition.

In summary, plant residues contain chemical components that can be relatively easy to decompose by all soil microorganisms, or relatively difficult to decompose (only a few soil microorganisms have the ability to break down these materials).

Focus Box 8.2. Litter decomposition dynamics.

Litter decomposition is a continuous process that can be divided into five major stages.

1. In the first hours to days after organic residues fall to the soil surface, loss of matter is governed by the dissolution and leaching of inorganic nutrients and small organic substances to deeper layers of soils or by runoff. Simple phenolic compounds, sugars and amino acids are susceptible to leaching and dilution by rainfall and soil water. Up to 30% of the mass of green leaves is soluble in cold water. This fraction moves to deeper soil layers where it can be assimilated by microbes or plant roots.
2. The very first degradative biochemical reactions, particularly in dead animal residues, are due to the activity of the enzymes released after the breakdown of cell membranes. The products of these autolytic reactions in recently dead tissues produce simple, easily mineralizable substances also susceptible to leaching through the soil profile. Processes in (1) and (2) could be responsible for as much as 20% loss of fresh litter within the first days, especially after intense rainfall events.
3. During the first stages of decomposition, simpler constituents (water-soluble compounds, mostly simple sugars and peptides) are assimilated and respired. Next, more complex compounds begin to be attacked. A generalized sequence of decomposition for water-insoluble constituents could be: simple sugars > starch > hemicellulose > pectin and proteins > unprotected cellulose > lignin and protected cellulose > suberin and cutins.
4. The physical and biochemical activity of detritivorous meso- and macro-fauna plays an extremely important role at these stages. The burial, mixing and comminution of the remaining litter by the detritivorous fauna liberate organic compounds that were protected within the physical structure; these compounds now become available for microbial attack. Detritivorous fauna assimilate only a small part (much less than 30%) of the energy they ingest, but directly and indirectly accelerate litter decomposition.
5. Depletion of simple substances leads to a shift from opportunistic and generalist microbial populations to specialist decomposers, which are able to take advantage of more complex compounds by producing extracellular enzymes to hydrolyse specific substrates. For instance, cellulolytic enzymes break down cellulose to smaller units; since many organisms can produce cellulases, this leads to relatively fast decomposition. Sometimes extracellular enzymes accumulate, leading to increased degradative activity in favourable microsites. The lignin-protected cellulose soon becomes the biggest energy reservoir in decomposing litter, as lignin breakdown depends on the activity of enzymes released by a few specialized soil fungi.

Composition and concentration of the different organic compounds in the decaying organic matter is modified as it is continuously mineralized to CO_2 by decomposers. There is a strong link between the quality of the organic matter, decomposer community structure and biomass, and time. As the availability of resources diminishes, the microbial biomass decreases in size. But the degradative activity (loss of carbon as CO_2) does not decrease at the same rate because the dominant microbial population shifts from fast-growing species to others in succession as the resource quality changes.

Once the recalcitrant compounds (lignin, suberin, etc.) are biochemically disrupted, other simpler compounds shielded within their three-dimensional structure become available for decomposers. Thus, we could find peaks of soluble compounds derived from the depolymerization of more complex substances. Opportunistic bacteria make use of all the intermediate compounds resulting from the processes described above to obtain carbon substrates for energy and consequently mineralize organically bound N, P and S.

Generalized model for litter decay

Empirical observations of litter decay (measured as loss of initial mass of organic matter (OM_{ini})) are usually close to a negative exponential function:

$$\% \text{ Remaining mass} = OM_t \,/\, OM_{ini} = e^{-k.t} \qquad [8.1]$$

Researchers have used exponential functions to model the empirical observations of organic matter decay in terrestrial ecosystems. Mathematically, the potential decomposition rate is a constant (k), and the observed mass loss is affected by the quantity of organic matter remaining (OM_t) after a period of time (t). Thus, the decomposition rate does not change through time. The exponential function predicts that organic matter loss occurs rapidly during the initial stages of decomposition, eventually reaching the lower asymptotic limit (Fig. 8.11).

The exponential function is not entirely accurate because it implies that all components of the organic matter decay at the same rate. As shown in Fig. 8.12, decomposition rates vary for the different fractions of organic matter due to their distinct chemical characteristics and susceptibility to degradation by enzymes of microbial origin. While the general exponential decay function does not describe the decomposition of each organic matter fraction, it is a good general model to predict litter decomposition.

Continued

Focus Box 8.2. *Continued*

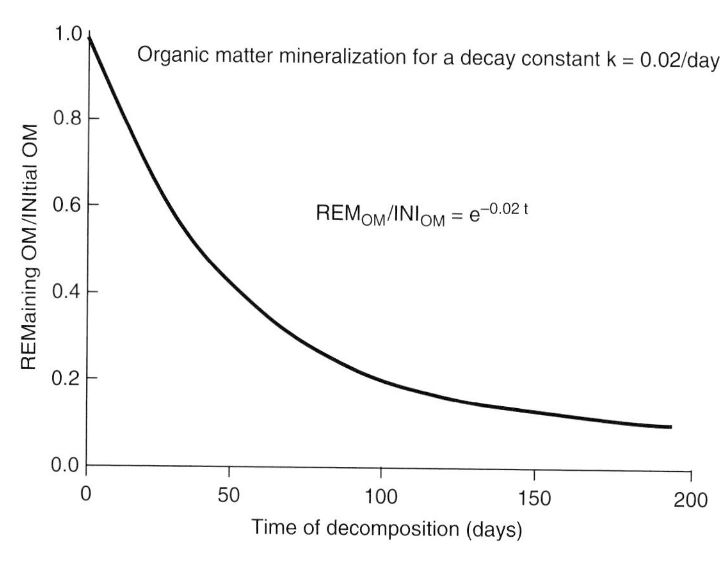

Fig. 8.11. General exponential function describing litter decay dynamics. In this model, decay constant is 0.02/day, which means that absolute loss of mass is as much as 2% of the remaining organic matter (OM) per day.

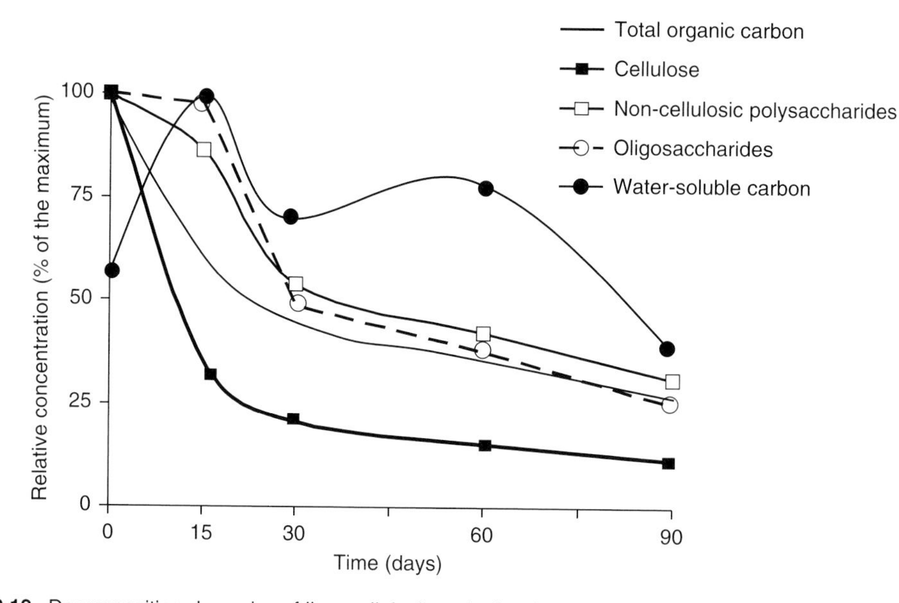

Fig. 8.12. Decomposition dynamics of lignocellulosic agricultural wastes, illustrating the changes in the relative concentration (relative to the maximum observed for each component) of organic matter fractions during a 90-day decay period. Decomposition of unprotected cellulose released simpler organic carbon compounds, particularly water-soluble carbon. (From Sampedro, 1999.)

8.4 Formation of Soil Humus

As shown in Fig. 8.12, readily decomposable substances such as cellulose are greatly diminished after 90 days of decomposition. Resistant compounds like lignin tend to be decomposed slowly in comparison, due to their size, irregular structure and the presence of non-hydrolysable bonds in the chemical structure. Thus, it has long been postulated that resistant lignin is converted into humic materials, organic matter with distinctive chemical properties (Table 8.4) whose origin can no longer be determined. The characteristic brown to black colour of detritus and soils is related to the concentration of humic materials, known collectively as humus.

Humus represents up to 80% of the organic matter contained in soils. By definition, the soil organic matter contains:

1. Residues that can be identified as originating from plants, animals or microorganisms.
2. Exudates (e.g. from plant roots).
3. Excreted materials and mucus secreted by soil organisms (e.g. earthworms).
4. Dissolved organic matter in the soil solution.
5. Humus: complex, carbon-rich materials that are highly resistant to decomposition.

Humus may be stored in the soil for decades, centuries and millennia (Table 8.4). In forest soils, the mean residence time of the soil organic carbon ranged from under 50 years to more than 3800 years (Table 8.5). Clearly, the formation of recalcitrant humus in soils is of great importance for carbon cycling. Due to the large size of the global soil carbon pool, a gain in stable, carbon-rich humic materials could reduce the CO_2 concentration in the atmosphere if all other carbon transformations remain constant. Therefore, understanding the mechanisms that control humus formation is essential for developing land management practices that allow soils to capture more carbon and offset global climate change.

Table 8.4. Characteristics of soil humic materials, compared with lignin.

Characteristic	Humic materials	Lignin
Colour	Black	Light brown
$-OCH_3$ (methoxyl) content	Low	High
C content (%)	50–58	~70
N content (%)	3–6	0
Carboxyl and phenolic hydroxyl	High	Low
Total exchangeable acidity (cmol/kg)	≥ 150	≤ 0.5
Amino N	Present	None
Vanillin content (%)	< 1	15–25
Mean residence time in soil (years)	500–1000+	15–25

Table 8.5. Quantity of soil organic carbon (SOC) and its age, as determined by ^{14}C radiocarbon dating, in two forest sites from northern Bavaria, Germany. (Modified from Marschner *et al.*, 2008.)

Soil type/ horizon	Depth (cm)	SOC (g/kg)	SOC age (years before present)
Podzol			
Oa	Surface detritus	340	n.d.*
EA	0–10	32	525
Bs	12–30	54	745
Bw	30–55	24	1570
C	55–70	2	3840
Cambisol			
Oa	Surface detritus	122	n.d.
A	0–5	44	< 50
Bw	5–24	6	< 50
Bg	24–30	1	655

*n.d.: not determined.

What evidence do we have that the humus comes from undecomposed plant lignin? There are two theories that would account for the transformation of lignin from plant residues into soil humus:

1. Plant Alteration Theory: lignin is incompletely decomposed by microorganisms and the residual compounds become part of the soil humus. Lignin is modified by hydroxylation and oxidation of $-OCH_3$ to $-COOH$, which reacts with amino compounds to yield humic acids. This is probably the major pathway for the formation of humic materials in poorly drained soils where peat is formed. Lipids originating from plant residues, soil organisms and organic fertilizers (including fats, waxes and resins) constitute about 3 to 20% of the soil humus stored in peaty soils.
2. Lignin-derived Quinone Theory: many bacteria possess the necessary enzymes to remove methoxyl side groups ($-OCH_3$) without altering the phenol

ring. The oxidation of the side chains and subsequent demethylation of lignin produces a polyphenolic compound. Next, the phenol undergoes enzymatic conversion with polyphenoloxidase enzymes to produce quinones, which condense with amino compounds to yield humus. This reaction is likely in aerobic soils, but it is not the only reaction pathway leading to humus formation.

Although lignin molecules can persist in soils for a few decades, they are not generally as stable as soil humus. The apparent lignin residence time in temperate grassland and agricultural soils ranges from 13 to 22 years (Table 8.6). The complete decomposition of a lignin molecule to CO_2 depends on its chemical structure – cinnamyl units decompose faster than syringyl units, which are decomposed more readily than vanillyl units – as well as the plant organ in which it is found. Rasse *et al.* (2005) provide evidence that lignin, suberin and other resistant compounds in plant roots decompose more slowly than recalcitrant molecules in plant shoots. However, lignin originating from plants does not appear to be stabilized for more than a couple of decades in soils, suggesting that lignin is not a major component of the soil humus.

There is growing evidence that much of the resistant soil humus is a by-product of microbial biosynthesis. Many bacteria produce polysaccharides, which may be capsular polysaccharides tightly associated with the cell surface or a slime that is released from the cell into the surrounding environment (extracellular polysaccharides). These polysaccharides may be composed of a single sugar monomer, but more commonly are composed of several types of monomers, as shown for *Rhizobium japonicum* (Table 8.7). In this N_2-fixing bacterium, the polysaccharides bind to a lectin (sugar-binding protein) receptor on the root surface of soybean, thus facilitating the first contact between the symbiotic bacteria and its host. Nearly all Gram-negative bacteria produce extracellular polysaccharides, with some bacteria investing more than 70% of their energy to coat the outer surface of their cells with capsular polysaccharides and secreting the remainder as slime. The major functions of bacterial extracellular polysaccharides are listed in Table 8.8.

Table 8.6. Lignin residence time in soils as determined from isotopic labelling and analysis of lignin monomers in temperate grassland and agricultural sites. (Data from Marschner *et al.*, 2008.)

Site	Crops studied	Lignin residence time (years)
Grassland	*Lolium perenne*	14
	Trifolium sp.	15
Bioenergy crop	*Miscanthus giganteus*	13
Agricultural	*Zea mays*	14–22

Table 8.7. Molar composition of the polysaccharides produced from *Rhizobium japonicum* strain I38 cultures, relative to a mannose concentration of 1.0 mol/l. (Modified from Mort and Bauer, 1980.)

Component	Capsular polysaccharides	Extracellular polysaccharides
Mannose	1.0	1.0
Glucose	2.35	2.30
Galacturonic acid	1.08	1.18
Galactose	0.22	0.32
4-O-Methyl galactose	0.78	0.68
Acetate	2.15	1.30
Ketodeoxyoctonate	trace	trace
Protein (by weight)	1–2%	1–2%

Proteins of microbial origin are also important in soil humus formation. Bacteria actively use proteins to attach to surfaces. Fungi secrete a broad class of surface-active glycoproteins, including glomalin, through their hyphae and spores. Tremblay and Brenner (2006) demonstrated that between 60 and 75% of the nitrogen in highly decomposed detritus was of bacterial origin, not from plants. Due to their small size, bacteria and fungi are in intimate contact with the soil minerals. Proteins on cellular surfaces or released from microbial cells come into direct contact with soil minerals, where they can be decomposed into amide compounds that are stabilized and stored as part of the humus fraction. Chitin (Fig. 8.13) is another N-rich substance that is a structural polysaccharide in fungal cells. When hyphae die, this compound may be decomposed by extracellular enzymes (chitinases) of fungal and bacterial origin. If the glucosamine products are not assimilated by a living cell, they can bind to soil minerals and become stabilized in the humus.

There are two theories that would account for the transformation of microbial by-products into soil humus:

Table 8.8. Major functions of bacterial extracellular polysaccharides. (Data from Weiner *et al.*, 1995.)

Function	Survival advantage(s)
Physical/protective barrier	Protection from desiccation and predation; provides immunity and resistance to toxins, antibiotics and poisons
Cell–cell recognition and interaction	Plant symbiosis (nodulation); formation of microcolonies; binding to protein receptors in the intestinal tract of invertebrate and vertebrate hosts
Response to environmental stress	Sequestration and import of charged ions into bacterial cell; creates a reducing environment around cell for redox reactions
Adhesion and biofilm formation	Immobilization on to nutrient-rich surfaces; dissociation from nutrient-depleted surfaces

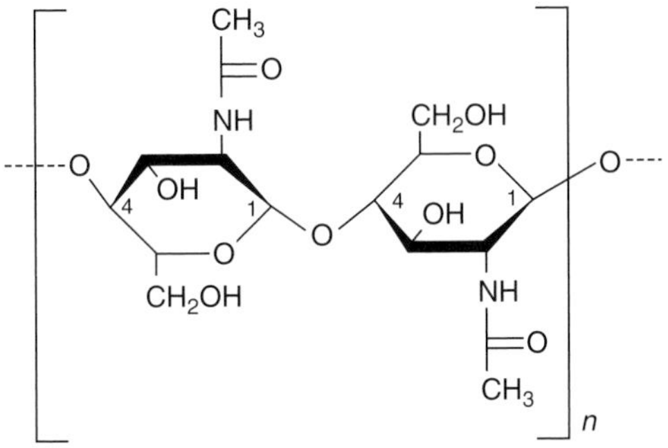

Fig. 8.13. Structure of the chitin molecule showing N-acetylglucosamine subunits bonded through a β-1,4 linkage. Hydroxyl groups (–OH) are bound to the glucose ring structure in positions that lack other side groups.

1. Reducing sugar theory (Maillard reaction): reducing sugars and amino acids formed as by-products of microbial metabolism undergo non-enzymatic condensation reactions to yield humus. In this reaction, the reactive carbonyl group of the sugar reacts with the nucleophilic amino group of the amino acid. The 'reducing sugar theory' is one of the earliest theories of humus formation, based on observation of brown, nitrogenous polymers formed during the dehydration of certain food products. This chemical reaction requires heat, low moisture content or alkaline conditions, and thus it probably does not occur in many soils, if at all.
2. Microbial synthesis theory: polyphenols are synthesized by fungi from non-lignin carbon sources such as cellulose. The polyphenols are then enzymatically oxidized to quinones that condense with amino compounds to yield humus. This reaction occurs predominantly in aerobic soils.

While the microbial synthesis theory appears to be valid, there is a sense that aliphatic carbon compounds with hydrophilic and hydrophobic characteristics may be an important component of the soil humus. These compounds could include the breakdown products of the extracellular polysaccharides, proteins and chitin coming from microbial cells, as discussed above, as well as hydrophobic long-chain fatty acids (n-$C_{21:0}$ to n-$C_{34:0}$) synthesized by microbial cells. In contrast, short-chain fatty acids (n-$C_{10:0}$ to n-$C_{20:0}$) in cellular membranes are quite susceptible to decomposition. The

Table 8.9. Mean residence time in long-term agricultural field experiments of bioproducts synthesized by soil microorganisms. (Data from Marschner *et al.*, 2008.)

Compound	Mean residence time (years)
Polysaccharides	44–161
Protein/Chitin	48–284
Unspecified	30–3880

average residence time of microbially synthesized compounds is given in Table 8.9.

The previous discussion has focused on the chemical composition and susceptibility of biochemical breakdown of various organic substrates that leads to the formation of soil humus. We cannot forget the role of soil minerals in humus stabilization. A recent model of organo-mineral interactions in soils has been proposed by Kleber *et al.* (2007), based on the adsorption of organic compounds in discrete zones (Fig. 8.14). These are:

1. The contact zone, where the strongest organo-mineral bonding occurs. This may be achieved when polar organic functional groups create ligand-exchange bonds with hydroxyl compounds on the mineral surface. The adsorption of proteins at the mineral surface leads to a denaturation of the three-dimensional protein structure, so that bind-

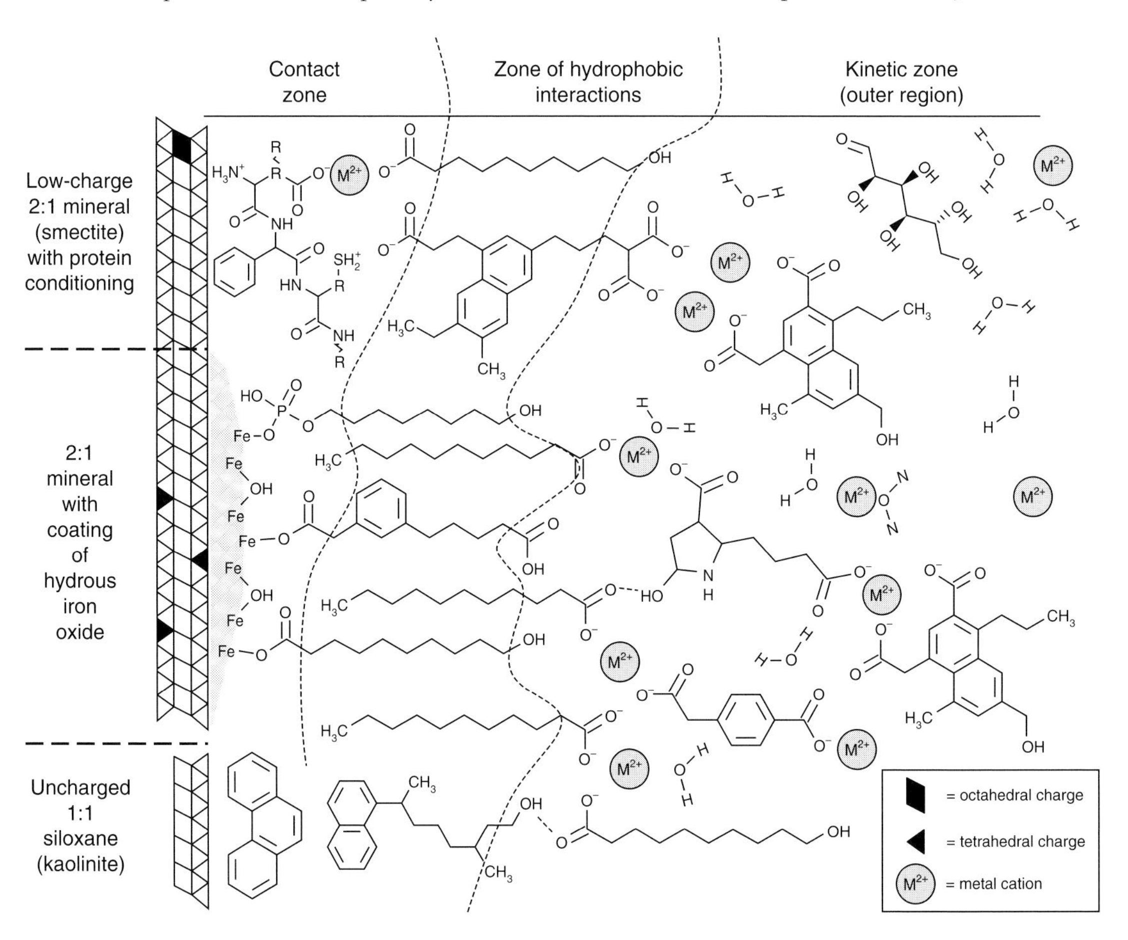

Fig. 8.14. The zonal model of organo-mineral interactions hypothesized for three types of mineral surfaces (low-charge 2:1 mineral, 2:1 mineral coated with hydrous iron oxide and uncharged 1:1 siloxane). Explanation of the bonding mechanisms in each zone is given in the text. (Reproduced with kind permission from Springer Science+Business Media©: *Biogeochemistry*, a Conceptual Model of Organo-mineral Interactions in Soils, vol. 85, 2007, M. Kleber, P. Sollins and R. Sutton, Fig. 2, p. 12.)

ing occurs through both hydrophobic interactions and electrostatic forces. Organic fragments bound to the mineral surface at the contact zone are probably bound irreversibly and are not susceptible to decomposition.
2. The hydrophobic zone, where the hydrophobic portions of molecules are bound to the contact zone through hydrophobic interactions. This creates a bilayer structure on the mineral surface. Organic molecules in this zone can be exchanged into the surrounding soil solution, but are still retained with considerable force.
3. The kinetic zone contains organic material that is loosely bound to hydrophilic moieties or to absorbed proteins by cation bridging, hydrogen bonding and other interactions. Organic materials in the kinetic zone may experience high exchange rates with the surrounding soil solution and thus have a shorter residence time than organic substances in zones that are in closer contact with the soil minerals. The adsorption and desorption rates are probably affected by molecular stereochemistry, the degree of amphiphilicity (e.g. the proportion of hydrophobic and hydrophillic moieties in the molecule), the cation content of the soil solution, pH, temperature and other factors that are not yet known.

Stabilization of soil organic matter

To this point, we have focused on the transformation of organic substances into humus, which is the most stable form of soil organic matter and has the longest residence time in the soil (decades, centuries or millennia). Yet, organic substrates of plant, animal and microbial origin can be stabilized in the soil in the short term (weeks or months) to medium term (years to decades). As long as these materials remain in the soil organic matter and are not completely decomposed to CO_2, then they count as part of the soil organic carbon pool. Soil carbon sequestration occurs when the rate of increase in carbon content in soil is greater than the carbon emitted from soil as CO_2. Even if carbon compounds entering the soil are not transformed into humus, they can still count as a 'soil carbon credit' if they are stabilized and not returned to the atmosphere as CO_2. There are three ways in which organic carbon compounds becomes stabilized: through biochemical, chemical and physical processes.

Biochemical stabilization

Most organic substrates entering the soil are complex polymers and must be decomposed by extracellular enzymes before the simple monomeric compounds can be absorbed and used to support the metabolic processes of soil microorganisms. Extracellular enzymes are specific in the chemical bonds that they can break. This means that the enzyme can only bind to certain sites on the organic substrate before beginning the hydrolysis process. If an organic substrate becomes saturated with enzymes at all available adsorption sites, then additional extracellular enzymes produced by the microbial cell will diffuse further away. As proteins, extracellular enzymes tend to bind to mineral surfaces and organo-mineral complexes if they do not encounter and bind to an appropriate substrate. Sometimes a bound enzyme becomes inactivated because the three-dimensional protein structure cannot fold or bend in the required manner. In other cases, enzymes bound to soil surfaces retain their catalytic ability long after the organism that produced them has passed away. Soil organic matter is considered to be biochemically stabilized when there is insufficient enzyme activity to completely degrade the fresh material to CO_2. If no organisms in a soil system are capable of producing cellulase enzymes required for cellulose depolymerization, for instance, then any cellulose present in the soil would be considered to be biochemically stabilized.

Another type of biochemical stabilization occurs due to polymerization and condensation reactions. In the microbial synthesis theory of humus formation, we see that it is possible for sugars, amino acids and quinones released from fungal metabolism to polymerize into large polyphenol conglomerates that are held together by hydrophobic interactions and hydrogen bonding. Melanin-type compounds could be formed through this process, and these are quite resistant to enzymatic breakdown.

Chemical stabilization

Organic matter can be stabilized for a short period of time through biochemical mechanisms, but sooner or later, fresh material will be consumed by soil fauna, which brings it into more intimate contact with microbial populations and their extracellular enzymes. Even the most recalcitrant plant compounds such as lignin, suberin and cutin will eventually be decomposed. The process of chemical stabilization refers to the formation of organo-minerals as shown in Fig. 8.14. Depending on the zone where binding occurs, the molecule can be bound very strongly to the organo-mineral surface or it can be readily exchangeable between the organo-mineral and the soil solution (kinetic zone). Factors that strongly affect the chemical stabilization of soil organic matter include the quantity and characteristics of clay minerals in a particular soil. The presence of polyvalent cations also facilitates chemical stabilization. Soils saturated with polyvalent cations such as Ca^{2+} or Mg^{2+} tend to have greater binding capacity (e.g. more ligand bridges) than soils with a high Na^+ or K^+ concentration. Since the amount of organic matter that can bind to mineral surfaces or be absorbed in the kinetic zone is limited by the number of sorption sites, it is possible for mineral surfaces to become saturated with carbon, such that any additional organic residues entering the soil will remain unprotected and quickly decomposed.

Physical stabilization

Organic matter can be physically protected from decomposition when it becomes concealed or covered by soil particles during the process of aggregate formation. These can be small micro-aggregates (< 20 μm) formed when clay and silt particles are bound together by humic material or extracellular polysaccharides of microbial origin. Micro-aggregates are bound together by fine roots or fungal hyphae to form macro-aggregates (> 250 μm). Aggregates provide physical protection of organic matter in three ways: (i) the organic matter is not accessible to the microbial population due to pore size exclusion, where the microorganisms are too large to fit into the pores leading to the organic matter; (ii) when bacterial cells and fungal hyphae enter pores, but predatory microfauna cannot due to pore size exclusion, decomposition will occur at a slower rate because grazing pressure is absent; and (iii) gas exchange is compromised due to the small size of the pores (generally < 10 μm) leading to the organic matter. This creates an anaerobic environment within the aggregate, causing the microbial activity to slow down significantly. Physical stabilization of organic matter occurs because the soil environment within aggregates is modified in a way that limits the decomposition rate.

Benefits of soil organic matter

We have already discussed the interest from the scientific community to find ways of storing more carbon in soils, by understanding how carbon compounds can be transformed into humus or stabilized in soil organic matter. In practice, it is land managers and farmers who are responsible for managing natural ecosystems and selecting agricultural practices that increase soil carbon sequestration. While the idea of obtaining 'carbon credits' for storing more carbon in a particular ecosystem may be of interest, a number of improvements in soil properties can be achieved by maintaining or increasing the soil organic matter content in terrestrial systems, as farmers around the world have known for generations. For example:

- Soil organic matter improves soil biological properties: partially decomposed residues and rhizodeposits in the soil organic matter are a source of energy for a large and diverse group of soil organisms, as described in Focus Box 8.1. Supporting the soil microbial and faunal life ensures a constant supply of nutrients and plant growth-promoting substances. Soil microorganisms also produce the extracellular polysaccharides, hyphae and cellular debris that serve as a 'glue' to hold soil particles together and create aggregates. Primary production in natural ecosystems and crop production in agroecosystems cannot be sustained without the activity of the soil microbial and faunal communities.
- Soil organic matter improves soil chemical properties: the cation exchange capacity of a soil is greatly increased by the presence of negatively charged organic molecules. Essential nutrients such as ammonium, potassium and magnesium (NH_4^+, K^+ and Mg^{2+}) can be adsorbed to cation exchange sites and then desorbed into the soil solution when needed by crops. Soil organic matter also possesses positively charged organic molecules that provide anion exchange sites for essential nutrients such as phosphate, sulfate and chloride ($H_2PO_4^-/HPO_4^{2-}$, SO_4^{2-} and Cl^-).The soil organic matter buffers changes in the soil pH. Organically bound nutrients such as nitrogen, phosphorus and sulfur are slowly released as soil organic matter decomposes. Small organic molecules act as natural chelating agents (e.g. oxalic acid), keeping essential micronutrients such as zinc, copper and iron (Zn^{2+}, Cu^{2+} and Fe^{2+}) in a soluble form that can be absorbed

by plant roots. Organic acids released during decomposition also aid in mineral soil weathering, which causes phosphate ions (HPO_4^{2-} and $H_2PO_4^-$) to enter the soil solution and become available for plant uptake. Finally, many persistent organic molecules (e.g. pesticide and chemical residues) adhere to the soil organic matter and organo-minerals until they are degraded by microorganisms.

- Soil organic matter improves soil physical properties: producers often note that a mineral soil with a high organic matter content is easier to till. The dark colour improves adsorption of solar energy, so these soils warm up more quickly in the spring and can be planted earlier. Soil organic matter is like a sponge that can absorb and retain water, releasing it slowly during dry periods in the growing season. Aggregate size is larger in mineral soils that receive organic fertilizers and when crop residues are retained. A well-aggregated soil has large, continuous macropores that permit water drainage during heavy rainfall events. Compaction tends to be lower in a well-aggregated soil than a soil with a more fragile structure.

It should be realized that soil organic matter is not equivalent to soil organic carbon. The method for determining soil organic matter is called 'loss on ignition' and involves placing a pre-weighed soil sample into a furnace, heating the sample to 360°C for at least 4 h, and then re-weighing the sample. As organic materials are volatilized at this temperature, all that remains are the soil mineral particles. Since the soil organic matter can include non-carbonaceous organic materials such as nitrogen-rich proteins, organic phosphorus and organic sulfur, it is necessary to use a conversion factor if the soil organic matter value is used to estimate soil organic carbon. Traditionally, a conversion factor of 1.724 has been used based on the assumption that organic matter contains 58% organic C (i.e. g organic matter/1.724 = g organic C) (Nelson and Sommers, 1996). However, there is no universal conversion factor because the organic carbon content varies from soil to soil, from soil horizon to soil horizon within the same soil, and will vary depending upon the type of organic matter present in the sample. Conversion factors range from 1.724 to as high as 2.5 (Soil Survey Laboratory, 1992; Nelson and Sommers, 1996). Alternatively, the soil organic carbon can be measured directly by high-temperature combustion of the soil at 900°C. This converts carbonaceous compounds into CO_2, which is then analysed by gas chromatography.

The size of the soil organic matter pool in a particular ecosystem is a delicate balance between the amount of organic residue entering the system and the amount of CO_2 lost during the decomposition process. Scientists and land managers are interested to know whether certain ecological or agricultural practices will lead to a decline in the soil organic matter, or maintain or increase this pool. A variety of mathematical models have been developed to describe decomposition and make predictions about what will happen in the future, based on various scenarios. One of the most well known is the CENTURY model (Focus Box 8.3), which has been cited in more than 500 scientific papers.

Focus Box 8.3. Estimating decomposition with the CENTURY model.

Modelling the soil organic matter dynamics is a top priority in soil ecology and management. Conceptual and mathematical models express our current understanding of litter decomposition, biotic interactions in soil, humus formation and other relevant processes. Mathematical models are the most powerful because they can be used to make projections about future responses to soil management changes, climate, response to major disturbances, and so on. For example, the projected increase in global temperatures may increase the rate of carbon emission from soil organic matter, causing a positive feedback between temperature and CO_2 emissions. Since we cannot make measurements of events that have not yet occurred, a model is the best way to determine the extent to which this hypothesized feedback will affect soil organic carbon pools.

Some decomposition models are more or less restricted to the soil compartment, whereas others include vegetation or herbivore modules, and can be extended to a broader biological framework. The University of Kassel and the GSF–National Research Center for Environment and Health have developed the Register of Ecological Models (REM), which is a

Continued

Focus Box 8.3. *Continued*

meta-database for existing mathematical models in ecology available on the internet (http://www. ecobas. org/www-server/). Despite great advances in modelling soil organic matter respiration as determined by climatic and pedological factors, a better understanding of the multiple interactions occurring among soil organisms in the context of decomposition still presents a challenge. The accuracy of predictions regarding the soil organic matter balance in future global scenarios is highly dependent on the quality of models describing decomposition processes.

Models of soil organic matter decomposition can be classified according to the procedure they use to estimate stocks of carbon and other nutrients in soil. The most common approach is empirical modelling of the processes responsible for moving matter between compartments, and the rate at which transfers occur. This kind of model is called a 'process model'. Another approach is to estimate the carbon and nitrogen cycling rates, explicitly accounting for the transfer rates due to the metabolic activity of soil microorganisms and fauna decomposing the litter resource. This is called a 'foodweb model' or 'metabolic model'.

Process models divide the soil organic matter into several compartments according to its physicochemical characteristics and decomposition rate. The decomposition rate in each compartment is regulated by a variety of factors. Soil temperature, soil moisture and nitrogen content are major drivers of catabolic processes. Other rate regulators included in models are the C:N ratio of vegetation, lignin content of fresh organic residues, concentration of other nutrients, soil pH and clay concentration, among others. All models begin with the general assumption that the soil organic matter integrated in the considered compartments has different residence times, leading to sequential degradation. This leads to algorithms of first-order decomposition rates for most soil organic matter compartments, meaning that the rate that organic matter flows from one compartment into another depends on the inputs to the first component. To avoid oversimplification that would occur in a model based exclusively on linear-unidirectional fluxes, certain feedbacks and loops are inserted to represent realistic decomposition rates. Among the most common soil organic matter models are CENTURY, Roth C, Daisy, Candy, NCSOIL and DNDC, which have been validated using data from existing field sites and also used to study decomposition rates and other parameters under future climate or perturbations scenarios.

The CENTURY model (Parton *et al.*, 1987; http:// www.nrel.colostate.edu/projects/century/ (accessed on 21 March 2009)) is a general model of plant–soil nutrient cycling, which has been used to simulate carbon and nutrient dynamics for different types of ecosystems including grasslands, agricultural lands, forests and savannahs. It includes three soil organic matter pools (active, slow and passive) with different potential decomposition rates, above- and belowground litter pools and a surface microbial pool, which is associated with decomposing surface litter (Fig. 8.15). Each litter pool has an intrinsic maximum decomposition rate. The actual rate of decomposition is then determined by the value of the modifiers, i.e. soil moisture, temperature and plant cover, operating on the maximum rate. The CENTURY model is composed of various submodels: the soil organic matter/decomposition submodel, the water budget model, the grassland/ crop submodel, a forest production submodel and management and events-scheduling functions.

Several driving variables are needed for the monthly calculations such as: average monthly air temperature (maximum and minimum values), monthly precipitation, soil texture, plant nitrogen, phosphorus and sulfur content, lignin content of plant material, atmospheric and soil nitrogen inputs, initial soil carbon and nitrogen content. A distinguishing feature of CENTURY is that it contains two plant production submodels: a grassland/crop submodel and a forest production submodel. Both plant production models assume that the monthly maximum plant production is controlled by moisture and temperature, and that maximum plant production rates are decreased if there are insufficient nutrient supplies, with the flexibility of specifying potential primary production curves for the specific plant community under consideration. The forest submodel simulates the growth of deciduous or evergreen forests in juvenile and mature phases. To simulate a savannah or shrubland, CENTURY uses both of these submodels with some additional code to simulate nutrient competition and shading effects. Simulation of complex agricultural management systems including crop rotations, tillage practices, fertilization, irrigation, grazing and harvest methods is also possible. Disturbances such as fire, harvest, grazing and cultivation can be simulated via the management and events-scheduling functions.

Continued

Focus Box 8.3. *Continued*

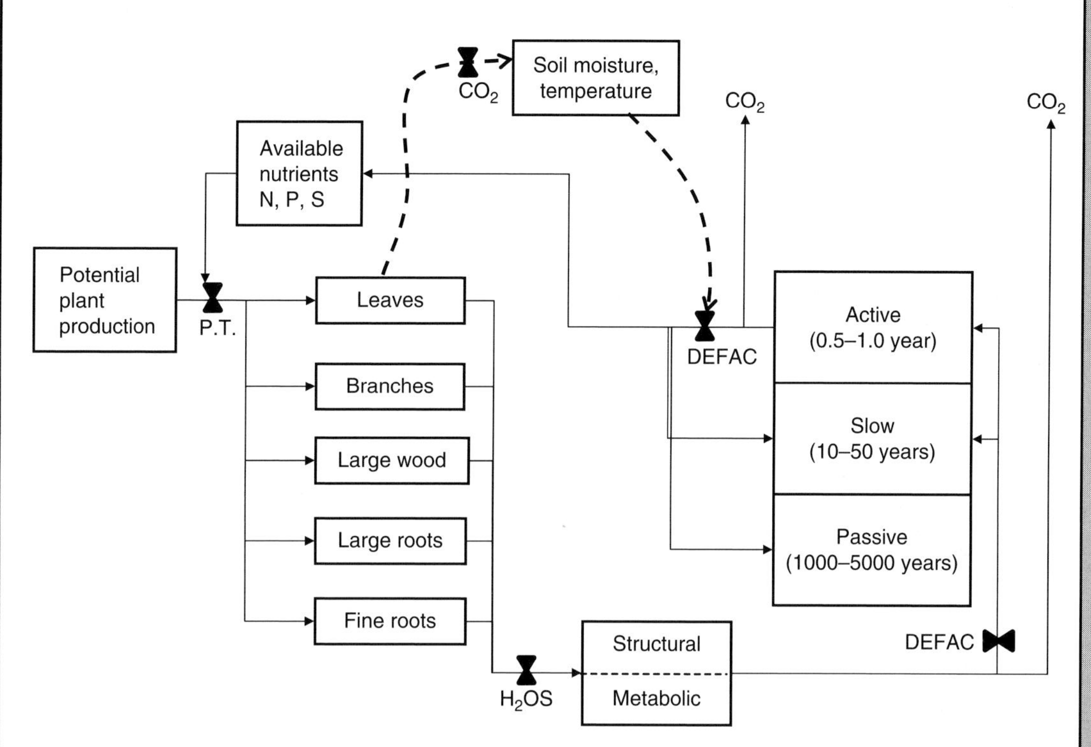

Fig. 8.15. Overview of the conceptual structure and compartments in the CENTURY model. Abbreviations: P.T., potential transpiration; DEFAC, abiotic decomposition factor based on temperature and moisture; H_2OS, rainfall and soil texture. (Redrawn from Cerri *et al.*, 2004.)

9 Nutrient Cycling

The decomposition of organic residues by soil foodweb organisms is of considerable importance for carbon cycling, but we cannot forget that the turnover of organic residues is also essential for releasing organically bound nitrogen, phosphorus and sulfur in soils. In natural systems and agroecosystems where most of the nutrients for primary production come from litterfall, organic fertilizers and similar residues, it is critical to understand and appreciate the factors that affect nutrient cycling.

In organic agriculture, for example, producers are encouraged to grow N_2-fixing legumes that can be ploughed into the soil and decompose to release nitrogen for subsequent crops. To be successful, an organic producer must understand how decomposition of organic residues occurs so that nutrients are mineralized and transferred to crops during the period in the growing season when plant nutrient requirements are high. In herbaceous plants, nutrient requirements are greatest during the vegetative growth stages, while woody plants need nutrients to support leaf growth and photosynthesis after new foliage emerges. There is little agricultural benefit to have the peak nutrient mineralization occur when crops have reached reproductive growth stages or when fields are bare, generally from late autumn to early spring in temperate regions. Any nutrients released in plant-available forms during this time are not used by crops and may be susceptible to export from the soil into aquatic ecosystems or the atmosphere. Agricultural producers would like to achieve synchrony between the nutrient mineralization from organic materials and the nutrient uptake by crops, for maximum crop production and environmental protection.

Organic residues of plant origin contain appreciable quantities of nitrogen and other essential nutrients (summarized in Table 7.4). Biosolids and animal manure contain all of these nutrients, as well as trace metals from contaminants in the environment that are accumulated and excreted from animals. Land-spreading or incorporation of organic residues of plant or animal origin in the soil leads to a rapid release of water-soluble nutrients into soil solution (Table 9.1), followed by re-establishment of an equilibrium between the nutrient concentrations in soil solution and in adsorption sites on organo-minerals.

Small quantities of inorganic nitrogen (NH_4^+, NO_3^-), phosphorus ($H_2PO_4^-$, HPO_4^{2-}) and sulfur (SO_4^{2-}) are present in plant sap, but most nitrogen, phosphorus and sulfur molecules are covalently bonded to carbon atoms, typically in the structural plant components. Their transformation into ionic forms requires microbial action. More than 90% of the nitrogen, phosphorus and sulfur in residues and soil is in organic forms, and each of these nutrients have a central role in the biological functioning of microbial, plant and animal cells (Table 9.2). Therefore, the nutrient cycles of these elements attract considerable scientific attention and are the subject of practical interest, especially in agricultural systems where the health, development and production of crops and livestock can be negatively impacted by nitrogen, phosphorus and sulfur deficiencies.

9.1 The Nitrogen Cycle

An overview of the nitrogen cycle is presented in Fig. 9.1. Biological and chemical reactions control the rate at which nitrogen is transferred from the atmosphere to terrestrial ecosystems, cycled through plants, animals and soils in the terrestrial ecosystem and emitted to aquatic systems (groundwater, surface water) or released back to the atmosphere. The following discussion will focus on the role of soil biota in the nitrogen cycle.

We begin by considering the pools of biological active nitrogen. The largest of these is in the form of dinitrogen gas (N_2), which constitutes 78% of the Earth's atmosphere (Table 9.3). Soil organic

©CAB International 2010. *Soil Ecology and Management* (J.K. Whalen and L. Sampedro)

Table 9.1. Water-soluble and organically bound nutrients essential for plant growth, and the ionic forms present in soil solution.

Water-soluble nutrients		Organically bound nutrients	
Nutrient	Form(s) in soil solution	Nutrient	Form(s) in soil solution
Potassium	K^+	Nitrogen	NH_4^+, NO_3^-
Calcium	Ca^{2+}	Phosphorus	HPO_4^-, $H_2PO_4^-$
Magnesium	Mg^{2+}	Sulfur	SO_4^{2-}
Copper	Cu^{2+}		
Iron	Fe^{2+}, Fe^{3+}		
Manganese	Mn^{2+}		
Nickel	Ni^{2+}		
Zinc	Zn^{2+}		
Boron	HBO_3^0, $H_4BO_4^-$		
Molybdate	MoO_4^{2-}		
Chlorine	Cl^-		

Table 9.2. Major biological functions of nitrogen, phosphorus and sulfur in plants.

Nutrient	Functions
Nitrogen	Required for formation of: - amino acids, peptides, proteins, enzymes and genetic material (DNA, RNA) - mitochondrial structure, chlorophyll formation and photosynthesis reactions Sufficient nitrogen ensures: - faster growth and maturation - high protein content in grain/seeds Other functions: - influences carbohydrate deposition (plants with low N content tend to be more succulent and have weaker stems)
Phosphorus	Essential for: - energy storage and transfer (ADP↔ATP) via phosphorylation reactions Processes or pathways involving ADP/ATP include: - membrane transport, cytoplasmic streaming, photosynthesis, protein biosynthesis, nucleic acid synthesis, lipid and phospholipid biosynthesis, generation of membrane electrical potentials, respiration, biosynthesis of cellulose, pectins, hemicelluloses and lignin, and isoprenoid biosynthesis Sufficient phosphorus ensures: - development of reproductive organs and seed formation - increased root growth - speeds the process of grain ripening
Sulfur	Required for formation of: - S-containing amino acids (cystine, cysteine and methionine) - disulfide bonds (-S-S-) between polypeptide chains needed for proper configuration of structural and catalytic proteins - synthesis of coenzyme A, involved in oxidation and synthesis of fatty acids, amino acids and reactions in citric acid cycle - chlorophyll formation - ferredoxins (Fe-S proteins) in chloroplasts, required for assimilation of nitrogen from N_2-fixing bacteria (symbiotic or free-living) Sufficient sulfur ensures: - more nitrogen transformed into protein, less non-protein NH_2 and NO_3^- in plant Other functions: - sulfur compounds are responsible for the characteristic taste and smell of plants in the mustard and onion families

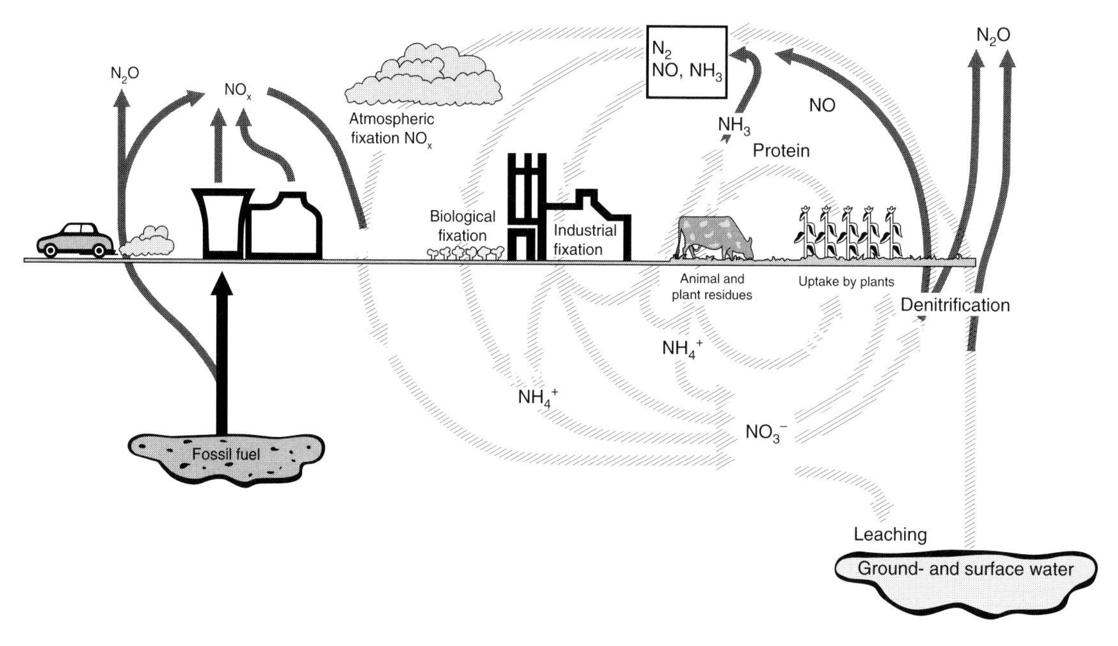

Fig. 9.1. The global (terrestrial) nitrogen cycle. (Redrawn with permission from Loegreid *et al.*, 1999.)

Table 9.3. Pool sizes of biologically active nitrogen in terrestrial ecosystems to a depth of 1 m. (Modified from Myrold, 1998.)

Pool	Typical size (range) (g N/m^2)
N_2 (Dinitrogen)	1,150 (230–27,500)*
Organic N	725 (100–3,000)
Plant N	25 (1–240)
NH_4^+ (Ammonium)	1 (0.1–10)
NO_3^- (Nitrate)	5 (0.1–30)

*N_2 gas contained in soil pore space and around vegetation in terrestrial ecosystems.

nitrogen is the second largest terrestrial pool, followed by plant nitrogen. The organic nitrogen includes nitrogen from plant and animal residues, in microbial biomass and associated with soil organic matter. Variation in the soil organic nitrogen pool size of terrestrial ecosystems is due to climatic variables, especially temperature and moisture, land use and soil pedogenesis. Soil microorganisms can transform N_2 and organic nitrogen into forms that can be used by plants, which is vital for primary production on Earth. Plant nitrogen is affected by the vegetation type, land management and plant-available nitrogen available within various soils and climatic zones. Forests and savannahs are expected to have more plant nitrogen than sparcely vegetated desert or tundra ecosystems. The inorganic nitrogen pools are very small, with concentrations often less than 1 mg N/kg soil in natural ecosystems and rarely exceeding 100 mg N/kg soil in fertilized agricultural soils.

Transformation of dinitrogen (N_2)

As discussed in Chapter 7, the N_2-fixing bacteria possess the nitrogenase enzyme complex that can convert N_2 into NH_3, which is subsequently protonated in the presence of water to yield NH_4^+. This enzyme complex is actually made of two enzymes: dinitrogenase reductase (Fe-containing protein), which gathers electrons for the reaction, and dinitrogenase (MoFe-containing protein), which uses these electrons to produce NH_3, which is then protonated to NH_4^+ and bonds to glutamate, yielding glutamine. The reaction is illustrated in Fig. 9.2 and outlined in equations 9.1 to 9.3.

Dinitrogenase reductase and dinitrogenase are Fe-S proteins, meaning that both iron and sulfur are required as cofactors; in addition, molybdenum or vanadium is needed as a cofactor in the dinitrogenase enzyme. Adequate phosphorus and magnesium are required for the production of

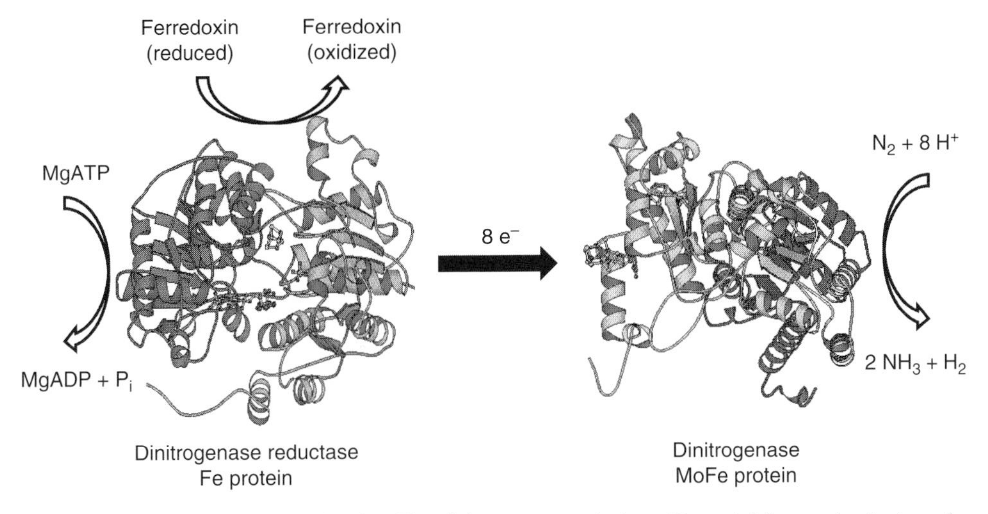

Fig. 9.2. Reactions involved in nitrogen fixation. The dinitrogenase reductase (Fe protein) accepts electrons from a low-redox donor such as reduced ferredoxin or flavodoxin and binds MgATP. Then, the dinitrogenase reductase binds to the dinitrogenase (MoFe protein), transfers one electron and hydrolyses two MgATP molecules to MgADP + P_i (inorganic phosphate). The two proteins dissociate, and the process is repeated. Once eight electrons have been transferred, the dinitrogenase is able to bind dinitrogen (N_2) and reduce it to NH_3. (Illustrations of enzyme structures were produced with the program MOLSCRIPT (Kraulis, 1991) from http://metallo.scripps.edu/promise/2MIN.html, accessed on 12 January 2009.)

MgATP. The nitrogenase complex is capable of reducing N_2 to NH_3, and can also reduce C_2H_2 to C_2H_4 and H^+ to H_2. The process of N_2 reduction is as follows:

$$N_2 + 6\ e^- + 12\ ATP + 8\ H^+ \xrightarrow{Mg} 2\ NH4^+ + 12\ ADP + 12\ P_i \quad [9.1]$$

Simultaneous reduction of H^+ in the enzyme complex occurs as follows:

$$2\ H^+ + 2\ e^- + 4\ ATP \xrightarrow{Mg} H2 + 4\ ADP + 4\ P_i \quad [9.2]$$

The overall equation then becomes:

$$N_2 + 16\ ATP + 8\ e^- + 10\ H^+ \xrightarrow{Mg} 2\ NH_3 + H_2 + 16\ ADP + 16\ P_i \quad [9.3]$$

and the addition of two H^+ ions (but no additional energy) yields two NH_4^+ molecules. The binding of NH_4^+ with glutamate yields glutamine, which is usually transformed in secondary reactions to other products before leaving the root via the xylem. Legumes that grow in temperate zones may transport glutamine or transform it into asparagine, whereas tropical legumes export the ureides allantoin and allantoic acid. Actinorhizal plants that form a symbiosis with N_2-fixing actinomycetes such as *Frankia* transport ureide citrulline.

In legume nodules, the energy (ATP) required for this reaction comes from the breakdown via glycolysis of photosynthates that were transferred from the host plant to the prokaryote symbiont. In theory, 16 ATP are required for the reaction to occur, but it could be more energetically costly under field conditions if additional energy must be expended during the repeated binding/dissociation of the Fe protein and MoFe protein in the nitrogenase complex, or if the symbiont requires another enzyme system such as hydrogenase to generate H^+ ions for the reaction. Naturally occurring ferrodoxin and flavodoxin synthesized within the prokaryote cell serve as low-redox reactants. Critical factors for this reaction include oxygen, which must be excluded because it destroys the enzyme (see Focus Box 9.1), and the presence of other substrates that can be catalysed by nitrogenase (Table 9.4). Most symbionts are mesophiles that grow optimally between 10°C and 37°C. They require adequate mineral nutrition for optimal N_2 fixation (Table 9.5).

Table 9.4. Substrates possessing triple bonds that can be hydrolysed by nitrogenase.

Substrate	Formula	Reaction product(s)
Hydrogen	H^+	H_2
Dinitrogen	$N \equiv N$	$NH_3 + H_2$
Nitrous oxide	$N \equiv N^+ - O^-$	$N_2 + H_2O$
Azide	$[N \equiv N^+ - N^-]$	$N_2 + NH_3 + N_2H_4$
Acetylene	$HC \equiv CH$	$H_2C = CH_2$
Cyanide	$[C \equiv N]^-$	$CH_4 + NH_3$
Carbon monoxide	$C \equiv O$	None. $C \equiv O$ binds to binding site but is not reduced, thus blocking the reaction of other substrates

Transformation of organic nitrogen

Soil organic nitrogen includes a variety of compounds that originate from microbial cells as well as plant and animal residues. Proteins and amino acids, microbial cell wall polymers and amino sugars, nucleic acids and metabolites such as vitamins, antibiotics and growth factors contain nitrogen. The common feature of all organic nitrogen compounds is that they are bound to carbon (-C-N- bonds). In plants, nitrogen-containing compounds represent between 0.2 and 1.5% of plant biomass. Among the organic nitrogen compounds in plant cells are amino acids, pyrimidines and purines, as well as porphyrin, the precursor to chlorophyll a. Plants tend to contain much more carbon (45%) than nitrogen, which leads to wide carbon:nitrogen ratios. For instance, the carbon: nitrogen ratio for tree leaves may range from 20 to 50 while branches and other parts rich in ligno-cellulosic material often have a carbon:nitrogen ratio in excess of 100.

Microbial cells are much richer in nitrogen-containing compounds than plants. Bacterial cells contain 6–9% nitrogen by weight. Much of this nitrogen is contained in the bacterial cell wall, which is composed of peptidoglycan (polymer of sugars and amino acids), and within the cell in proteins, peptides and amino acids. The cell wall of some fungi is composed of

Table 9.5. Essential nutrients required by symbiotic and free-living prokaryotes capable of biological N_2 fixation. (Modified from O'Hara, 2001.)

			Functions in rhizobia			Specific role in symbiosis	
Nutrient	Cell dry weight (%)	Cell structure	Enzyme function	Energy transfer	Signal/cell regulation	Nodulation	N_2 fixation
Macronutrients							
Carbon	50	+	+	+			
Oxygen	20	+	+	+			
Nitrogen	14	+	+	+			
Hydrogen	9	+	+	+			
Phosphorus	3	+	+	+	+	+	+
Sulfur	1	+	+				
Potassium	1		+		+	+	
Calcium	0.6	+	+		+	+	+
Magnesium	0.6		+				
Micronutrients							
Iron	0.3		+		+	+	+
Manganese	T		+				
Copper	T		+			+	+
Zinc	T		+				+
Molybdenum	T		+				+
Cobalt	T		+			+	+
Nickel	T		+			+	
Selenium	T		+				+
Boron	T					+	

T = trace amounts in microbial cells.

Focus Box 9.1. The 'oxygen paradox' in biological nitrogen fixation.

Many prokaryotes possess the nitrogenase enzyme complex and are capable of biological nitrogen fixation. This enzyme system is found in organisms covering the entire spectrum of heterotrophic (bacteria) and autotrophic (cyanobacteria) metabolism, in cells that function as aerobes, facultative anaerobes and strict anaerobes. One of the challenges for the nitrogen-fixing prokaryotes, especially free-living organisms, is that nitrogenase is extremely sensitive to molecular oxygen. Exposure to oxygen inactivates the enzyme, but oxygen is needed for the oxidative phosphorylation of ATP and production of the reductant (ferredoxin or flavodoxin). This creates a 'paradox', because the organism needs oxygen and wishes to avoid oxygen at the same time.

In symbiotic N_2-fixing organisms, it appears that both physical and metabolic mechanisms are important for protecting the nitrogenase enzyme. A physical barrier to oxygen diffusion exists in the nodule cortex, which reduces oxygen transport into sensitive areas of the cell. The membranes of certain cells have osmoregulatory functions that restrict oxygen conductance under stress conditions and in response to other biochemical reactions. It seems that the osmoregulation function is connected to nitrogenase activity or energy supply to the nitrogenase complex, but this is still under investigation (Schulze, 2004). The oxygen 'paradox' is an even greater challenge for free-living N_2-fixing prokaryotes because they are not protected within a plant host. A variety of unique strategies have been developed by these free-living organisms (diazotrophs) to protect the nitrogenase enzyme system, including:

- Avoidance: most anaerobic and facultative anaerobic bacteria fix N_2 in the absence of oxygen. One exception is *Klebsiella pneumoniae*, which can tolerate low levels of oxygen.
- Microaerophily: the N_2-fixing activity of most aerobic bacteria is optimal when the partial pressure of oxygen is low, about 0.7 kPa (30 times lower than the atmospheric oxygen concentration of 21 kPa). This minimizes the exposure of nitrogenase to oxygen.
- Respiratory protection: respiratory functions in aerobic bacteria divert oxygen away from nitrogenase. Bacteria of the genus *Azobacter* are well known for their high respiration rate, which scavenges oxygen from the cell and surrounding environment and protects nitrogenase from becoming inactivated. However, to maintain such high respiration, the bacteria must process large amounts of substrates and hence N_2 fixation is quite inefficient due to the amount of carbon that must be consumed to keep the nitrogenase complex functional.
- Conformational protection: when *Azobacter* species cannot metabolize substrates quickly enough to scavenge oxygen that could damage the nitrogenase complex, they produce a protein that binds to nitrogenase and alters its conformation (shape) to protect it from oxygen. As a result, these bacteria can stop N_2 fixation abruptly when the oxygen concentrations are too high; as oxygen concentrations decline to lower levels, the protein is released from the nitrogenase complex and N_2 fixation resumes.
- Production of specialized cells: many diazotrophic cyanobacteria produce specialized thick-walled cells called heterocysts that contain nitrogenase. These cells are not involved in photosynthesis and thus do not generate oxygen; in addition, their thick walls are a physical barrier to oxygen diffusion. The actinomycete *Frankia* produces vesicles that protect the nitrogenase from oxygen. When the oxygen concentration surrounding the cell increases, the vesicle wall becomes thicker and serves as a physical barrier to the entry of oxygen molecules.
- Temporal or spatial separation of N_2 fixation and oxygen evolution processes: cyanobacteria that are not capable of producing heterocysts solve the oxygen problem by fixing N_2 primarily during the dark phase of growth, when the cell is not photosynthesizing. During this phase, there is no oxygen production and the oxygen concentrations within the cell are depleted through respiration. This is an example of temporal separation of N_2 fixation and oxygen evolution. Other cyanobacteria achieve spatial separation of these processes by forming aggregates of cells that act as a physical barrier to oxygen diffusion. The nitrogenase complex functions inside the barrier at low oxygen concentrations (microaerophilic environment).
- Slime: aerobic diazotrophs growing on nitrogen-free agar frequently produce large, slime-covered colonies by secreting extracellular polysaccharides through their cellular membranes. The slime acts as a diffusion barrier, preventing the free flow of oxygen into the colony so that cells at the interior of the colony are not exposed to oxygen and therefore can catalyse N_2-fixation reactions. This strategy is not thought to be very efficient under field conditions where colonies are readily disrupted by predators and abiotic fluctuations in the soil environment.

Continued

Focus Box 9.1. *Continued*

Dinitrogen is one of the most stable diatomic molecules known. The triple bond in this molecule requires a lot of energy (945 kJ/mole = 226 kcal/mole) to break. In addition, the reaction that converts N_2 into NH_3 must also consider the energy needed to transform H_2 into the ionic H^+ form. This is the major challenge for any biological organism or chemical process that aims to catalyse N_2 fixation.

The chemical fixation of dinitrogen is accomplished through high-pressure catalysis, known as the Haber–Bosch process, named after the German scientists who developed the process in 1914. Originally, the NH_3 generated from this chemical reaction was converted to ammonium and nitrates for manufacturing explosives during World War I. The process has been adapted to large-scale industrial production of ammonium- and nitrate-based fertilizers, important for global agricultural production. The reaction for industrial N_2 fixation is:

$$N_2 + 3\,H_2 \xrightarrow[K_2O,\,Al_2O3]{\text{heat, pressure}} 2\ NH_3(aq) \qquad \Delta G = -53\ \text{kJ}\ (12.7\ \text{kcal}) \qquad [9.4]$$

There are several features of the chemical reaction that warrant attention. At the industrial scale, high pressures (about 20 MPa) and temperatures of 400 to 500°C are used to catalyse the reaction in the presence of inorganic catalysts (K_2O, Al_2O_3). Natural gas (methane) is often used as a source of H_2 for the reaction and to heat the reaction vessel. The energetic costs of industrial N_2 fixation were very high in the 1950s, requiring more than 80 GJ/t of NH_3 produced. By the late 1990s, the energetic costs had declined to 27 GJ (i.e. 0.645 t of oil equivalent) per tonne NH_3 due to improvements in the design and energy efficiency of fertilizer plants (Smil, 2001). Since this process depends on non-renewable energy, there is a sense that farming systems that rely less on industrial N_2 fixation as the primary nitrogen source would be more sustainable in the long term. Globally, fertilizer consumption increased from 10.8 million t N/year in 1960 to 99.8 million t/year in 2007, emphasizing our dependence on chemical N_2 fixation to support agricultural production (Maene, 2007).

Biological N_2 fixation accounts for about 20% of the nitrogen input in agricultural systems, with about 50–70 million t N/year captured by symbiotic and free-living prokaryotes. If both agricultural and natural ecosystems are considered, the estimated global biological N_2 fixation is probably in the range of 122 million t N/year (Herridge *et al.*, 2008). Just as chemical N_2 fixation requires a large amount of energy to complete the reaction, so do symbiotic and free-living prokaryotes. The principal difference between chemical and biological N_2 fixation is that biological organisms have the ability to produce NH_3 under ambient pressures and temperatures, relying on photosynthates supplied by a host or by undertaking photosynthesis themselves, as is the case with phototrophic diazotrophs such as cyanobacteria. Symbiotic organisms use 3 to 25% of the net energy captured by the host plant during a growing season and require 20–80 kg carbon/t NH_3 (Silvester and Musgrave, 1991). Free-living prokaryotes are less efficient, using 200–1000 kg carbon/t NH_3 (Zuberer, 1998).

nitrogen-rich chitin, a polysaccharide synthesized fromN-acetylglucosamine, but other species may have carbon-rich cell walls composed of cellulose-glycogen, cellulose-β-glucan and mannan-β-glucan. This leads to more variation in the nitrogen content of fungal cells, which may range from 2 to 5%. Since microbial cells contain about 45% carbon, the carbon:nitrogen ratio of bacterial cells ranges from 5 to 8, while fungal cells have a carbon:nitrogen ratio of 9 to 22. Some nitrogen-containing compounds of microbial origin are illustrated in Fig. 9.3.

Until the 1980s, most of the research pertaining to the chemical properties of soil organic nitrogen relied on chemical fractionation methods. These methods indicated that about 50% of the organic nitrogen was in amino acids, amino sugars and purine/pyrimidine derivatives, but the rest of the organic nitrogen could not be characterized. In the past 20 years, advanced chemical analysis using methods such as thermochemolysis and nuclear magnetic spectroscopy has demonstrated that 60 to 90% of the soil organic nitrogen is present as amides. It is believed that these amides are derived from microbial proteins and peptides, suggesting that organic nitrogen coming from fresh organic residues is rapidly transformed by soil microorganisms, in conjunction with the associated soil fauna, to the amide form (Table 9.6).

Since the vast majority of soil microorganisms produce amidase enzymes capable of degrading amides according to equation 9.5, it is surprising

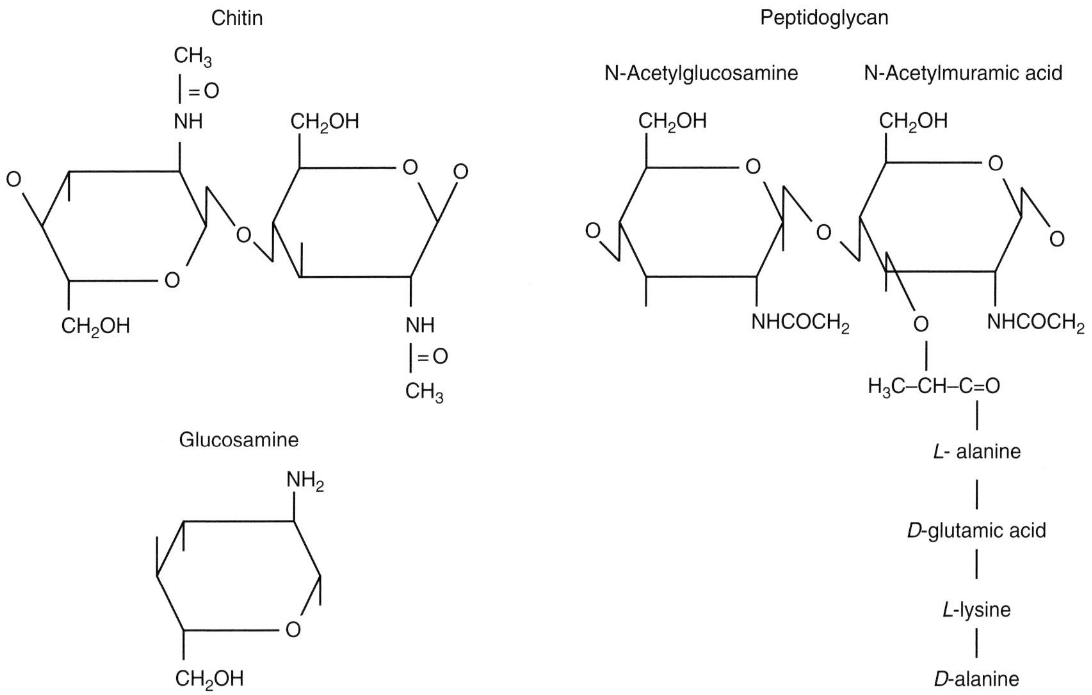

Chitin

CH_3
=O
NH
CH_2OH
O
O
O
O
O
CH_2OH
NH
=O
CH_3

Glucosamine

NH_2
O
CH_2OH

Peptidoglycan

N-Acetylglucosamine
N-Acetylmuramic acid
CH_2OH
CH_2OH
O
O
O
O
O
NHCOCH$_2$
O
NHCOCH$_2$
$H_3C{-}CH{-}C{=}O$
L- alanine
D-glutamic acid
L-lysine
D-alanine

Fig. 9.3. Chemical structure of chitin, a nitrogen-containing compound found in the fungal cell wall, peptidoglycan from Gram-negative bacteria and glucosamine, a precursor in the biochemical synthesis of glycosylated proteins and lipids.

Table 9.6. Proportion of amides in degraded plant substrates and soils. (Adapted from Knicker et al., 1995; Knicker, 2000; Smernik and Baldock, 2005.)

Sample	Carbon:nitrogen ratio	Peptide carbon (% of total carbon)	Amide nitrogen (% of total nitrogen)
Fungal melanins	8–31	6–18	> 70
Degraded algae	7.0	55	84
Degraded wheat (58 days)	30.2	13	85–88
Degraded wheat (4 years)	8.3	29	53–60
NaOH extract of forest soil	8.5	39	82–88
Pine forest soil	13	13	> 70
Mollisol, andisol, oxisol alfisol (0.2–2.0 μm fraction)	9.4–10.9	48–56	> 90
Peats	20–25	10–11	> 70

that such a large proportion of the soil organic nitrogen pool is composed of amides.

$$\text{Monocarboxylic acid amide} + H_2O \rightarrow \text{Monocarboxylate} + NH_3 \qquad [9.5]$$

It appears that only a fraction of the amide-nitrogen in soils can be hydrolysed by amidase enzymes, based on hydrolysis with a strong acid (Table 9.7). The remaining amides seem to be stabilized against biological and chemical degradation, possibly by steric hindrance or in chemically refractory biopolymers. Physical protection of amides is also likely. Most of the soil organic nitrogen is associated with the clay fraction, probably due to the binding of hydrophobic protein moieties to the silicate sheets of clay minerals through

Table 9.7. Concentrations of hydrolysable amino acids in soils treated with 6M HCl, expressed as a percentage of the soil organic nitrogen pool. (Data from: Friedel and Scheller, 2002; Dieckow *et al.*, 2005; Paul and Williams, 2005.)

Soil	Land use	Carbon:nitrogen ratio	Amino acid nitrogen (% of total nitrogen)
Acrisol	Grassland and arable	8.4–9.0	36–38
Luvisol	Arable	7.3–9.1	32–37
Fluvisol	Arable	7.5	28
Regosol	Grassland	8.3	48
Histisol	Abandoned	12.0	50
Planosol	Coniferous forest	16.1	47
Podsol	Coniferous forest	11.1	40
Humus iron podzol	Arable/moorland	18	18
Peaty podzol	Moorland	16	17
Peat	Peat	31	11

hydrophobic interactions as well as electrostatic and hydrogen bonding, as shown in Fig. 8.14. Physical protection of amides within microaggregates, particularly in micropores (< 10 μm diameter) that are too small to allow entry of microorganisms or degradative enzymes, could protect these organic nitrogen compounds.

Another component of soil organic nitrogen is protein complexes that form through association with tannins and other plant-derived polyphenol compounds, including lignin, that are capable of precipitating proteins. These compounds accounted for about 15–20% of the soil organic nitrogen in boreal forest (Adamczyk *et al.*, 2008) and are probably important in forests and aquatic ecosystems such as blackwater streams that receive a considerable input of plant detritus each year. Tannins are classified into two subgroups: (i) condensed tannins, mixtures of polymers of flavan-3-ol units with different degrees of polymerization and hydroxyl substitution; and (ii) hydrolysable tannins, consisting of gallic acid, its dimers (hexadydroxydiphenic acid) and its derivatives.

Tannins are found in the vacuoles and cuticular tissue of gymnosperms and angiosperms. Woody plants have more condensed tannins than hydrolysable tannins in their foliage, seeds, root and stem tissues. Tannins are released from leaves and other senescent tissues after cell breakdown and have an important role in soil chemistry. They reduce aluminium toxicity by binding Al oxides and hydroxides in the soil solution, and increase phosphate availability by preventing its adsorption on to clay minerals. Tannins also precipitate proteins through hydrophobic bonding and van der Waals interactions. This renders the protein temporarily insoluble, thus reducing the susceptibility of organic nitrogen to degradation and loss through leaching or gaseous emissions. In temperate forests and alpine ecosystems, the formation of tannin–protein complexes is believed to be a key process for sequestering organic nitrogen during the autumn and winter months, when trees are not actively taking up nitrogen (Northup *et al.*, 1995). When soils thaw in the spring, free-living saprophytic fungi, ectomycorrhizal and ericoid mycorrhizal fungi secrete extracellular enzymes that break down the tannin–protein complex, liberating inorganic nitrogen for fungal and plant growth. As far as we know, bacteria and actinomycetes are not able to hydrolyse tannin–protein complexes. The resistance of the tannin–protein complex to biochemical degradation is influenced by the chemical composition of the tannin, the length of polymer and the chemical properties of the bound protein.

Most soils do not contain much detectable heterocyclic aromatic nitrogen, with the exception of soils that contain 'black carbon' or 'char'. Between 10 and 35% of the organic nitrogen is found as pyrrole-type heterocyclic nitrogen (Fig. 9.4) in soils that contain charred materials, thermally altered plant materials and in the deep layers of anaerobic peat soils. The origin of these materials may be indole- and pyrrole-containing biopolymers, as well as porphyrins from chlorophyll or pigments. Yet, heteroaromatic nitrogen is not found in any significant quantity in young peats or in soils that were not subject to fire during their development. This suggests that chemical reactions, rather than biological processes, are responsible for the polymerization

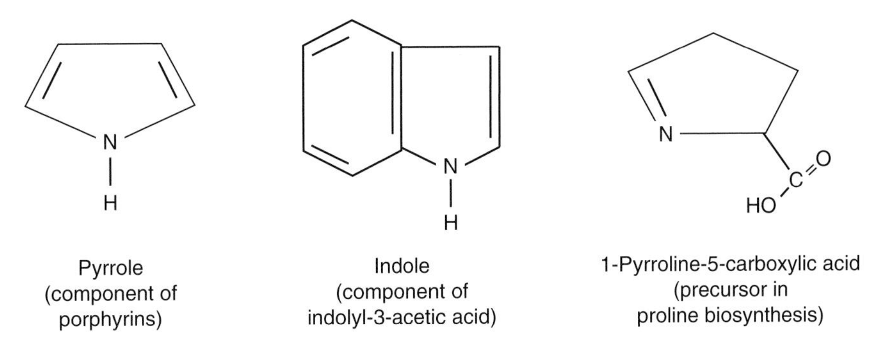

Fig. 9.4. Heteroaromatic nitrogen compounds of plant origin.

and condensation of preserved peptide-like material. The chemical structure of pyrroles and other heterocyclic nitrogen (chlorophyll, purines/pyrimidines, indoles and imidazoles) makes them difficult to decompose, so heteroaromatic nitrogen is considered to represent a sink of refractory soil organic nitrogen.

Nitrogen mineralization

The microbially mediated processes leading to the breakdown of complex organic nitrogen into simpler substances is referred to as nitrogen mineralization. The breakdown of organic nitrogen compounds is achieved through the action of extracellular proteolytic enzymes such as proteases or peptidases that are released into the soil solution by soil microorganisms. Smaller peptides such as amino acids, amines, amides, urea and nucleic acid fragments can be transported into the microbial cell, where intracellular enzymes further hydrolyse the material. This process is considered to be a two-step reaction, consisting of sequential aminization and ammonification reactions, with ammonium (NH_4^+) as the final product (Fig. 9.5).

The reaction rate can be limited by a number of factors, including the rate of extracellular enzyme production by soil microorganisms; diffusion of extracellular enzymes to the substrate and possible inactivation due to binding with soil organo-minerals; hydrolytic capacity of the extracellular enzyme; speed with which the product diffuses to microbial cell and is actively transported through the cell membrane; and the hydrolytic capacity of intracellular enzymes. In most cases, the rate-limiting step in nitrogen mineralization is the conversion of high-molecular weight proteins into low-molecular weight products (step 1 in Fig. 9.5). Measuring the activity of extracellular enzymes can provide an idea of the speed at which nitrogen mineralization occurs in a particular soil. Extracellular enzymes vary in their specificity, and may either hydrolyse a broad array of proteins or target specific proteins. Some examples of the extracellular enzymes important for nitrogen mineralization are listed in Table 9.8.

As proteins, extracellular enzymes are susceptible to interactions with clays and organic matter, as shown in Fig. 8.14. Depending on how the enzyme is bound to organo-minerals, it may retain its catalytic ability for a long period of time. Soil sterilization by steam treatment (pasteurization), heat and pressure (autoclaving) and gamma irradiation do not destroy its ability to hydrolyse substrates, indicating the extreme stability of extracellular enzymes bound to some organo-mineral surfaces. However, if the catalytic site(s) and protein conformation are affected during the binding process, the enzyme will have a lower hydrolytic capacity, if any at all.

Following the breakdown by extracellular enymes, low-molecular weight products are transported or diffuse through the microbial cell membrane for further hydrolysis by intracellular enzymes. Amino acids possess amine ($R\text{-}C\text{-}NH_2$) and amide ($R\text{-}NH_2\text{-}C{=}O$) functional groups that undergo degradation. Nitrogen in the amine compounds is released by amino-acid dehydrogenase and amino-acid oxidases in a process called deamination. The breakdown of an amide group is shown in Fig. 9.5, and involves an amidohydrolase that is specific for a particular amino acid (e.g. L-aspartine is hydrolysed by L-aspartase, L-glutamine is hydrolysed by L-glutaminase).

Amino sugars are metabolized in two steps. First, the amino sugar is phosphorylated by a kinase and then ammonia is released through deamination.

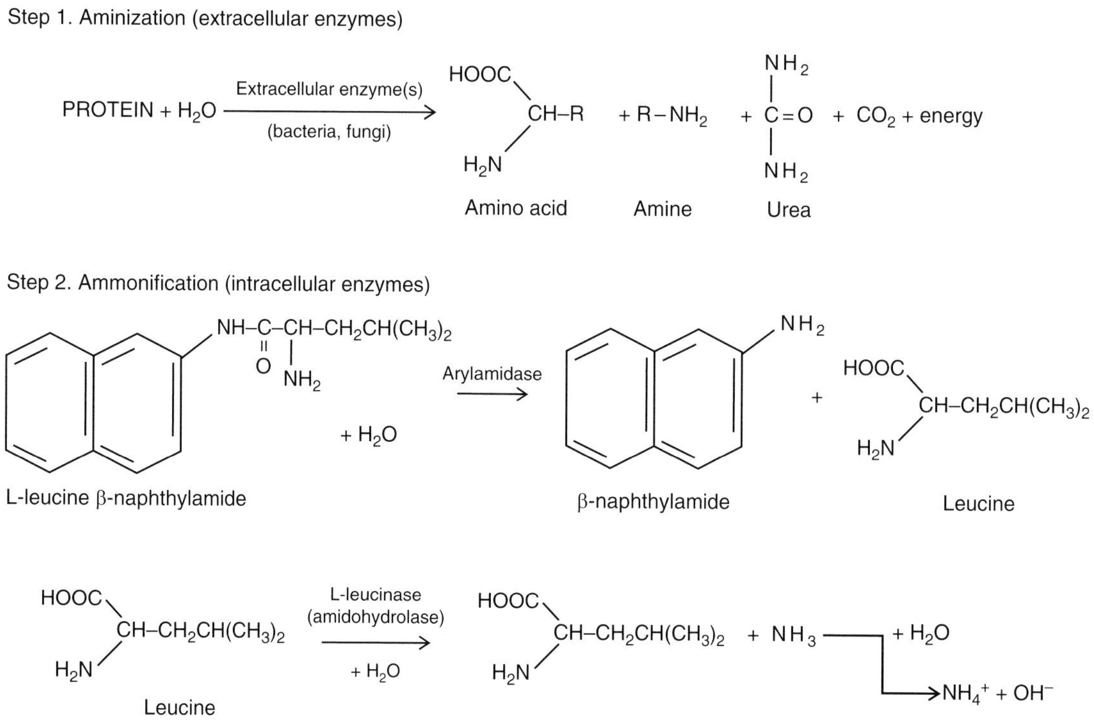

Step 1. Aminization (extracellular enzymes)

$$\text{PROTEIN} + H_2O \xrightarrow[\text{(bacteria, fungi)}]{\text{Extracellular enzyme(s)}} \underset{\text{Amino acid}}{\text{HOOC–CH(R)–}NH_2} + \underset{\text{Amine}}{R–NH_2} + \underset{\text{Urea}}{NH_2–C(=O)–NH_2} + CO_2 + \text{energy}$$

Step 2. Ammonification (intracellular enzymes)

$$\underset{\text{L-leucine } \beta\text{-naphthylamide}}{C_{10}H_7–NH–C(=O)–CH(NH_2)–CH_2CH(CH_3)_2} + H_2O \xrightarrow{\text{Arylamidase}} \underset{\beta\text{-naphthylamide}}{C_{10}H_7–NH_2} + \underset{\text{Leucine}}{\text{HOOC–CH(}NH_2\text{)–}CH_2CH(CH_3)_2}$$

$$\underset{\text{Leucine}}{\text{HOOC–CH(}NH_2\text{)–}CH_2CH(CH_3)_2} \xrightarrow[+ H_2O]{\text{L-leucinase (amidohydrolase)}} \text{HOOC–CH(}NH_2\text{)–}CH_2CH(CH_3)_2 + NH_3 \xrightarrow{+ H_2O} NH_4^+ + OH^-$$

Fig. 9.5. The process of nitrogen mineralization. Aminization requires extracellular enzymes such as protease of bacterial or fungal origin to break bonds in the three-dimensional protein structure, transforming protein into simpler peptides such as amino acids, amines and amides (e.g. urea). Small nitrogen-containing compounds are transported inside the cell for step two of the process (ammonification). The illustration shows arylamidase catalysing the hydrolysis of the N-terminal amino acid from an amide compound, L-leucine β-naphthylamide. The amino acid, leucine, is then hydrolysed by L-leucinase, a specific amidohydrolase enzyme. The NH_3 molecule is protonated to yield NH_4^+.

Table 9.8. Extracellular enzymes involved in nitrogen mineralization and selected soil microbial groups/species known to produce the enzyme.

Substrate	Enzyme(s)	Product(s)	Microbial groups/species producing the enzyme(s)
Protein	Proteinases, endo-protease	Peptides, amino acids	*Caulobacter*, *Microbacteria*, *Streptomycetes*, *Trichoderma*, *Bacillus megaterium*
Peptides	Aminopeptidase Carboxypeptidase	Amino acids	*Pseudomonas fluorescens Bacillus cereus*, *Bacillus mycoides*, *Flavobacterium-Cytophaga*
Chitin	Chitinase	Chitobiose	*Burkholderia*, *Pseudomonas*, *Trichoderma*, *Agrobacterium rhizogenes*, *Bacillus amyloliquefaciens*
Chitobiose	Chitobiase	N-acetylglucosamine	*Streptomycetes*, *Colletotrichum gloeosporioides*
Peptidoglycan	Lysozyme	N-acetylglucosamine and N-acetylmuramic acid	Any microorganism capable of degrading bacterial cell walls
DNA and RNA	Endonucleases, exonucleases	Nucleotides	Archaea, bacteria and fungi
Urea	Urease	NH_3 and CO_2	Some ammonia-oxidizing bacteria (*Nitrosomonas, Nitrosococcus, Nitrosospira*)

Nucleotide degradation requires several steps. First, nucleotides are hydrolysed to produce nucleosides and inorganic phosphate. Then, nucleosides are further hydrolysed to purine or pyrimidine bases and pentose sugars. The catabolism of purine bases produces a urea intermediate and finally ammonia. Pyrimidine bases undergo ring cleavage, and the usual end-products of catabolism are beta-amino acids plus ammonia and carbon dioxide.

Once NH_4^+ is produced in the microbial cell, it can be used for microbial growth and reproduction. The metabolic processes that permit microbial cells and other organisms, including plants, to convert ammonium into amino acids and eventually proteins are the glutamate dehydrogenase pathway and the glutamine synthetase–glutamate synthase (GOGAT) pathway. If the microbial cell has a surplus of ammonium, it will release the NH_4^+ ions into the soil solution. Other factors that lead to nitrogen release from microbial cells include predation, starvation, fluctuations in water potential or freezing/thawing cycles. In these situations, NH_4^+ and organic nitrogen compounds are released into the soil solution upon lysis of the cell membrane.

Environmental factors controlling the nitrogen mineralization rate

Nitrogen mineralization is a biological process controlled by the activity of soil microorganisms. Thus, any change in the environment that is favourable for microbial growth and metabolism should increase the nitrogen mineralization, and the converse. Some of the key environmental factors in soils are listed below:

- Aeration: well-aerated soils have a higher mineralization rate than water-logged soils.
- Soil moisture content: between 50 and 80% of field capacity is probably optimal for nitrogen mineralization.
- Temperature: from 25 to 35°C is optimal for microbial activity.
- Soil pH: neutral to slightly alkaline conditions (pH 6.5–8.0) are most favourable for nitrogen-mineralizing bacteria, whereas slightly acidic conditions (pH 5.5–6.5) are ideal for fungi.
- Physical size and location of organic residues: smaller particles have a larger surface area that permits microbial colonization and breakdown. Organic particles on the soil surface are more slowly decomposed than those mixed in the soil profile. At the aggregate scale, organic molecules that are bound within aggregates are less accessible and more resistant to decomposition than free organic matter (i.e. dissolved organic compounds in the soil solution) or adsorbed organic matter (i.e. found on soil surfaces).
- Organic residue chemistry: organic residues with high lignin content are harder to decompose and thus slow nitrogen mineralization. The total nitrogen content and C:N ratio of the material are also very important.

Microbial cells retain the NH_4^+ they need to fulfil their metabolic requirements, which is referred to as nitrogen immobilization or ammonium consumption. When the NH_4^+ concentration in the cell exceeds its metabolic requirements, the excess is released into the soil solution. This is referred to as nitrogen mineralization or ammonium production. The general principle is that microorganisms will retain NH_4^+ in their cells when their growth is limited by nitrogen availability. In a situation where there is more nitrogen than actually needed to support microbial growth, then the microorganisms will release NH_4^+ into the soil solution. The second scenario is more common, and indicates that the microbial cell has sufficient nitrogen, but is limited by another essential growth substance, generally carbon. The NH_4^+ released from microbial cells becomes available for uptake by plants, which is essential for the primary production of all non-leguminous plants and benefits plants that get some, but not all, of their required nitrogen from biological N_2 fixation.

When plant or animal residues are applied to soils or enter the soil through natural litterfall, we can predict the net effect on the NH_4^+ concentration in the soil solution. For instance, adding a carbon-rich material such as maize cobs that contain very little nitrogen will cause the microbial community to absorb the soluble nitrogen needed for its growth from the soil solution at a rapid rate, reducing drastically the concentration of available NH_4^+ in the soil solution. If a nitrogen-rich material such as barnyard manure is added to the soil, the microorganisms will break down the proteins to obtain the carbon they need to support their growth, and the excess nitrogen will be released into the soil solution, increasing the NH_4^+ concentration.

There is a critical carbon:nitrogen ratio in organic materials that determines whether microorganisms

Table 9.9. C:N ratios of selected organic materials.

Organic substance	C:N ratio*	Organic substance	C:N ratio*
Soil bacteria	5:1–8:1	Straw (cereal crops)	80:1
Biosolids (aerobic)	6:1	Timothy hay	80:1
Soil fungi	9:1–22:1	Bitumens, asphalts	95:1
Soil organic matter	10:1	Maize cobs	104:1
Clover (immature)	12:1	Coal liquids, shale oils	125:1
Municipal garbage	15:1	Red alder sawdust	135:1
Biosolids (anaerobic)	16:1	Paper (newspaper)	172:1
Lucerne hay	18:1	Oak residue	200:1
Barnyard manure (rotted)	20:1	Hardwood sawdust	250:1
Grass clippings (fresh)	20:1	Douglas fir bark	295:1
Clover residues (mature)	23:1	Crude oil	400:1
Green rye	36:1	Douglas fir sawdust	728:1
Peat moss	58:1	Pine sawdust	729:1
Maize/sorghum residues	60:1		

*The C:N ratio is the carbon content (by mass) divided by the nitrogen content (by mass) in an organic material. The value can be presented as a ratio, e.g. C:N ratio = 5:1 or as a number, e.g. C:N ratio = 5.

will absorb NH_4^+ from the soil solution (nitrogen immobilization) or release NH_4^+ into the soil solution (mineralization). Since most fresh organic materials contain 45% carbon by mass, the carbon:nitrogen ratio is affected primarily by the nitrogen content of the material. The carbon:nitrogen ratio (C:N ratio) of common organic materials is provided in Table 9.9.

Decades of research have shown that NH_4^+ is released into the soil solution when organic amendments with a C:N ratio less than 20 are added to the soil. As seen in Table 9.10, soil mixed with plant residues having a C:N ratio ≤20 provided nitrogen that was absorbed by young tomato seedlings. The nitrogen content in the tomato seedlings was from 0.31 to 2.25 mg nitrogen, which was more than the nitrogen content of tomato seedlings from the control soil (no residues added), and the mass of the tomato plants was up to four times greater than in the control. When plant residues with a C:N ratio > 20 were added, tomato seedlings had less nitrogen and plants did not grow very well, compared to the control. This indicates that residues with a C:N ratio ≤ 20 supply NH_4^+ to plants through nitrogen mineralization, and that residues with a C:N ratio > 20 do not because most of the ammonium is retained in microbial cells (immobilized).

Why do we see such dramatic nitrogen immobilization when residues with a C:N ratio > 20 are applied to soils? Typically, the growth and activity of heterotrophic microorganisms (bacteria, actinomycetes and fungi) is limited by the carbon supply, so adding organic materials that contain a lot of carbon, relative to nitrogen and other nutrients, promotes the growth of the heterotrophic microorganisms as well as soil fauna whose diet consists of carbon-rich organic residues and microorganisms. At some point, the microorganisms will not

Table 9.10. Nitrogen mineralized from vegetable residues mixed with soil, determined by the nitrogen uptake by tomato seedlings after 4 weeks of growth. (Data from Iritani and Arnold, 1960.)

Plant residue	C:N ratio of residues*	Total N uptake (mg)	Tomato mass (g, dry weight)
Control soil	–	0.294	0.34
Tomato stems	45	0.051	0.10
Maize roots	48	0.007	0.07
Maize stalks	33	0.038	0.09
Maize leaves	32	0.020	0.08
Tomato roots	27	0.029	0.09
Collard roots	20	0.311	0.33
Bean stems	17	0.823	0.76
Tomato leaves	16	0.835	0.72
Bean leaves	12	1.209	0.95
Collard stems	11	2.254	1.50
Collard leaves	10	1.781	1.33

*Residues above the dashed line have a C:N ratio > 20. Those below the dash line have a C:N ratio ≤20.

be able to support further growth because they will run out of nitrogen in their bodies and there is little nitrogen to be gained from a carbon-rich residue. This leads to nitrogen immobilization, which reduces NH_4^+ concentration in the soil solution and can have a negative impact on the growth of young plants, as shown for tomato seedlings in Table 9.10.

How long is nitrogen immobilized when a plant residue or any organic material with a C:N ratio > 20 is added to the soil? The typical pattern of nitrogen immobilization and nitrogen mineralization is illustrated in Fig. 9.6.

For simplicity, the graph in Fig. 9.6 can be examined at five stages:

1. The initial soil conditions show a measurable quantity of plant-available NH_4^+ in the soil solution, as well as CO_2 produced from soil microbial respiration.

2. A quantity of maize residue having a C:N ratio of 60 (Table 9.9) is mixed with the soil. Normally producers do not deliberately collect and add maize residues to a field; this is the residue left after harvest that is ploughed down in the field, either in the autumn or the spring. The maize residue is rich in carbon, an energy source for the soil microorganisms. There is a rapid increase in the CO_2 level as soil microorganisms break down the maize residue and use the energy released during decomposition for cellular growth and reproduction.

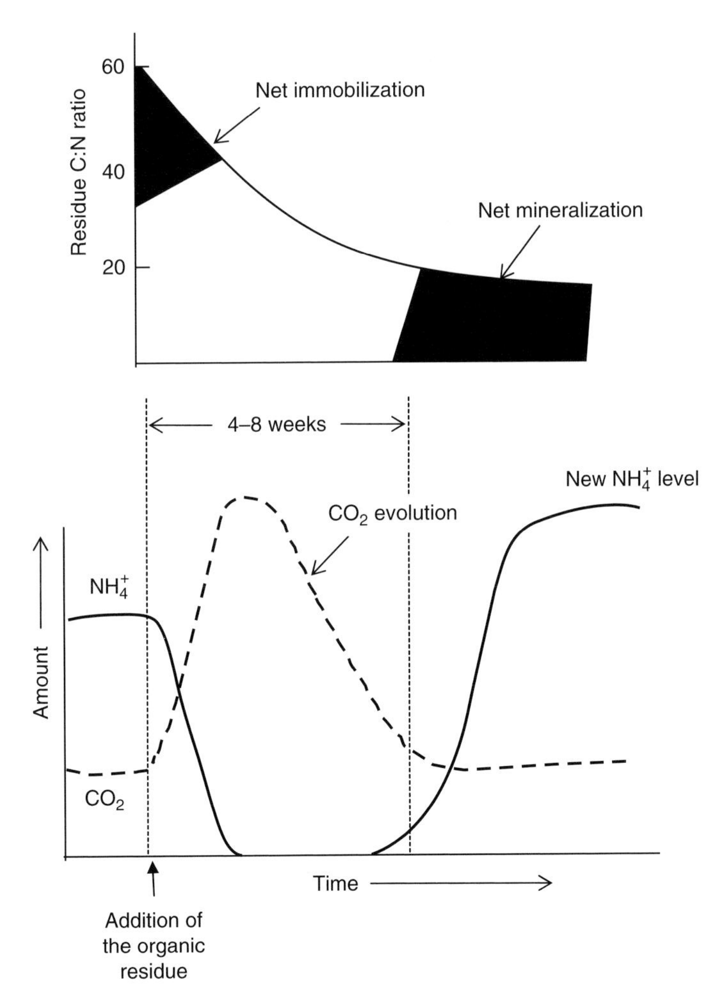

Fig. 9.6. General trend of N mineralization and immobilization following addition of organic residues to soil in temperate agroecosystems.

3. Microorganisms, like all living creatures, need nitrogen for amino acid production, protein synthesis and to build DNA/RNA. Cell walls of bacteria consist of peptidoglycan and the hyphae and spores of many fungi contain chitin, a nitrogen-rich substance. Thus, microbial growth will be limited by a lack of nitrogen. There is not enough nitrogen in the maize residue to meet the demands of the microbial community. They acquire the rest of the nitrogen they need from the soil solution, depleting the NH_4^+ concentration to a very low level. When the nitrogen is used by microorganisms (immobilized), it is not available to plants. Due to their high metabolic activity, microorganisms are much more efficient at taking nitrogen from the soil solution than plant roots.
4. After a period of 4–8 weeks, the soil microorganisms have used up much of the carbon that was in the maize residue. The C:N ratio of the residue declines from C:N = 60 to about C:N = 20 or less. As the food supply runs out, the microbial cells are less active (CO_2 level declines) and some die.
5. The death and lysis of microbial cells releases NH_4^+ into the soil solution (mineralization). The new NH_4^+ level reflects the original NH_4^+ immobilized from the soil solution plus NH_4^+ that was released through the breakdown of proteins, peptides and amino acids from the maize residue. Microbial activity has fallen to initial levels, so microbial cells do not have a high requirement for soluble nitrogen. The NH_4^+ in the soil solution is available for plant uptake.

Agricultural producers need to be able to predict the plant-available nitrogen supply when plant residues and organic fertilizers are added to the soil. From an agronomic point of view, it is not a very good idea to add a residue that will immobilize the NH_4^+ from soil solution when young seedlings are growing or transplanted into the field. Such a practice would lead to a nitrogen deficiency in the crop during vegetative growth, which is the most critical phase of crop development. To ensure the production of strong, healthy plants with sufficient nitrogen and a good grain yield at the end of the growing season, it is an excellent idea to add an organic residue with a C:N ratio $\leq$ 20 just before planting the crop or in the early vegetative growth stages. The organic nitrogen contained in the residue will be mineralized, releasing NH_4^+ into the soil solution.The sample calculation in Focus Box 9.2 gives an idea of how much nitrogen will be immobilized when residue with a high C:N ratio is mixed with an agricultural soil.

Focus Box 9.2. Sample calculation for N mineralization/immobilization.

A producer ploughs down 6000 kg/ha (dry matter) of maize residues to the soil after harvesting grain maize. The residues contain 45% C and 0.75% N. In the next year, soil microorganisms will decompose the residues, using 60% of the C in the residues for growth (biomass production) and 40% of the C in the residues as a source of energy (respired as CO_2).

a) What is the C:N ratio of the residues?

45% C ÷ 0.75% N = 60 C:N ratio = 60 : 1

b) How much C is contained in the residues?

6000 kg residues/ha × 45% C = 2700 kg C/ha

c) How much C is used by microorganisms for microbial growth?

2700 kg C/ha × 60% for growth = 1620 kg C/ha

d) How much N is needed for microbial growth (assume microbes have a C:N ratio of 8:1)?

1620 kg C/ha = 8 units C
? kg N/ha 1 unit N
Answer = 202 kg N/ha

e) How much N is contained in the residues?

6000 kg residues/ha × 0.75% N = 45 kg N/ha

f) Calculate the amount of N immobilized by soil microorganisms.

Microorganisms require 202 kg N/ha and residues contain 45 kg N/ha
202 kg N/ha – 45 kg N/ha = 157 kg N/ha

In the sample calculation, we assumed that the soil microorganisms had a C:N ratio of 8:1, although this

Continued

Focus Box 9.2. *Continued*

varies among microbial groups. If we apply glucose, a simple sugar, to a soil sample, we can use the microbial C:N ratio to determine how much nitrogen each group of microorganisms will immobilize. Clearly, low-nitrogen fungi will require much less nitrogen than high-nitrogen bacteria due to differences in their C:N ratio (Fig. 9.7).

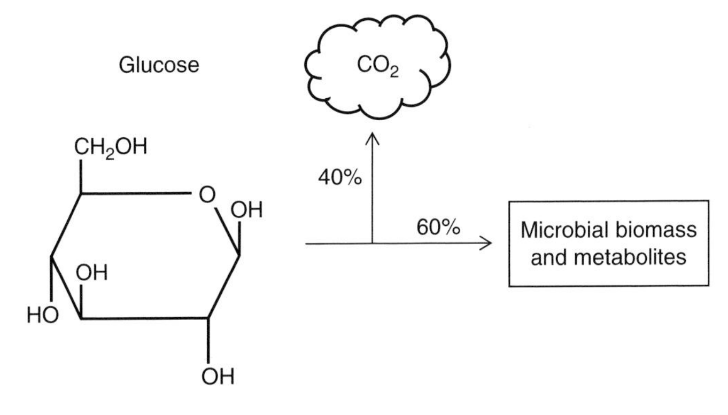

Inputs	High-N bacteria	Soil microorganisms	Low N fungi
Carbon from glucose	40 g	40 g	40 g
Microbial carbon	24 g	24 g	24 g
Microbial C:N ratio	4:1	8:1	15:1
Nitrogen needed	6 g	3 g	1.6 g
Preferred substrate C:N ratio	6.7	13.3	25

Fig. 9.7. Nitrogen requirements for growth of microorganisms with different C:N ratios, given 100 g of glucose, containing 40% C. The main assumptions are that 40% of the carbon will be used for microbial respiration and 60% of the carbon will be allocated for microbial biomass and metabolites. The preferred substrate C:N ratio refers to the properties of a glucose plus nitrogen mixture that would be needed to support the production of microbial biomass and metabolites. (Based on Paul and Clark, 1989.)

Focus Box 9.3. Solving the global nitrogen problem: it's a gas!

The importance of nitrogen for plant and animal production on Earth is undeniable. Historically, terrestrial ecosystems with low soil fertility, especially due to a low concentration of plant-available nitrogen, tended to have lower primary production than those with a large nitrogen supply. Animals that consumed nitrogen-deficient plants may have lacked some of the dietary protein needed for optimal growth and functioning. However, the situation has changed dramatically in the past century. Combustion of fossil fuels, industrial activities including the Haber–Bosch process and intensive livestock production have accelerated the nitrogen cycle and increased the amount of 'reactive' nitrogen on Earth.

Continued

Focus Box 9.3. *Continued*

'Reactive' nitrogen refers to forms of nitrogen that are chemically and biologically reactive (Table 9.11). The gaseous forms of reactive nitrogen are of particular concern because they are rapidly transported through the atmosphere and are deposited in aquatic and terrestrial ecosystems that may be far from their point of origin. Elevated nitrogen deposition in the landscape is linked to the acidification of soils and lakes, and can alter the balance from low nitrogen to 'nitrogen saturation' conditions. In forests, the decline of some woody species and changes in herbaceous plant communities have been linked to excess nitrogen deposition. Excess nitrogen from terrestrial ecosystems and lakes is eventually transported into rivers and coastal waters. The eutrophication of water bodies with nitrogen as well as phosphorus has been linked to uncontrolled algal blooms, anoxic conditions in the water column and a decline in the populations of fish and other marine organisms affected by low oxygen concentrations. The 'dead zone', a large, hypoxic saltwater body in the Gulf of Mexico discovered in the 1980s off Lousiana (USA), is the result of huge amounts of reactive nitrogen that have been transported from the agricultural fields of the USA through the Mississippi river and its tributaries to offshore water.

To solve the global nitrogen problem, it is clear that we need to reduce the quantity of reactive nitrogen on Earth. This could be achieved by cycling reactive nitrogen back to its inert form, N_2 gas, through the process of denitrification. Unfortunately, past attempts to estimate the efficiency of achieving this at scales relevant to pollution and health problems have been fraught with uncertainties, arising mainly from large variability in denitrification rates, difficulty in detecting N_2 fluxes from denitrification above the background level of 78% N_2 already present in the atmosphere, and the involvement of multiple reactants and products in other nitrogen-cycling processes. As Kulkarni *et al.* (2008) have pointed out, researchers are developing methods to quantify large-scale denitrification that address some of these issues. Some of the recent advances include: (i) novel techniques that allow detection of small changes in N_2 concentrations due to denitrification; (ii) models that can identify 'hot spots' of denitrification in the landscape; and (iii) remote-sensing technologies that provide better data for denitrification models.

These developments, and others, hold promise for advancing our understanding of denitrification and its potential to mitigate the environmental impacts of reactive nitrogen.

Table 9.11. Forms of reactive nitrogen generated from human activities, and their impact on the environment. (Adapted from Kulkarni *et al.*, 2008.)

Reactive nitrogen	Chemical formula	Source	Environmental impacts
Nitric oxide	NO (g)	Combustion of primarily fossil fuels	Component of smog, precursor to tropospheric ozone (reduces visibility, negative effect on human health)
Nitrous oxide	N_2O (g)	Biological transformation of NO_3 and NO_x via denitrification, especially from soils	Greenhouse gas (about 300× more potent than CO_2), destroys stratospheric ozone layer
Nitrogen oxide	NO_x (g)	Manure, N fertilizer and soils	Precursor to N_2O in biological denitrification process
Ammonia	NH_3 (g)	Industrial fixation (Haber–Bosch process), animal manure	Wet and dry deposition contains NH_3. Causes acidification of soils and aquatic systems. Nitrogen saturation of natural ecosystems (including eutrophication)
Nitrate	NO_3^- (aq)	Manure, N fertilizer and soils	Easily leached from terrestrial ecosystems, reduces water quality for human and animal consumption. Contributes to eutrophication of waterways. Precursor to N_2O in biological denitrification

Nitrification

Not all of the NH_4^+ produced from the mineralization of organic residues is immobilized by microorganisms or absorbed by plants. Ammonium can be volatilized through chemical reaction (especially at alkaline soil pH values) and thus lost from the soil, adsorbed to cation exchange sites or fixed in clay lattices. In most agricultural soils, the NH_4^+ concentrations in soil solution are quite low because most of the excess ammonium (NH_4^+) is rapidly converted into nitrate (NO_3^-) through the nitrification process. The three key reactions involved in this process are shown in Fig. 9.8.

The biochemical reactions in Fig. 9.8 are mediated by chemo-autotrophic microorganisms, which derive energy from the reaction. The carbon needed to support cellular growth in these organisms comes from carbon dioxide. Two groups of autotrophs are involved: ammonia oxidizers, which convert ammonia to nitrite (reactions 1 and 2), and nitrifiers that transform nitrite to nitrate (reaction 3).

The most well known of the ammonia oxidizers is the autotrophic bacterium *Nitrosomonas*. Although it has been studied extensively, *Nitrosomonas* is not the most abundant ammonia oxidizer in most soils (Table 9.12). Ammonia oxidation requires a membrane-bound ammonia monooxygenase enzyme produced by the ammonia monooxygenase (*amo*) gene cluster. This enzyme has a broad catalytic ability and can also oxidize methane (CH_4), albeit at a lower rate than methane-specific enzymes, and is involved in the co-metabolism of several small organic compounds such as trichloroethylene, chlorinated ethanes and chloroform.

Ammonia oxidizers possess the hydroxylamine oxidoreductase enzyme, which is found in the periplasmic fluid between the inner and outer membranes of bacteria. Hydroxylamine oxidation is required to transform hydroxylamine into nitrite, but it does not produce NO_2^- exclusively. Important by-products of this reaction are nitroxyl radicals (HNO), nitric oxide (NO) and nitrous oxide (N_2O). The production of the gaseous by-products NO and N_2O is of interest because it can lead to nitrogen losses from soils during nitrification. While the quantity of nitrogen lost through this process is not expected to be very large compared with the nitrogen requirements of plants, it represents a small, continual loss of nitrogen from aerobic soils.

Ammonia-oxidizing autotrophic bacteria remain an important group capable of catalysing the first step of this reaction, but they are not the only microorganisms that possess the ammonia monooxygenase enzyme (Table 9.12). Amplification of ammonia-oxidizing genes from environmental samples revealed that the Crenarchaeota possess homologous genes as do ammonia-oxidizing bacteria, although there are

Reaction 1. Ammonia oxidation

$$NH_3 + O_2 + 2\,H^+ + 2\,e^- \xrightarrow{\text{Ammonia monooxygenase}} NH_2OH + H_2O + \text{energy}$$

Reaction 2. Hydroxylamine oxidation

$$\underset{\text{Hydroxylamine}}{NH_2OH} + H_2O \xrightarrow{\text{Hydroxylamine oxidoreductase}} NO_2^- + 5\,H^+ + 4\,e^-$$

Reaction 3. Nitrite is oxidized to nitrate

$$NO_2^- + 5\,H^+ + 4\,e^- \xrightarrow{\text{Nitrite oxidoreductase}} NO_3^- + 2\,H^+ + 2\,e^- + \text{energy}$$

Fig. 9.8. The nitrification process involves the progressive oxidation of ammonium (NH_4^+, oxidation state −3) to hydroxylamine (NH_2OH, oxidation state +1), nitrite (NO_2^-, oxidation state +3) and finally nitrate (NO_3^-, oxidation state +5).

Table 9.12. Examples of microorganisms capable of catalysing ammonia oxidation/hydroxylamine oxidation reactions. (Adapted from: Myrold, 1998; Hayatsu *et al.*, 2008.)

Microbial group/species	Characteristics
Autotrophic bacteria	
Nitrosomonas europaea	Obligate chemoautotroph, well-studied, widely distributed in soils, sewage and fresh water
Nitrosococcus nitrosus	Obligate chemoautotroph, produces urease, found in soil and marine environments
Nitrosolobus multiformis	Obligate chemoautotroph, most abundant genus in many soils
Nitrosospira briensis	Obligate chemoautotroph, produces urease, large populations in acid soils
Nitrosovibrio tenuis	Obligate chemoautotroph, grows as slender, curved rods that are morphologically distinct from *Nitrosolobus* (lobate cells) and *Nitrosospira* (spiral cells)
Archaea	
Nitrosopumilus maritimus	Marine organism. This chemolitho-autotroph uses NH_3 as its sole energy source
Unidentified *Crenarchaeota*	Soil and marine systems. A survey of 12 soils showed that more amoA gene copies (up to 3000x more) were produced by *Crenarchaeota* than ammonia-oxidizing bacteria
Heterotrophic bacteria	
Paracoccus denitrificans	Soil bacterium. Facultative aerobe, but can also grow as a chemolitho-autotroph. Possesses ammonia- and hydroxylamine-oxidizing enzymes
Alcaligenes faecalis	Found in soil, water and sewage. Obligate aerobe. Possesses ammonia- and hydroxylamine-oxidizing enzymes
Pseudomonas putida	Found in soil. Obligate aerobe capable of decomposing complex organic compounds, including toluene and petroleum. Possesses ammonia- and hydroxylamine-oxidizing enzymes

some distinctions between the way that the genes are clustered in Crenarchaeota versus Proteobacteria. A number of heterotrophic bacteria also possess the ammonia- and hydroxylamine-oxidizing enzymes and thus have the ability to catalyse the transformation of the first two reactions shown in Fig. 9.8.

Nitrite is the end product of the ammonia- and hydroxylamine-oxidation reactions. Many plants, animals and microorganisms exhibit NO_2^- toxicity at low concentrations, below 5 parts per million (ppm). Fortunately, NO_2^- does not accumulate in soils because it is rapidly transformed to NO_3^- by nitrifiers. These microorganisms possess nitrite oxidoreductase or other enzymes that can catalyse the transformation of nitrite to nitrate. The autotrophic bacterium *Nitrobacter* is often mentioned as a typical nitrifier. However, nitrifiers are a very diverse group that includes autotrophic bacteria as well as heterotrophic bacteria and fungi. At present, no archaea capable of oxidizing nitrite have been found, but this still remains to be investigated since evidence of archaeal involvement in the nitrogen cycle is a recent discovery (Venter *et al.*, 2004).

The autotrophic nitrifiers can be obligate chemoautotrophs that obtain the carbon needed for growth from CO_2, or facultative chemo-autotrophs that are capable of heterotrophic growth under some conditions (Table 9.13). When nitrite is not available, *Nitrobacter winogradskyi* can use organic carbon sources as its main substrate and nitrate as an electron acceptor (thus reducing nitrate to NO_2^-, NO and N_2O). This is not considered to be its preferred metabolic pathway, as growth of this organism is much slower when it relies on heterotrophic rather than autotrophic metabolism.

Heterotrophic nitrifiers include a variety of bacteria and fungi (Table 9.13). Bacteria in the genus *Alcaligenes* have the enzyme nitrite oxidoreductase, whereas *Arthrobacter* spp. may possess a hydroxylamine-nitrite oxidoreductase enzyme that is capable of converting NH_2OH to NO_2^- before the nitrite is oxidized to nitrate. However, there are other enzymes that can oxidize nitrite. For instance, a catalase enzyme produced by *Bacillus badius* can convert NO_2^- to NO_3^-. The heterotrophic nitrifier *Burkholderia cepacia* uses nitric oxide dioxygenase

Table 9.13. Examples of microorganisms capable of catalysing the nitrite oxidation reaction. (Adapted from: Myrold, 1998; Hayatsu *et al.*, 2008.)

Microbial group/species	Characteristics
Autotrophic bacteria	
Nitrobacter winogradskyi	Facultative chemoautotroph, well-studied, widely distributed in soils, fresh water and salt water
Nitrospina gracilis	Obligate chemoautotroph and halophile. Marine organism
Nitrococcus mobilis	Obligate chemoautotroph and halophile. Aerobic metabolism. Found in marine environments
Nitrosospira marina	Obligate chemoautotroph, found in soil and aquatic environments
Heterotrophic bacteria	
Alcaligenes spp.	Soil bacterium. Facultative aerobe, but can also grow as a chemolitho-autotroph. Capable of oxidizing NH_2OH and NO_2^-
Arthrobacter spp.	Soil bacterium. Obligate aerobe. Capable of oxidizing NH_2OH and NO_2^-
Bacillus badius	Soil bacterium. Belongs to a group of aerobic, spore-forming bacteria that are versatile chemoheterotrophs capable of respiration using a variety of simple organic compounds (sugars, amino acids, organic acids)
Burkholderia cepacia	Found in soil and water. Catalase-producing bacterium
Heterotrophic fungi	
Aspergillus wentii	Soil fungus. Obligate aerobe with heterotrophic metabolism. May use amino acids as a substrate in nitrification
Penicillium spp.	Soil fungus. Obligate aerobe with heterotrophic metabolism. Capable of transforming peptone and ammonium into NO_3^-. Involved in nitrification in acidic forest soils

to produce nitrate under aerobic conditions. Several fungal species have been shown to oxidize NO_2^- to NO_3^- in pure culture and in acidic forest soils, but the biochemical mechanisms involved remain unclear.

A consequence of nitrification that should be mentioned is the impact of this process on soil pH. The hydroxylamine oxidation and nitrite oxidation reactions release H^+ ions into the soil solution, so the net effect of nitrification is localized acidification where microorganisms are active. Over a period of time, this process can lower the pH of natural and agricultural soils, depending on the ability of soils to buffer those changes.

The nitrification process is vital for primary production in many ecosystems of the world. While NH_4^+ tends to be rapidly adsorbed to cation exchange sites on organo-minerals via adsorption reactions and is generally adsorbed by plants through root interception, NO_3^- is not very tightly bound to soil surfaces and moves through mass flow. As mentioned in Chapter 7, root interception requires contact between the root surface and the soil surface where the NH_4^+ ion is bound. In contrast, mass flow permits NO_3^- to move readily in the soil solution throughout the root zone, and it is carried to the plant with water as a result of the evapotranspiration process. The NH_4^+ concentration in soil solution is generally quite low, in the order of 1–10 ppm, whereas the NO_3^- concentration can easily exceed 100 ppm in fertilized agricultural soils. The high mobility and often high concentrations of NO_3^- in some soils can pose an environmental risk because an NO_3^- molecule that is not absorbed by plants or immobilized in the biomass of microorganisms and soil fauna could be susceptible to leaching below the root zone and eventually make its way to ground- and surface-waters. Another possibility is that the NO_3^- molecule will be used as a substrate in the biological reactions catalysed by denitrifying organisms (Focus Box 9.4).

In most natural ecosystems, the loss of NO_3^- through these processes is low because this limiting nutrient is quickly absorbed by plants and other organisms through a process called assimilatory nitrate reduction. The assimilatory pathway involves the movement of NO_3^- from the soil solution into the cell, followed by the reduction of NO_3^- to NH^+, which is the precursor for the biosynthesis of macromolecules. Assimilatory nitrate reduction requires greater energy expenditure by the plant or microbial cell, relative to NH_4^+ uptake. Plants vary in their preference and ability to assimilate NH_4^+ and NO_3^-, but they generally have sufficient energy from photosynthesis to take advantage of the ionic form that is most abundant in the soil solution.

Agricultural soils that receive large inputs of nitrogen-rich fertilizer are more prone to have excess NO_3^- that can be lost from the system via leaching and gaseous emission. In the last century, agricultural scientists developed models to predict the risk of NO_3^- loss from soils and advised farmers on the most environmentally friendly techniques related to nitrogen fertilizer use in agroecosystems. However, the logistics of applying the correct amount of nitrogen fertilizer, in the right place (e.g., in the crop root zone) and at the right time to achieve maximum crop yields, still remain challenging. The biggest constraint to achieving better nitrogen fertilizer efficiency in agricultural soils is related to our inability to forecast weather accurately. Rainfall is the major factor contributing to NO_3^- losses from soils due to the high solubility of NO_3^- and its susceptibility to denitrification when soil pores are filled with water.

Environmental factors controlling the nitrification rate

Nitrification is a biological process controlled by the activity of soil microorganisms and is affected by many of the same processes that control the nitrogen mineralization rate, including:

- Aeration: soils must be well aerated for nitrification to occur.
- Soil moisture content: between 50 and 80% of field capacity is probably optimal for nitrification.
- Temperature: from 25 to 35°C is optimal for nitrifying bacteria.
- Soil pH: slightly acidic to neutral conditions (pH 6.0–7.0) are most favourable for autotrophic nitrifying bacteria. Some heterotrophic bacteria and the nitrifying fungi are active in acidic soils (pH 5.0).
- NH_4^+ supply: the reaction is limited by the amount of NH_4^+ released from nitrogen mineralization. Ammonia oxidizers and nitrifiers transform simple substrates and thus can rapidly deplete the NH_4^+ supply in soil solution.
- Nitrification inhibitors: certain bacteriocides such as nitrapyrin (N-Serve), dicyandiamide (DCD) and etradiazol (terrazole) are toxic to *Nitrosomonas* and other ammonia oxidizers. Nitrification inhibitors are typically applied as a coating on inorganic NH_4-based fertilizers. This blocks the conversion of NH_4^+ to NO_3^- for a period of several weeks or longer. Inhibitors were first developed to conserve autumn-applied nitrogen fertilizer in maize production systems, but have also been used to slow the nitrification rate of fertilizers applied in the early spring. Fertilizers treated with nitrification inhibitors are more expensive than untreated fertilizer, so their use is generally limited to high-value speciality crops. Nitrification inhibitors are not permitted in organic agriculture.

Focus Box 9.4. Transformations of nitrate: dissimilatory nitrate reduction, including denitrification and anammox reactions.

Nitrate is readily produced through nitrification. Most of the NO_3^- entering the soil solution is probably assimilated by plants and microorganisms. Some is adsorbed weakly to anion exchange sites on soil surfaces, leached below the root zone or undergoes reduction by dissimilatory processes.

There are several nitrate reduction processes known to occur in soils and sediments. The major pathway is respiratory denitrification, normally referred to as denitrification, a microbial process wherein NO_3^- is transformed to N_2 under anaerobic or hypoxic conditions according to the overall reaction:

$$2\,NO_3^- + 5\,H_2 + 2\,H^+ \rightarrow N_2 + 6\,H_2O \qquad [9.6]$$

The denitrification pathway involves the sequential reduction of NO_3^- from a +5 oxidation state to N_2 (oxidation state = 0) and requires four reductase enzymes: (see Equation 9.7 at bottom of page)

$$\underset{+5}{2\,NO_3^-} \xrightarrow[\text{reductase}]{\text{dissimilatory nitrate}} \underset{+3}{2\,NO_2^-} \xrightarrow{\text{nitrite reductase}} \underset{+2}{NO_{(g)}} \xrightarrow[\text{reductase}]{\text{nitric oxide}} \underset{+1}{N_2O_{(g)}} \xrightarrow[\text{reductase}]{\text{nitrous oxide}} \underset{0}{N_{2(g)}} \qquad [9.7]$$

Continued

Focus Box 9.4. *Continued*

The reduction of nitrate generates energy via oxidative phosphorylation, and is an example of anaerobic respiration. Denitrification is the most important reductive process that occurs when aerobic soils under forests, grasslands and agricultural management become waterlogged after rainfall, irrigation or flooding events. Anaerobic soil conditions are highly favourable for facultative anaerobes that can use nitrate as an alternative electron acceptor for oxygen.

Dissimilatory nitrate reductase (Nar) refers to enzymes that can reduce nitrate to nitrite. In bacterial cells, this includes the membrane-bound Nar enzymes (bound to the inner cell membrane) and the periplasmic Nar enzymes. These enzymes contain a metal co-factor (molybdenum, iron, copper or zinc) and labile sulfur groups. The activity and *de novo* synthesis of the Nar enzyme is inhibited by oxygen, and stimulated by NO_3^-. The reaction generates energy through oxidative phosphorylation.

Nitrite reductase (Nir) is a key enzyme in the denitrification process. Two structurally different Nir enzymes are found in the periplasm of denitrifying bacteria. The first contains copper (Cu-Nir) and the second contains cytochromes *heme c* and *heme d1* (cd1-Nir). A few strains in the genera *Pseudomonas* and *Alcaligenes, Bacillus, Rhizobium, Nitrosomonas, Thiosphaera* and others possess the Cu-Nir enzyme. The majority of denitrifying bacteria catalyse the reduction of NO_2^- to NO with the *cd1-Nir* gene, including most of the *Pseudomonas, Alcaligenes, Paracoccus, Thiobacillus* and *Azospirillum*. No functional difference in these enzymes has been reported, and the *de novo* synthesis of both is repressed by oxygen, but induced by NO_2^-.

Nitric oxide reductase (Nor) is a membrane-bound enzyme that contains cytochromes *b c* in its catalytic centre. It is believed that electron transport associated with Nor activity is linked to ATP synthesis. This enzyme is common among denitrifying bacteria, but there is variation in the enzyme structure in nature. As of late 2007, there were eight distinct structures reported for Nor enzymes that are capable of reducing NO to N_2O. Synthesis of Nor is suppressed by oxygen and induced by NO.

The nitrous oxide reductase (Nos) enzyme is located in the periplasm of denitrifier cells. This enzyme requires copper as a co-factor. In some bacterial cells, cytochrome *c* is also needed for catalysis. Nos activity is regulated by oxygen and nitrogen oxides. The Nos enzyme is also sensitive to acidic pH conditions and inhibited by sulfide and acetylene.

Until recently, it was believed that denitrification was catalysed by soil bacteria only, but current research shows that fungi and archaea possess the same or analogous enzymes and can catalyse dissimilatory nitrate reduction. A summary of enzymes and their characteristics, as well as representative microorganisms, is provided in Table 9.14. Soil fungi known to

Table 9.14. Enzymes and representative microorganisms involved in respiratory denitrification under anaerobic soil conditions.

Enzyme	Enzyme location* and characteristics	Representative organisms
Dissimilatory nitrate reductase (Nar)	Membrane-bound	Bacteria: *Paracoccus denitrificans* Fungi: *Fusarium oxysporum*
	Periplasmic	Bacteria: *Pseudomonas* sp. and *Rhodobacter sphaeroides* f. sp. *denitrificans*
Nitrite reductase (Nir)	Periplasmic, containing copper (Cu-Nir)	Bacteria: *Alcaligenes, Bacillus* and others Fungi: *Fusarium oxysporum, Leptosphaeria maculans*
	Periplasmic, containing cytochromes (cd1-Nir)	Bacteria: *Alcaligenes, Flavobacterium, Pseudomonas* and others
Nitric oxide reductase (Nor)	Membrane-bound, has cytochromes b c	Bacteria: all denitrifying bacteria
	Membrane-bound, cytochrome P450 (P450-Nor)	Fungi: all denitrifying fungi
Nitrous oxide reductase (Nos)	Periplasmic	Bacteria: all denitrifying bacteria Fungi: all denitrifying fungi

*In bacterial cells, the membrane-bound enzymes are located in the inner cell membrane and the periplasmic enzymes are found between the inner and outer cell membranes. The fungal denitrification system is located in the mitochondria, and enzymes are either bound to the internal membrane or in the periplasmic space between the inner and outer mitochondrial membranes.

Continued

Focus Box 9.4. *Continued*

reduce nitrite and simultaneously release N_2O or N_2 include organisms belonging to the Ascomycota: the plant pathogens *Fusarium oxysporum* and *Fusarium solani*, as well as *Cylindrocarpon tonkinese* and *Gibberella fujikuroii*. The Basidiomycete *Trichosporon cutaneum* is another denitrifying fungus.

Several archaea, such as the hyperthermophile *Pyrobaculum aerophilum* and the halophile *Haloferax denitrificans*, are capable of reducing NO_3^- to N_2 using the same dissimilatory nitrate reduction pathway as bacteria. However, there are clear differences in the organization of denitrifying enzyme genes, enzyme structure and enzyme regulation in archaea and bacteria. Practically nothing is known at present about the denitrifying capabilities of soil archaea, and this remains an open research field.

Other reactions leading to dissimilatory nitrate reduction

Chemical reduction of nitrate to nitrogen gases (mostly NO, with trace amounts of N_2) through dismutation of nitrite has been reported in acid soils. The reaction of NO_2^- with amino groups from organic nitrogen compounds can also lead to the production of N_2. However, these reactions are not very common and are much less important than biological processes.

In contrast to respiratory denitrification described above, microorganisms are also involved in non-respiratory denitrification that produces N_2O under aerobic soil conditions. This process does not generate energy for the organism. While this has been demonstrated for bacteria, fungi, algae, plants and animals under laboratory conditions, the ecological significance of this reaction under natural soil conditions is not known

Another process that produces N_2O or N_2 is called co-denitrification and is accomplished by fungi. The reaction involves combining NO_2^- and other nitrogen compounds (co-substrates) under anaerobic conditions. It appears that the NO_2^- induces the P450Nor enzyme, which has a broad catalytic ability. In *Fusarium oxysporum*, the P450Nor enzyme was able to form a hybrid N_2O molecule from NO and a co-substrate (azide or NH_4^+) without an electron donor. Normally, the reduction of NO to N_2O by P450Nor requires NADH or NADPH as a direct electron donor, so the co-denitrification process is a unique reaction.

Dissimilatory nitrate reduction to ammonium (DNRA) is achieved by nitrate-respiring bacteria, which convert NO_3^- to NO_2^- under anaerobic conditions to gain energy via oxidative phosphorylation. Facultative anaerobes and other bacteria can further reduce the NO_2^- to NH_4^+, although they do not tend to gain energy from this reaction. It seems likely that the reduction of NO_2^- is done to prevent build-up of NO_2^- concentrations to levels that would be toxic to the bacterial cell. The overall reaction is:

$$NO_3^- + 4\ H_2 + 2\ H^+ \rightarrow NH_4^+ + 3\ H_2O \qquad [9.8]$$

The DNRA process is very common in carbon-rich, anaerobic environments such as sediments and sewage sludge, but rare in soils.

Soil fungi also catalyse a DNRA reaction called ammonia fermentation, which involves the reduction of NO_3^- to NH_4^+ with the simultaneous oxidation of ethanol to acetate to generate ATP. In the reaction, NO_3^- is the terminal electron acceptor for fermentation. When environmental conditions are very anoxic, the soil fungus *Fusarium oxysporum* switches from denitrification to ammonia fermentation to generate energy. In fact, *F. oxysporum* can grow in an aerobic enrivonment using oxygen respiration, in a moderately anaerobic environment by switching to denitrification (nitrate respiration) and in an extremely anaerobic environment by changing to ammonia fermentation. The broad metabolic capacity exhibited by *F. oxysporum* may be common among soil fungi. Zhou *et al.* (2002) suggest that at least some of the soil fungi should be properly labelled as 'facultative anaerobes' rather than 'obligate aerobes', given their abilities as denitrifiers and ammonia fermenters. It seems that fungi have a much greater role in the soil nitrogen cycle than previously realized.

The anammox reaction (anaerobic ammonium oxidation)

In wastewater treatment facilities and marine environments there are anaerobic bacteria (*Brocadia anammoxidans, Kuenenia stuttgartiensis* and *Scalindua sorokinii*) that possess hydroxylamine oxidoreductase-like proteins and can obtain energy from the 'anammox' process:

$$NH_4^+ + NO_2^- \rightarrow N_2 + 2\ H_2O \qquad [9.9]$$

Although these bacteria are not ammonia oxidizers in the conventional sense, they have some similar genes and enzymes to organisms that live in aerobic environments. The most interesting feature of this reaction is that it rapidly transforms 'reactive' nitrogen into inert dinitrogen gas without producing harmful intermediates such as NO (g) and N_2O (g). It has been suggested that anammox bacteria are responsible for 50% of N_2 emission from oceans (Dalsgaard *et al.*, 2003; Brandes *et al.*, 2007). This unusual process is being used to remove N from wastewater treatment plant effluents and return it directly to the atmosphere as N_2 gas, which reduces the potential for environmental pollution from N-rich effluents.

9.2 The Phosphorus Cycle

Phosphorus is not as abundant in soil as nitrogen, but it is also less susceptible to loss via leaching and is not transformed into gaseous forms. The total phosphorus in soils ranges from about 0.005 to 0.15%. The primary mineral form of phosphorus is rock phosphate (apatite), which must be weathered before the soluble phosphate ions $H_2PO_4^-$ and HPO_4^{2-}, collectively referred to as orthophosphate, are released into the soil solution. Generally, the orthophosphate concentration in soil solution is very low due to the fact that these soluble forms are rapidly assimilated by biota or quickly react with soil surfaces and minerals. The phosphorus cycle can be divided into biological pools (plants, soil microorganisms, labile and stable organic phosphorus) that are distinct from geochemical pools (primary minerals, secondary minerals, adsorbed- and occluded-phosphorus) for simplicity, but the reality is that biological processes can affect the chemical reactions, and vice versa (Fig. 9.9).

The loss of phosphorus from soils through erosion and leaching was thought to be a minor process of little significance to the environment, but it is now understood that certain land management practices, especially manure disposal associated with intensive livestock production, can oversaturate the soil's capacity to adsorb phosphorus. Consequently, some of these phosphorus-saturated soils become a source of diffuse pollution because phosphates are lost from the land through drainage water, in eroded sediments and in water runoff from the soil surface. Primary production in many freshwater lakes and rivers is limited by phosphorus availability, so the phosphate input from terrestrial systems can cause eutrophication, which can produce 'blooms' of blue-green algae (cyanobacteria). An estimated 40 genera are involved, but the main ones are *Anabaena, Aphanizomenon,*

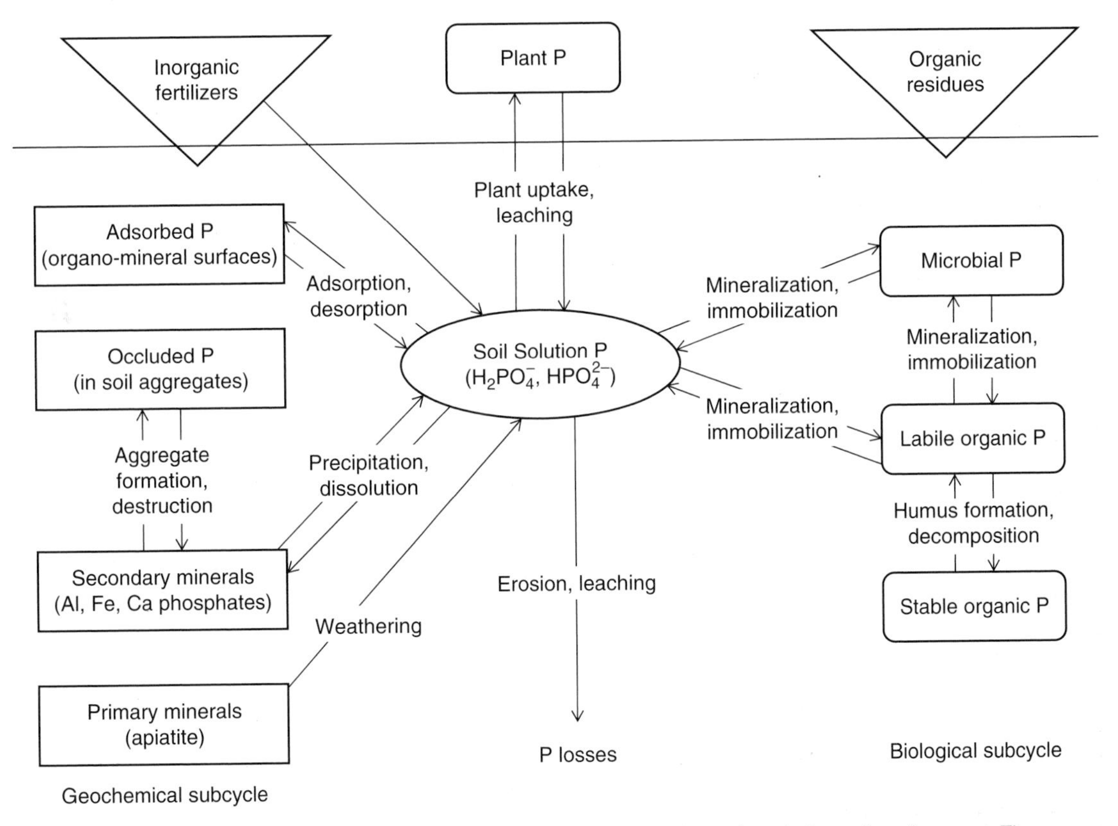

Fig. 9.9. The phosphorus cycle, showing inputs, losses and major transformations in the soil environment. The phosphorus cycle consists of two subcycles: geochemical and biological.

Cylindrospermopsis, Lyngbya, Microcystis, Nostoc and *Oscillatoria (Planktothrix)*. Not only are cyanobacterial blooms unsightly, but the decomposition of dead biomass reduces the dissolved oxygen level in the waterway and has been linked to fish kills. In addition, cyanobacteria produce toxins such as cytotoxins and biotoxins that are poisonous to wild/domestic animals and humans. It is expensive but necessary to eliminate these toxins from drinking water sources. In many regions of North America and Europe, a great deal of effort is being expended to calculate phosphorus inputs and monitor soil phosphorus build-up on agricultural lands. Riparian management programmes have led to the installation of erosion controls and construction of wetlands that can adsorb phosphates and thus prevent their entry into sensitive waterways.

The orthophosphate concentration in the soil solution is generally very low, often less than 0.1 mg soluble P/kg soil. In addition, orthophosphate is an immobile ion – it moves through the soil solution by diffusion and is prone to reaction with other ions and soil surfaces encountered. Plants require concentrations of 0.003–0.3 mg soluble P/kg soil for optimal growth. Most of the orthophosphate assimilated by plants is absorbed by unsuberized cells near root tips. Chemical equilibrium in the soil solution leads to replenishment of the orthophosphate pool as it is depleted by plant uptake, which is essential in areas where young roots are rapidly absorbing orthophosphate.

The chemical reactions affecting the orthophosphate concentration in soil solution are seen in the geochemical subcycle. The weathering process releases orthophosphate from primary minerals at a very slow rate. Secondary minerals are sparingly soluble, and may not contribute much orthophosphate to the soil solution, while occluded phosphorus is both chemically and physically protected; even if the chemical forms are dissolved, the orthophosphate is unlikely to remain soluble as it diffuses from within aggregates to the soil solution. The main source of orthophosphate in most soils probably comes from phosphates that are adsorbed on to organo-mineral surfaces through electrostatic forces (e.g. in anion exchange sites).

Orthophosphate supplied from the geochemical subcycle is not generally sufficient to meet the phosphorus requirements of plants and other biota, especially in natural ecosystems that rely on organic residue recycling as the major phosphorus input. Organic residues from plants, animal manure and dead microbial biomass contain organic phosphorus compounds that are mineralized, releasing orthophosphate into the soil solution. The pools of labile and stable organic phosphorus are distinguished by the relative ease with which the compounds in each pool can be mineralized to release orthophosphate.

Given the importance of phosphorus for energy storage and transfer, plants and microorganisms have evolved various strategies to ensure that there is sufficient orthophosphate in the soil solution to support their growth and survival. Therefore, it is important to understand the transformations of phosphorus compounds in the geochemical and biological subcycles.

Transformations of chemically bound phosphorus

As illustrated in Fig. 9.9, orthophosphate from soil solution can be adsorbed to organo-mineral surfaces or precipitated in secondary minerals. Collectively, these reactions are called phosphorus fixation. Adsorption reactions are the dominant process when there is a low orthophosphate concentration in the soil solution. The electrostatic bonds between negatively charged orthophosphate and positively charged organo-mineral surfaces are easily broken, so adsorbed phosphorus is easily desorbed and released back into the soil solution. This pool is considered to be more 'labile' than the phosphorus associated with primary and secondary minerals.

Precipitation with hydroxides and oxides of aluminium and iron (Al-P and Fe-P compounds), in alumina-silicate minerals and with calcium carbonate (Ca-P compounds) occurs when the soil solution contains a high orthophosphate concentration, such as shortly after the application of water-soluble P fertilizer. Surface precipitation and chemisorption reactions lead to the formation of compounds that are chemically stable. Once precipitated in secondary minerals, orthophosphate is released very slowly through dissolution (Fig. 9.10).

Factors influencing phosphorus fixation reactions and the phosphorus fixation capacity of a soil include:

1. Soil minerals: in acidic soils, more orthophosphate is adsorbed by 1:1 clays such as kaolinite than by 2:1 clays (e.g. montmorillonite) due to the greater quantities of aluminium and iron oxides

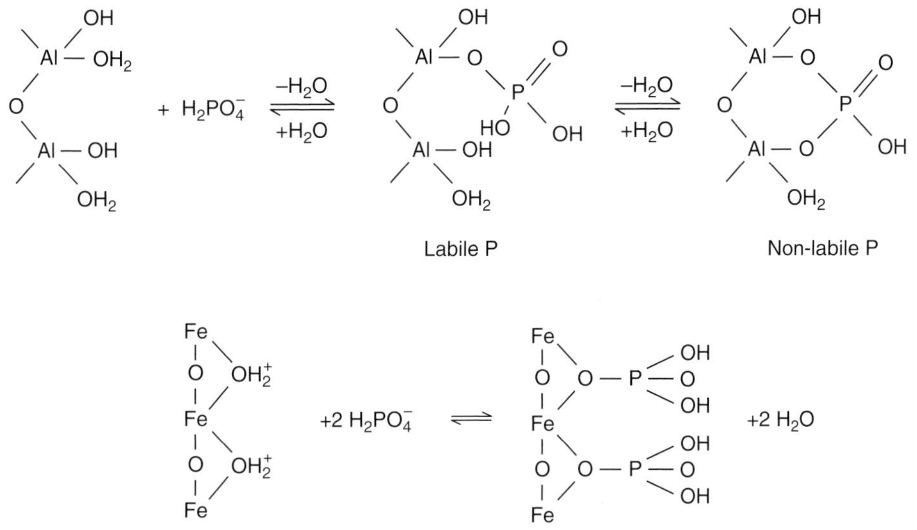

Fig. 9.10. Mechanism of phosphorus adsorption to aluminium and iron oxide surfaces. Orthophosphate bonding through one Al–O bond is considered to be labile, as it can be desorbed into the soil solution. Bonding through two Fe–O or two Al–O bonds produces a stable structure that is non-labile and unlikely to be readily desorbed.

associated with 1:1 clays. Highly weathered soils also tend to have more exposed hydroxyl groups on the alumino-silicate minerals, which will bind and fix orthophosphate. In alkaline soils, the amount and reactivity of calcium carbonate control the amount of orthophosphate adsorption and precipitation. Secondary minerals containing orthophosphate are shown in Table 9.15.

2. Soil pH: the soil pH affects the ionic form of orthophosphate in soil solution. In very acidic solutions, orthophosphate exists as H_3PO_4, but this form does not exist in solutions with pH ≥ 4.0. The form PO_4^{3-} is found in very alkaline solutions, when the pH ≥ 10.0. Since the soil solution is generally not so acidic or alkaline, the major forms of orthophosphate are $H_2PO_4^-$ and HPO_4^{2-}. These ionic forms are present in equal quantities at pH 7.2.

The adsorption of orthophosphate by aluminum and iron oxides declines with increasing soil pH because the $H_2PO_4^-$ is replaced by OH^- and HCO_3^- ions from soil solution. Adsorption of orthophosphate by calcium carbonate compounds increases as soil pH increases because HPO_4^{2-} becomes the dominant orthophosphate form in soil solution, and it readily precipitates with Ca^{2+}. The application of agricultural lime to adjust the soil pH to 6.5 will reduce precipitation of orthophosphate, leaving more in soil solution for plant uptake. The

Table 9.15. Secondary minerals containing phosphate found in acidic, neutral and calcareous soils. Minerals are listed in order of decreasing solubility, with the most soluble at the top of the column.

Acidic soils		Neutral and calcareous soils	
Name	Chemical form	Name	Chemical form
Variscite	$AlPO_4 \cdot 2\ H_2O$	Dicalcium phosphate dehydrate	$CaHPO_4 \cdot 2\ H_2O$
Strengite	$FePO_4 \cdot 2\ H_2O$	Dicalcium phosphate	$CaHPO_4$
NH_4-Flatt's salt	$Al_2NH_4(PO_4)_2OH \cdot 2\ H_2O$	Octacalcium phosphate	$Ca_4H(PO_4)_3 \cdot 2.5\ H_2O$
K-Flatt's salt	$Al_2K(PO_4)_2OH \cdot 2\ H_2O$	β-tricalcium phosphate	$Ca_3(PO_4)_2$
NH_4-taranakite	$Al_5(NH_4)_3H_6(PO_4)_8 \cdot 18\ H_2O$	Hydroxyapatite	$Ca_5(PO_4)_3OH$
K-taranakite	$Al_5K_3H_6(PO_4)_8 \cdot 18\ H_2O$	Fluorapatite	$Ca_5(PO_4)_3F$

rhizosphere tends to be more acidic than the surrounding bulk soil because roots release H^+ into the soil solution when they absorb cations. Organic acids of plant and microbial origin are another source of acidity in the rhizosphere. In alkaline soils, these acids could contribute to the dissolution of Ca-P compounds in the rhizosphere.

3. Cation effects: divalent and trivalent cations have more capacity to adsorb and precipitate orthophosphate than monovalent cations. Cations with a larger hydrated radius have a greater adsorption capacity than those with a smaller hydrated radius. The adsorption capacity of soil cations on mineral surfaces and in soil solution for orthophosphate is: $Al^{3+} > Fe^{3+} > Fe^{2+} > Ca^{2+} > Mg^{2+} > K^+ > NH_4^+ \geq Na^+$

4. Anion effects: both inorganic and organic anions can compete with orthophosphate for adsorption sites, resulting in decreased adsorption of orthophosphate. The strength of bonding of the ion to the mineral surface determines the competitive ability of that anion. Among the inorganic anions, the relative adsorption order would be: $OH^- > H_2PO_4^- > H_3SiO_4^- > HPO_4^{2-} > MoO_4^{2-} > SO_4^{2-} > HCO_3^- > NO_3^- \geq Cl^-$.

Negatively charged organic acids from plants, soil microorganisms and released from decomposing organic residues also compete with orthophosphate for binding sites on organo-mineral surfaces. Among the soil microorganisms, the bacterial genera *Bacillus* and *Pseudomonas*, as well as fungi belonging to *Penicillium* spp. and *Aspergillus* spp., are often termed 'phosphate-solubilizing microorganisms'. Organic acids of biological origin (Table 9.16) such as citrate, oxalate, tartrate and malate are well known for their ability to displace orthophosphate from anion exchange sites. Another role could be to lower the pH in the soil solution surrounding the microbial cell, which would alter the chemical equilibrium and possibly release orthophosphate into soil solution. Organic acids dissolve Al-P and Fe-P complexes, accelerate the weathering of primary minerals (including apatite-rich rock phosphate fertilizer), and can form stable complexes with hydrous aluminium and iron, which reduces orthophosphate adsorption and precipitation.

Table 9.16. Sources of organic acids in soils.

Source	Organic acids
Plant roots	Acetate, aconitate, citrate, fumarate, glycolate, isocitrate, lactate, malate, oxalate, succinate, tartaric acid
Bacteria	Acetate, adipic acid, butyrate, formic acid, fumarate, gluconate, glyconate, 2-ketogluconate, lactate, malate, malonate, propionate, succinate
Ectomycorrhizal fungi	Citrate, oxalate
Free-living fungi	Citrate, fumarate, malate, oxalate

5. Extent of P saturation: more orthophosphate adsorption occurs in soils with little orthophosphate adsorbed to organo-mineral surfaces or precipitated in secondary minerals. Soils possess a finite capacity to bind orthophosphate through phosphorus fixation reactions. Repeated applications of phosphorus-rich materials such as animal manure can increase the orthophosphate concentration to such an extent that all accessible binding sites are filled. Phosphorus-saturated soils are expected to have a much higher orthophosphate concentration in soil solution than soils with low phosphorus fertility, and they pose a risk for environmental pollution.

6. Soil water content (flooding): soils that are waterlogged and anoxic for a period of time are likely to see Fe^{3+} transformed into Fe^{2+} through redox reactions. The conversion of Fe^{3+}-P compounds, which are relatively insoluble, to Fe^{2+}-P compounds (more soluble) leads to orthophosphate release in the soil solution. Similarly, flooded soils exhibit accelerated weathering/dissolution of orthophosphate from primary and secondary minerals, and greater diffusion of orthophosphate.

Transformations of organically bound phosphorus

The organic phosphorus pool represents about 50% of the total phosphorus in soils, with values between 15 and 80% representing the natural variation in soil due to parent materials and pedogenesis. Organic phosphorus forms account for about 1–3% of the soil organic matter. This is much less than the amount of carbon and nitrogen contained in the soil organic matter. The ratio of organic carbon:organic nitrogen:organic phosphorus in most soils is, on average, 140:10:1.3. Hence, there is more than 100 times more organic carbon than organic phosphorus, and about eight times more organic nitrogen than organic phosphorus in soils.

Soil organic phosphorus constitutes a diverse group of compounds of plant, animal and microbial origin, in various states of decay (Table 9.17).

Table 9.17. Soil organic phosphorus compounds. (Modified from Quiquampoix and Mousain, 2005.)

Compound	Percentage of organic phosphorus pool
Nucleic acids	< 3
Sugar phosphates	< 1
Monophosphorylated carboxylic acids	< 1
Phosphoamides and phosphoproteins	< 5
Phospholipids (including phosphoglycerides)	> 7
Organophosphates (from pesticides)	Variable
Inositol phosphates	< 80
myo-inositol phosphate	
scyllo-inositol phosphate	
D-*chiro*-inositol phosphate	
neo-inositol phosphate	

The chemical nature of the soil organic phosphorus fraction is studied with advanced chemical methods such as ^{31}P-nuclear magnetic resonance spectroscopy and high-performance chromatic separation-mass spectroscopy detection. The most abundant compound is inositol phosphate, which exists in several stereoisomeric forms (Fig. 9.11). The *myo*-inositol phosphate form is found in soils, plants, animals and microorganisms, while D-*chiro*-inositol phosphate is generally reported in soils and plants (seeds and leaves). The *scyllo*- and *neo*-inositol forms are thought to be exclusively of microbial origin. The inositol forms are relatively stable and resistant to decomposition, compared with other organic phosphorus compounds, which explains their abundance in the soil organic matter. Inositol phosphates are associated with clay minerals, humus, proteins and metallic ions, suggesting that they may be stabilized in organo-mineral complexes in a similar fashion to organic carbon and nitrogen (Fig. 8.14).

Most plants and microorganisms absorb orthophosphate from the soil solution, although there are a few exceptions. The alga *Ochromonas danica* was able to assimilate glucose 1-phosphate

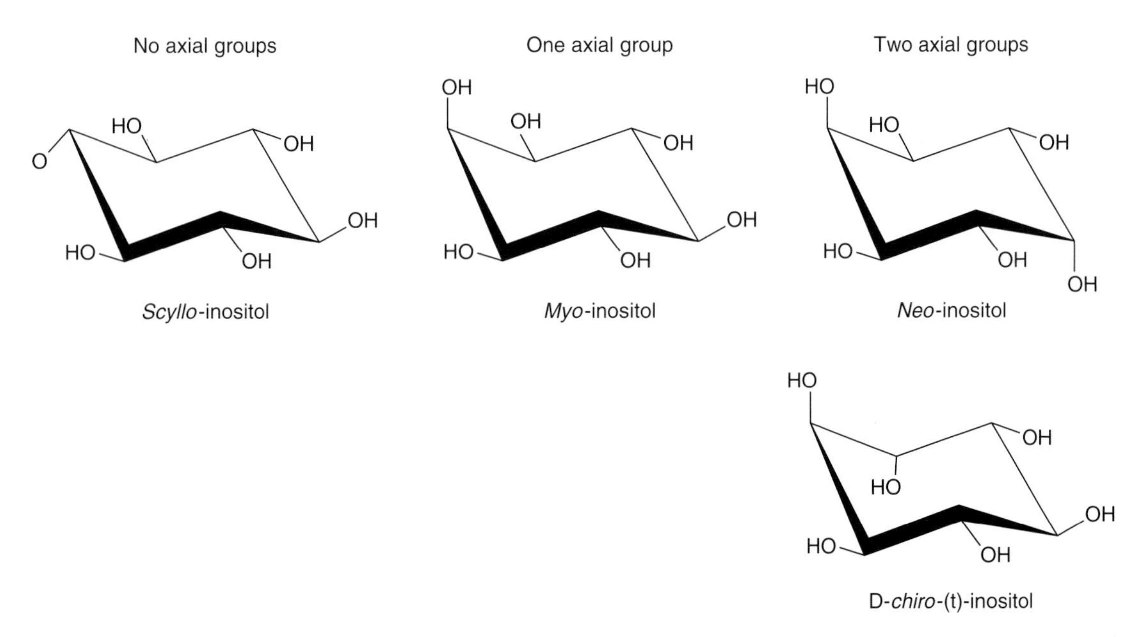

Fig. 9.11. The four inositol stereoisomers that occur in phosphorylated forms in soil. The four isomers differ only by the orientation of a single hydroxyl group. It should be noted that *scyllo*-inositol has no axial groups (all are more or less in the plane of the ring), which seems to confer resistance to enzymatic hydrolysis and accounts, in part, for its persistence in soils. Phosphate binds to the ring by displacing the H atom of the –OH side chain. Theoretically, up to six P atoms can bind to each ring (redrawn from Turner, 2007).

and glucose 6-phosphate, although this did not support optimal growth. Certain mycorrhizal fungi are able partially or totally to utilize soluble sodium and calcium phytates, but not iron phytate, under laboratory conditions (Quiquampoix and Mousain, 2005). Normally, organic phosphorus must be hydrolysed enzymatically to release orthophosphate, according to the general reaction:

$$R\text{-}O\text{-}PO_3 + H_2O \xrightarrow{\text{phosphohydrolase}} H\text{-}O\text{-}PO_3^- + R\text{-}OH \qquad [9.10]$$

All organic phosphorus compounds contain ester linkages (C-O-P) and thus can be hydrolysed by a class of enzymes broadly called phosphohydrolase enzymes. These are extracellular enzymes, which can bind to the outer cell wall of the microorganism after they are secreted or released into the soil solution (from microbial cells and plant roots). Often, the phosphohydrolases bind to soil organominerals and remain stable and active for an extended period. A variety of phosphohydrolases exist, including:

- Phosphomonoesterases hydrolyse the $H_2PO_4^-$ from monoester forms of phosphorus, such as those in nucleotides or phospholipids. These enzymes can break down a wide range of substrates under acid and alkaline conditions.
- Phosphodiesterases hydrolyse the $H_2PO_4^-$ from diester forms of phosphorus, such as those in nucleic acids.
- Phytases hydrolyse the $H_2PO_4^-$ from inositol phosphates. Two types are known: the 3-phytase, which cleaves the phosphate at the 3-position of the *myo*-inositol ring, and the 6-phytase that initiates dephosphorylation at the 6-position of the *myo*-inositol structure.

Phosphatase enzyme activity is affected primarily by soil temperature (35°C is optimal) and soil pH. Acid phosphatases have a high hydrolytic potential between pH 5.0 and 5.5, whereas alkaline phosphatases function optimally from pH 7.5 to 8.0. Phytases have a much broader range of pH optima, ranging from pH 2.2 in the yeast *Pichia farinose* to pH 7.5 in the bacterium *Bacillus subtilis*. Inhibitors of phosphatases and phytases include fluorine, polyvalent anions (phosphate, molybdate, arsenate, etc.), metal ions (silver, zinc, mercury, etc.) and chelating agents such as oxalate and tartrate. These enzymes can be activated by divalent metal ions such as calcium, magnesium and cobalt.

Mineralization of organic phosphorus in soils is affected by the ratio of organic carbon to organic phosphorus (C:P ratio) in soils and organic residues. Organic residues with a C:P ratio less than 200 will release $H_2PO_4^-$ into the soil solution. When the C:P ratio is between 200 and 300, there is no net gain or loss of inorganic phosphate. Net phosphorus immobilization in microbial cells occurs when organic residues with a C:P ratio greater than 300 are mixed with the soil. The quantities of organic P mineralized during a growing season in temperate climates range from < 1 to 20 kg P/ha/year.

9.3 The Sulfur Cycle

Soils contain between 0.002 and 10% total sulfur, with the highest concentrations found in tidal flats, saline soils, acid sulfate and organic soils. Organic sulfur is a relatively small component of the soil organic matter, compared to organic carbon and nitrogen. The ratio of organic carbon:organic nitrogen:organic sulfur in most soils is, on average, 120:10:1.4. This means there is about 85 times more organic carbon than organic sulfur, and about seven times more organic nitrogen than organic sulfur in soils. About 90% of the total sulfur is in organic sulfur forms, which must be mineralized into sulfate (SO_4^{2-}) before it can be absorbed by plants and microorganisms (Fig. 9.12). Many soils receive considerable inputs of dissolved sulfur and sulfate from wet and dry deposition. In addition, plants can absorb some SO_2 through their stomata. Sulfur deficiencies are much less widespread in agriculture than nitrogen or phosphorus deficiencies, although some arid and semi-arid regions do benefit from sulfur fertilizer applications.

Soils contain a variey of organic and inorganic forms of sulfur, with oxidation states ranging from −2 to +6. The organic sulfur forms can be grouped into two classes:

1. Organic sulfates: this group constitutes 30–75% of the total organic sulfur and includes sulfate esters (C-O-S bonds), sulfamates (C-N-S) and sulfate thioglycosides (N-O-S bonds). In forest and peat soils, the ester sulfates are 14–39% of the total sulfur (Table 9.18).
2. Carbon-bonded sulfur: this includes sulfur in amino acids, proteins, polypeptides, heterocyclic compounds, sulfinates, sulfones, sulfonates and fuloxides. The major carbon-bonded sulfur groups

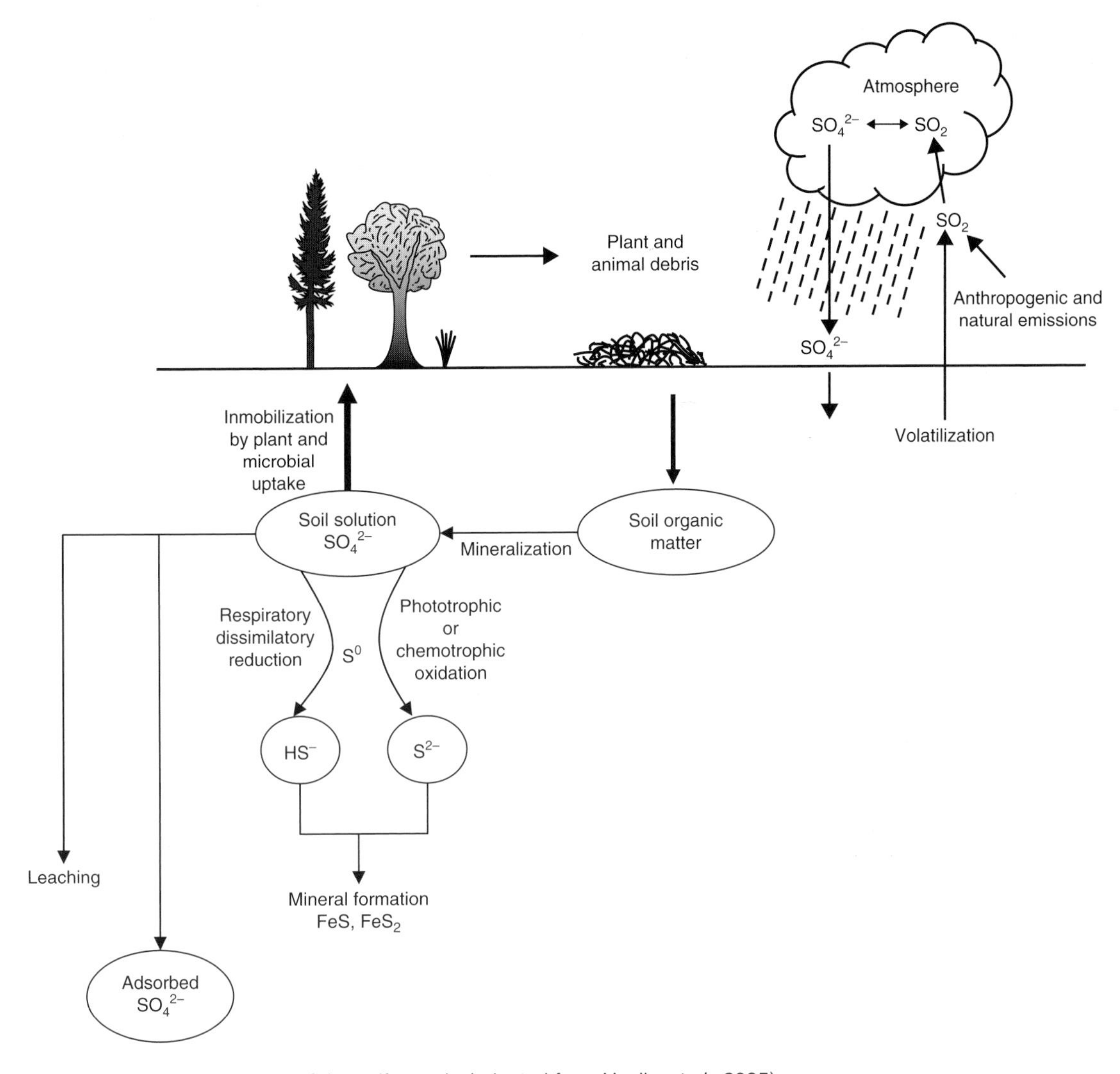

Fig. 9.12. Simplified scheme of the sulfur cycle (adapted from Havlin *et al.*, 2005).

in forest and peat soils are the organic disulfide/thiol/organic monosulfide group and sulfones (Table 9.18).

Transformations of organically bound sulfur

Organic sulfur from plant residues, animal manure and lysed microbial cells must be mineralized to release sulfate, according to the general reaction:

$$R\text{-}O\text{-}SO_3 + H_2O \xrightarrow{\text{sulfohydrolase}} HSO_4^- + R\text{-}OH \qquad [9.11]$$

The sulfatase enzymes responsible for this reaction, technically classified as sulfohydrolases, are extracellular enzymes. The best studied is arylsulfatase, secreted from plant roots and microbial cells into the soil solution. They can be stabilized on soil particles and function for many months or perhaps years to break down ester sulfates. Carbon-bonded sulfur compounds (C-S bonds) are broken down by proteases and other proteolytic enzymes. This releases amino acids containing sulfur, such as methionine and cysteine, that are then absorbed by microbial cells.

Table 9.18. Sulfur compounds found in soils, grouped by the oxidation state of the sulfur atom and percentage of total S in each class in forest and peat soils (Adapted from Prietzel *et al.*, 2007).

Sulfur species	Chemical structure	Example	Oxidation state of S atom	Percentage of total S (%)
Reduced				
Inorganic sulfide	S^{2-}	Troilite	–2	0
Elemental sulfur	S°		0	–
Organic polysulfide	R-S-S-S-R'		+0.15	–
Organic disulfide	R-S-S-R'	Cystine	+0.2	27–52
Thiol	R-SH	Cysteine	+0.5	
Organic monosulfide	R-S-R'	Methionine	+0.5	
Intermediate				
Sulfoxide	R-S=O-R'	Methionine sulfoxide	+2	10–14
Sulfite	SO_3^{2-}	Sodium sulfite	+3.7	5–7
Sulfone	R-S=O_2-R'	Phenyl sulfone	+4	15–27
Sulfonate	R-SO_2-O-X	Cysteinic acid	+5	0
Oxidized				
Ester sulfate	R-O-SO_3	Sodium dodecyl sulfate	+6	14–39
Inorganic sulfate	SO_4^{2-}		+6	–

Factors affecting sulfur mineralization by arylsulfatase are similar to those that affect nitrogen and phosphorus mineralization. The optimal soil temperature range for sulfatase is 20–40°C. The optimal soil moisture content is about 60% of field capacity. Soil pH has little effect on arylsulfatase activity. The presence of plants stimulates arylsulfatase production by microbial cells, and plant roots also secrete arylsulfatase. In general, cultivation and addition of organic fertilizers increases sulfur mineralization, although it depends on the carbon to sulfur ratio (C:S ratio) of the organic residues mixed or added to the soil. Organic residues with a C:S ratio less than 200 will release SO_4^{2-} into the soil solution. Net sulfur immobilization in microbial cells occurs when organic residues with a C:S ratio greater than 400 are mixed with the soil.

10 Soil Structure

One of the fundamental characteristics of soil is the manner in which primary particles, organic matter and pore space are arranged within three-dimensional space (soil structure) and the ability of the soil to maintain this arrangement when exposed to different stresses (soil stability). Soil structure determines how easily water infiltrates and drains through a soil and controls the rate of gas diffusion between soil and the atmosphere. Plant roots must be able to penetrate the soil fabric to gain a strong foothold that will anchor the plant and permit it to acquire water and nutrients from the soil solution. Within the soil profile are microsites where bacteria can thrive without risk of predation. Macropores allow larger fauna such as earthworms to rapidly move from the soil surface to depths of 1 m or more in the soil profile.

Soil structure can be described at various scales, ranging from the microscale to the macroscale. The size of a bacterial cell is less than 10 μm, hence soil bacteria are affected by microscale changes in soil structure. In contrast, architects and engineers responsible for building high-rise apartments, highways or dams would be concerned with soil structural stability at the macroscale. This chapter examines factors that lead to the formation of a stable soil structure, and the implications for survival of plants and soil organisms.

10.1 Soil Structure

Structure is a physical characteristic of soils that describes the mass and volume occupied by solid particles, water and gas. Primary particles, derived from the weathering of parent material or deposited as sediments following transport by wind and water, are essential building blocks in the soil physical structure. The most important particles are clays, which have a relatively large surface area, flocculate readily and are often highly reactive. The primary particles are bound together into aggregates of various sizes by inorganic and organic binding agents, described in more detail in subsequent sections. A soil aggregate is a group of primary particles that are held together by binding agents with enough strength that it can be separated from adjacent aggregates in the soil matrix. The space between adjacent aggregates constitutes the pore space, containing water and the soil air. Soil aggregates and primary particles usually account for about 50% of the soil volume, and the rest is the pore space.

The shape and orientation of aggregates determines the porosity and controls water transport, gas diffusion, the movement of mobile soil food-web organisms and the growth of plant roots. The soil mass is composed of natural aggregates, called peds, and artificial aggregates called clods that are formed by tillage and other soil manipulation. There are five types of soil structure (Fig. 10.1):

1. Platy structure: the units are flat and platelike, with a horizontal orientation. A platy structure is common in subsurface soils that have been leached, compacted by machinery or trampled by grazing animals. Plates often overlap and impair permeability, blocking water flow and plant root growth.
2. Prismatic structure: column-like peds without rounded caps (the top is flat or sub-rounded). The outer surface is flat, with long vertical surfaces. This structure is found in subsurface soil and B horizon. Vertical cracking permits water and root infiltration. Sometimes prismatic aggregates break into smaller, block-like peds.
3. Columnar structure: column-like peds that resemble prismatic structure, except that the top and vertical surfaces tend to be rounded. This structure is characteristic of the B horizon in solonetz (high sodium) soils. Columnar structure is very dense and difficult for plant roots to penetrate. These soils can be improved by very deep tillage with a subsoiler.

©CAB International 2010. *Soil Ecology and Management* (J.K. Whalen and L. Sampedro)

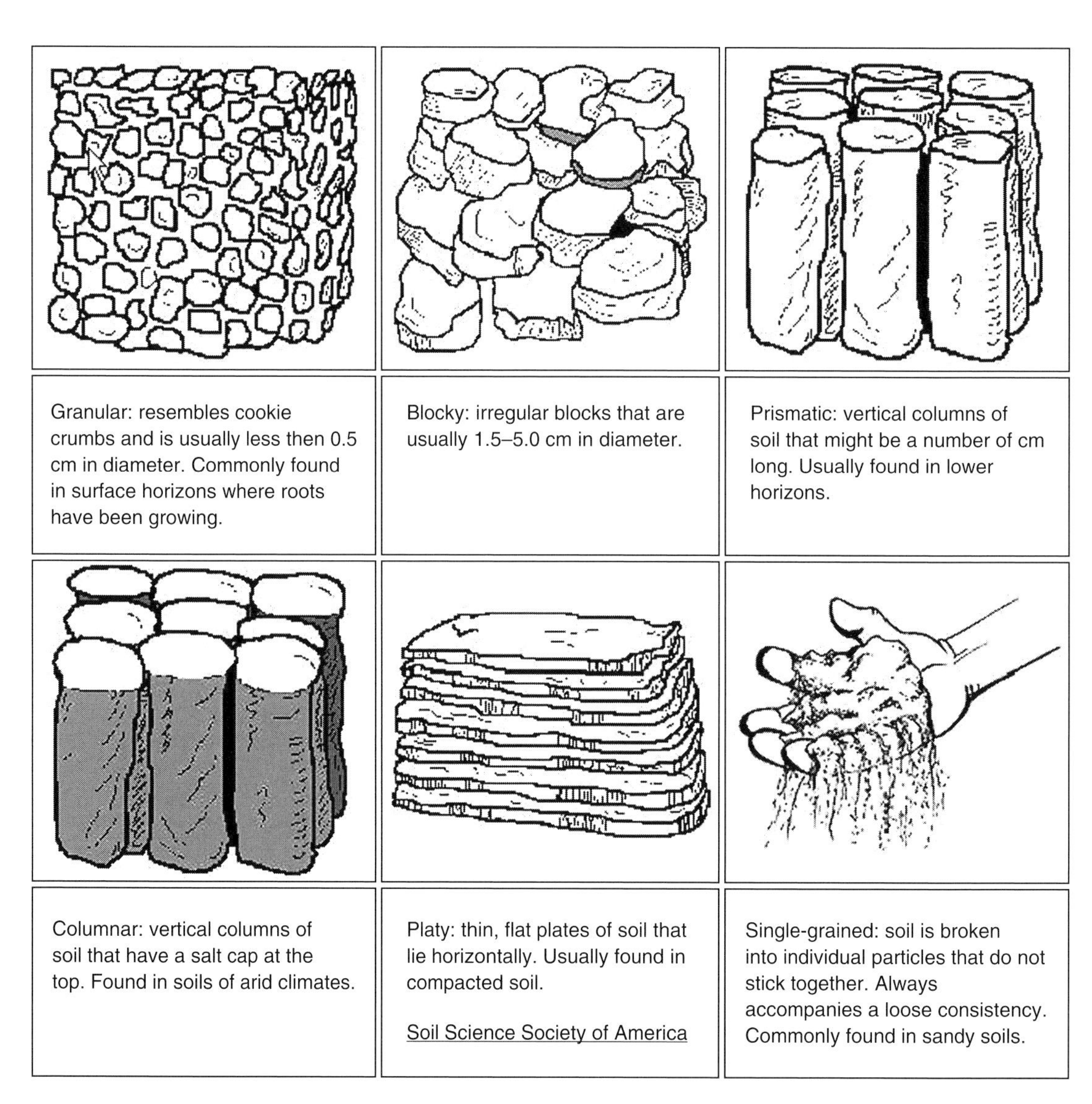

Fig. 10.1. Shape of peds (naturally occurring aggregates) responsible for soil structure. (Image courtesy of Dr Elissa Levine, The GLOBE Program Teachers Guide, Soil Characterization Investigation (http://www.globe.gov). From: http://soil.gsfc.nasa.gov/pvg/prop1.htm (accessed February 2009).)

4. Blocky structure: block-like (polyhedral) peds with flat surfaces, surrounded by other aggregates with distinct angular surfaces. If the surfaces are a mixture of flat and slightly round faces and the corners tend to be round, the peds have a subangular blocky structure. This structure is most common in the subsoil (B horizon), but can also occur in surface soils with a high clay content. Blocky structure forms due to the swelling and shrinkage of expandable clay minerals. During dry periods, cracks between peds permit water movement through preferential flow.

5. Granular structure: aggregates are approximately spherical or polyhedral, with curved surfaces, and distinct from adjacent aggregates. This is sometimes referred to as 'crumb' structure because the surface resembles cookie crumbs. Granular structure is common in the surface soils of many grasslands, forests and agroecosystems. Organic residues mixed with these soils through the action

of earthworms or cultivation produce a loose, friable soil with good porosity that facilitates water transport and gas exchange. Soils with granular structure are easy to cultivate (good tilth), and the structure can be maintained by retaining crop residues or regularly applying organic materials.

Some soils lack visible aggregation or an orderly arrangement of peds. These are referred to as structureless soils. A single-grained structureless soil possesses many particles that function individually, with little cohesion or attachment to other particles, such as in a very sandy soil. In a massive structureless soil, there is very high cohesion between particles. When heavy clay soils are tilled or trampled when very wet by grazing animals, the clay particles tend to fill the pore spaces, making the soil very dense and leading to the formation of large, hard clods upon drying.

Aggregates can be classified in a hierarchical model, based on their size and stability. The smallest and most stable aggregates are flocculated clays (< 2 μm), often with a coating of microbial gums, humic debris and polyvalent cations. The agglomeration of flocculated clays, in combination with extracellular polysaccharides of microbial origin, leads to the formation of microaggregates (< 0.25 mm diameter), which are somewhat less stable to dispersive forces than clays. The binding of microaggregates to the surfaces of dead roots and fungal hyphae leads to the formation of macroaggregates (> 0.25 mm diameter), which are even less stable (Fig. 10.2). This model of aggregate formation and stability is highly relevant for soils where organic matter is the major binding agent, but less important in soils where inorganic binding agents predominate. This has led to the current view of aggregate formation, illustrated in Fig. 10.3. Microaggregates may form directly from primary particles, but they may also form within macroaggregates and be released upon the breakdown of macroaggregates. The concept of 'microaggregate formation within macroaggregates' is supported by several authors, notably Angers *et al.* (1997), who demonstrated that ^{13}C from decomposing wheat straw was initially incorporated into macroaggregates, and later redistributed into microaggregates.

Stability refers to the ability of aggregates to withstand mechanical stresses that cause the particles to break apart. Aggregates are not solid: as illustrated in Fig. 2.12, they possess pores of varying size and connectivity, depending on the manner with which primary particles and binding agents aligned during their formation. In general, macroaggregates are most susceptible to breakdown

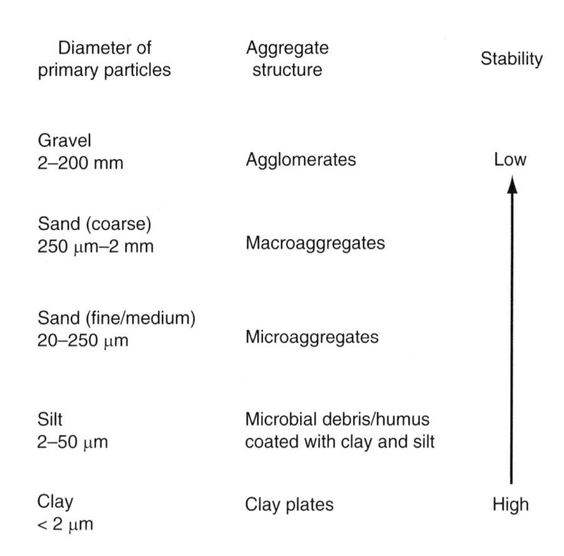

Fig. 10.2. The relative size of primary particles found in aggregate structures, and their relative stability to dispersive forces. (Adapted from Tisdall and Oades, 1982.)

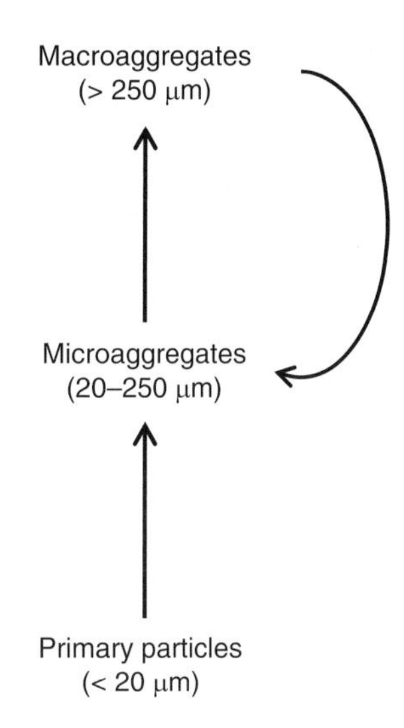

Fig. 10.3. The modified hierarchical model of aggregate formation. Microaggregates are formed from the agglomeration of primary particles (clay and silt particles), but they may also form around a core of organic matter (plant residues and fungal hyphae) within macroaggregates and are released as the macroaggregates disintegrate. Adapted from Oades and Waters (1991).

because they contain larger intra-aggregate pores, which constitute planes of weakness within the structure. Hence, the smallest aggregates are the most stable to dispersive forces because they have greater contact between particles, stronger bonds between particles and higher tensile strength.

It is important to distinguish between intra-aggregate pores, which are found within an aggregate, and inter-aggregate pores, which occur between aggregates. Whereas the pores between aggregates permit easy movement of water and dissolved ions, the intra-aggregate pores tend to collect and store water. Gas diffusion is expected to be more rapid in macropores than the smaller intra-aggregate pores. The integrity of pores is strongly influenced by hydration pressure, especially as the water content changes during wetting/drying cycles. The expansion and contraction of water molecules that accompanies phase changes during freezing/thawing cycles is another natural force that can cause the expansion or collapse of soil pores. The disruption of aggregates through these natural processes, as well as anthropogenic forces, has a major impact on pore connectivity, affecting water movement and gas diffusion through the soil matrix. Disturbance of the soil structure also impacts organisms living in this milieu, such as plant roots and soil biota.

Soil structure and soil foodweb organisms

Plant roots and soil biota inhabit the soil pores, hence the survival and proliferation of biological organisms in terrestrial ecosystems is strongly influenced by the soil structure. Porosity controls the spatial distribution of organisms within the soil matrix, as the microorganisms and fauna are often found in pores that correspond to their body size (Fig. 10.4). The soil-dwelling earthworms excavate or push aside soil to create burrows, while spiders, centipedes and other large macrofauna seek refuge in soil cracks and fissues. Enchytraeids and micro-arthropods lack the musculature to dig burrows, and find a home in the large macropores between the macroaggregates. These organisms may also be found in soil fissures, old root channels and unoccupied earthworm burrows.

Nematodes, protists and fungal hyphae can inhabit the small macropores within macroaggregates, while soil bacteria are expected to colonize the micropores within microaggregates. This is

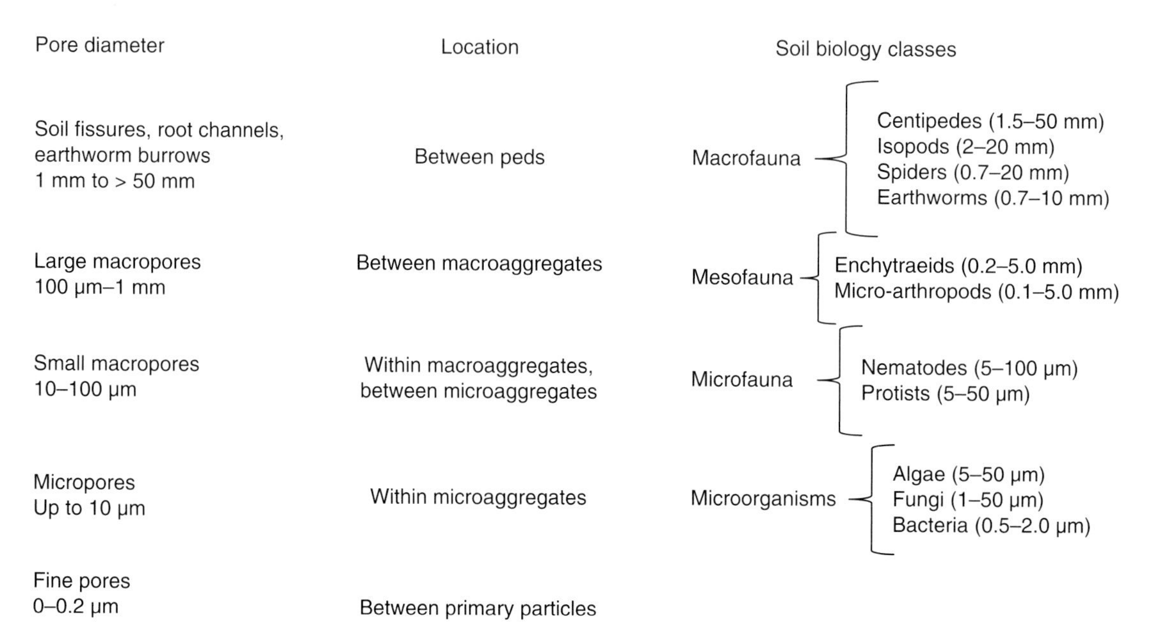

Fig. 10.4. Soil foodweb organisms inhabit pore spaces relative to their body size. Biologically relevant pore spaces range from micropores (within microaggregates) to soil fissures (between peds). (Adapted from Elliott and Coleman, 1988.)

not to imply that the smaller soil organisms cannot live elsewhere in the soil matrix – for instance, bacterial colonies grow abundantly along earthworm burrows, in root channels and in macropores. However, these bacteria are susceptible to predation by bacterivorous protists and nematodes moving through macropores in search of food, and may also be consumed by omnivorous meso- and macrofauna. It is believed that bacteria are protected from predation in micropores, which their predators cannot enter because their body size is larger than the pore diameter. Because micropores rapidly fill with water when it rains or the soil is irrigated, the bacteria living in this habitat may be facultative anaerobes, capable of switching to anaerobic metabolism when the pore space becomes saturated and there is insufficient oxygen for aerobic activities.

The presence of facultative anaerobic bacteria in micropores provides an explanation for an observation that has long puzzled soil scientists. There are many reports indicating that denitrification, an anaerobic process, occurs in soils that are apparently aerobic (e.g. not waterlogged and containing sufficient oxygen for aerobic metabolism). It seems that oxygen diffuses slowly from the external surface of an aggregate through the intra-aggregate pores, particularly under saturated conditions when the pores are filled with water. This leads to the occurrence of aggregates with an anaerobic core. An aerobic soil can contain many anaerobic microsites that support the activity of denitrifying bacteria. Using microsensors, Sexstone *et al.* (1985) measured a gradient of oxygen concentrations, ranging from atmospheric levels (21% oxygen) at the surface and less than 1% at the centre of waterlogged soil aggregates. Some of these aggregates produced gaseous nitrogen through denitrification, but others did not. It may be that denitrification was limited by other factors than available oxygen concentration in some aggregates. As discussed in Focus Box 9.4, there are many factors that control the respiratory denitrification process: denitrifying bacteria must be present, nitrate is required as an electron acceptor, a carbon source is needed for energy, the necessary co-factors for the enzyme systems must be available, soil pH must be in the optimal range for enzyme activity, and so on.

Soil pore space is a dynamic environment. The proportions of water and gases vary in the pore space, depending on climate, weather and water management such as irrigation and drainage. The volume of water-filled pore space is discontinuous due to the fact that macropores drain quickly and micropores release water more slowly. The quantities of dissolved carbon and nutrients in soil water fluctuate spatially, depending on the proximity to plant roots, decomposing residues and fertilizer bands. The soil atmosphere tends to be oxygen-rich near the surface, with lower oxygen concentrations within microaggregates, around actively growing roots and microbial communities, and with increasing depth in the soil profile. The soil structure is also subject to change: the shape, size, orientation and connectivity of pores are altered when aggregates form or break down in response to natural processes and anthropogenic management. Understanding soil biological functioning within structured soils requires some discussion of the abiotic and biotic factors contributing to aggregate formation, as well as the processes that cause aggregate breakdown.

10.2 Factors Controlling Aggregate Formation

Aggregate formation is controlled by environmental variables, which affect soil water, inorganic binding agents, plant roots, soil microorganisms and earthworms (Fig. 10.5). Environmental variables such as climate, land use and management practices and fluctuations in the soil water during wetting/drying and freezing/thawing cycles can lead to aggregate formation or breakdown (see Section 10.3). Fundamentally, aggregate formation is affected by the presence of binding agents, such as the inorganic oxydroxides and polyvalent cations, along with organic substances of plant and microbial origin. Although root growth can mechanically disrupt the soil structure and leaves channels following the decomposition of dead roots, stable macroaggregates often contain fungal hyphae and fine plant roots, illustrating the importance of these materials for aggregation. Earthworms are well known for their ability to ingest soil particles and organic residues, which are mixed during passage through the digestive tract and deposited as casts that dry and harden into stable macroaggregates.

Inorganic binding agents

Binding agents contribute to the agglomeration of primary particles into aggregates of various shapes

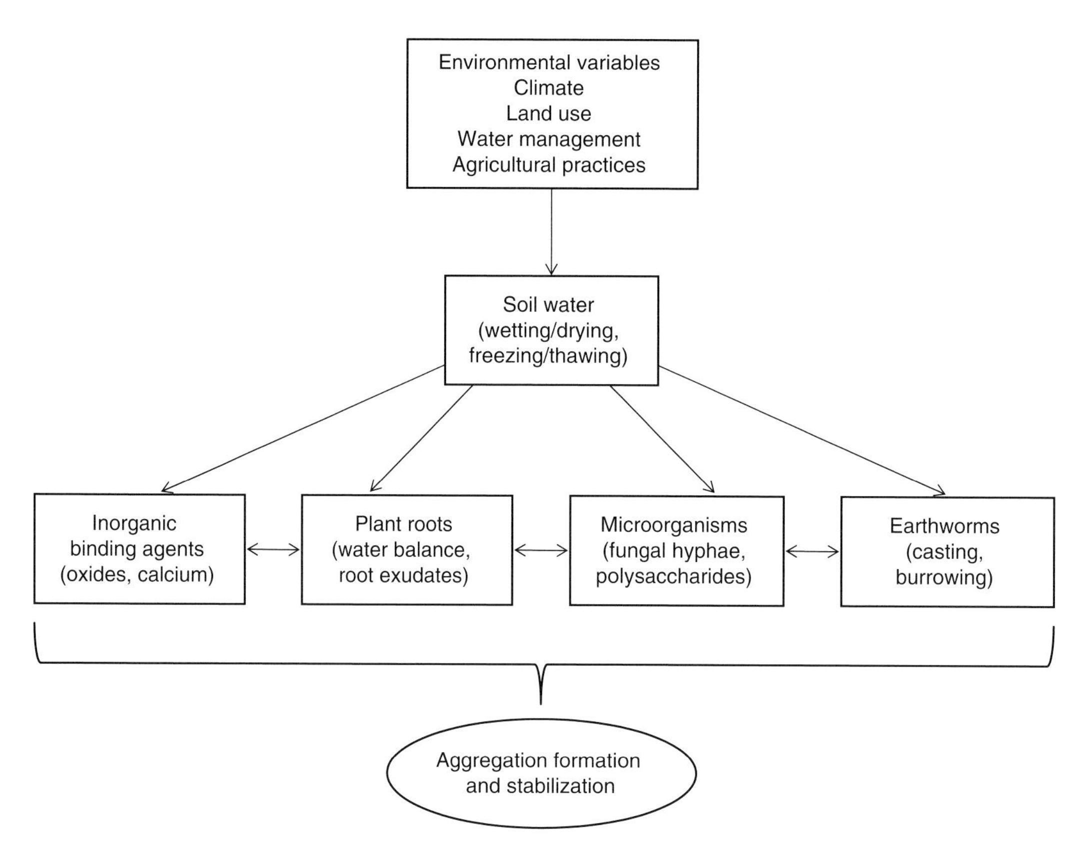

Fig. 10.5. Interactions between the major factors influencing aggregate formation and stabilization. (Adapted from Six *et al.*, 2004.)

and sizes. Inorganic binding agents such as aluminosilicates, iron oxides and calcium carbonates are quite important for the production of stable microaggregates. Some key points about each group are provided below:

- Aluminosilicates: the geochemical processes in some soils, mainly acidic weathered soils, lead to the formation of highly reactive Al compounds and poorly ordered aluminosilicates. Aluminium and other metallic ions, such as iron, magnesium and calcium, are adsorbed on to the surface of amorphous silicate, creating a stable silicate coating. This increases the surface reactivity of layered silicate clays such as kaolinite, halloysite, montmorillonite and vermiculite. The highly weathered silicate clays and modified clays such as allophane ($Si_3Al_4O_{12} \bullet nH_2O$), imogolite ($Si_2Al_4O_{10} \bullet 5\ H_2O$) and ferrhydride ($Si_3Fe_5HO_2 \bullet 4\ H_2O$) form the most stable aggregates, especially in the presence of organic matter. In addition, Al has the capacity to bind humus into Al–humus complexes, resulting in strong stabilization of organic matter.
- Iron oxides: the major binding agent in oxide-rich soils (oxisols). Microaggregate formation is strongly influenced by the presence of oxides. They may be important for macroaggregation in some soils. The mechanism for oxide binding may be: (i) electrostatic binding between positively charged oxides and negatively charged clay minerals; (ii) oxide coating on mineral surfaces forms bridges between primary and secondary particles; and (iii) oxides on mineral surfaces bind and adsorb organic materials, contributing to the formation of organo-mineral complexes.
- Calcium carbonates: the formation of clay–polyvalent cation–organic matter complexes critical for the stabilization of aggregates. Polyvalent cations such as Ca^{2+} reduce the

repulsive forces between the negatively charged clay particles, facilitate flocculation and confer strength to the aggregate. This stabilization effect occurs mostly at the microaggregate level, and can be considerable. In one study, stable Ca-organic matter in aggregates represented more than 20% of the soil organic carbon content (Clough and Skjemstad, 2000). In acidic soils, the application of calcium carbonate that alters the soil pH and stimulates microbial activity can contribute to macroaggregation.

Organic binding agents

In many soils, aggregation is strongly related to the presence of organic binding agents. Polysaccharides, proteins, lipids and humic materials are effective at binding clays and other primary particles into aggregates. These organic substances originate from microbial cells and plant roots, or are by-products of organic matter decomposition. Polysaccharides have many hydroxyl groups and also contain carboxyl groups in uronic acids and acetylamino groups in amino sugar units. The amino groups of proteins and hydrophobic chains of lipids are also believed to be important binding agents. Binding occurs through Van der Waals bonds and weak H^+ bonds between charged groups and the siloxane oxygen on clay surfaces. Similarly, the negatively charged surfaces of clays bind with carboxyl groups in single-ring aromatic and aliphatic compounds of humus, as well as the hydroxyl groups contained in phenols and alcohols of humic compounds.

Polysaccharides have long been recognized as important binding agents that contribute to aggregate formation and stability. Extracellular polysaccharides of microbial origin, discussed in Chapter 8 (Section 8.4), possess a macromolecular structure that is readily adsorbed on to clay surfaces and acts as glue to bind particles together. Hydrophilic polysaccharides are especially effective at binding clay and silt particles within microaggregates. The polysaccharides (mucilages) secreted by plant roots are effective at forming aggregates, and are often concentrated in the macroaggregate fraction. Adsorption of these binding agents on to clay surfaces is affected by many factors, including the chemical composition and structural complexity of the macromolecule, the surface charge of the clay mineral, pH and the presence of polyvalent cations.

Glomalin is an insoluble glycoprotein secreted by arbuscular mycorrhizal fungi. This hydrophobic compound contains N-linked oligosaccharides and iron, which may facilitate its binding to soil particles. Indeed, a number of studies indicate that glomalin has a role in aggregate formation and stabilization, perhaps much more than the fungal hyphae. The turnover time of glomalin ranges from 6 to 42 years, which is longer than the persistence of fungal hyphae (Rillig *et al.*, 2001). Lipids may also contribute to aggregate formation and stabilization, as these hydrophobic materials increase the resistance of aggregates to slaking (Paré *et al.*, 1999).

Roots and soil microorganisms

Root growth and expansion in the pore space exerts a mechanical force that disrupts aggregates within 50–200 μm of the root surface (Dorioz *et al.*, 1993). This is accompanied by an increase in the bulk density around roots, bringing particles into closer contact and leading to aggregate formation. Changes in the soil water status, specifically soil drying due to plant water uptake, promote the binding of root exudates on clay particles.

Fine roots are readily coated with clays, which stick to the mucilage and dead cells sloughed off during root penetration into the soil. Fragments of dead roots are often found at the core of newly formed macroaggregates. In addition, root exudates stimulate the growth of many free-living soil microorganisms. These organisms produce organic substances that bind to clays, outside of the cell or on the surface of living and dead microbial cells. Root architecture (e.g. degree of branching, root thickness, proportion of fine roots, rooting depth), the amount and chemical composition of exudates released from roots, and the root turnover rate are variables that affect aggregate formation.

The fungal mycelium is another sticky surface where clays adhere. Fungal hyphae grow throughout the soil in an extensive, intertwining network that entangles soil particles and cements them in place through glomalin and extracellular polysaccharide production. A great number of plant roots are colonized by mycorrhizal fungi, which makes it difficult to separate the effects of roots and fungal hyphae on aggregate formation. The formation of macroaggregates with a diameter > 2 mm was due to entanglement of particles by fine roots plus hyphae, with little effect of soluble carbohydrates on macroaggregate formation (Jastrow *et al.*, 1998). Thus, the physical entanglement of soil particles by roots and fungal hyphae can generate stable macroaggregates.

Earthworms

Earthworms have long been recognized for their ability to modify the soil structure. Due to the large quantities of organic litter and soil that they consume, Darwin (1881) referred to them as the 'intestines of the earth'. When soil particles and fresh/partially decomposed organic residues pass through the earthworm's digestive tract, the original structure is destroyed by grinding in the gizzard and a new aggregate structure generated as the soil particles are reoriented and organic residues are partially decomposed. The undigested material is deposited in the soil profile or at the soil surface in earthworm casts, which can be transformed into stable macroaggregates upon drying. In addition to casting, earthworms also affect the soil structure by burrowing and through interactions with plant roots and microorganisms.

Studies of earthworm casts have revealed that the stability of earthworm casts and their eventual transformation into aggregates depends on the quality of ingested organic residues, as well as the available food supply. Casts appear to be macroaggregates with a core of stable, carbon-rich microaggregates. This stability may result from the cementing of soil particles in the earthworm gut by calcium humates formed from decomposing organic material and calcium secreted by the earthworm's calciferous glands. Extracellular polysaccharides and other organic substances secreted by the intestinal soil microflora or released during the decomposition of organic residues could serve as binding agents in the organo-mineral complex. The continued growth of free-living bacteria and fungi within newly deposited casts could generate additional organic substrates that strengthen bonds with mineral particles.

In the ferruginous tropical soil of Côte d'Ivoire, Blanchart *et al.* (1997) reported the presence of 'compacting' and 'decompacting' earthworms that had opposing effects on soil bulk density and aggregation. The earthworms *Pontoscolex corethrurus* and *Millsonia anomala* produce large, compact casts (diameter > 5 mm), which are deposited on the soil surface. Decompacting earthworms of the Eudrilidae feed on the large casts and leave smaller, delicate casts (0.5–5.0 mm diameter) on the soil surface. The complementary activity of compacting and decompacting earthworms maintains good soil tilth for agriculture. However, sites with compacting earthworms alone can develop a thick, compacted layer that is impermeable to water.

Burrowing is another earthworm activity that impacts soil structure. The endogeic earthworms push aside and excavate soil to construct burrows that extend horizontally in the topsoil (< 20 cm depth). Most of the horizontal burrows are short-lived, as they get backfilled with soil as the earthworms move from one area to another, presumably in search of food. Anecic earthworms create vertical burrows that can extend to 1 m or deeper. These burrows have a typical 'J' shape, as the earthworm appears to leave a small cavity or resting place at the bottom. Earthworms appear to maintain these vertical burrows and deposit unwanted soil and other materials, such as fresh crop residues, at the mouth of the burrow in a midden. Earthworm burrows constitute a very large macropore, and the burrow lining is stabilized by radial pressure from the earthworm body as well as the mucus secreted from the earthworm body and deposited in the burrow wall.

Earthworm interactions with plant roots and soil microorganisms can indirectly affect aggregation. Earthworm casts are relatively rich in carbon and plant-available nutrients (ammonium, nitrate, phosphate and potassium) compared with bulk soil, making them a 'hot spot' for root growth and microbial activity. Soil microorganisms are abundant in the lining of burrows inhabited by earthworms, perhaps because mucus secreted from the earthworm body is an appropriate organic substrate for microbial growth, while roots may be found growing in abandoned earthworm burrows. Earthworms are omnivores that consume primarily dead and decomposing organic residues, but may also consume small quantities of living roots, fungal hyphae and bacterial cells. The soil microbial biomass tends to be smaller, but more active, based on CO_2 respiration measurements, in the presence of earthworms (Schindler Wessells *et al.*, 1997).

10.3 Factors Affecting Aggregate Breakdown

Aggregation is a dynamic process – factors leading to aggregate formation are balanced by forces that cause aggregate breakdown. Root penetration, the soil-working and feeding activities of earthworms, fluctuations in soil water content and environmental variables can lead to aggregate disruption at scales ranging from the rhizosphere to many hectares (e.g. soil drying and aggregate breakdown caused by wind erosion). Therefore, aggregate breakdown can be a small-scale, localized phenomenon or a large-scale process that leads to the redistribution of soil, organic matter and nutrients in the landscape.

Aggregate stability and breakdown at the microscale

Small-scale processes leading to the breakdown of aggregates could reduce microaggregates (< 0.25 mm) to primary particles, but this probably does not occur very often. Strong cohesive forces within microaggregates are attributed to the strength and stability conferred by inorganic and organic binding agents. The following discussion will focus on processes that reduce macroaggregates (> 0.25 mm) to microaggregates, which is quite common.

Inorganic binding agents are important for macroaggregate stabilization in tropical soils. Subjecting these macroaggregates to wetting/drying cycles can cause clay plates to shrink and swell, generating mechanical forces that can weaken the aggregate by disrupting electrostatic and chemical bonds. On the other hand, soil water carries soluble cations such as Al^{3+} and Fe^{3+} oxides that can coat newly exposed silicate clay surfaces, increasing the surface reactivity and binding strength within micro- and macroaggregates.

Organic binding agents, as well as cellular debris originating from plants and microorganisms, are important in macroaggregate formation. At the same time, the intra-aggregate pore space of macroaggregates is a favourable habitat for bacteria, free-living fungi and microfauna. A macroaggregate with a core of plant residue or fungal hyphae is susceptible to breakdown because microorganisms living within the macroaggregate will decompose the organic materials after some period of time (weeks, months or years). Any physical disturbance of the weakened macroaggregate structure would lead to its collapse. The modified hierarchical model of aggregate formation in Fig. 10.3 suggests that macroaggregates disintegrate into microaggregates.

In the context of soil organic-carbon dynamics, the length of time that organic residues are protected from decomposition within macroaggregates is of great importance. Organic matter within intact macroaggregates is physically protected from consumption by larger soil fauna, compared with organic residues that are not incorporated in aggregates. The decomposition rate of residues at the core of a macroaggregate will be slow if the centre of the aggregate is an anaerobic zone. The chemical composition of organic residues such as chitinous fungal hyphae within macroaggregates may be difficult to decompose, compared with cellulose-rich plant residues. This contributes to humification of organic residues within macroaggregates. Even when the macroaggregates break down, some of the humic substances remain protected within microaggregates and adsorbed to the surface of clay and silt particles.

Monnier (1965) proposed a model to explain the effect of organic residues on aggregate dynamics, showing that aggregate stability is related to the chemical composition of organic residues and their gradual decomposition to humic substances (Fig. 10.6). This model was tested by Abiven *et al.* (2009), who conducted a meta-analysis of data from nearly 50 experiments with a wide range of organic residues. Their results indicate that mucilage and simple labile compounds (probably converted to microbial by-products such as polysaccharides) rapidly produce highly stable aggregates, whereas compost, peat and lignin are more slowly incorporated into aggregates, and those formed are not very stable (Fig. 10.7).

The physical rearrangement of soil particles due to plants, microorganisms and earthworms occurs in the soil space within a few micrometres of microbial cells and within a millimetre of the rhizosphere. Due to their mobility, earthworms may impact aggregate stability and contribute to aggregate breakdown within centimetres to perhaps a metre or more. Hence, the impact of these biological organisms on aggregate stability and breakdown occurs at the microscale.

Aggregate stability and breakdown at the macroscale

Soils experience continual fluctuations in soil water content due to precipitation and evapotranspiration. The soil water content is strongly controlled by environmental variables such as climate, land use, water management and agricultural practices. Since these environmental variables operate in a

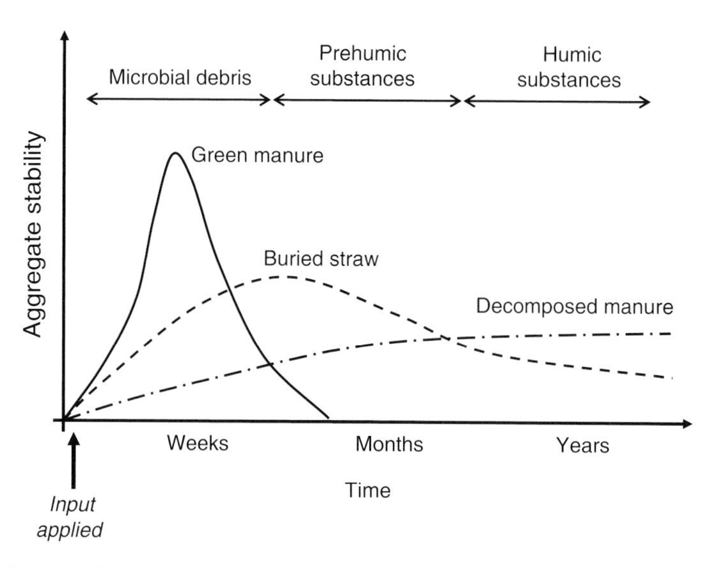

Fig. 10.6. Conceptual model of aggregate stability dynamics after the addition of three organic inputs (green manure, straw and decomposed manure). The factors contributing to aggregate stabilization include microbial debris (short-term), prehumic substances (medium-term) and humic substances (long-term). (Adapted from Monnier, 1965.)

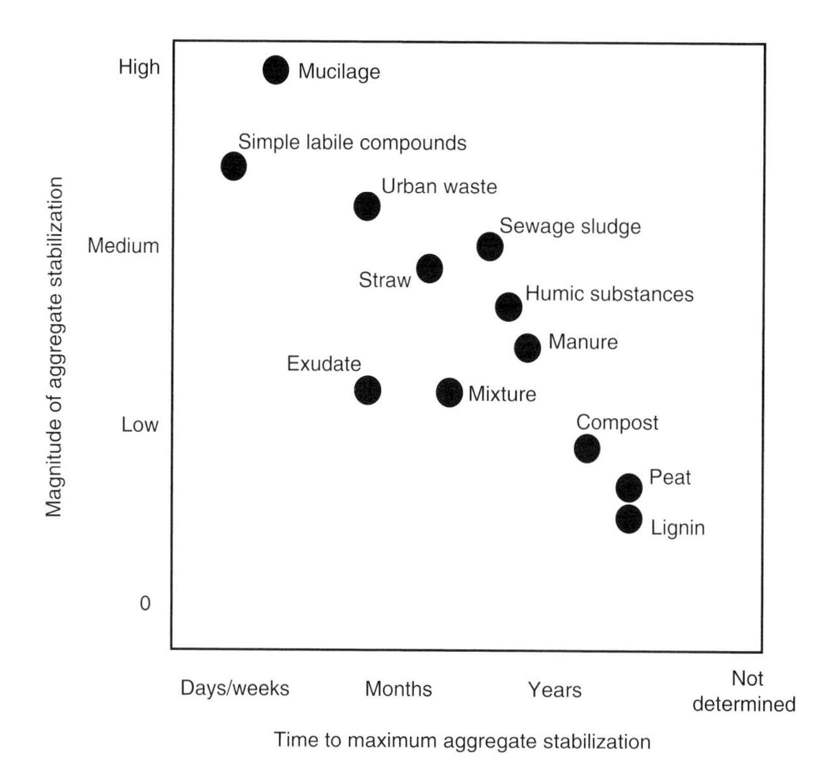

Fig. 10.7. Relationship between the time to achieve maximum aggregate stabilization and the magnitude of aggregation in soils treated with a variety of organic residues. (Modified from Abiven *et al*., 2009.)

large geographical context (within and across ecosystems, landforms and regions), they are considered to affect aggregate stability and breakdown at the macroscale.

Soil wetting and drying are counteracting forces that can cause a decrease or increase in aggregation, depending on soil properties and management. When a water droplet strikes the soil surface, it possesses sufficient force to physically disrupt aggregates and detatch particles at the soil surface. Heavy rainfall or irrigation events that destabilize aggregates at the soil surface can cause erosion, in which the soil mass is transported from the original position to another part of the landscape. The translocation of soil in runoff water can be categorized as: (i) sheet erosion, where soil is removed uniformly from every part of the slope; (ii) rill erosion, where soil is transported preferentially through tiny gullies, especially on bare surfaces; and (iii) gully erosion, where the volume of water moving over the soil surface becomes concentrated and undercuts the soil in some parts of the surface, forming a small or large ravine.

Intense wetting events could lead to erosion at the soil surface and also within the soil, as aggregates are broken apart and the dislodged particles are transported through soil fissures, earthworm burrows, root channels and macropores with gravitational water. The breakdown of aggregates within the profile can occur when: (i) silicate clays are rapidly and unevenly hydrated, which causes them to swell; and (ii) the rapid filling of soil pores with water displaces air and creates internal pressure in the aggregate. Rehumidification of aggregates through capillary wetting is far less destructive. In addition, an air-dry aggregate is more likely to be disrupted by rewetting than an aggregate that was already partially moist.

Soil drying occurs through drainage and evapotranspiration. In arid and semi-arid regions, dry aggregates at the soil surface are susceptible to wind erosion. The abrasive action of the wind leads to detachment of tiny particles (minerals and organic fragments) from the aggregate surface. These small particles can then be transported through the following processes:

- Saltation: bouncing movement of medium-sized aggregates (2.5–3.75 mm diameter). These large particles bounce short distances, to a height of less than 30 cm, along the soil surface. Between 50 and 75% of the particles transported by wind erosion move through saltation.
- Soil creep: the rolling and sliding along the surface of smaller aggregates (> 0.84 mm diameter). The force of large bouncing particles provides the momentum that pushes these smaller aggregates along, accounting for 5–25% of the particles eroded.
- Suspension: particles smaller than the size of fine sand (< 0.2 mm) become dislodged from the soil surface and enter the turbulent wind stream. Although most particles fall back to the ground within a few metres, some are carried into the atmosphere and moved hundreds of kilometres from their point of origin. These particles return to the earth when the wind subsides or when it rains. Suspension accounts for 15–40% of particles transported by wind erosion.

In some soils, repeated wetting and drying cycles tend to induce aggregate stabilization because particles tend to settle into packed configurations that lead to greater cohesion. This depends somewhat on soil mineralogy, since aggregates derived from swelling silicate clays are continually disrupted and will be less stable than aggregates built from non-expanding clays.

Similarly, freezing and thawing cycles are expected to induce aggregate breakdown. The force generated by freezing water is equivalent to about 1465 Mg/m^2 (Brady, 1984). This is sufficient to crack rocks during the weathering process, so it is not surprising that particles are dislodged and aggregates break apart when they are frozen and thawed. This effect is generally limited to the surface soil layers, which are also the most prone to erosion. Aggregates that have high internal cohesion because they have been formed from non-expanding clays and bound together with inorganic and organic binding agents should be stable during freezing/thawing cycles.

Climate is an important predictor of the frequency, duration and intensity of rainfall events, as well as freezing and thawing events. Soil temperature is another relevant parameter that is affected by climate, since temperature controls the rates of chemical and biochemical reactions related to adsorption–desorption and decomposition–humification processes.

Land use influences the vegetation and management practices within ecozones. For instance, the use of fire for slash-and-burn agriculture in tropical regions contributes to a significant loss of soil organic matter from surface horizons, while promoting soil dehydration and the cementing of soil particles by thermally transformed aluminium and iron oxides. The net effect is the formation of inorganically stabilized macroaggregates, at the expense of soil organic carbon loss (Six *et al.*, 2004). Water management affects soil aggregates by influencing the soil water balance, as discussed above.

Agricultural practices such as the application of organic residues can increase aggregation, but this depends on the quantity and frequency of organic matter additions, and the type of organic residue applied (Fig. 10.7). Tillage can cause aggregate formation and breakdown. On one hand, tillage implements break up clods, incorporate organic matter into the soil and prepare the seedbed for plant growth. However, tillage implements mix and abrade soil particles, leading to the mechanical breakdown of macroaggregates. Tillage also stimulates decomposition, which reduces the content of organic binding agents that stabilize macroaggregates. Heavy machinery crushes aggregates and contributes to soil compaction in the surface layers, especially when farm equipment is driven over wet soils. Careful timing of tillage operations is essential to avoid compaction. The mechanical force associated with ploughing can produce a compacted plough-pan in the subsurface soils, but this can be alleviated by periodically growing deep-rooting crops such as lucerne in the rotation or by very deep tillage with a subsoiler.

Aggregate formation and breakdown is of particular interest in agricultural systems because of the importance of aggregation for crop production. Sandy soils have loosely packed particles, good aeration and drainage, and are easily tilled. However, they lack water-holding capacity, so they are often droughty with a low nutrient supply. Applying organic residues to sandy soils increases the proportion of stable macroaggregates that can store water and release it gradually during the growing season. The organic residues can also decompose to release nutrients for crop uptake.

A different problem is encountered in clay soils. Many are high-fertility soils with excellent agricultural productivity due to the cation exchange capacity of clay surfaces and their inherent ability to produce stable aggregates. However, fine-textured soils tend to be compacted if driven on when wet. When these soils dry, they exhibit a hard, densely packed structure. It is difficult for plant roots to penetrate through a densely packed substratum and for water to drain through small soil pores. Mixing organic residues with clay soils can produce macroaggregates and generate a structure with larger pore spaces, but care must be taken to avoid compaction. Rotating with deep-rooting perennial crops and switching to less intensive tillage practices such as zone tillage or even no-tillage can optimize the soil structure of clay soils for agricultural production.

Focus Box 10.1. Measuring aggregate size and stability.

From the previous sections, it should be clear that aggregate size and stability are fundamental physical characteristics that impact a number of soil functions, including: (i) the ability of a soil to resist erosive forces (wind, water) and mechanical disruption during wetting/drying and freezing/thawing cycles; (ii) soil water storage within the intra-aggregate and inter-aggregate pore space; (iii) plant growth, due to the importance of porosity for root penetration and growth; (iv) soil organic matter turnover, due to the intimate connection between primary particles and organic matter in organo-mineral complexes; and (v) soil foodweb organisms: the population size, diversity and activity of soil organisms are regulated by the soil porosity within and between aggregates.

How does one measure aggregate size and stability? There are two common techniques: wet-sieving and dry-sieving. Briefly, the wet-sieving method involves placing soil clods or coarsely sieved, field moist soil on a nest of sieves with diameters ranging from 4 mm (top sieve) to less than 0.25 mm (bottom sieve) and repeatedly immersing the sieves in tap water for about 10 min. The material remaining on each sieve is collected, dried and weighed. The mass of aggregates on each sieve, relative to the total soil mass, represents the proportion of aggregates of various sizes in the soil. The effect of land management or soil type on aggregation can be assessed by comparing the average aggregate size in each soil. While this provides some information on aggregate

Continued

Focus Box 10.1. *Continued*

formation and stabilization under field conditions, this can be examined further by subjecting aggregates from each size class to dispersive forces. Aggregates can be dispersed with sodium hexametaphosphate solution or subjected to dispersive energy with an ultrasonic probe. Stable aggregates are those that resist disruptive forces associated with wetting, drying and dispersion.

The dry-sieving method is similar, although it uses air-dry soil samples and a rotary sieving device to separate aggregates and calculate the average aggregate size. In wind erosion research, the aggregates < 0.84 mm diameter are considered to be susceptible to erosion. Aggregates collected with the dry-sieving method can be rewetted and subjected to dispersion to gain additional insight into their stability.

11 Biocontrol

The ecological role of soil foodweb organisms that function as plant symbionts and contribute to organic residue decomposition and nitrogen mineralization is well known. Modification of soil structure and aggregation is an important pedological function of the soil foodweb. This chapter focuses on the ability of certain biota to inhibit or suppress soilborne parasites and pathogens that infect plant roots and cause disease. Pathogenic soil viruses, bacteria, fungi, oomycetes, protists and nematodes are responsible for many economically important diseases affecting crops and natural vegetation. In some cases, disease outbreaks can be controlled by naturally occurring predators and competitors. For example, some soil microorganisms release chemical substances that inhibit fungal pathogen growth, while others parasitize nematodes and destroy their eggs. Rhizobacteria and mycorrhizal fungi can rapidly colonize plant roots more effectively than fungal pathogens, essentially out-competing the pathogen for space on the root surface and preventing or reducing root infection.

There is considerable potential for these biological interactions to reduce disease outbreaks by reducing pathogen populations to low, manageable levels. We will examine the cultural practices and environmental conditions that support soil organisms that provide biological control ('biocontrol') of crop pests. Although biocontrol is relevant in all terrestrial ecosystems, most of the examples provided in this chapter will come from studies conducted in agroecosystems. The global demand for inexpensive and nutritious food compels us to manage plant health for high crop yields. Examples of biocontrol agents that can control diseases in cereals, vegetable crops, fruit and nut trees and ornamental plants (trees, shrubs and flowers) will be discussed as case studies. We will also highlight the biological organisms and plant growth-promoting substances that are permitted as crop production aids in organic agriculture.

11.1 Fundamental Principles of Biological Control

Any organism that interfers with another organism's metabolic processes in a manner that weakens the host and causes disease is called a pathogen. The soilborne pathogens can cause disease in microorganisms, plants and animals, including humans. This chapter focuses on the plant pathogens, particularly root pathogens, that live on or inside the host plant and obtain their nutrition from the host. Yet, the presence of a pathogen does not necessarily lead to plant disease. The pathogen may persist in a dormant state, in a non-pathogen life stage or by living on a plant where it causes marginal damage and no disease. In agricultural fields, weeds are often a secondary host for pathogens, indicating the importance of an appropriate weed management programme to control pathogen populations. Not all cultivars of a particular crop are susceptible to injury by a specific pathogen – the non-susceptible hosts are called disease-resistant cultivars. Even the infection of a susceptible cultivar by a pathogen does not guarantee that disease will occur. The environment must be favourable for the establishment of the disease-causing organism. The relationship between these factors that leads to plant disease is known as the disease triangle (Fig. 11.1).

The occurrence of a plant disease epidemic indicates that one or more of the following conditions exist:

1. The pathogen is highly virulent or has a high inoculum density.
2. The soil environment or climatic conditions are more favourable for the pathogen than for the host.
3. The host plant is genetically homogeneous, highly susceptible and continuously or extensively grown in a field or broader area.
4. There are no biological organisms capable of reducing the pest population. Perhaps such a

©CAB International 2010. *Soil Ecology and Management* (J.K. Whalen and L. Sampedro)

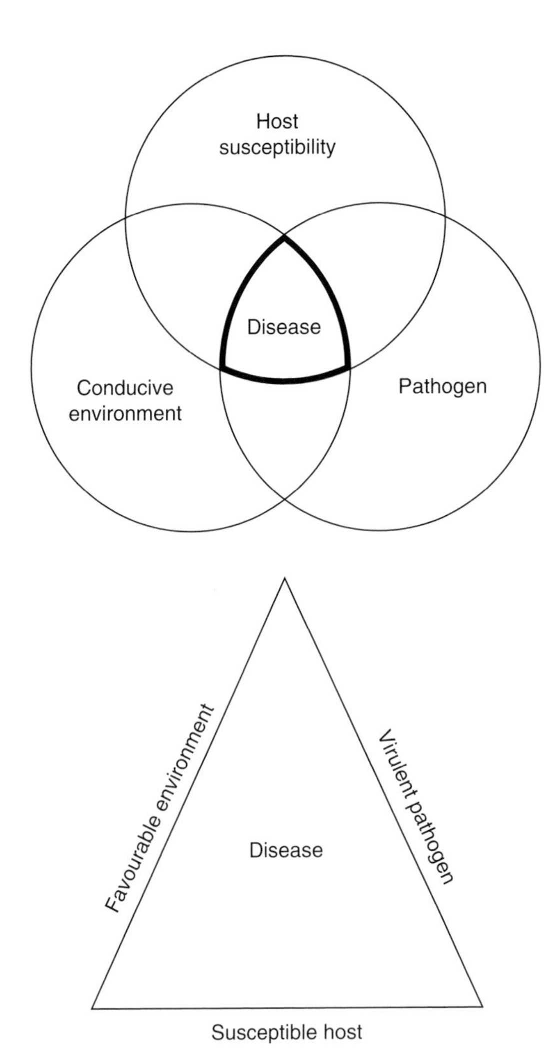

Fig. 11.1. The disease triangle. The interaction of a virulent pathogen with a susceptible host in a conducive environment leads to disease outbreak. (Redrawn from http://www.apsnet.org, accessed on 2 February 2009.)

biological agent does not exist in the area. Another possibility is that the biological agent is limited in growth due to unfavourable environmental conditions or inhibited by other biological groups. In such a case, the population of the biological agent is too small to effectively reduce the pest population.

The biocontrol of plant disease occurs when the inoculum of a pathogen or its ability to produce disease in a plant is reduced by other biological organisms, known as antagonists. Biocontrol can occur in the bulk soil, in the rhizosphere or within the plant root. Most of the examples and case studies will focus on the fungi pathogens and nematode parasites that attack plant roots, due to their widespread distribution and economic importance. There are three recognized mechanisms whereby antagonists can reduce the inoculum or activity of these pathogens:

1. Antibiosis: the antagonist produces substances that inhibit the activity of the pathogen or cause its death. These products include lytic agents that break apart the pathogen's cells, enzymes, volatile compounds, antibiotics and other toxic substances.
2. Competition: the antagonist reduces the pathogen's activities by successfully competing for and using resources needed for the pathogen to grow, reproduce and survive. This may include organic and inorganic nutrients, energy, growth factors, oxygen and space.
3. Hyperparasitism: the antagonist invades the pathogen by secreting enzymes such as chitinases, cellulases, glucanases and other lytic enzymes.

There are a number of methods that can be employed to control crop diseases, and it is helpful to review them and gain a perspective on the potential use of biocontrol agents. Chemical controls such as fumigants, fungicides and nematicides can be applied to conventional agroecosystems, but are not generally permitted in organic agriculture. Soil fumigants tend to have a broad-spectrum effect and act against the target pathogen as well as other foodweb organisms. Methyl bromide (bromomethane) is perhaps the most well known soil fumigant. Effective against pathogenic bacteria and nematodes, but having limited fungicidal properties, methyl bromide was in common use until the beginning of the 21st century. Due to its ozone-depleting properties, the use of methyl bromide has been phased out under the Montreal Protocol, an international agreement developed in 1987 and last ratified in 1999. Fumigants and pesticides that may be applied to control fungal pathogens and nematodes causing plant disease are listed in Table 11.1. Fumigants are injected in a gaseous form through the soil pore space, where they can reach target and non-target organisms. Liquid pesticides can be injected into the soil using a shank injector or drip-line method, which permits control of root pathogens *in situ*. In addition, fungicides are often applied as a soil drench or root treatment during transplanting.

Table 11.1. Selected chemical controls to protect plant roots from fungal pathogens and plant-parasitic nematodes. (From: Nel *et al.*, 2007; Murillo-Williams and Pedersen, 2008; Sanchez-Moreno *et al.*, 2009.)

Chemical control	Comments
Fumigants	
Chloropicrin	Broad-spectrum fumigant, effective against bacteria, fungi and nematodes. Low toxicity to non-target organisms, animals and humans. Moderate persistence in the environment
Metam sodium (sodium N-methyl dithiocarbamate)	Broad-spectrum fumigant and herbicide, controls soilborne plant pathogens, parasitic nematodes and weeds. In moist soil, it decomposes to the biocide methyl isothiocyanate. Low toxicity to non-target organisms, animals and humans. Moderate persistence in the environment
Pesticides	
1,3-dichloropropene	Nematicide. Shank injection or drip-line application. Low toxicity to non-target organisms, including humans. Moderate persistence in the environment
Carbosulfan	Nematicide and insecticide. Low toxicity to non-target organisms, including humans. Low persistence in the environment
Mefenoxam	Fungicide. Can be applied as a seed treatment. Low toxicity to non-target organisms, including humans. Low persistence in the environment
Benomyl	Fungicide. Root treatment or drip-line application. Low toxicity to non-target organisms, including humans. Moderate persistence in the environment

Fungicide coatings on seeds are effective in protecting the newly germinated seedling from fungal attack, particularly in cool, humid soils.

Cultural controls include sanitation and crop rotations. Sanitation refers to the inactivation of the disease agent in the soil environment. This can be achieved by sterilizing or pasteurizing the growth substrate, as is common in the greenhouse industry. Solarization is the exposure of soil or growth substrates to heat and ultra-violet radiation. Excavating soil and spreading it on a surface where it can dry under direct exposure to sunlight is an effective way to kill pathogenic nematodes. Burning the plant residues that contain the disease-causing organism is another effective sanitation method. When burning is not recommended, chopping and burying the residue may be effective, depending on the ability of the pathogen to resist such treatment. The stem rot fungal pathogen *Sclerotinia sclerotiorum* overwinters on surface residues and can be disrupted by chopping and incorporating surface residues (Gracia-Garza *et al.*, 2002). However, root rot pathogens belonging to the fungal genus *Phytophthora* spp. can survive in moist field soil for up to 6 years (Hwang and Ko, 1977). Composting involves the chopping of infected residues and heat treatment due to the activity of thermophilic bacteria during the process, and is an effective means of reducing or eliminating pathogens. Crop rotation involves alternating a host crop with a non-host crop to 'break' the disease cycle. In fields that are severely infested, it may take a number of years of growing the non-host crop before the pathogen population is reduced sufficiently that the host crop could be included in the rotation.

Biological control (biocontrol) methods include plant breeding initiatives to develop resistant cultivars and the introduction of antagonists. Usually, biocontrol methods are less rapid and not as immediately effective as the other methods of controlling crop diseases. However, they are generally more stable and provide a longer-lasting effect than chemical controls and cultural techniques. Since a lag time is expected when biocontrol methods are used to control disease, it is advisable to adopt a combination of chemical, cultural and biocontrol techniques to minimize the negative consequences of crop diseases in the short term.

11.2 Manipulating the Soil Foodweb to Achieve Biocontrol

The initial objective of biological control is to maintain a population of resident antagonists that will be able to suppress the disease organism. Due to crop or soil management practices, some soils possess microorganisms and fauna that coexist with the pathogen and the crop. They may be able to partially suppress the activity of the pathogen. Their influence is observed in pathogen-suppressive soils, which could have the following characteristics: (i) the pathogen does not establish or persist; (ii) the pathogen establishes, but causes little or no damage; or (iii) the pathogen causes disease, but the disease is not severe even when the pathogen persists in the soil.

While the development of soil suppression remains poorly understood, the basic principle is that disease incidence and severity is lower in suppressive soils than in conducive (non-suppressive) soils due to a combination of soil conditions and biotic interactions between resident antagonists and pathogenic organisms that are detrimental to the pathogen. The activity of resident antagonists is affected by cultural controls and other agricultural practices, discussed later in this chapter. Relying on the inherent ability of suppressive soils to control pathogens such as those listed in Table 11.2 is complementary to traditional methods of pest control, i.e. chemical control with fungicides and the breeding of resistant host cultivars.

General and specific suppression of pathogens

The agricultural practices described in the last paragraph alter the biotic interactions between resident antagonists and soilborne pathogens, leading to partial suppression of the pathogen population. This phenomenon is called general suppression and is achieved by competition for resources during critical periods in the life cycle of the pathogen. In these soils, a large, active microbial community and associated soil fauna compete for energy, nutrients and other substances required for pathogen survival and growth. Free-living soil microorganisms that live in the rhizoplane (at the plant root surface) and in the rhizosphere (the soil immediately surrounding plant roots) will occupy favourable habitats and thus block pathogens from inhabiting these regions (Fig. 11.2). Additional control may be achieved by microbial predators such as protists, bacterial-feeding and fungal-feeding nematodes that are attracted to the rhizosphere by root exudates and microbial activity.

Fungal pathogens and plant-parasitic nematodes need to attach to root surfaces before they can pierce the cell wall and extract cellular debris as an energy source, or enter the roots. The colonization of root surfaces by symbiotic nitrogen-fixing bacteria and mycorrhizal fungi of all types (arbuscular mycorrhizae and ectomycorrhizae are the most common) will prevent pathogens from attaching to the root surface. The physical barrier generated by mycorrhizal fungi can be extensive, since the pathogens form a network of mycelium within the root and extend their hyphae into the surrounding soil. Roots that are extensively colonized by mycorrhizal fungi can effectively block a pathogen from entering the plant root (Fig. 11.3).

Another suppression mechanism is called specific suppression. This refers to individual antagonists or groups of soil foodweb organisms that are antagonistic to a pathogen during one or more

Table 11.2. Soilborne pathogens and parasites that are controlled in suppressive soils. Examples of diseases or crops affected by each organism are provided. (Adapted from Weller *et al.*, 2002.)

Fungal pathogen	Bacterial pathogen
Gaeumannomyces graminis var. *tritici* (causes take-all disease)	*Ralstonia solanacearum* (causes bacterial wilt)
Fusarium oxysporum (causes plant wilt)	*Streptomyces scabies* (actinomycete) (causes potato scab)
Thielaviopsis basicola (causes black root rot)	Protists
Rhizoctonia solani (causes root rot and damping off)	
	Plasmodiophora brassicae (causes club root disease)
Oomycetes	Plant-parasitic nematodes
Aphanomyces euteiches (causes foot and root rot)	
Pythium splendens (causes blast disease on oil palm)	*Heterodera avenae* (cereal cyst nematode)
Pythium ultimum (causes root rot and damping off)	*Heterodera schachtii* (sugarbeet nematode)
Phytophthora cinnamomi (causes root rot and dieback)	*Criconemella xenoplax* (ring nematode on grapes)
Phytophthora infestans (causes late blight on potatoes and tomatoes)	

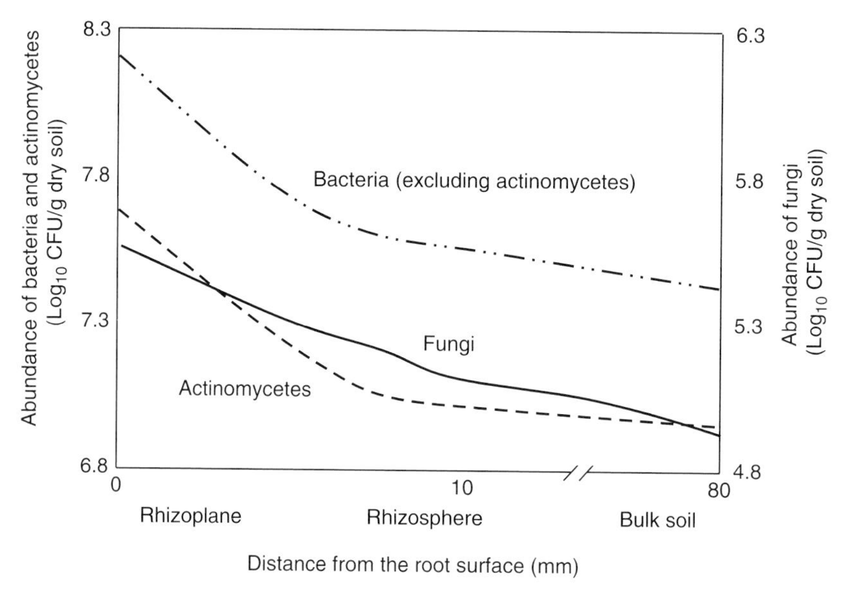

Fig. 11.2. Distribution of naturally occurring actinomycetes, fungi and bacteria on the rhizoplane and rhizosphere of spring wheat and bulk soil (CFU = colony-forming units). (Adapted from Graham and Mitchell, 2005.)

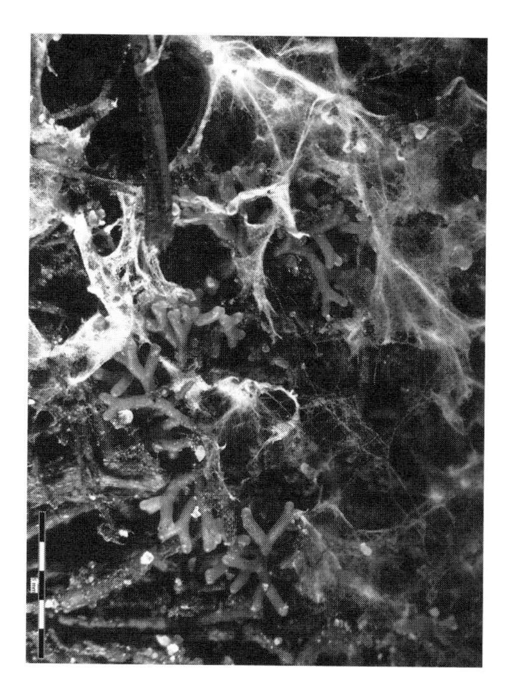

Fig. 11.3. The root tips of this Scots pine seedling are colonized by an ectomycorrhizal fungus. This results in formation of the pale brown dichotomously branched ectomycorrhizae. The white filaments radiating from the ectomycorrhizae are the fungal hyphae: they form the extraradical mycelial network. (Photo courtesy of Martin Vohník, Institute of Botany ASCR, Czech Republic.)

stages of its life cycle. The organisms involved in specific suppression often achieve biocontrol through antibiosis and hyperparasitism, described in Section 11.1. A soil that exhibits specific suppression could be studied more closely to isolate the key antagonists, which could then be cultured and transferred to other soils. In contrast, general suppression is achieved through competition between a resident soil foodweb and a pathogen. Management practices are helpful in generating an active and healthy soil foodweb that can act against pathogens through general suppression.

How can we determine whether a soil has the ability to suppress certain pathogens? Most suppressive soils maintain their activity even after drying and sieving, which means we can measure a soil's suppressive ability in the greenhouse or laboratory. This permits isolation of key microbial groups that contribute to soil suppression in the field. The first step is to determine whether suppression can be destroyed by soil fumigation (the soil is exposed to chloroform, methyl bromide or another substance that kills microbial cells), soil pasteurization (the soil is exposed to aerated steam heated to 60°C for 30 min) or with biocides such as novobiocin and chloropicrin. Soils can also be tested under harsher conditions such as autoclaving (the soil is steam-sterilized at 105°C under high pressure for 45 min) and gamma irradiation.

In most cases, general suppression is reduced but not eliminated by soil fumigation and pasteurization (Hoitink and Fahy, 1986). Transfer experiments are often conducted to verify the biological nature of antagonism against a particular pathogen. This involves making a soil–water mixture of the suppressive soil and adding it to a conducive soil (containing the pathogen and known to be susceptible to disease outbreaks). Less severe disease in a host plant growing in the conducive soil that received the soil–water mixture may confirm that the suppressive soil did indeed possess a resident population of antagonistic soil foodweb organisms (Fig. 11.4).

Agricultural practices to promote soil suppression of pathogens

Perhaps the simplest way to achieve biological control of pathogens is to manage the soil in a way that promotes an active soil foodweb containing resident antagonists. General suppression is often enhanced by adding organic matter, since this is an important food source for most soil bacteria, actinomycetes and fungi. Building soil fertility is also extremely important because a moderate to high level of soil fertility will benefit crops and support the growth of beneficial microorganisms inhabiting the rhizosphere. Producers can select agricultural practices that favour pathogen suppression, such as:

- select certain cultivars of a host plant that favour the growth of resident antagonists in the rhizosphere;
- design crop rotations with non-susceptible plants, especially those that produce allelopathic substances that inhibit pathogens;
- in some cases, a fallow period would be recommended so plant hosts are not present and the pathogen propagules die from starvation;
- apply compost and other organic fertilizers to support a large, active soil foodweb capable of pathogen suppression;
- tillage accelerates the decomposition of organic residues and can reduce pathogens that overwinter in crop residues;

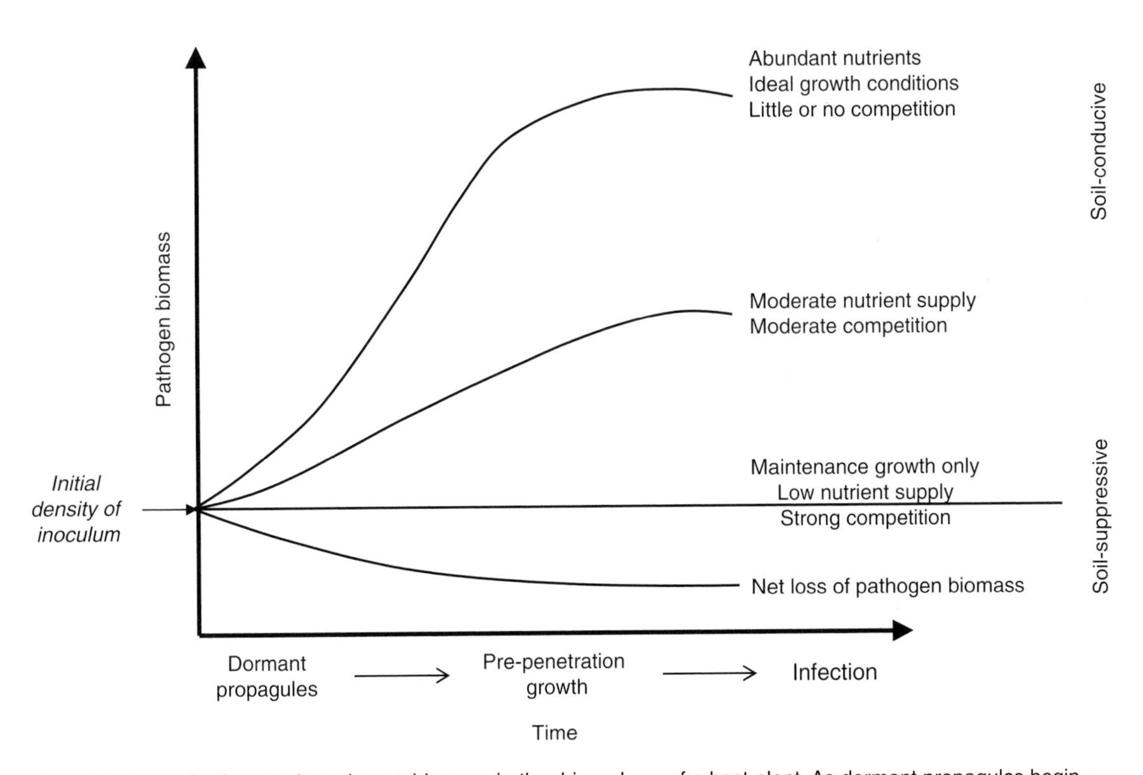

Fig. 11.4. Possible changes in pathogen biomass in the rhizosphere of a host plant. As dormant propagules begin pre-penetration growth and enter an infective state, the pathogen biomass may increase or decrease depending on whether the soil is conducive or suppressive to disease development. (Adapted from Graham and Mitchell, 2005.)

- add lime, install drainage and follow other good soil management practices that favour a large, active community of resident soil foodweb organisms that are antagonistic to plant pathogens; and
- soil treatments with steam, biocides or solarization that suppress or eliminate the pathogen, but not the resident antagonists, can be used.

Solarization involves covering soil or growth media with transparent polyethylene plastic sheets, thereby trapping incoming radiation from the sun for several days, until the material is completely dry. The combination of ultra-violet radiation and drying is expected to reduce the soilborne pathogen populations because it kills pathogen propagules. Weed seeds are also killed during solarization.

Case study: *Phytophthora*-suppressive soils

Phytophthora cinnamomi is a widespread soilborne fungal pathogen that infects woody plants causing root rot and cankering. It needs moist soil conditions and warm temperatures to thrive, but is particularly damaging to susceptible plants (e.g. drought-stressed plants in the summer). *P. cinnamomi* poses a threat to forestry, ornamental and fruit industries and infects over 900 woody perennial species (Fig. 11.5; Plate 2).

In avocado groves of Queensland, Australia, a severe and economically significant outbreak of root rot due to *P. cinnamomi* was documented (Broadbent and Baker, 1974). Researchers investigating this disease outbreak discovered suppressive soils in a neighbouring avocado grove that exhibited only minor damage from root rot. The disease suppression in this grove was attributed to good soil management practices: the producer applied poultry manure and limestone annually to maintain a soil pH above 6.0, and the space between tree rows was continually cropped with a legume–maize cover crop. This management promoted the growth of an active soil microbial population that

Fig. 11.5. Little leaf disease caused by *Phytophthora cinnamomi* infection of sand pine *Pinus clausa*. (Photo source: Edward L. Barnard, Florida Department of Agriculture and Consumer Services, Bugwood.org. Licensed under Creative Commons.)

was not eliminated with steam treatment (60°C for 30 min), but more intensive heat treatment (100°C for 30 min) was lethal to the resident antagonists. The temperature effect suggests that spore-forming bacteria and actinomycetes were the biological antagonists acting against the *Phytophthora* fungal pathogen. Further work demonstrated that the resident antagonists were producing celluloase and laminarinase enzymes that degraded cellulase in the hyphae of *P. cinnamomi*. No single organism was capable of suppressing the fungal pathogen, which led the researchers to conclude that the control mechanism was general suppression, because it was related to the total amount of microbial activity in the soil at a critical time in the pathogen's life cycle.

Case study: suppression of take-all disease in wheat

Take-all disease of wheat is caused by the fungus *Gaeumannomyces graminis* var. *tritici*. Root infection by the fungus blocks the conductive tissue and reduces water uptake. Early signs of the disease include stunting and yellowing. Plants mature earlier, may have fewer tillers and can be recognized by the characteristic 'white heads' that appear within a healthy crop (Fig. 11.6; Plate 3). The roots of diseased plants are black in colour, and the base of the stems may also exhibit this symptom. Affected plants are easy to pull out of the ground due to poor root development.

In many parts of the world, producers gain partial control of the disease by rotating wheat and barley crops with other non-host crops. However, the disease is transferable from field to field, so it can be spread by machinery to large areas. As discussed by Weller *et al.* (2002), there are two biological control mechanisms that can reduce take-all severity in wheat: (i) competition from bacteria and fungi in root fragments, which reduces the inoculum density of the pathogen (general suppression); and (ii) antagonistic microorganisms in the rhizosphere and young root lesions attack the pathogen and limit its infection and spread through roots (specific suppression).

General suppression of the take-all pathogen is partially attributed to the activity of non-pathogenic, saprophytic *Fusarium* spp. The take-all pathogen is able to grow and reproduce on crop residues (root fragments and stems) when the crop is not present in the field. The fungus *Fusarium*

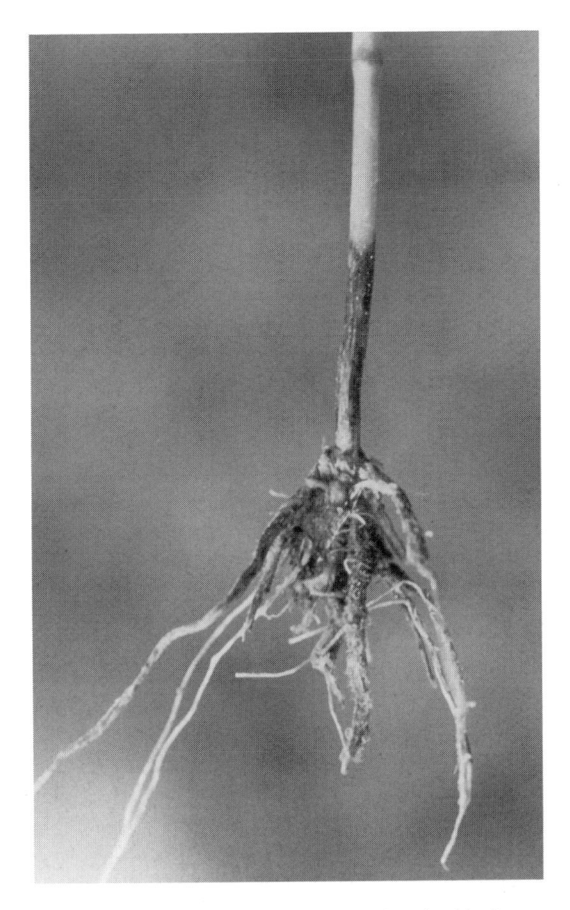

Fig. 11.6. Take-all disease in wheat, showing black root and stem of an affected plant. (Photo courtesy of Kansas State University.)

oxysporum is more efficient at obtaining energy and nutrients from decomposing residues than the take-all pathogen, thus reducing the amount of *G. graminis* var. *tritici* inoculum that could infect the next wheat or barley crop. The competitive activity of *Fusarium* spp. is improved by tillage, which accelerates the breakdown of crop residues and promotes soil aeration. Delayed seeding is another beneficial agricultural practice because this allows more time for the *Fusarium* spp. to displace the *G. graminis* var. *tritici* from crop residues.

Specific suppression of the take-all pathogen by antagonistic fluorescent pseudomonads is an important means of biological control. Pseudomonad characteristics include: (i) they are well adapted to the rhizosphere environment; (ii) they use a variety of organic substrates; (iii) they synthesize antifungal compounds that inhibit the growth of *G. graminis* var. *tritici*; and (iv) they grow rapidly on roots with take-all lesions.

The most effective biological control strains of *Pseudomonas* spp. produce 2,4-diacetylphloroglucinol (2,4-DAPG). This is an antibiotic synthesized by *Pseudomonas* spp. living in the rhizosphere and is effective at reducing the growth of the take-all pathogen. This is a clear example of specific suppression. Researchers have also found that *G. graminis* var. *tritici* growth on crop residues is inhibited by *Pseudomonas* spp. The green, fluorescent cells of *Pseudomonas fluorescens* are the most easily viewed under high magnification, as shown in Fig. 11.7 and Plate 4. They enter the epidermis and cortex of plant roots, where they can directly attack pathogens that have infected the root.

Case study: suppression of root-knot nematode (*Meloidogyne* spp.)

Root-knot nematodes of the genus *Meloidogyne* are a serious pest in crop production because they infect a variety of crops, including tomato, banana, grape, carrot, strawberry, lettuce, pumpkin and beans. The life cycle involves four juvenile stages separated by moults (Fig. 11.8). Infection occurs when second-stage juvenile nematodes invade the

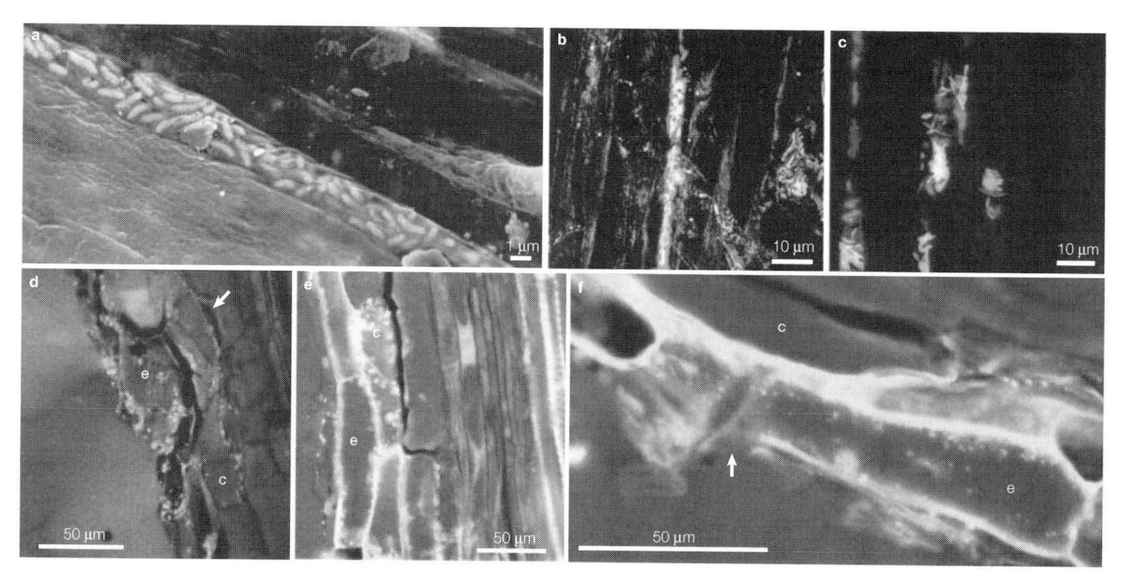

Fig. 11.7. Microscopic photos of: (a) microcolony of *Pseudomonas fluorescens* strain WCS365 on tomato root; (b) tomato root colonized by *P. fluorescens* WCS365 expressing autofluorescent proteins (red, yellow and green colours); (c) a mixture of autofluorescent proteins produced by strain WCS365; (d) and (e) *P. fluorescens* strain CHA0 cells associated with the cortex and epidermis of tobacco roots. The arrow in d points to bacteria found between cortical cells; (f) a damaged cell wall (see arrow) of epidermal cells colonized by *P. fluorescens*. In images (e) and (f), ‘c’ indicates the cortex and ‘e’ an epidermal cell. (Reprinted with permission from Macmillan Publishers Ltd, *Nature Reviews Microbiology* 3, 307–319, Biological control of soil-borne pathogens by fluorescent pseudomonads, Haas, D. and Défago, G.,® 2005. Panel a reproduced, with permission, from Chin-A-Woeng, T.F.C., de Priester, W., van der Bij, A. and Lugtenberg, B.J.J. (1997) Description of the colonization of a gnotobiotic tomato rhizosphere by *Pseudomonas fluorescens* biocontrol strain WCS365, using scanning electron microscopy. *Molecular Plant–Microbe Interactions* 10, 79–86,© (1997) American Phytopathological Society. Panels b and c reproduced, with permission, from Bloemberg, G.V., Wijfjes, A.H.M., Lamers, G.E.M., Stuurman, N. and Lugtenberg, B.J.J. (2000) Simultaneous imaging of *Pseudomonas fluorescens* WCS365 populations expressing three different autofluorescent proteins in the rhizosphere: new perspectives for studying microbial communities. *Molecular Plant–Microbe Interactions* 13, 1170–1176© (2000) American Phytopathological Society. Panels d–f reproduced, with permission, from Troxler, J., Berling, C.-H., Moënne-Loccoz, Y., Keel, C. and Défago, G. (1997) Interactions between the biocontrol agent *Pseudomonas fluorescens* CHA0 and *Thielaviopsis basicola* in tobacco roots observed by immunofluorescence microscopy. *Plant Pathology* 46, 62–71© (1997) Blackwell Publishing.)

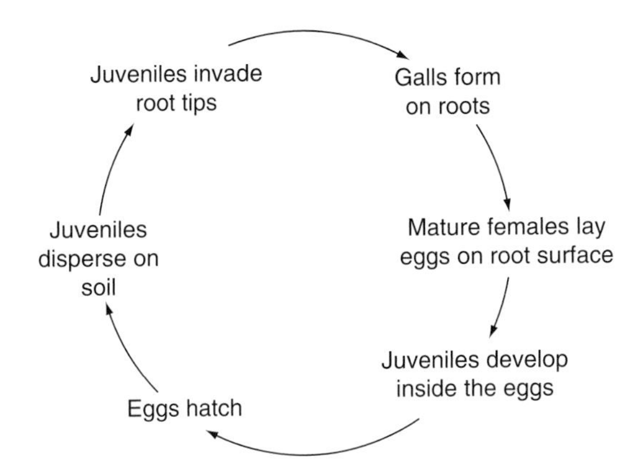

Fig. 11.8. Life cycle of the root-knot nematode. When soil temperatures are optimal (25–30°C), the entire life cycle takes 3 to 4 weeks (adapted from: http://www2.dpi.qld.gov.au/horticulture/4855.html, accessed on 6 February 2009).

root tip and migrate towards the vascular tissue to establish a permanent feeding site. Once feeding begins, the juvenile becomes sedentary and undergoes its three successive moults, developing into adults. While females remain non-motile during adult stages, the males regain motility during the third moult and leave the root.

At the feeding site, the infective juvenile triggers the development of five to seven giant cells within the root cortex that eventually developed into galls. Within the galls, the female produces hundreds of eggs and releases them on to the root surface. The voracious feeding by root-knot nematodes results in misshapen root crops, an economic loss for producers. Damage to the root system caused by nematodes reduces the root's capacity to absorb water and nutrients necessary for crop production. Wilting, chlorosis and other visual nutrient deficiencies may be a sign of root nematode infestation, and often lead to a decline in the yield and quality of crops.

Biological control of *Meloidogyne* spp. can be achieved through specific suppression with *Paecilomyces lilacinus*, a nematophagous fungus capable of reducing root-knot nematode populations by 76%. The nematode-trapping fungus *Monacrosporium lysipagum* is even more efficient, showing a 93% reduction in root-knot nematode populations (Khan *et al.*, 2006). The nematophagous fungus *Verticillium chlamydosporium* can reduce root-knot nematode populations by parasitizing nematode eggs on plant roots. However, the efficacy of *V. chlamydosporium* depends on both nematode population density and plant species. There seems to be a tripartate interaction between plant roots, nematodes and nematophagous fungi, whereby the nematode affects root exudation by the host plant, and the root exudates from some plants act as a chemical attractant for the nematophagous fungi. Not all plant species produce exudates that stimulate the growth of fungal hyphae towards their root surfaces (Bourne *et al.*, 1996). The bacterium *Bacillus penetrans* proved to be as effective as chemical nematicides in controlling *Meloidogyne javanica* on tomatoes and grapes (Stirling, 1984).

General suppression of root-knot nematode is also possible and relies on crop rotations and cover-cropping with legumes. Intercropping susceptible crops and non-host crops is another technique that has been used with some success. Soil solarization is helpful in some cases, but nematodes encyst when soil conditions become too hot or dry for their activity, so solarization is not a reliable method of reducing root-knot nematode populations to economically acceptable levels.

11.3 Biocontrol with Introduced Antagonists

If indigenous biological organisms (resident antagonists) are not present in a particular field or their numbers are inadequate to achieve biological control, then it may be necessary to add biocontrol organisms. Introduced antagonists may be added once, occasionally or repeatedly to maintain an appropriate level of biocontrol. Pathogens that cause plant disease in the soil, such as seed rots, damping-off of seedlings, root rots and wilts, are the best candidates for control with introduced antagonists. A small amount of inoculum can be applied to seeds, cuttings or spread in the furrow before planting seeds or transplants. It is important to introduce the antagonist early in the growing season to protect plant roots during the crucial establishment phase.

To be effective biocontrol agents, introduced antagonists must have rhizosphere competence, that is, they are capable of colonizing the root surface (rhizoplane) or growing in the rhizosphere in the presence of competitors and predators. This applies to any introduced antagonist, whether applied directly to the seed, on the planted end of cuttings or applied to the soil in the vicinity of seeds or transplants. There are numerous biotic and abiotic factors, listed in Table 11.3, that determine whether an introduced antagonist will be able to colonize the rhizosphere of a plant. Briefly, the microorganism must be able to acquire energy and nutrients and defend itself from competitors and predators. Plant factors affect the chemical signals sent to potential colonizers; in addition, there must be unoccupied binding sites or niches for the introduced antagonist on the plant root. Finally, environmental factors may affect the survival of the introduced biocontrol agent (Table 11.3).

Rhizosphere-competent organisms must be capable of growing and proliferating along a developing root for optimal protection. In this way, the root is protected from infection by a pathogen, since the root surface is covered by beneficial organisms. As well as protecting against pathogens, these organisms may produce plant growth-promoting substances like growth hormones, stimulate nitrogen mineralization, secrete chelating compounds that bind and transport essential nutrients to the plant root, or acquire water for plant growth. The rhizobacteria are perhaps the best known group of rhizosphere-competent bacteria, although some actinomycetes and fungi such as *Trichoderma* are also rhizosphere-competent. Further information on the rhizobacteria was provided in Focus Box 7.1.

Rhizobacteria are extremely effective introduced antagonists because they can attach and distribute themselves along the growing root surface. Some of the bacterial traits that make them well suited for this role include:

- Production of extracellular polysaccharides: these substances surround and bind bacterial cells together to form microcolonies. The extracellular polysaccharides protect the cells in the colony from desiccation, antibiotics secreted by microbial competitors, and predators. This protection helps the introduced antagonist to bind to the root and resist displacement by indigenous microorganisms.
- Production of fimbriae: these proteinaceous, filamentous appendages aid microbial attachment to roots. Several bacteria possess these appendages, including free-living nitrogen-fixing bacteria belonging to *Klebsiella* and *Enterobacter* species and the free-living *Pseudomonas fluorescens* that suppresses take-all disease in wheat and barley (specific suppression agent).
- Production of flagella: certain bacteria possess flagella that permit them to propel themselves for some distance along a growing plant root to sites that are not yet colonized by other organisms.
- Chemotaxis: the ability to receive and interpret chemical signals from plant roots may guide rhizobacteria through soil pores and the soil solution to the root surface. This is especially important if rhizobacteria are not placed in direct contact with the seed or the end of a cutting (e.g. placed in the furrow, not in the immediate vicinity of a growing root).
- Tolerance of dry soil conditions: rhizobacteria that can form spores or enter into very small soil pores during unfavourably dry soil conditions are more likely to survive in the long term.
- Ability to use complex carbohydrates as an energy source: although rhizobacteria are expected to be most active in the rhizosphere, where they receive simple carbohydrates and amino acids excreted by the plant root, a few of them are able to use more complex carbon compounds such as cellulose, hemicelluloses and pectin. This ensures their survival when plant roots are not present at the site of introduction (e.g. in the bulk soil, not in the immediate vicinity of a growing root).

Table 11.3. Factors or characteristics influencing rhizosphere colonization by introduced antagonists. (Adapted from Graham and Mitchell, 2005.)

Microbial characteristic	Plant factor	Soil characteristic
Nutritional versatility	Species	Texture
Rapid growth rate	Age	Moisture
Cellulase production	Altered genetics	Soil air
Antibiotic tolerance or production	Foliar treatments	Temperature
Siderophore production	Root surface (binding sites)	Fertility
Unique physiological attributes	Root exudates	Applied pesticides
Tolerance to fungicides or other chemicals		

The effectiveness of rhizobacteria or any other introduced antagonist in biocontrol depends on their survival when applied to the soil, the establishment and growth of the organism in the soil and its proliferation in root lesions and infection sites created by the pathogen. Weller (1988) described the colonization of a root by an introduced antagonist as follows:

1. Phase I – the introduced antagonist attaches to the elongating root tip and is transported with the growing root.
2. Phase II – the introduced antagonist spreads locally and proliferates to the limits of its ability, considering competition with indigenous organisms, and survives.

In Phase I, the introduced antagonist, added to the soil on seeds, crushed grain or other materials, comes in contact with emerging roots. It can be assumed that the introduced antagonists are rhizobacteria. Viable cells attach to the elongating root and are nourished by root exudates. As the root elongation continues, some rhizobacteria are carried along the root tip, whereas others are left behind as a source of inoculum on older root parts. Transport of rhizobacteria with the root tip continues as long as bacterial populations keep multiplying at the same pace as root tips grow. The primary limitation for the introduced antagonist may be that the rhizobacteria cannot multiply quickly enough to keep pace with the growth of the root tip (extending through soil at a rate of 2–9 cm/day). Other reasons that the rhizobacteria may not fully colonize the root are that they can be physically removed from the root tip and adsorbed on to soil particle surfaces, and they can be outcompeted for space on the root surface by indigenous microorganisms.

In Phase II, the rhizosphere-competent bacteria will grow and reproduce on the root, whereas less fit bacteria are rapidly displaced. The fate of introduced microorganisms is ultimately governed by their ability to compete for resources with indigenous microorganisms. Nutrient limitation is thought to be the main reason that introduced antagonists fail to survive in some soils. Bacteria tend to congregate at cell junctions where nutrients are abundant. If indigenous strains of bacteria are more efficient at acquiring nutrients from these 'hot spots', then the introduced bacteria may not survive.

The survival of introduced antagonists during Phase I and Phase II is profoundly affected by environmental factors, including the soil water potential, temperature, pH, soil texture, salinity and so on. The optimal temperature for the growth of *P. fluorescens* and *P. seudomonas putida* is 25–30°C in the laboratory, but root colonization is greatest when soil temperatures are < 20°C. Thus, it is important to understand the characteristics of introduced antagonists so they can be added at an appropriate time or soil can be prepared in a manner that favours their survival. Researchers have also found that some plant genotypes are more receptive to colonization by an introduced antagonist than others. This is not unusual – it is well known that choosing an appropriate *Rhizobium* strain for a particular legume is necessary to get the best nodulation and achieve a high level of N_2 fixation. For instance, the recommended inoculum for lucerne is *Rhizobium meliloti*, while red clover should be inoculated with *Rhizobium trifoli*. Researchers must consider the microbial and plant characteristics, as well as environmental factors (Table 11.3), when assessing the efficiency of introduced antagonists in the field.

Focus Box 11.1. Earthworms improve plant tolerance to nematode parasites.

As discussed in Chapter 10, earthworms modify the soil environment by mixing and redistributing organic residues with soil mineral particles. The consumption of organic residues by earthworms accelerates the decomposition and nutrient recycling from these materials. Plant growth often benefits from the change in soil structure and nutrient availability due to the presence of earthworms. This has been shown many times in controlled laboratory and greenhouse studies.

Another way that earthworms may interact positively with plants is through their effects on soil microbial communities. Free-living microorganisms are redistributed in the soil profile by earthworms, into nutrient-rich casts that support rapid root proliferation. Consequently, the earthworm may indirectly

Continued

modify the soil environment in a manner that brings roots and saprophytic microbiota into closer contact. Some of these microorganisms could be effective antagonists against root pathogens.

The plant-parasitic nematode *Heterodera sacchari* forms external cysts on rice roots, leading to serious damage in upland rice fields in Africa. A greenhouse study by Blouin *et al.* (2005) showed that the presence of the earthworm *Millsonia anomala* in nematode-infected soils under rice production led to a rapid improvement of rice growth (Fig. 11.9).

Within 3 months, rice grown in pots without any nematode control exhibited less growth, lower photosynthetic rates and greater expression of plant stress genes. It appeared that the nematodes had caused irreversible damage to the photosynthetic system of the rice plant. In pots where the rice plants were exposed to nematodes and earthworms, there was very little damage: stress genes were in the normal range, the photosynthesis system was not damaged and the rice plants grew to a similar size as those in the control soil (no earthworms or nematodes

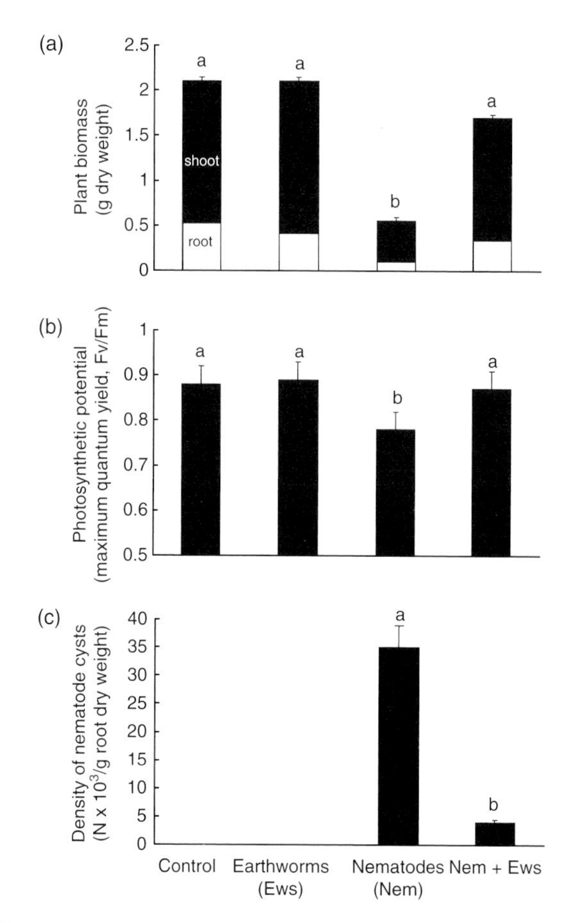

Fig. 11.9. Earthworms (*Millsonia anomala*) living in the experimental pots did not enhance significantly (a) rice biomass or (b) plant health indicators, compared with the control plants without earthworms. Plants experimentally infected with the plant-parasitic nematode *Heterodera sacchari* showed reduced performance, plant growth and photosynthetic rates. However, when earthworms were present simultaneously with plant-parasitic nematodes (Nem + Ews), shoot and root dry biomasses were similar to the control, although (c) there was an 80% lower nematode infection rate. Columns with the same letter were not significantly different. (Source: Blouin *et al.*, 2005.)

Continued

1 3

2

Plate 1. The symbiotic association between nitrogen-fixing bacteria and leguminous plants. The nodules on the root of the soybean plant illustrated contain *Rhizobium* bacteria. Within the nodule, the bacteria fix atmospheric nitrogen, which they share with the plant. In exchange, the plant supplies the bacteria with a source of carbon and energy for growth. (Photo courtesy of David M. Dennis.)
Plate 2. Little leaf disease caused by *Phytophthora cinnamomi* infection of sand pine, *Pinus clausa* (courtesy of Edward L. Barnard, Florida Department of Agriculture and Consumer Services, Bugwood.org. Licensed under Creative Commons).
Plate 3. Take-all disease in wheat, showing affected roots and stem (courtesy of Kansas State University).

4

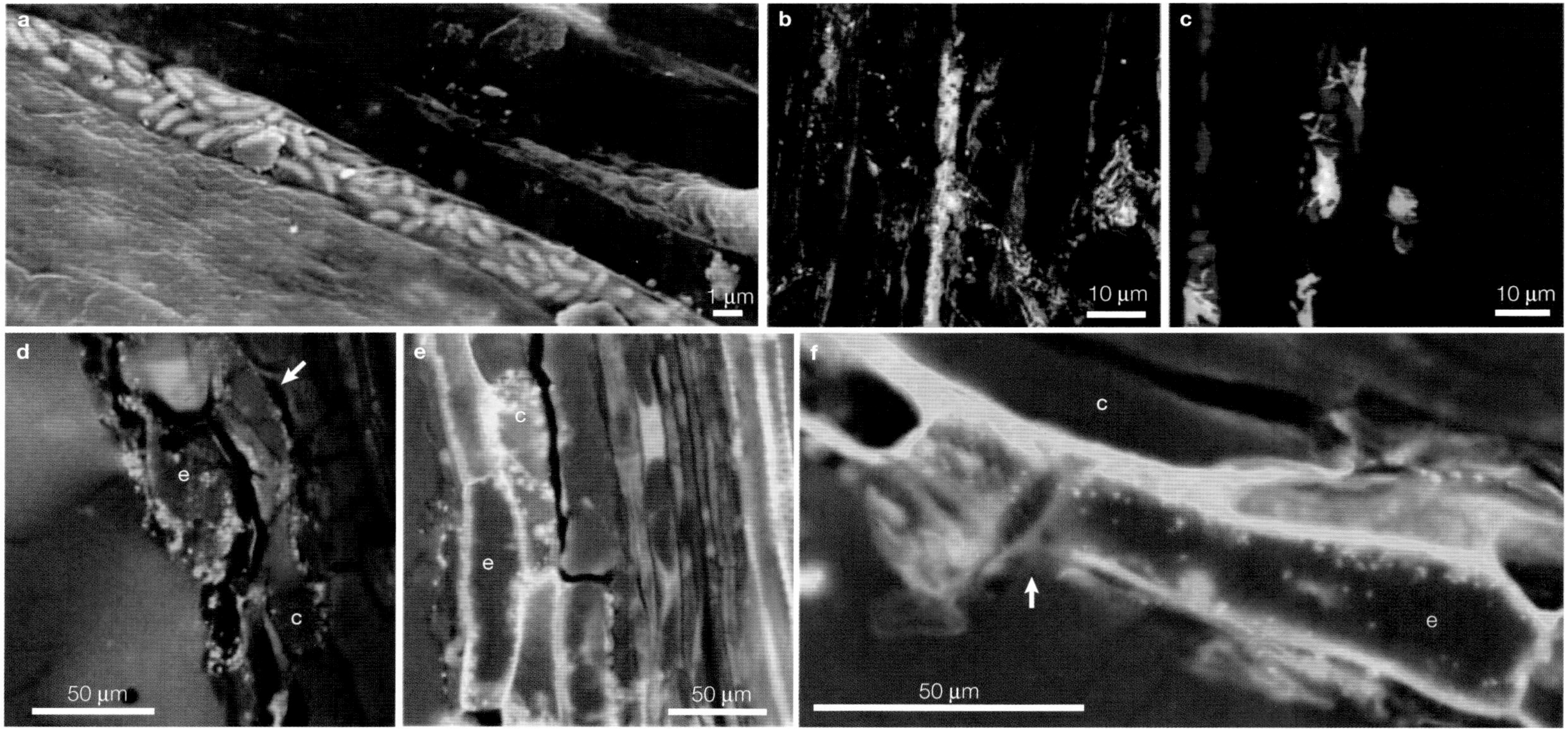

a b c d e f

1 µm 10 µm 10 µm 50 µm 50 µm 50 µm

Plate 4. Microscopic photos of (a) microcolony of *Pseudomonas fluorescens* strain WCS365 on tomato root; (b) tomato root colonized by *P. fluorescens* WCS365 expressing autofluorescent proteins (red, yellow and green colours); (c) a mixture of autofluorescent proteins produced by strain WCS365; (d) and (e) *P. fluorescens* strain CHA0 cells associated with the cortex and epidermis of tobacco roots. The arrow in d points to bacteria found between cortical cells; (f) a damaged cell wall (see arrow) of epidermal cells colonized by *P. fluorescens*. Reprinted by permission from Macmillan Publishers Ltd, *Nature Reviews Microbiology* 3, 307–319, Biological control of soil-borne pathogens by fluorescent pseudomonads, Haas, D. and Défago, G., ®2005. Panel a reproduced, with permission, from REF. 90 © (1997) American Phytopathological Society. Panels b and c reproduced, with permission, from REF. 105 © (2000) American Phytopathological Society. Panels d–f reproduced, with permission, from REF. 107 © (1997) Blackwell Publishing.

Focus Box 11.1. *Continued*

present). Earthworms did not reduce the number of nematode cysts, but they did lower the root infection rate by 80%.

Blouin *et al.* (2005) concluded that earthworms enhanced plant tolerance to nematodes in a systemic manner. Earthworms did not directly reduce the nematode population, nor did they alter the soil foodweb in a way that caused resident antagonists to attack the parasite. Thus, the reduction in root infection was probably due to an indirect effect of the earthworms on the soil environment that improved plant health or that physically blocked the nematode from adhering to the rice roots. These results emphasize the complexity of below-ground interactions in suppressive soils.

Methods of introducing antagonists for biocontrol

A common method of introducing antagonists for biocontrol is to place the antagonist on a seed, which delivers it into the soil environment. Biological control agents are often applied as a slurry on the seed with binding agents such as methyl cellulose and polyvinyl acetate that create a seed coating with viable bacteria or fungal cells. Many seeds undergo controlled hydration or dehydration treatments known as seed priming to improve their germination after planting. Yet, these treatments can be detrimental to biological control agents. A procedure known as solid-matrix priming was developed to overcome this challenge and has been successfully used to prepare seeds coated with *Trichoderma harzianum*, providing effective and reliable biological seed treatment for a number of crops.

Seed treatment with biological control agents is an expensive proposition for a number of reasons. Fungicides are more effective in seed protection and less expensive than biological agents. The shelf life of chemicals is superior to biological agents, and chemicals protect the seeds under a wider range of temperature and environmental conditions than biological control agents. However, there are opportunities to use introduced biological control agents for: (i) organic agriculture applications where chemical pesticides are not approved for use; (ii) applications to control crop root diseases that are inadequately controlled with chemical pesticides, due to difficulties in applying the pesticide or the development of pesticide-resistance pests; (iii) replacement or reduction of chemical pesticides in sensitive environments such as greenhouses where worker safety is an important consideration; and (iv) replacement of chemical pesticides that are no longer permitted due to stricter regulations for human and environmental safety.

Granulated products that contain bacterial cells or fungal spores are another option. The inoculant is mixed with water and sprayed on the seed just before it is planted. Cuttings and tubers can be dipped in the inoculant before they are planted. This method is relatively cheaper than purchasing coated seeds. Fungal biocontrol agents are often grown on pieces of grain, which can then be broadcast on a field site or applied in a furrow before the seeds are planted.

It is possible to apply introduced antagonists as biocontrol agents, and there have been some very successful cases, such as the biocontrol of crown gall (see Case study: biological control of crown gall, below). Producers must be aware that the record for other introduced biocontrol treatments and biopesticides is far less consistent. There are several reasons for this:

1. The marketplace has focused almost exclusively on single organisms, even though a composite of introduced antagonists (several antagonistic organisms in a single mixture) is nearly always more effective against plant disease because multiple organisms can attack the pathogen through different mechanisms (antibiosis, competition, hyperparasitism). This is due to the fact that it is easier to patent and market a single biocontrol organism than a group of biocontrol organisms.
2. For marketing reasons, biocontrol agents are designed to be effective for a short period of time, which guarantees the need for repeated applications and enhances sales.
3. The technology needed to produce and deliver individual biocontrol agents is similar to the existing technology for mass production of microorganisms. There are production issues that limit our

ability to produce formulations containing several microorganisms to generate a composite of introduced antagonists.

There are at least two other reasons why introduced antagonists perform in an inconsistent manner in the field. The most common explanation is that the soil environmental conditions are not favourable for the introduced antagonist and it is not able to adapt to these conditions (i.e. it has lost rhizosphere competence). Secondly, the biocontrol agent may produce antibiotics or other inhibitory compounds too late for effective disease control.

Despite these drawbacks, a number of potential biocontrol agents have been identified and have either been developed commercially or are under development (Table 11.4).

Case study: biological control of crown gall

The best example of plant disease control achieved with an introduced biocontrol agent is crown gall. This disease is caused by the soilborne bacterium *Agrobacterium tumefaciens*. The disease causes uncontrolled cell division in the host plant, usually on the roots or around the crown (Fig. 11.10). A wide variety of dicotyledonous plants are susceptible to crown gall, especially stone fruits, roses and grapes.

A highly successful biological control for crown gall was developed in Australia. The method involves inoculating the planting stock with non-pathogenic *Agrobacterium radiobacter* strain K84 that produces the bacteriocin agrocin 84. A bacteriocin is a chemical that inhibits or kills closely related bacterial species. Biological control is achieved by dipping the planting material in a water suspension of strain K84 containing 10^7 to 10^8 cells immediately before planting. Strain K84 has performed better than commercial chemical treatments in preventing crown gall. It can provide 100% control in naturally infested soil. Strain K84 has been sold commercially in Australia since the early 1970s and is now permitted for use in many countries.

Case study: biological control of root-knot nematode

The bacterium *Pasteuria penetrans* has several characteristics that make it a potentially useful biological control agent against the root-knot nematode *Meloidogyne* spp. as well as other nematodes. Bacterial populations are parasitic on root-knot nematode, and prevent reproduction and reduce plant root infection by juvenile nematodes. The spores of *P. penetrans* have morphological biochemical features of endospores, and consequently they can tolerate environmental extremes. Endospores can be stored in air-dried soil without loss of infectivity, have a considerable tolerance to temperature extremes and are relatively small and therefore redistributed by percolating water. The last feature is particularly useful because the spores could be applied at or near the surface of sandy soils and will move through the soil profile with rainfall or irrigation

Table 11.4. Examples of microorganisms that have shown potential as biocontrol agents when inoculated in the soil or rhizosphere. (Adapted from Graham and Mitchell, 2005.)

Microorganism	Application(s) in biocontrol
Agrobacterium tumefaciens	Controls crown gall in grapes
Alcaligenes faecalis	Reduces *Rhizoctonia solani* root rot in tomatoes
Bacillus subtilis	Increases germination and growth of cabbage
Azospirillum brasilense	Produces phytohormones, specifically the auxin indole-3-acetic acid (IAA), which promotes crop growth and reduces the impact of crown gall and leaf blight
Pseudomonas fluorescens	Increases growth of seeds and cuttings of many flowers
	Reduces damping-off of cotton seedlings
	Suppresses 'take-all' disease of wheat
Pseudomonas putida	Reduces *Pythium* damping-off of sugarbeet
Pseudomonas syringae	Controls mould and rot on apple and citrus fruits
Trichoderma harzianum	Reduces the populations of several soilborne fungal pathogens that cause root diseases, notably *Aphanomyces cochlioides, Rhizoctonia solani* and *Fusarium oxysporum*

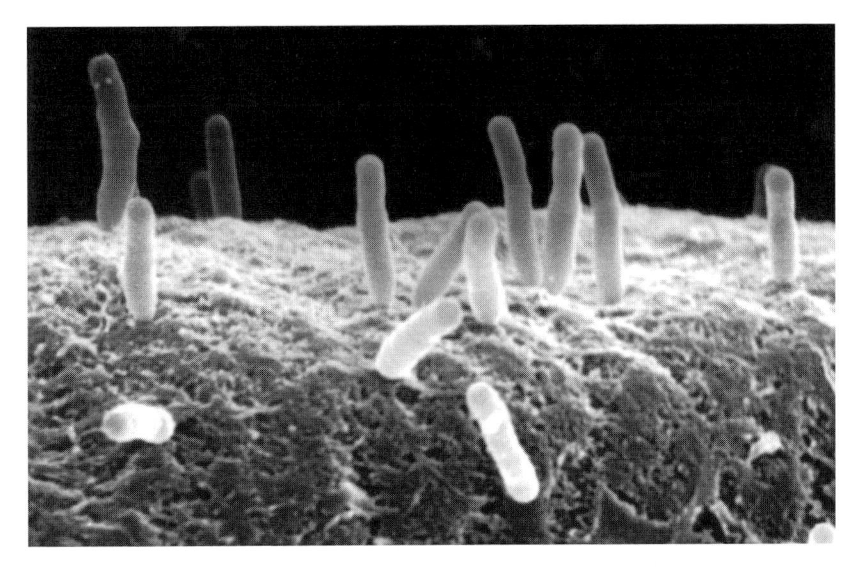

Fig. 11.10. *Agrobacterium tumefaciens* cells attached to the root tip of a susceptible host. (Photo courtesy of Martha C. Hawes.)

water to the root zone, where nematodes are present. One drawback is that *P. penetrans* is an obligate parasite, meaning that it will not grow or reproduce unless it comes in contact with a nematode (Fig. 11.11). Thus, it may not be very effective in soils with a low number of pathogenic nematodes.

11.4 Plant–Microbial Interactions to Promote Disease Resistance

We have examined resident antagonists that can be manipulated with cultural practices to suppress pathogen populations. The importance of crop rotations, resistant crops and organic fertilizer applications was mentioned. Introduced antagonists can be purchased and added to agroecosystems to reduce and control pathogens causing economically important crop diseases.

We cannot neglect to discuss the plant itself. Plants with a deep, well-developed root system that received adequate amounts of water and essential nutrients during their life cycle are generally less susceptible to disease than plants that were stressed during growth. However, many plants have the ability to resist invading pathogens through constitutive and inducible defences. To understand the plant defence strategies, we will examine how plants respond when material and energy are transferred from the plant to another organism. In many

Fig. 11.11. Nematode parasitized by *Pasteuria penetrans*. (Photo courtesy of Drs Brian Kerry and Keith Davies, Rothamsted Research, Hertfordshire, UK.)

ways, plant pathogens are similar to herbivores (Fig. 11.12). Large vertebrate herbivores such as cows and bison graze upon plants, biting off or chewing upon the leaves and stems. Cutting insects such as grasshoppers and ants also bite and cut leaves and stems. Sap-sucking insects (e.g. aphids) insert a stylet into the leaf or stem and suck out the juices. Soil-borne pathogens and parasites disrupt plant root tissues, breaking apart cell walls or burrowing into roots, so they obtain energy for growth and reproduction directly from the host plant. Unlike other herbivores, which generally do not kill the plant, pathogens alter the normal physiological functions of a plant in a manner that can weaken or destroy cells and tissues, manifesting in plant disease.

There are a number of ways that plants may repel herbivores. Constitutive defences include external barriers, such as waxy cuticles, spines, thorns and other physical deterrents so that herbivores have difficulty in penetrating the epidermis or would suffer harm if they consumed the plant. The bark growing on woody plants is an example of a constitutive defence, since it protects the plant from boring insects during its long lifespan. The constitutive defences are produced in all plants of the same phenotype. Thus, all woody plants have bark and all cacti have spines, regardless of whether they are grazed upon by herbivores during their life or not.

Inducible defences are generated by the plant in response to an environmental stress such as herbivory or infection by a pathogen. There are two major categories of inducible defences: (i) induction of cell elongation and division; and (ii) induction of chemical defences (Fig. 11.13).

How does the induction of cell elongation and division manifest in plants? In response to herbivory and other stresses, some plants respond by allocating more energy towards flowering and seed production. If the plant cannot survive the stress from the herbivore or pathogen, it will die, but its genes will live on through its offspring. Other plants allocate energy to root and rhizome production. If the leaves and stems are consumed, the plant will have enough energy reserves in the roots to regrow once the herbivore has moved to another area.

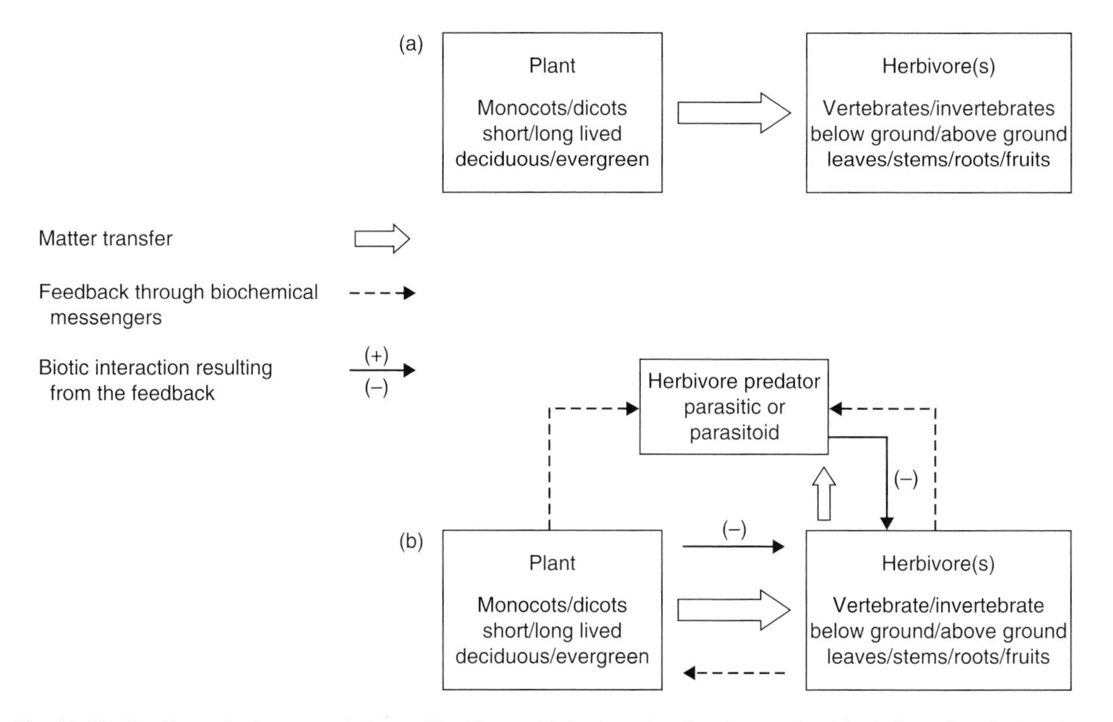

Fig. 11.12. Two theoretical representations of herbivory: (a) the transfer of carbon and nutrients from the plant to the herbivore occurs in a unidirectional manner, or (b) there is a biochemical feedback between herbivores and plants, which permits the plant to respond and defend itself against the loss of matter and energy. Illustration (b) is probably the more accurate representation of plant–herbivore interactions.

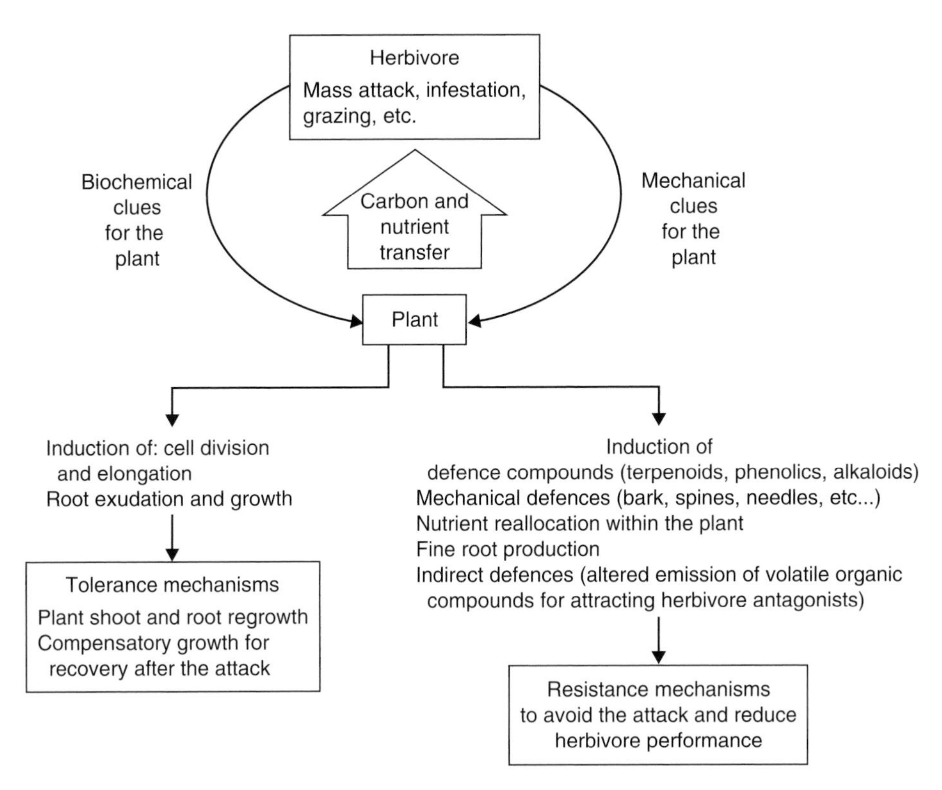

Fig. 11.13. The plant–herbivore interactions shown here suggest two possible lines of defence. When herbivory occurs, the plant receives a signal from the herbivore that stimulates cell division and elongation (positive feedback). A second possibility is that the plant responds by allocating energy towards the production of chemical defences (negative feedback).

This strategy may also be used to defend against root grazing or root infection by pathogens. More energy allocated to the root system is expected to stimulate production of root exudates. As listed in Table 7.8, plant roots exude chemical compounds that are a mixture of carbohydrates (simple sugars and polysaccharides), amino compounds, organic acids, nucleotides, flavones, enzymes and growth factors. These substances stimulate the growth of plant growth-promoting rhizobacteria that contribute to biological control and a number of other ecological functions in the rhizosphere (Focus Box 7.1).

The plant growth-promoting rhizobacteria seldom act directly against pathogens (most are not antagonists) but promote biological control by stimulating induced systemic resistance in the plant. In other words, the rhizobacteria send chemical signals to the plant, which prompt the plant to activate biochemical pathways that lead to physical barriers or chemical compounds that impede the pathogen (Fig. 11.14).

Chemical defences are another type of inducible defence that is observed when plants receive chemical signals from herbivores or pathogens. Chemical signals may be exogenous, generated when the invading organism mechanically damages the plant tissue or creates a wound in a root or stem to gain entry into the plant. Endogenous chemical signals are generated when an invasive organism infects the plant and begins to acquire energy from the plant for its metabolic requirements. In either situation, plants initiate biochemical pathways and manufacture compounds in response to these chemical signals. Plants can produce lignin, which confers structural rigidity to cells, or tannins, which render the plant tissues less palatable to a grazing organism; but also specific deterrent proteins, greater amounts of latex or terpenoid resin, or specific chemical compounds that are very toxic.

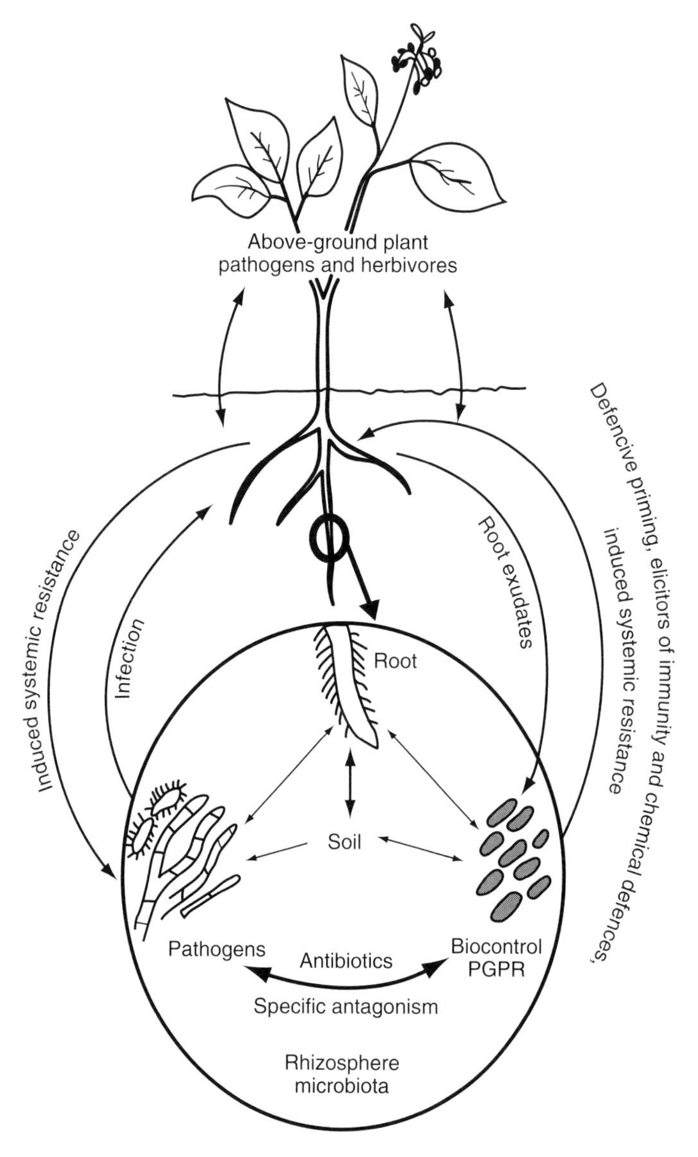

Fig. 11.14. Interactions between biocontrol plant growth-promoting rhizobacteria (PGPR), plants, pathogens and soil. (Adapted from Haas and Défago, 2005.)

These compounds are generally produced in stems and leaves, as a deterrent to herbivores that consume these tissues. Plants produce allelopathic compounds that inhibit the growth of other plants and some plant pathogens (Table 11.5). Some may be secreted directly through the root system, or could be extracted from leaves, tubers and stems and applied to soils.

Allelopathy is an inducible defence that is initiated when a plant is under stress, due to competition for resources, stress from disease, temperature extremes, moisture deficit or herbicide damage. Allelopathic chemicals are produced from the shikimate pathway, which is used by plants and bacteria to generate aromatic amino acids (phenylalanine, tyrosine and tryptophan) or the acetate pathway, which transforms acetyl-CoA to acetate and energy under anaerobic conditions. A diverse group of reactive compounds are formed as byproducts of these reactions (Table 11.5). Some act

Table 11.5. Allelochemicals produced through the shikimate pathway and the acetate pathway that may serve as inducible plant defences. (Adapted from Popa *et al.*, 2008.)

Shikimate pathway	Acetate pathway
Cinnamic and benzoic acids/aldehydes → coumarins	Long-chain fatty acids
Gallic acid → hydrolysable tannins	Various terpenes
Hydroxamic acids	Sesquiterpene lactones
Cyanogenic glycosides	Unsaturated lactones
Sulfides	Alcohols, aldehydes and ketones
Purines	Organic acids
	Certain quinines
Polypeptides*	Polypeptides*
Alkaloids*	Alkaloids*
Flavonoids*	Flavonoids*

*Compounds produced by both biochemical pathways.

as growth stimulants for plants, symbiotic microorganisms and growth-promoting rhizobacteria. Others act as herbicides, allowing plants to deter the growth of weeds and other undesirable plants in close proximity, increasing the competitive advantage of the plant species producing this compound. Finally, some allelochemicals act as chemical barriers to herbivores and root pathogens, providing a degree of protection against grazing pests and agents of plant disease.

Allelochemicals may also serve as chemical signals to resident antagonists. Roots secrete pheromone-like molecules that attract antagonists living in the bulk soil to the rhizosphere. There are at least two groups of soil organisms that are attracted to roots through this signalling mechanism. The first group are the nematode-trapping fungi, predators known for their ability to capture and consume plant-parasitic nematodes. The second is the rhizobacterium *Bacillus penetrans*. This organism produces lytic enzymes that break down the cell walls of some root parasites, causing cell lysis and eventually death. It is evident that plant interactions with these soil foodweb organisms lessen the severity of pathogenic attacks on the host plant.

Biological control with compost

Compost-amended media are often used in greenhouse container production and the floriculture industry as an alternative to biocides such as fungicides. Consistent and sustained biological control of fungal pathogens of the genera *Fusarium*, *Pythium*, *Phytophthora* and *Rhizoctonia* has been achieved using compost-amended media produced from wood by-products. Similar results have been obtained in field soil, although with less consistency.

In practice, composts do not contain a complete suite of soil foodweb organisms, which is needed for suppression of a broad spectrum of pathogens. The following mechanisms of biological control have been reported in compost-amended substrates:

1. General suppression of *Pythium* and *Phytophthora* spp. appears to be achieved in composts that contain a wide diversity of biocontrol agents, particularly bacteria belonging to the *Pseudomonas*, *Pantoea* and *Bacillus* spp.
2. Specific suppression of *Rhizoctonia solani* and *Sclerotium rolfsii* can be achieved if biocontrol agents are added after the peak heating period of the composting process, before colonization by mesophilic microorganisms has occurred (Fig. 11.15).
3. Release of fungitoxic compounds during organic matter decomposition can cause a general reduction of all fungi, including pathogens.
4. Induced systemic resistance in plants can be achieved when compost containing certain microorganisms is applied in and around the root zone. Microorganisms capable of inducing plant resistance in the rhizosphere include the non-pathogenic fungus *Trichoderma* spp. and the bacterium *Pantoea agglomerans*. However, fewer than 10% of composts stimulate resistance to plant diseases through this mechanism.

There are three phases in the decomposition of organic residues during composting (Fig. 11.15):

1. Phase I (thermophilic stage): the immature or non-stabilized compost is decomposed by thermophilic organisms at extremely high rates. The temperature within the pile is often too high for microorganisms other than thermophiles (Table 11.6). High temperatures cause heat stress and are effective at killing most of the pathogens, parasites and weed seeds that may be present in the undecomposed organic waste.
2. Phase II (mesophilic stage): the temperature of the pile cools to 20–40°C. This supports decomposition

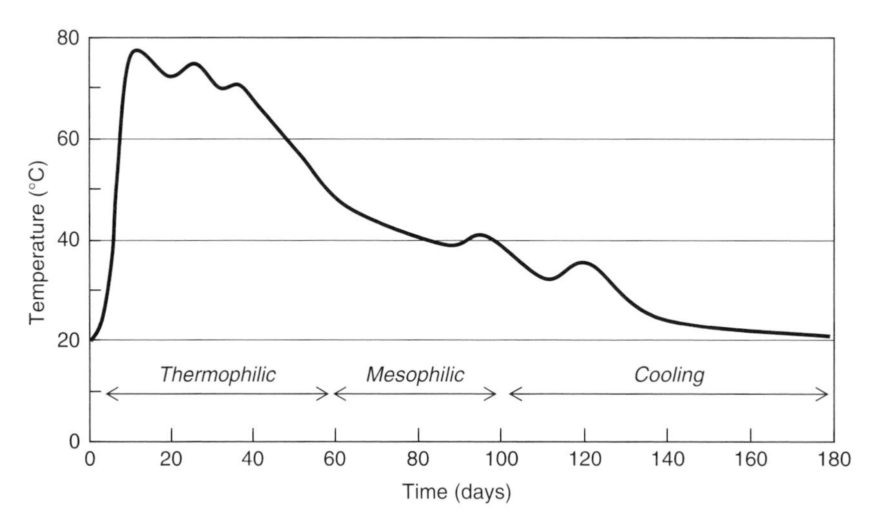

Fig. 11.15. Typical time–temperature relationship for composting organic wastes in aerated piles. Within several days, or even hours, the internal pile temperature rises rapidly from the ambient mesophilic (20–40°C) to the thermophilic (> 40°C) stage of the composting. It should be maintained by appropriate aeration close to 65°C for optimum microbial performance. If inefficient aeration occurs, the microbial populations will be deactivated when temperatures reach 70–80°C and the pile may collapse due to uneven heat diffusion. When resources become less abundant, microbial metabolic heat is easily diffused, pile temperature reduces and thermophilic populations are substituted by mesophilic ones, and microbial activity is sustained at a lower temperature for some time. The cooling phase is related to the depletion of available substrates for microorganisms, and to the secondary decomposition process leading to organic matter stabilization.

Table 11.6. Representative bacterial and fungal genera/species found in the thermophilic and mesophilic stages of composting. Adapted from: Boulter *et al.*, 2002; Zibilske, 2005; Ashraf *et al.*, 2007; Adams and Frostick, 2008.)

Composting stage	Bacteria	Fungi
Thermophilic	*Bacillus* spp.	*Aspergillus fumigatus*
	Geobacillus spp.	*Chaetomium thermophile*
	Streptomyces spp.	*Humicola lanuginosus*
	Thermoactinomyces spp.	*Paecilomyces* spp.
	Thermobrachium spp.	*Rhizomucor pusillus*
	Thermomonospora spp.	*Scytalidium thermophilum*
	Thermosinus spp.	*Thermoascus aurantiacus*
Mesophilic	*Achromobacter* spp.	*Alternaria* spp.
	Bacillus spp.	*Aspergillus* spp.
	Chitinophaga spp.	*Cladosporium* spp.
	Clostridium spp.	*Coccidioides* spp.
	Cytophaga spp.	*Helminthosporium* spp.
	Flavobacterium spp.	*Humicola* spp.
	Klebsiella spp.	*Penicillium* spp.
	Lactobacilli spp.	*Scedosporium* spp.
	Micrococcus spp.	*Trichoderma* spp.
	Pseudomonas spp.	
	Serratia spp.	
	Streptomyces spp.	

Table 11.7. Dynamics of microbial populations during composting of poultry waste. (Adapted from Zibilske, 2005.)

Organism	Thermophilic[a] stage	Mesophilic[a] stage	Maturation[a] stage	Number of species found
Bacteria				
Thermophilic	10^{17}	10^{8}	10^{7}	15
Mesophilic	10^{6}	10^{8}	10^{11}	6
Fungi				
Thermophilic	10^{7}	10^{3}	10^{3}	16
Mesophilic	10^{3}	10^{6}	10^{5}	18

[a]Colony-forming units/g compost material.

by a large, diverse community of bacterial and fungal species within compost piles (Table 11.7). Strong chemical breakdown of organic residues is also achieved during this stage.

3. Phase III (maturation/cooling): temperatures cool to ambient levels, which slows decomposition but allows for further processing of the waste by microorganisms and fauna. The final product is a finely divided, nutrient-rich compost (C:N ratio < 20) with good physical structure. It will have a near neutral pH, low salinity level and contain humus. Bacterial communities are generally dominated by Gram-negative genera such as *Pseudomonas*, *Serratia* and *Klebsiella*, while *Bacillus* spp. is often the most abundant Gram-positive bacterium in mature composts (Boulter *et al.*, 2002).

For biological control, the best results are generally obtained when plants are treated with compost or compost tea obtained during mesophilic cooling before maturation stage, and before the organic residue is completely decomposed (C:N ratio < 20). Compost tea is a water-based soil amendment solution created by steeping mature compost in order to promote beneficial bacterial growth. According to most definitions, mature compost has undergone thermophilic decomposition (Phase I) and then decomposed for about 6 weeks (Phase II). Depending on the decomposition rate and cooling of the compost pile, the compost may reach the appropriate stage for making compost tea near the end of the mesophilic stage or the beginning of the maturation stage (Phase III). This period is accompanied by an increase in mesophilic bacteria and a stabilization of mesophilic fungi (Table 11.7). Presumably, some of these organisms can act as antagonists when they come in contact with pathogens.

While compost is easy to work with and appropriate for greenhouse and floriculture production, there is an interest in using compost and other organic amendments to control pathogens under field conditions. A review of more than 2000 experiments by Bonanomi *et al.* (2007) showed that suppression of soilborne pathogens followed the order: organic waste = compost ≥ crop residues > peat. Up to 82% of studies reported suppression of pathogens with compost and organic waste, with an increase in disease incidence in 3–23% of cases (Table 11.8). Crop residues provided disease control in a maximum of 74% of cases, but also boosted pathogen populations in up to 48% of cases. Peat was effective at suppressing disease in 12% or fewer cases. The organic wastes considered in these studies were a diverse group of undecomposed animal manure, papermill wastes, fish, meat and bone meal, and olive mill residues. Crop residues were considered to be non-decomposed materials such as green manure and non-harvested plant remains such as leaves, stems and roots. Organic amendments such as compost, partially decomposed and fresh organic residues hold promise for controlling pathogens and plant disease in agroecosystems. Due to inconsistent results, further work is needed to improve the formulation of composts and management of organic amendments to achieve better control of pathogen populations and improve disease suppression in soils.

Table 11.8. Effect of organic amendments on diseases caused by soilborne pathogens (*Fusarium* spp., *Phytophthora* spp., *Pythium* spp., *Rhizoctonia solani* and *Verticillium dahliae*). (Adapted from Bonanomi *et al.*, 2007.)

Organic amendment	Suppressive[a]	No effect[a]	Conducive[a]
Compost	32–74	21–48	3–20
Organic waste	41–82	19–46	8–23
Crop residues	29–74	16–41	3–48
Peat	4–12	11–50	38–60

[a]Percentage of cases classified as suppressive, no effect or conducive to plant disease (range for five soilborne pathogens).

Further Reading and Web Sites

Carbon Dioxide Information Analysis Center http://cdiac.ornl.gov

The flow of energy in ecosystems http://www.globalchange.umich.edu/globalchange1/current/lectures/kling/energyflow/energyflow.html

The Montreal Protocol on Substances that Deplete the Ozone Layer. Available at: http://www.theozonehole.com/montreal.htm (accessed 6 February 2009).

Yadav, V. and Malanson, G. (2007) Progress in soil organic matter research: litter decomposition, modelling, monitoring and sequestration. *Progress in Physical Geography* 31, 131–154.

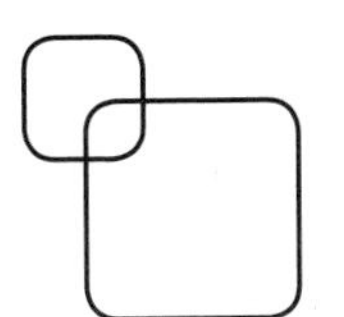

PART IV

Ecosystem Management and the Soil Foodweb

12 Ecosystem Management and Soil Biota

Humans exploit the Earth's resources to obtain sustenance, shelter and energy for homes and cities. Terrestrial ecosystems are especially susceptible to human activities because they provide much of the food, fibre, fuel, ores and other materials that we demand for living. The sustainable use of terrestrial resources underlies much of the discussion related to management of agricultural lands, grasslands and forests. There is general agreement that land should be used in a manner that does not lead to its degradation or pollution. Careful use of our land is necessary to ensure that harvests and yields of renewable resources can be maintained at the level required by society now and in the future. Since terrestrial ecosystems do not exist in isolation, it is understood that improper land use could pollute aquatic systems or the atmosphere, which may impair ecological processes in these components of the biosphere.

Ecosystem management relies on the use of ecological principles to understand the status of a particular ecosystem, to promote the long-term sustainability of this ecosystem and to ensure delivery of essential ecosystem goods and services to society. The concept of ecosystem management was described by Christensen *et al.* (1996) as follows:

1. Long-term sustainability is a fundamental value.
2. There are clear, operational goals for ecosystem use.
3. These goals are predicated on a sound ecological understanding.
4. The connectedness and complexity of the ecosystem has been considered.
5. The dynamic character of the ecosystem has to be considered.
6. The scale and context of the management activity is stated.
7. Humans are included as a component of the ecosystem.
8. Flexible, adaptive approaches are stressed.

Thus, ecosystem management allows us to make decisions or recommendations about the present and future use of lands. A simple example is the preservation of soil organic matter in agroecosystems. The depletion of soil organic matter due to intensive cultivation is a major problem because this causes a loss in soil fertility and reduces the soil water-holding capacity, leaving crops more susceptible to nutrient deficiencies and drought. Prolonged depletion of soil organic matter is expected to reduce crop yields. Agricultural producers could implement a number of changes in their operations to halt soil organic matter loss and reverse the trend to return the soil to a more productive state. Reducing tillage intensity, diversifying the crop rotation and applying organic fertilizers are agricultural practices that can build soil organic matter, and will be discussed further in the next section.

However, ecological management should go beyond improving crop yield and consider the ecological functions more broadly. Understanding the role of soil organic matter in carbon cycling, for instance, suggests that there is a benefit in terms of the global carbon balance when carbon is stored in soils rather than when carbon dioxide enters the atmosphere and contributes to the greenhouse effect. Enhancing water storage within aggregates (organo-mineral complexes) in a well-structured soil slows the hydrologic flow through terrestrial ecosystems, and reduces the nutrient and sediment loads entering waterways. This could reduce the risk of eutrophication and siltation in aquatic systems that receive water from agricultural lands. Soil organic matter is the major food source of many free-living saprophytic organisms, so increasing the soil organic matter content or changing the management of agricultural debris could contribute to biodiversity. In this way, an increase in soil organic matter may affect biogeochemical cycling and biodiversity, as well as crop yields.

©CAB International 2010. *Soil Ecology and Management* (J.K. Whalen and L. Sampedro)

From the previous example, it is clear that the policies guiding the practice of ecosystem management require a holistic view. Ecosystems are complex, with interactions and feedbacks occurring between the abiotic and biotic components. It is not possible to have a detailed knowledge of all of the interactions or even all of the components in the system. By necessity, the system has to be described in a simplified manner, based on the following assumptions:

1. Extremely complex processes can be understood more easily when one considers a number of very simple processes rather than a few, complicated processes. For example, crop yield is affected by many climatic factors: air temperature, rainfall (quantity and distribution), relative humidity, light (quantity, intensity, duration), altitude/latitude, wind (velocity, distribution), carbon dioxide and ozone concentration, etc. For simplicity, researchers may use the mean, median and range values for a crop grown in a specific climatic region to integrate the effect of climatic factors on crop yield.
2. Complex processes are dynamic, so we must choose an appropriate timescale when examining cause and effect relationships. The lifespan of an agricultural crop such as soybean is about 140 days from seeding to harvest. Measuring the hourly photosynthetic rate of the crop is not likely to be of much use in predicting the harvestable yield of soybean. However, hyperspectral data collected by remote sensing satellites at critical growth stages, combined with monthly averages for rainfall, temperature and soil moisture, can accurately predict soybean yields (Prasad *et al.*, 2006).
3. The central aim of ecosystem management is to optimize processes, i.e. make the best choice from an array of alternatives at different periods of time. The processes to be optimized can be purely utilitarian (e.g. optimize crop yield or biofuel production from a limited land base), economic (e.g. optimize profitability of an ecological processes) or aesthetic (e.g. optimize the well-being of humans, animals, plants and microorganisms that are part of an ecosystem). Ecosystem management is thus concerned with the multiple functions of ecosystems, thresholds in processes and trade-offs between ecosystem users. The focus then becomes on *optimizing* processes so they meet the needs of multiple users, rather than *maximizing* the process. Imagine that all trees from an old-growth forest were harvested – this would maximize the yield of wood products and provide economic benefits to a few users in a short period of time. However, this would be detrimental for individuals who like to visit the forest and to the wildlife that live in this forest. There is a threshold to the extent of forest harvesting that can be undertaken in the context of ecosystem sustainability, and compromise (trade-offs) must be made to consider the needs of various users.
4. There is feedback control in ecosystems that causes processes to deflect towards equilibrium (steady state). Any natural disturbance or human-induced perturbation of the ecosystem can cause a deviation in an ecosystem process from the normal range. Ecosystems are considered to be resistant to change when processes are not affected by a disturbance, whereas those that exhibit a marked fluctuation followed by a rapid return to the steady state are resilient (Fig. 12.1). However, some controlling factors have such a strong influence on ecosystem processes that they can cause irreversible or long-

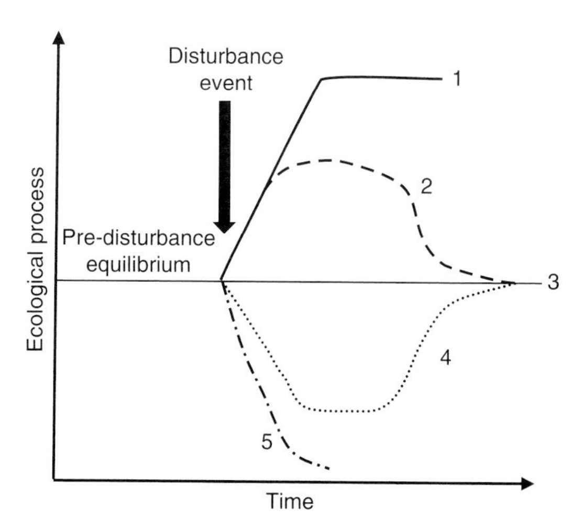

Fig. 12.1. The possible responses of an ecological process (e.g. primary production) to a disturbance event. (1) A positive feedback resulting from the disturbance causes a shift to a new steady state; (2) a temporary increase in the ecological process occurs following disturbance, but the process eventually returns to the steady-state conditions (resilience); (3) the ecosystem is unaffected by the disturbance (resistant); (4) a temporary decline in the ecological process following disturbance, but the process eventually returns to the steady-state conditions (resilience); (5) the ecological process cannot persist following the disturbance, resulting in the complete loss of this ecological function.

term change in ecological functions. For instance, an intense forest fire that burns a large area and destroys all of the mature trees will create space for younger trees and understorey plants that survived the blaze. The soil foodweb, plant community and animal life in this forest may not resemble that existing in the pre-fire period for many years.

In the context of ecosystem management, identifying and fostering negative feedbacks is considered to be the most appropriate strategy to constrain ecological processes and maintain the ecosystem in steady-state conditions. For example, the use of biological control agents to control plant pathogens and parasites is an example of a negative feedback (discussed in Chapter 11). As the pathogen population gets larger, so does the biocontrol population because it has more food. As the pathogen population declines, so does the biocontrol population because it runs out of food. In contrast, positive feedbacks will tend to push the ecosystem towards a new steady state. For instance, practices that promote nitrogen fixation will increase the primary production in an ecosystem to a higher level than would be achieved if plant growth relied on soil organic nitrogen alone. While a positive feedback is not necessarily detrimental, it implies that the ecosystem process is not constrained and thus will not resemble the old steady-state condition.

5. Many of the processes that occur in ecosystems involve interactions among biotic and abiotic components. As discussed in a previous chapter, the predation of soil bacteria by protists and bacterial-feeding nematodes is controlled by the physical location of the bacteria. Colonies located in the inner surfaces of microaggregates are physically protected from predators for a period of time. If those microaggregates are consumed and structurally reorganized in the intestinal tract of an earthworm, then the bacteria could become exposed to their predators. The interplay between biotic and abiotic components contributes to the complexity of studying predation in the soil ecosystem.

6. There is repetition in the processes that occur in systems, through space and time. If a researcher understands the nutrient cycling process in temperate grasslands of North America, they are likely to have a pretty good idea of how nutrient cycling occurs in the steppes of Russia, due to the similarity in vegetation, climate and soil-forming processes in each region. For this reason, it is possible to make some generalizations about ecological processes such as primary production, biogeochemical cycling, population dynamics and community interactions in soils within ecoregions that have similar geographic features, climate, plant and animal communities.

These assumptions provide a framework for developing an ecosystem management plan, which should be flexible and adaptive (Christensen *et al.*, 1996). An adaptive policy is one that is designed from the outset to test hypotheses about the ways that ecosystem processes respond to human interventions. If one set of actions fails, it provides an opportunity to learn so that better methods can be applied in the future. A flexible, adaptive approach accepts the high degree of uncertainty associated with ecological processes in the real world. Rather than delay action until one is certain about the current state of the ecosystem or the expected outcome of ecosystem restoration, an adaptive approach permits individuals to take steps based on reasonable hypotheses about how the ecosystem functions and how it will respond to interventions.

Once an ecosystem management plan has been implemented, it should be followed by monitoring. There is a variety of indicators that can be used to determine the overall health and functioning of an ecosystem. In an agroecosystem, the biomass and nutrient content of crops is carefully monitored and compared to average values for a particular field or region. Annual fluctuations in crop yield due to weather conditions are not unusual, but a sustained decline over a period of years could be a sign of a change in the steady state of the agroecosystem. However, there are economic and social consequences of leaving an agroecosystem in decline for several years. Monitoring other components in the agroecosystem could be informative when quick decisions are needed to implement corrective actions, before there are negative consequences for people or the environment.

The soil foodweb is expected to be a sensitive indicator of changes in terrestrial ecosystems because it is intimately associated with the soil at various scales (micrometres to metres) and soil organisms have shorter lifespans than most plants or wildlife. Regular monitoring of the soil biota in agroecosystems, grasslands and forests could detect subtle changes in ecosystem functions before severe problems become apparent. The use of ecological indicators such as soil quality indicators (see Focus Box 12.1) permits a proactive approach to ecosystem management. Just as it is wise to inspect your house regularly and make repairs (proactive) rather than waiting until the house has collapsed (reactive), it makes sense to monitor the most sensitive indicators of soil function and adjust management practices to favour their continued survival before a complete loss of ecosystem function occurs.

While the soil foodweb is a relevant indicator for ecosystem management, there are a few challenges in assessing the diversity and activity of soil foodweb organisms. As discussed in previous sections, the soil foodweb exhibits a spatially aggregated structure across ecosystems, and populations display temporal fluctuations due to environmental conditions and disturbances. Some organisms are difficult to isolate and identify due to their small size and the fact that the taxonomy is still being revised. However, general ecological theories about community structure and function seem as applicable to the soil foodweb as to plants and animals with limited mobility, albeit at a smaller scale. This chapter provides insight into how management practices and disturbances in agroecosystems, grasslands, forests and urban ecosystems affect the soil foodweb, which has implications for the sustainability and long-term functioning of terrestrial ecosystems.

Focus Box 12.1. Soil quality indicators as a measure of soil and ecosystem health.

Soils are a fundamental component of ecosystems that support diverse and highly productive communities of plants and animals, clean and recycle water, capture and exchange gases with the atmosphere and meet human demands for food, fibre and fuel. Yet, how do we assess the health of our soils, to know whether the soil can sustain all of these important ecological functions? In some cases, it is evident that a soil has been degraded. Salt crusting on the soil surface following years of excessive irrigation on saline soils, deep gouges in a field where the soil has been eroded by wind or water and compacted tracks from heavy machinery – all of these visual symptoms indicate that the soil has been degraded and its function impaired. Scientists and land managers need indicators to proactively assess soil health before such symptoms of soil degradation appear. In this way, alternative management practices can be implemented before costly rehabilitation is needed. Indicators also provide insight into whether changing of management practices leads to an improvement in the soil quality in a particular ecosystem.

Soil quality indicators were first developed as an index of soil health in agroecosystems because there is a clear link between soil conditions and crop yields. Changes in the soil structure due to compaction by heavy farm equipment hinder root development, thus reducing the quantity of soil nutrients and water that can be accessed by crops. A decline in soil organic matter following intensive tillage can reduce the water-holding capacity of a soil, making the crop more susceptible to water deficits and drought during the growing season. Farmers realize that a soil with good physical and chemical qualities should also be able to produce higher crop yields and generate more income than a poor-quality soil.

According to agricultural producers in the USA, a good-quality soil has a deep, dark topsoil layer that is easy to cultivate. The soil acts as a sponge for water, but also drains early in the spring after snow melting, allowing for the timely planting of crops. Residues left after crop harvest in the autumn are rapidly decomposed, many earthworms are found in the soil and the soil has a sweet, fresh smell. The organic matter content is high, and there is little erosion on good-quality soils. Finally, less fertilizer is required and crop yields are higher; there are also more weeds, but fewer problems with insect pests and diseases. These descriptions reveal that soil quality assessment must consider the physical, chemical and biological characteristics of soils (Doran *et al.*, 1996). Some of the most common indicators of soil quality are listed in Table 12.1.

Although soil quality indicators are related to crop production, they are also broadly relevant to other ecosystem services. A good-quality soil should also be able to filter and purify water before it is released to waterways. Inorganic and organic pollutants can be adsorbed and some can be degraded by a good-quality soil. Finally, a good-quality soil should serve as a buffer for climate change by promoting the growth of plants that sequester CO_2 from the atmosphere and contributing to the humification and physical protection of carbon from plants and other organic residues.

Soils support a diverse set of ecosystem services, so it is challenging to select one test or a few tests that can gauge the soil's contribution to overall ecosystem function. Some researchers suggest that the soil organic matter content can be used as a universal indicator of soil quality. As discussed in Part III of this book, organic matter is implicated in soil structure and water-holding capacity, as a

Continued

Focus Box 12.1. *Continued*

Table 12.1. Physical, chemical and biological soil indicators that may be included in a minimum data set for assessing soil quality. (Adapted from Wienhold *et al.*, 2004.)

Physical	Chemical	Biological
Texture	Soil organic carbon	Microbial biomass carbon
Topsoil depth	Total nitrogen	Microbial biomass nitrogen
Water infiltration rate	pH	Potentially mineralizable nitrogen
Bulk density	Electrical conductivity	Soil respiration
Water-holding capacity	Extractable nutrients (N, P, K)	

reservoir of organic carbon, nitrogen and other essential elements, as a chemical buffer and as a resource for the detrital foodweb. Yet, the total soil organic matter pool is not homogeneous across ecosystems nor can changes in soil organic matter due to anthropogenic and natural processes be detected quickly. The search for chemical or physical fractions of the soil organic matter pool that could be relevant indicators of soil quality is ongoing (De Bona *et al.*, 2008).

12.1 Agricultural Systems Management

Agricultural systems management began thousands of years ago when Neolithic people began to settle in sedentary, agrarian societies. The development of methods for food production by these people and subsequent civilizations is documented in archaeological studies and surviving written records, revealing a long history of innovation and technologies for agricultural systems management. Modern agroecosystems are perhaps the most intensively managed ecosystems on Earth due to their importance in supplying food for humans and animals. To achieve high levels of food production, agricultural lands are often tilled, irrigated, fertilized and sprayed with pesticides to control weeds and other undesirable organisms. One or more crops may be harvested per year, or the land may be left fallow for a period of time between crops. All of these anthropogenic interventions are expected to affect the diversity and activity of soil biota. The following sections examine how some key agricultural management practices affect the soil foodweb.

Tillage

Agriculture is considered to have begun when our ancestors first began cultivating land for crop production, using digging sticks and other rudimentary tools. The first plough was developed around 3000 BC and pulled by people, but was soon modified to be pulled by draught animals such as oxen and horses. Paintings on the walls of ancient Egyptian tombs dating back 5000 years show oxen yoked together by the horns, drawing a plough made from a forked tree. The Greeks improved the Egyptian plough by adding a metal point. By Roman times, tillage tools and techniques had advanced to the point where tillage was a recommended practice for crop production. The British author Jethro Tull, inventor of the first seed drill, expounded on the need for thorough cultivation in his book *Horse-hoeing Husbandry* published in 1731 because he believed (erroneously) that fine soil particles were absorbed by plants and improved crop growth.

Until the 1800s, most ploughs were constructed from a wooden frame with metal strips and a metal ploughshare. The casting of the metal mouldboard plough provided farmers with a durable, long-lasting tool that was essential for cultivating prairies and cleared forests, permitting the settlement of North America. With the introduction of tractors and modern farm machinery in the early 1900s, farmers were able to till agricultural lands more intensively and deeper than was previously possible. This intensive cultivation left soils more vulnerable to erosive forces and depleted the soil organic matter reserves by up to 40% in the former grasslands of North America. The detrimental effects of over-tillage were seen vividly during the 'Dust Bowl' years from 1930 to 1936, when a combination of drought and wind storms in the southern Great Plains of North America, including Texas, Oklahoma, New Mexico, Colorado and Kansas, led to widespread crop failure and loss of topsoil through wind erosion from millions of hectares of farmland. Dust clouds were seen as far away as New York and Washingon, DC. By 1940, about 2.5 million people had been displaced from the affected areas. The US Government responded by forming the Soil Erosion Service by 1933, which was reorganized as the Soil Conservation Service in 1935 and is now the Natural Resources Conservation Service. Globally, governmental organizations and non-governmental agencies are involved in devising and disseminating information about appropriate tools and techniques to conserve soils. The concepts of minimum tillage, contour cultivation, strip cropping, terrace agriculture, windbreaks and other erosion controls have been promoted by extension personnel, through the media and most recently via the Internet.

There are many reasons why agroecosystems are tilled. Principally, tillage is an excellent way of preparing the land for seeding – the mechanical mixing of soil aerates, loosens the soil and permits water infiltration. Tilled soils warm more quickly in the spring, making the seedbed more hospitable for germination and early seedling growth. Weedy plants are disrupted mechanically by tillage implements. Dead plants and other residues on the soil surface are incorporated by tillage, which speeds their decomposition and the rate of nutrient recycling from these organic substrates. Tillage systems can be grouped into three categories:

1. Conventional tillage is the most intensive operation. Typically, the soil is cultivated in the autumn with a mouldboard plough, which turns the entire soil mass in a furrow, generally to a depth of 17–20 cm. In the spring, the seedbed is further prepared by cultivating the field to a depth of 10–15 cm with a disk harrow. Multiple passes with a harrow may be required to adequately control weeds before the field is planted.
2. Reduced tillage can include many operations, such as ridge tillage, zone tillage and strip tillage. Normally, the field is tilled to a shallower depth (perhaps the top 10 cm) and the entire soil mass is not completely mixed. This might be accomplished by using a chisel plough or spring-tooth harrow. Another type of reduced tillage system is ridge tillage, in which semi-permanent ridges are created and planted with row crops. The space between rows is generally not tilled. Similarly, zone tillage and strip tillage are reduced-tillage systems where a limited area of the field (generally the seed row only) is tilled. In zone and strip-tillage systems, the space between planted rows is not cultivated.
3. No-tillage is also referred to as direct seeding, since it uses equipment that makes a very shallow groove, pushes the seed directly into the ground and firms the earth over the seed. Thus, the disturbance of the soil mass is localized within the planted row. Fertilizer is often drilled into a band a few centimetres away from the seed, either at planting or during the early vegetative growth of the crop. Plant residues are left on the soil surface and weeds are controlled by chemical pesticides.

It is obvious that crops can be grown without much tillage, but crop production is enhanced in some agroecosystems when the soil is tilled. The mouldboard plough is an appropriate tool for burying crop residues, particularly in humid regions where residues decompose slowly, and keeps the soil from warming quickly in the spring (seed germination is slower in cold, wet soils). The mouldboard plough may be helpful in breaking up compact subsurface layers called traffic pans that form when heavy equipment is driven repeatedly over a field. Soil compaction can be avoided to some extent by not driving on fields with wet soils. However, mouldboard ploughing can also generate a compacted layer called a plough pan in some poorly structured soils. This compacted layer may inhibit root penetration and water infiltration. Plough pans and other naturally occurring compacted layers (hardpans) can be broken apart by deep tillage with a subsoiler to a depth of 25–30 cm. If a soil is susceptible to developing hardpans, another option is to switch to perennial crop production such as a grass-based hayfield to minimize the amount of traffic and ploughing on the land.

Many agricultural producers are interested in reducing the amount of tillage on their farm because it is costly to purchase the fuel and pay for the labour needed to support frequent tillage operations. Switching to reduced tillage or a no-till system requires an initial investment in the appropriate machinery, but should be economically feasible if the system gives good crop yields, reduces soil erosion and builds soil organic matter. There is a tendency for nutrient stratification in no-till systems because lime and fertilizers are applied to the soil surface or in a shallow band near the seed row. In contrast, soil amendments and fertilizers are well mixed throughout the entire plough layer of conventionally tilled soils. For this reason, it is recommended that farmers build up the soil fertility and adjust the soil pH of acidic soils to pH 6–7 before beginning the transition to no-tillage. If the seedbed tends to be wet and cold in the spring, which could slow seed germination or favour plant diseases such as damping-off caused by a fungal pathogen (*Fusarium*), a reduced-tillage system may be more appropriate than no-tillage. If root growth is constrained by soil compaction or hardpans, this would be another good reason to choose reduced-tillage rather than no-tillage.

About 3 to 5 years after reducing tillage intensity producers will begin to notice differences in the physical and chemical characteristics of no-till soils, compared with conventionally tilled soils (Fig. 12.2). When the soil is not disturbed, it tends to settle and thus will have a greater bulk density than a tilled soil, which is more fluffy and has a lower bulk density. In a no-till soil, the macropores formed by root channels and earthworm burrows become more important for water infiltration and gas exchange. Macropores may also act as preferential flow pathways for the transport of agrochemicals and nutrients from the soil surface into subsurface drains. Generally, no-till soils have greater structural stability, meaning that they are less likely to be eroded by wind and water. This is partly due to the presence of surface residues that protect the surface soil layers from disturbance. In addition, aggregates tend to be larger and contain more undecomposed or physically protected organic residues of plant or microbial origin. Thus, no-till soils may store more organic carbon than their conventionally tilled counterparts, although this is very much dependent on site-specific factors such as soil texture, the types of crops grown and the activity of soil organisms, which are responsible for transforming organic materials into more stable, humified carbon compounds. Soils with greater organic carbon content also have more organic nitrogen, phosphorus and sulphur as well as greater cation exchange capacity for cations such as potassium, calcium and magnesium.

The decrease in soil disturbance from tillage implements and greater retention of crop residues on the soil surface makes a no-till system a favourable habitat for soil microorganisms and fauna. Fungi tend to grow more extensive hyphal networks in undisturbed soils. Litter-dwelling soil fauna such as collembola and epigeic earthworms are more abundant in soils with a surface litter layer. Any fluctuation in the number, biomass and activity of soil organisms due to changes in tillage practices is

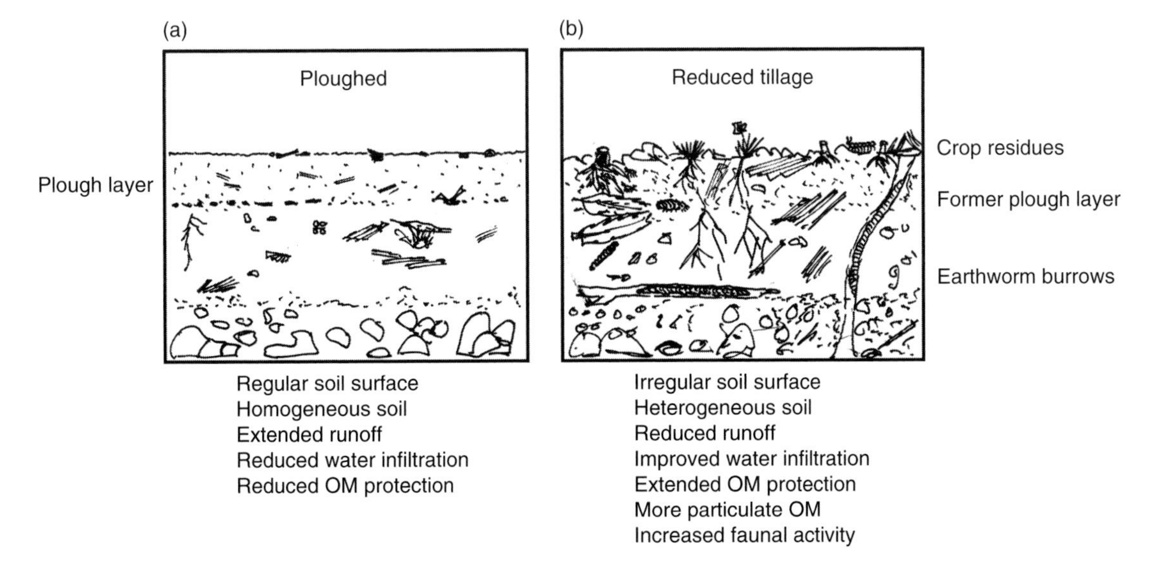

Fig. 12.2. A comparison of the biological, physical and chemical properties of a soil under (a) conventional tillage (ploughed) and (b) reduced tillage (no-till). Adapted from Stinner and Blair (1990).

Table 12.2. The effects of tillage functions on soil biological groups and root growth. (Modified from Giller *et al.*, 1997.)

Ecological/pedological function(s)	Biological group(s)
Residue comminution/decomposition	Microorganisms, microfauna, mesofauna, macrofauna
Carbon sequestration	Microorganisms, especially fungi; macrofauna that build structures
Organic matter/nutrient distribution	Roots, mycorrhiza, macrofauna
Nutrient cycling, mineralization/immobilization	Soil microorganisms, microfauna
Soil porosity, bioturbation	Roots, soil macrofauna
Soil aggregation	Roots, fungal hyphae, mesofauna, macrofauna
Population control	Predators/grazers, parasites, pathogens

expected to produce a feedback on soil ecological and pedological functions, as shown in Table 12.2.

Root growth is also affected by tillage practices. As discussed in earlier chapters, plant roots interact with the soil environment in ways that affect soil biological activity. Tillage has the potential to affect the way that roots assimilate water and nutrients from the soil solution, the quantity and distribution of carbon-rich exudates and allelochemicals released from roots, and the physical changes in soil structure caused by growing roots. The roots of cereal crops, perennial grasses and legumes are not disturbed by crop harvest, and when the soil is not disturbed by tillage they will be left to gradually decay and be transformed into humified soil organic carbon. During their lifespan, fine roots secrete mucilages and other adhesive substances that attract and bind mineral particles, leading to the formation of macroaggregates. When roots are decomposed, they leave spaces (root channels) that contribute to soil porosity. Since roots account for about 20% of the net primary production of annual crops, and an even greater proportion of the biomass in perennial crops, the contribution of roots to soil ecological and pedological functions cannot be overlooked.

The effect of tillage on soil biological groups is not the same in every field, but some general trends have been found. Wardle (1995) reviewed more than 100 studies from the scientific literature that compared the soil biota in conventional tillage and no-till agroecosystems. He calculated an index (V), which represents the degree to which each group of organisms was inhibited by tillage (Fig. 12.3). Generally, organisms with a larger body size are more inhibited by tillage than smaller organisms. There are two reasons for this observation: (i) any decrease in the food availability for smaller organisms induces a trophic cascade that results in a diminished food supply for the top-level detritivores and predators; and (ii) smaller organisms have a rapid life cycle that permits them to recover quickly from soil disturbance due to tillage, whereas larger organisms are more susceptible to physical damage and it takes longer for them to repopulate when they are injured or their activity is impaired by tillage operations.

Tillage effects on soil microorganisms

Generally there is less microbial biomass (bacteria plus fungi) in conventional tillage than in no-till soils. The primary reason is disturbance of the soil environment after tillage, as well as damage of the fungal hyphal network. Since fungi have a greater biomass than bacteria, the overall effect is a decline in the soil microbial biomass. Bacterial colonies are sufficiently small that they avoid damage by tillage equipment, but soil inversion probably exposes some bacterial colonies to desiccation and higher temperatures at the soil surface. Soil inversion during ploughing also disrupts the pore continuity of the soil, changing the water balance and airflow through macro- and micropores until these are reformed by soil biota.

The location and integrity of the organic residues that serve as a food supply for soil biota change after a soil is cultivated. No-tillage systems are considered an excellent habitat for mycorrhizal fungi that colonize plant roots and also saprophytic fungi that grow on plant residues within the soil (e.g. dead roots) and at the soil surface. In contrast, bacteria tend to dominate the microbial community in conventional tillage systems, probably because chopped residues that are fragmented and thoroughly mixed with the soil provide a large surface area for bacterial colonization and decomposition.

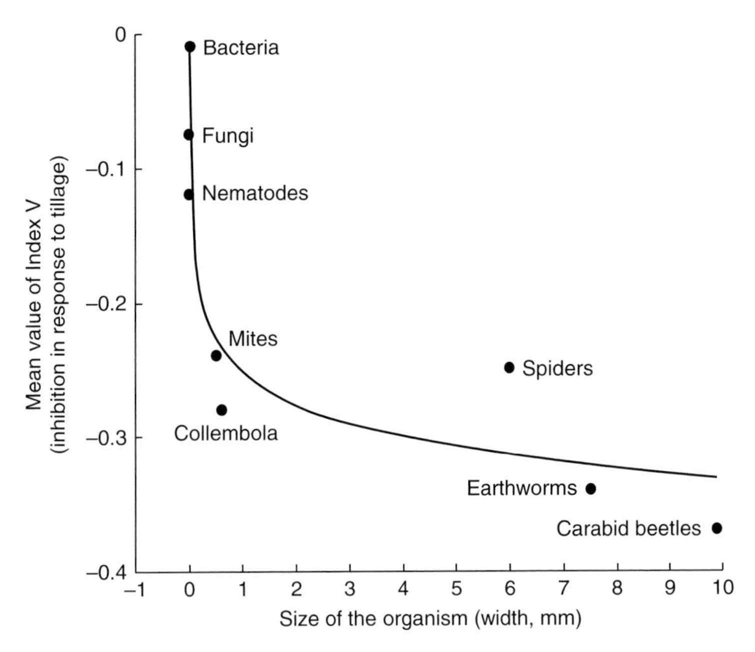

Fig. 12.3. The degree of inhibition of soil foodweb organisms, given as the mean value of the index V, in relation to the organisms' body size. The relationship follows a negative logarithmic function. (Adapted from Wardle, 1995.)

There is often a burst of soil respiration after soils are ploughed or harrowed, which points to a rapid growth of soil bacteria. Decomposition occurs more quickly in tilled soils than in no-till soils. Ammonia-oxidizing bacteria respond positively to tillage, which leads to a peak in nitrate concentrations following tillage operations.

Table 12.3 shows the composition and diversity of bacterial communities in conventional tillage and no-tillage agroecosystems, in a successional forest and an old-growth forest that were all on the same soil type (sandy-loam, Typic Rhodic Kanhapludult) (Upchurch *et al.*, 2008). The major finding of this study was that the bacterial communities in the agricultural area were more diverse than those in the old-growth forest. The agroecosystem possessed a dynamic ensemble of crops and weeds that varied seasonally, supporting the development of a diverse bacterial community. In addition, cropland is more exposed than forest to bacterial dispersal agents (wind, animals, humans), which could permit the establishment of novel species.

The no-tillage soils at Horseshoe Bend had more bacterial phylogenetic groups, and according to the Chao1 estimator, had the greatest bacterial diversity (Table 12.3). At this site, the no-till agroecosystems have similar organic carbon content to conventional agroecosystem and successional forest, so bacterial diversity was probably not related to organic substrates. The no-till agroecosystem contains larger populations of protists, nematodes and microarthropods than the cultivated agroecosystem, which could cause selection pressure through predation, although this remains to be confirmed.

Tillage effects on soil microfauna and mesofauna

Since bacteria are the main food supply for protists and bacterivorous nematodes, the tillage system that stimulates the most bacterial growth should favour these populations. The effects of tillage on soil porosity (continuity, orientation) and the persistence of water films are also expected to be important for the microfauna. Bacterial feeding nematodes are abundant in many soils, often comprising more than 50% of the soil nematode community. A long-term tillage and residue management study in Australia (Rahman *et al.*, 2007) demonstrated that bacterial feeding and omnivorous nematodes were more abundant in a no-till than conventionally tilled soils when the crop residues were burned (Table 12.4). However, when the crop

Table 12.3. Soil bacteria grouped into phylogenic groups as determined from 16S rRNA clones, and diversity indices for the bacterial communities in 16S rRNA gene libraries. Bacterial rRNA was isolated from soil taken from the 5–10 cm depth of a conventionally tilled agroecosystem, a no-till agroecosystem, a successional oak–sweetgum forest or an old-growth oak–sweetgum forest at Horseshoe Bend, Georgia, USA. (Selected data from Upchurch *et al.*, 2008.)

Phylogenetic group	Conventional tillage	No-tillage	Successional forest	Old-growth forest
Actinobacteria	4	16	10	7
Bacteroidetes				
Flavobacteria	1	0	0	0
Sphingobacteria	3	2	2	0
Other	0	0	1	0
Chloroflexi	1	0	0	0
Firmicutes	1	1	4	1
Gemmatimonadetes	5	8	2	1
Nitrospira	1	0	1	1
Planctomycetes	0	1	2	3
Proteobacteria				
α-Proteobacteria	20	28	17	12
β-Proteobacteria	8	5	4	3
γ-Proteobacteria	2	3	3	1
δ-Proteobacteria	1	1	0	1
Unclassified	0	2	4	8
Verrucomicrobia	4	6	4	3
Unclassified bacteria	42	43	45	46
Total (all taxa)	93	116	99	87
Diversity index	**Conventional tillage**	**No-tillage**	**Successional forest**	**Old-growth forest**
Evenness (H/log S)	0.98	0.96	0.97	0.95
Richness (N-1)/log N	0.77	0.74	0.71	0.63
Shannon index (H)	0.92	0.90	0.90	0.86
Chao1 estimator	211	452	201	170

Table 12.4. Effects of tillage and crop residue management on the number of free-living nematodes per square metre in a 22-year experiment at the Wagga Wagga Agricultural Institute, New South Wales, Australia. The free-living nematodes were the bacterial feeding Rhabditidae (53–99% of the community) and omnivorous Dorylaimidae (< 2% of the community). (Adapted from Rahman *et al.*, 2007.)

	Conventional tillage	No-tillage
Crop residue burned	4266	6310
Crop residue retained	7413	4169

residues were retained, the conventionally tilled system supported more free-living nematodes than the no-till system. These results suggest that the nematode populations were strongly controlled by the food supply (particulate organic matter and soil bacteria) and by predators, particularly in the no-tillage system where crop residues were retained.

The impact of tillage on soil mesofauna has not been studied extensively. Tillage operations can kill some microarthropods because they are mechanically abraded by tillage implements or they become trapped in soil clods when the soil is inverted during ploughing. Changes in the soil physical environment (pore continuity, soil temperature and moisture) and the mixing or fragmentation of crop residues are probably the most important factors controlling soil mesofauna populations. Collembola, cryptostigmatid and mesostigmatid mites tend to be impacted negatively by tillage. The abundance of prostigmatid mites may decrease or increase, depending on how their preferred food resources (fungal biomass, nematodes and smaller arthropods) are affected. In some cases, tillage may expose some of these smaller foodweb organisms, making them easy prey for the predatory mites. Tillage has a variable effect on the number of astigmatid mites, which consume organic residues. Although the astigmatid mites may decline in number temporarily after tillage, there is typically a more rapid recovery of the population to pre-tillage levels than observed for other mites.

In about 60% of studies, enchytraeid populations are positively stimulated by tillage. Enchytraeids are ecologically adaptable and able to tolerate soil disturbance, profile inversion and soil drying better than larger annelids (earthworms). The fragmentation and incorporation of organic residues is beneficial to enchytraeids, which preferentially consume partially decomposed residues within the soil profile rather than surface residues. In many agroecosystems, there appears to be competition for food resource between earthworms and enchytraeids, so a decline in earthworm populations due to tillage reduces the competitive pressure on enchytraeid populations.

Tillage effects on soil macrofauna

In agroecosystems that are ploughed, rototilled or harrowed, there tend to be fewer macroarthropods, earthworms and termites. These large organisms are often cut or physically damaged by tillage implements. Soil inversion buries macroarthopods such as carabid beetles that live in the soil litter layer. The reduction in surface residue cover is also highly unfavourable for beetles, opiliones and spiders. These macrofauna thrive in fields with surface residues such as decaying plant materials, mulch or weeds – these residues provide a cool, moist habitat where the organism can avoid desiccation, and where they can hide while waiting for prey or avoiding predation by larger organisms.

Termite nests in the plough layer are destroyed by tillage. If termites have retreated into deeper soil layers during tillage operations, they may find themselves trapped and have to expend considerable energy to rebuild nests and passages to the soil surface, which can weaken the organism so it is less able to defend against diseases, parasitism or predation. In many African countries, subterranean termites are considered to be crop pests because they damage the roots and stems of crops, causing crops to 'lodge' (fall over). This represents an important yield loss in some areas. Deep tillage to a depth of 30 cm with a mouldboard plough or to 25 cm with a chisel plough is an effective way to disturb subterranean nests and reduce termite populations, although it usually does not lead to termite extinction.

Earthworms are soft-bodied and highly susceptible to physical damage that causes death. Soil inversion during tillage brings earthworms to the soil surface, where they are consumed by seagulls, plovers and other birds. Burrows in the plough layer are destroyed by tillage and need to be rebuilt by deep-dwelling species. The incorporation of surface litter reduces the primary food source for epigeic and anecic earthworms. Populations of epigeic earthworms, whose habitat is disrupted when the litter layer is chopped and incorporated, and anecic earthworms such as *Lumbricus terrestris* L. that travel through vertical burrows to feed and reproduce at the soil surface, are expected to be negatively affected by tillage. Yet, tillage operations that alleviate soil compaction or incorporate crop residues serving as a food source for earthworms can maintain or enhance earthworm populations. Boström (1995) reported that rotary cultivation and ploughing reduced earthworm populations by 73–77%, but 1 year later, there were five times more large adult earthworms and earthworm biomass was similar to pre-tillage levels. The incorporation by tillage of lucerne and meadow fescue residues, which are readily decomposed and consumed by the dominant endogeic earthworm *Aporrectodea caliginosa* (Savigny), was considered to be a major factor in earthworm population recovery.

Cropping systems

The term 'cropping system' refers to the type of crop species and the sequence in which they are grown in agricultural fields. The key considerations in cropping systems that may be relevant to the soil foodweb are: (i) the space occupied by the crop(s) and other plants in the field, related to the distance between planted rows and the space occupied by plant roots; and (ii) temporal aspects of cropping systems, including fallow periods and the sequence of crops grown in a field during a crop cycle (rotation).

Spatial arrangement of crops

An agricultural field that contains one crop species during a growing season is called a monoculture.

Large-scale production of annual row crops such as cereals, maize, cotton and potatoes is generally achieved using monocultures because it is simpler for the agriculture producer to care for, harvest and market a large volume of a single crop per field. Row crops are planted with precise spacing so the farmer can pass through the field with machinery to apply fertilizer when it is needed during crop development, till between the rows with a mechanical weeder, apply herbicides and other agrochemicals for weed and pest control if necessary, and harvest the crop efficiently. In potato production, a hilling machine must enter the field once or twice during the growing season to ensure that the potatoes are adequately covered with soil for optimal growth and to avoid sunburn.

Once the row spacing has been determined, based on machinery specifications, the crop is planted at a population density to achieve maximum light capture and photosynthesis, and minimize competitive interactions between desirable crop plants. Maize is often planted in 76 cm-wide rows at a density of 75,000 to 80,000 plants/ha, while soybeans and cereals are typically planted in narrower rows (12–20 cm between rows) with more plants per area (about 300,000 soybean seeds/ha and up to 400,000 plants/ha in cereal production systems). The populations of forage plants may be even higher, leading to complete coverage of the soil surface. Forages are grasses, legumes or grass–legume mixtures that are grown for pasture and hay production. Annual and perennial forages are an essential component of livestock diets, particularly for ruminant animals.

More complex cropping systems are polycultures that contain two or more crop species within an agricultural field. When two or more crops are grown together, each must have adequate space to maximize cooperation and minimize competition. This involves consideration of the spatial arrangement of the crops, plant density, maturity dates of the crops and plant architecture.

The spatial arrangement of crops must be practical and meet the production goals of the producer. The major types of intercropping systems are:

1. Strip intercropping, in which two or more crops are grown in strips wide enough to permit crop production using machinery, but close enough for the crops to interact. This is a common cropping system on sloped fields that are susceptible to soil erosion. Cereal grains are grown in alternating strips with forages across the slope, which provide stability and trap sediments and water moving down the slope. In some areas, strip cropping involves alternating densely seeded crops such as cereals and hay with widely spaced row crops such as maize, soybean, cotton, potatoes or sugarbeet. The strips must be designed so they are wide enough to accommodate farm equipment used for planting, spraying and harvesting of the crops. As mentioned, strips should be established across the slope (not up and down the slope) and they should be oriented perpendicular to the direction of prevailing winds in areas that are susceptible to soil loss from wind erosion.

2. Row intercropping, where two or more crops are grown at the same time and at least one crop is planted in rows. It is important to choose crops that can be grown together successfully. If one crop is taller than the other, the second crop should be tolerant to shading, or the taller crop should be planted with a wide row spacing to minimize the impact on the understorey crop. To avoid competition for water and nutrients, it is preferable that the root system of each crop occupies a separate niche within the soil profile. Tree-based intercropping is a type of agroforestry system that meets these criteria because the trees are planted in wide rows, about 12–15 m apart. The space between the trees is used to produce annual crops such as maize, soybeans and cereal crops. Tree roots tend to grow much deeper and within a limited zone (e.g. tree row), compared with the annual crops. Although shading can be an issue with older trees, they also act as a windbreak and create a warmer microclimate for the intercrop.

3. Relay intercropping is a system in which a second crop is planted into an established crop during vegetative or reproductive stages. Growing a legume and non-legume together is often a good strategy in intercropping. An extension bulletin published by the University of Wisconsin illustrates a successful maize–kura clover intercrop (http://www.uwex.edu/ces/crops/uwforage/kurademo.pdf, accessed on 11 February 2009). In this system, the kura clover (perennial) was established in year one of the rotation. Using zone tillage, rows were cultivated for maize planting in the following spring. Clover residues are nitrogen-rich and contribute to the nutrient requirements of the maize crop. The clover was suppressed with herbicides during the early maize growth, to minimize competition, and thereafter grew under the maize crop. After the maize crop was harvested, clover was left in the

field over winter to protect against soil erosion. In year three of the rotation, the producer could let the clover regrow for forage production or plough down the clover as a green manure before planting the next annual crop.

4. Mixed intercropping involves growing two or more crops together with no distinct row arrangement. Companion crops grown together in flower and vegetable gardens are an example of mixed intercropping. Due to the absence of distinct rows, weeding and crop management are likely to be accomplished by hand rather than with farm machinery.

The optimal plant density of each crop in an intercropping system depends on the goals of the agricultural producer. The appropriate row spacing and plant populations to achieve maximum light interception, water and nutrient use in monoculture systems are well known. If each crop in a polyculture system were planted at these densities, neither one would do well due to competition for light and other resources. Therefore, the number of seeds will need to be adjusted to a level that can accommodate both crops. Intercropping systems generally feature crops with different maturity dates to minimize competition. The most extreme example of different maturity dates is seen in temperate tree-based intercropping systems. Each year, an annual crop is planted and harvested after a 4–5 month growing period, but the trees are not harvested from the system for 20–40 years, depending on the species. It is necessary to consider the architecture of the above-ground plant components and rooting systems when choosing intercrops – to optimize resource use, it is preferable that the understorey plant can tolerate shading and the rooting systems of the crops do not have much overlap in the soil profile.

Intercropping and other agricultural practices such as maintenance of shelterbelts, crop rotations and cover cropping increase the diversity of plants grown in agroecosystems, while low-input tillage and organic fertilization increase the amount of organic residues in the agroecosystem. Together, these practices provide habitat for a broader array of insects and soil organisms, birds and other animals than might be present in a monoculture system (Fig. 12.4; Table 12.5). If some of these animals are the natural enemies of crop pests, their predatory nature is expected to suppress pests.

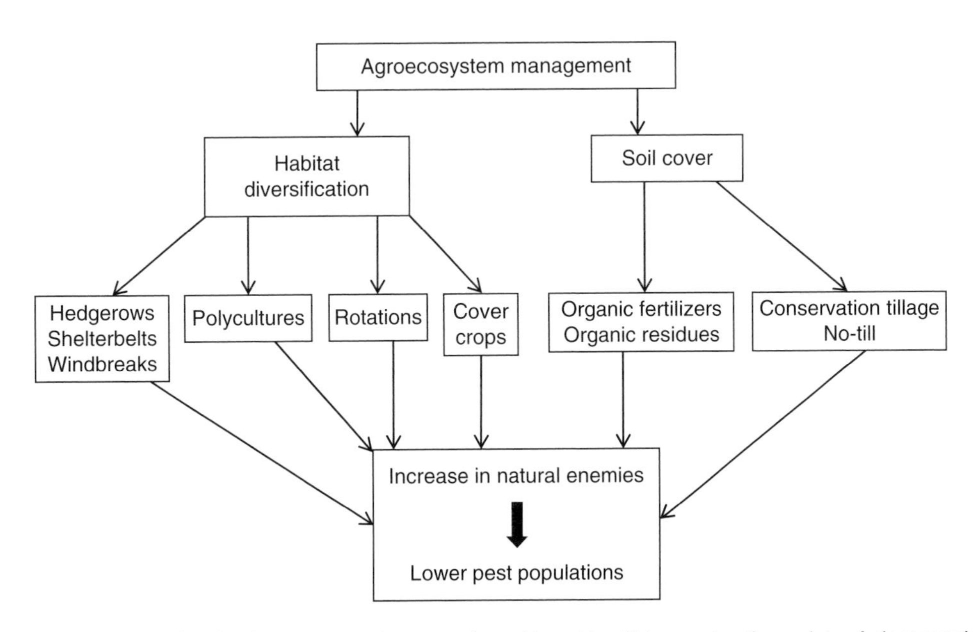

Fig. 12.4. Habitat diversification in agroecosystems can be achieved by: (i) increasing the variety of plant species present with hedgerows, polycultures, crop rotations and cover crops; and (ii) increasing the amount of organic residues at the soil surface through reduced tillage and organic fertilizer applications. These agricultural management practices often increase the diversity of arthropods, some of which may be the natural enemies of crop pests. (Adapted from Altieri, 1999.)

Table 12.5. Comparison of arthropod abundance in agroecosystems with diverse cropping and conventional cropping practices. More than 250 studies from the scientific literature were considered in the analysis. (Adapted from Attwood *et al.*, 2008.)

Taxon	Studies where arthropods were more abundant with diverse cropping practices (%)	Taxon	Studies where arthropods were more abundant with diverse cropping practices (%)
Predators		Decomposers	
Araneae	79	Acari	69
Carabidae	73	Collembola	50
Chilopoda	40	Herbivores	
Coccinellidae	84	Chrysomelidae	40
Formicidae	80	Curculionidae	72
Neuroptera	80	Dermaptera	64
Scarabaeidae	71	Homoptera	43
Staphylinidae	71	Orthoptera	63
		Thysanoptera	62

Diversifying the cropping system to create a more heterogeneous environment could limit the foraging activity and reproduction of pests that specialize on a particular crop species. There are a number of mechanisms whereby crop pests can be controlled through the careful selection of a companion crop:

1. Trap crops are often closely related species that are similar in form and growth habit to the main crop. They may secrete pheromones that are attractive to the pest, thereby attracting these insects before they land upon the main crop. The diamondback moth is strongly attracted to yellow rocket (*Barbarea vulgaris*), but larvae do not survive on this intercrop because the plant produces triterpenoid saponins that are toxic when consumed by the larvae. This affords some protection to the main cabbage crop from diamondback moth, although the intercrop needs to be planted at a sufficiently high density and seeded earlier than the cabbage crop to be effective (Badenes-Perez *et al.*, 2005).
2. Predators/parasites are the natural enemies of an insect pest. They benefit from the presence of an intercrop because it provides alternative resources (habitat, food) during periods that the insect pest is absent from the field. Maize planted into established lucerne or kura clover benefits from the presence of ground beetles (Coleoptera: Carabidae) that are abundant in these undisturbed, perennial forages. Ground beetles prey upon the pupae of the European corn borer, which provides some biological control of the corn borer, although not complete suppression (Prasifka *et al.*, 2006).
3. Physical interference refers to the fact that intercrops may affect the ability of the insect pest to find and colonize the host plant. When insects have to expend more energy to move and search for the preferred host, they may have fewer reserves for reproduction. Some intercrops may effectively camouflage the host plant. Hooks and Johnson (2003) noted that the presence of wheat stubble in no-till plots served as a physical barrier that blocked the flea beetle (*Phyllotreta cruciferae*) from colonizing and laying eggs on Brassica oilseed crops.
4. Chemical repellents are substances produced by a plant that repel insects, thus reducing their colonization in an agroecosystem. Tomato produces a volatile substance called rutin that deters the oviposition of diamondback moth on neighbouring cabbage and collard plants. This suggests that the diamondback moth uses olifactory stimuli to locate and colonize its preferred hosts, rather than visual images. Hooks and Johnson (2003) reported that glucosinolates produced by Brassicae deter the feeding activity of the grey field slug, but stimulated feeding by the cabbage stem flea beetle (*Psylliodes chrysocephala*). While chemical repellents hold promise for controlling crop pests, further work is needed to determine their effectiveness under field conditions.

Temporal dynamics of crops

A crop rotation refers to the sequence of crops that are grown in a particular field and is a well-established agricultural practice. Crop rotations have been practised for centuries. Written documents from ancient Roman authors report that certain crops, which we now know to be legumes, should be grown before the field was planted with cereals and other non-leguminous crops. Among the benefits observed were: an inprovement in soil fertility due to the legume crop; less water stress when crops with deep and shallow root systems were alternated in the rotation; and a reduction in pathogens and pest populations.

A crop rotation provides information on the frequency that crops of various types are planted in a particular field and the length of the fallow period, if any. Until recently, most crop rotations included a fallow period. From the Middle Ages until the 20th century, many European farmers relied on a rotation that featured rye or winter wheat in the first year, followed by spring oats or barley in the second and a fallow period in the third year. During the fallow period, the weeds growing in the field were consumed by livestock, which deposited manure on the field and contributed to the overall soil fertility. Weed consumption by livestock and mechanical weeding (harrowing) also provided some control over weeds, which was necessary because chemical herbicides did not become widely available until after the Second World War. Another popular rotation in Europe from the 1800s to the early 20th century was a four-year rotation of wheat–turnip–barley–clover. The turnip and clover were fed to livestock, while wheat and barley grain were for human consumption.

In semi-arid regions, many agricultural producers rely on the fallow period to preserve water, as well as to control weeds. Until the 1980s, a common crop rotation in the Great Plains of North America was a 2-year wheat–fallow rotation. The widespread adoption of no-tillage practices permits continuous cropping because soils tend to remain wetter and cooler when crop residues are retained on the soil surface. Another practical innovation is to increase the height of the cutting bar on the combine harvester, which leaves more intact stubble in the field and increases the amount of snow and rainfall that is trapped. A fallow period does not result in much primary production because only weeds grow in the field, so the elimination of the fallow period can improve the profitability of the farm and also contributes more crop residues that can be transformed into soil organic carbon via carbon sequestration.

Cropping system effects on soil foodweb organisms

Crop species and sequence of crops grown in rotations are selected by agricultural producers to achieve economically profitable yields, provide food for livestock on the farm, to achieve biological control of pests and diseases and to maintain soil and environment quality in a sustainable manner. Due to the variety of crop rotations that are practised in different parts of the world, it is not possible to provide an exhaustive list of cropping systems that could affect the activity and diversity of the soil foodweb. This section aims to provide a few examples and general observations.

Plants interact with symbiotic soil organisms and with free-living soil organisms. As the rhizosphere is an important zone of biological activity, it is expected that the soil foodweb should be more active and exhibit larger populations in the planted row than in the space between rows, particularly in wide-spaced crops such as maize or sugarcane. As predicted, parasitic nematodes were concentrated along the planted row of a sugarcane plantation, although they were not evenly distributed along the entire row (Fig. 12.5). This indicates that other factors, such as soil texture, bulk density and hydrology, are important in controlling the distribution of soil foodweb organisms.

Forage crops, intercrops and surface mulches are highly favourable for soil foodweb organisms, although for a variety of reasons. The beneficial effect of the rhizosphere on soil biota is not confined to the planted row when forages such as grass/legume mixtures are grown in a field, or when an intercrop grows between the rows of a main crop. When perennial forages or intercrops are maintained, soil disturbance due to tillage is expected to be minimal and some of the detritus left by these crops serves as a food resource for the soil foodweb. Surface mulches are also favourable for soil organisms because they are potentially a food resource. In addition, a soil that is covered with mulch or other organic residues may remain cooler and moister because precipitation can be trapped by the mulch and evapotranspiration is slowed as long as the surface is covered.

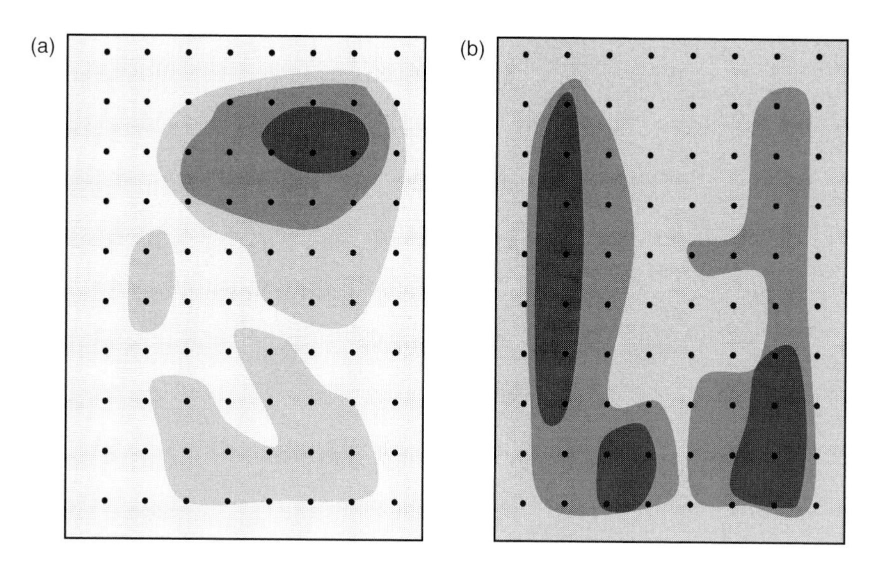

Fig. 12.5. Distribution of plant-parasitic nematodes in a sugarcane crop in north-east Martinique. Nematode species were (a) *Criconemella onoensis* and (b) *Helicotylenchus erythrinae*, *Hemicriconemoides cocophillus* and *Pratylenchus zeae*. Darkness in the grey scale indicates increasing density. (Adapted from Rossi *et al.*, 1996.)

In a study from New Zealand, Fraser *et al.* (1996) demonstrated that earthworm populations are larger in long-term pastures (> 9 years) than in agroecosystems that were cultivated for annual crop production. Earthworm diversity declined from five species in the long-term pasture to two species in the long-term arable cropping system (Fig. 12.6). An intercropping system where cereals were directly drilled into perennial clover supported 572 earthworms/m^2, whereas directly drilled cereal monoculture without the intercrop had 280 earthworms/m^2 and a conventionally tilled cereal monoculture cereal system had

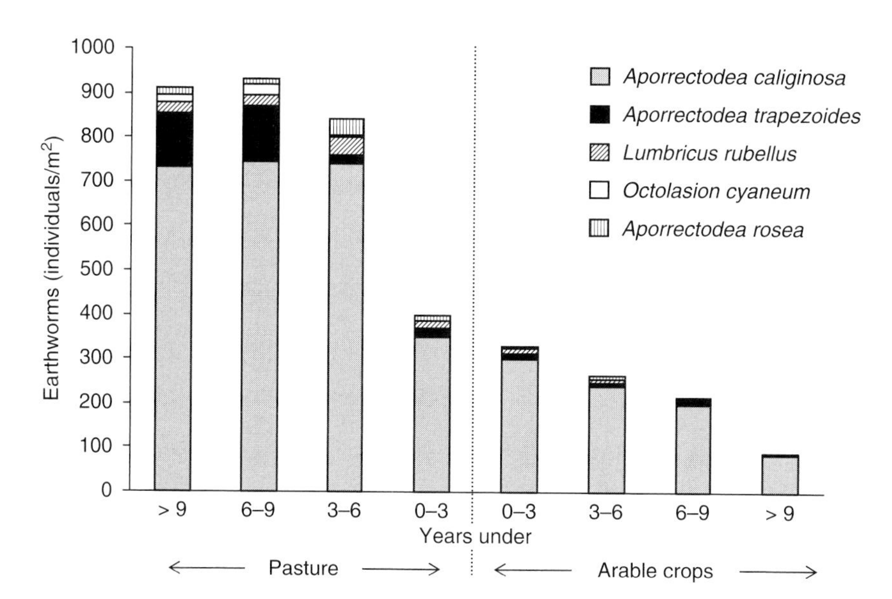

Fig. 12.6. Effect of cropping intensity on the diversity and number of earthworms in the Canterbury Plains, New Zealand. Sampling locations were grouped by cropping system (pasture, arable crops) and by the number of years under pasture or arable crops. (Adapted from Fraser *et al.*, 1996.)

211 earthworms/m^2 (Schmidt *et al.*, 2003). However, earthworm diversity was not affected in these agroecosystems, as all cropping systems had eight or nine earthworm species. These results indicate that the combination of crop species grown together or in rotations has a marked effect on the size and activity of earthworm populations, and probably on other organisms within the soil foodweb.

Focus Box 12.2. Genetically modified crops and soil biota.

Genetically modified crops are plants that have one or more genes from another organism added deliberately. Their development has been controversial due to the fact that genes from dissimilar organisms (bacteria and animals) can be incorporated into plants using recombinant DNA technology. The gene transformation is achieved by using gold particle bombardment or via horizontal gene transfer using *Agrobacterium tumefaciens*, a soil bacterium, to move the engineered plasmid vector or selected genes into the target plant species. The most well-known genetic modification of crop plants is with *Bacillus thuringiensis* (Bt), a Gram-positive bacterium that is common in soil. This bacterium was discovered independently by a Japanese scientist (1901) and a German scientist (1911) while investigating the cause of mortality in Lepidoptera. Natural pesticides containing Bt were available in the 1920s and 1930s, but they were prone to inactivation by ultraviolet radiation when sprayed on plants and also less effective than the chemical insecticides developed after the Second World War. Environmental concerns and insect resistance to chemical insecticides led to a renewed interest in Bt in the 1980s. Today, it is found in transgenic plants such as Bt maize, Bt potato and Bt cotton and food and fibre crops that are highly susceptible to insect damage. Before the release of Bt maize, crop losses and insect control costs in the USA were estimated to exceed $1 billion annually (University of Michigan Extension, 2008).

Maize hybrids that contain the Bt gene express the insecticidal lepidopteran-active crystalline protein (Cry1Ab) endotoxin for the control of European corn borer (ECB, *Ostrinia nubilalis* Hubner). These Bt maize hybrids have been available to maize growers in North America since 1996. Another type of Bt maize hybrid expresses the Cry3Bb1 endotoxin that is effective in controlling Western corn rootworm *Diabrotica virgifera virgifera* LeConte (Coleoptera: Chrysomelidae). Some hybrids have stacked genes, which provide control for both corn borer and rootworm. The 'Roundup Ready' (*RR*) gene provides protection to crops against glyphosate, a systemic broad-spectrum herbicide. In no-tillage agroecosystems, planting RR crops allows the producer to apply a post-emergence spray that will kill broadleaf and grass weeds without affecting the crop. This reduces the quantities of other herbicides that are needed for weed control in no-till systems. However, any producer who plants genetically modified maize (single or stacked hybrids) must follow a resistance management programme, which involves leaving refuges for corn borer and corn rootworm. Refuges are areas of the field where non-Bt maize is planted, allowing a pest like corn rootworm to survive without exposure to Bt maize. If a corn rootworm happened to survive exposure to Bt maize, it would most likely mate with a susceptible corn rootworm from the refuge. Corn rootworms produced from this mating would only be partially resistant, and should not be able to survive if they consumed Bt maize. It would be highly undesirable for resistant corn rootworm to mate with each other, as the offspring are likely to be fully resistant to the Bt maize (University of Michigan, 2008).

It is important to control European corn borer, corn rootworm, weeds and other pests without damaging existing, beneficial soil fauna. The crystalline proteins produced by Bt maize can enter soils and are stabilized by binding to clay surfaces and humic substances. The activity of the Cry1Ab protein is detectable in some soils for more than 200 days after purified toxin is added (Tapp and Stotzky, 1998), and it could enter the soil through root exudation or in crop residues. However, the Cry1Ab toxin released from Bt maize root exudates and crop residues had no apparent effects on bacteria, fungi, protists, nematodes and earthworms in laboratory experiments (Saxena and Stotzky, 2001). Field trials in several locations also reveal that the Bt maize producing the Cry1Ab toxin has no effect on microbial biomass, collembola, soil mites, diplura, symphyla and macrofauna (earthworms, spiders, millipedes and centipedes), even after Bt crops were grown in the same plot for 3 years (Frouz *et al.*, 2008). Under field conditions, Bt maize expressing the Cry3Bb1 protein had no effect on nematodes, collembola and soil mites (Al-Deeb *et al.*, 2003), but this protein does not persist in soil for more than a few days (Prihoda and Coats, 2008). Studies on Bt cotton also reveal that there was no effect in the functional bacteria in the rhizosphere of Bt and non-Bt cotton, even after 5 years of continuous cultivation of Bt cotton (Hu *et al.*, 2009). These results suggest that transgenic crops containing the Bt gene should not affect the structure and functions of the soil foodweb in agroecosystems.

Fertilizers

Fertilizers have been used for centuries to improve soil fertility and crop production. Early farmers recognized that some soils have greater inherent fertility than others, as the first agricultural settlements were established on rich alluvial soils in the Tigris and Euphrates river valleys and along the Nile, Yangtze, Hwang Ho and Indus rivers. Written records from ancient Greece report the benefit of animal manure, night soil (sewage sludge), limestone, marl, seaweed, guano, blood and bone dust for crop growth. These and other fertilizers are effective because they supply essential nutrients for plant growth, but this was not realized until the 1800s. Early experiments by Theodore de Saussure (1767–1845) showed that plant roots are selectively permeable to salts, which allows more rapid uptake of water than salt, and that plants require nitrogen, potassium, calcium and magnesium for growth. Justus von Liebig (1803–1873), a German chemist, reported that phosphorus was essential for seed formation. In 1830, Jean Baptiste Boussingault established the first agricultural field station at Bechelbronn in Alsace, France. Through field plot experiments and chemical analysis of soils and crops, he proved that legumes obtained N from a source that was not available to other plants. The discovery that N_2 fixation was due to symbiotic bacteria within nodules in the legume root was was reported by Hellriegel and Wilfarth in 1886.

The first inorganic fertilizer produced on a commercial scale was superphosphate fertilizer, using the manufacturing process patented by John Bennet Lawes in 1842. The son of an English lord, he became interested in chemistry during his studies at Eton and Oxford. After hearing that bone meal increased turnip production on some of the local farms, but not on others, he became interested in the problem of how to increase the solubility of phosphates in the bone meal so they could be absorbed by plants. He achieved a better fertilizer product by adding sulfuric acid to crushed bones.

Commercial production of single superphosphate fertilizer began in 1843, using rock phosphate. This apatite mineral contains about 12–18% total phosphorus and is mined from natural deposits in Morocco (50% of the world's reserves), the USA, Russia, South Africa and China. Phosphorus fertilizer is produced by the following reaction: (see Equation 12.1 on the bottom of page)

Single superphosphate contains 7.0–9.5% total phosphorus and is 90% water soluble, so essentially all of the phosphates in this fertilizer are available for plant uptake. Further developments in fertilizer technology led to the development of triple superphosphate, which is made by treating rock phosphate with sulfuric acid to generate H_3PO_4 (wet process acid) and then reacting the wet process acid with rock phosphate. Triple superphosphate is a granular fertilizer containing 17–23% total phosphorus and is nearly 100% water soluble. The market for single superphosphate and triple superphosphate is small, as most producers prefer the mixed nitrogen–phosphorus fertilizers such as monoammonium phosphate and diammonium phosphate.

The Haber-Bosch process for production of anhydrous ammonia (equation 9.4, Chapter 9) was even more critical for crop production on a global scale. The application of inorganic nitrogen fertilizers at appropriate agronomic rates is highly beneficial to most non-leguminous crops. Although the process for manufacturing anhydrous ammonia was developed in 1914, it was not adapted for large-scale industrial production of ammonium- and nitrate-based fertilizers until after the Second World War. The first commercial use of anhydrous ammonia (NH_3) involved injecting it into irrigation water in California in the 1930s, producing the following equilibrium reaction:

$$NH_3\ (aq) + H_2O \longleftrightarrow NH_4^+ + OH^- \qquad [12.2]$$

One of the first difficulties encountered in this process is that the hydroxide (OH^-) ion increases the pH of the water, causing a decline in the solubility of dissolved salts such as calcium and magnesium. This can cause salt precipitation, clogging the nozzles of sprinkler systems and restricting water flow. This problem can be reduced by adding sodium hexametaphosphate, which sequesters the salts and reduces precipitation. Another difficulty is that NH_3 (aq) must be stored in a cool, pressurized vessel if it is to remain in a liquid state (the

$$\text{Rock phosphate} + H_2SO_4 \rightarrow \underset{\text{superphosphate}}{CaHPO_4} + \underset{\text{gypsum}}{CaSO_4} \qquad [12.1]$$

boiling point of NH_3 is −33°C at atmospheric pressure), thus it is prone to volatilize and be lost to the atmosphere. Although anhydrous ammonia is still used and directly injected into the soil in some areas, it is generally transformed into a variety of nitrogen-rich fertilizer products (Fig. 12.7).

There are several reasons why inorganic nitrogen and phosphorus fertilizers are widely used for agricultural production:

1. They are water soluble, so most of the nutrients applied go directly into the soil solution where they can be absorbed by plant roots. In contrast, organic residues need to be decomposed by soil organisms before the organic nitrogen and organic phosphorus are transformed into inorganic forms (NH_4^+, NO_3^-, $H_2PO_4^-$, HPO_4^{2-}) that are absorbed by the plant.
2. They are nutrient rich – only a small quantity is needed to achieve an improvement in crop yield and it is economical to transport fertilizer over long distances. Diammonium phosphate (18–52–0) contains about ten times more nitrogen and phosphorus than fresh cattle manure (2.3–1.6–2.2). Cattle manure is also wet (45–70% moisture content) and odorous. It is clear that a much larger volume of cattle manure must be applied to the field to supply the same amount of plant-available nitrogen and phosphorus as the inorganic fertilizer. Thus, it is generally not cost-effective to transport cattle manure further than 20–30 km from the source to fertilize fields, based on the value of nitrogen and phosphorus in the manure (Whalen *et al.*, 2002). However, it would be beneficial to transport and apply cattle manure to more distant fields for other reasons (i.e. to build soil organic matter, to improve soil aggregation, to supply essential micronutrients for plants).
3. They are well characterized – inorganic fertilizers come with a guaranteed analysis, stating the total nutrient content and percentage of water-soluble nutrients and trace elements. The nutrient content of organic fertilizers is highly variable. Livestock manure from neighbouring farms can have a substantially different nutrient content depending on the age of animals, their diet, the barn cleaning system and the manure storage method. Periodic analysis of livestock manure and other organic residues is recommended as part of the farm nutrient management programme.

Since crops grow best when they obtain the right amount of nutrients, at the right time during the growing season, agricultural producers need to

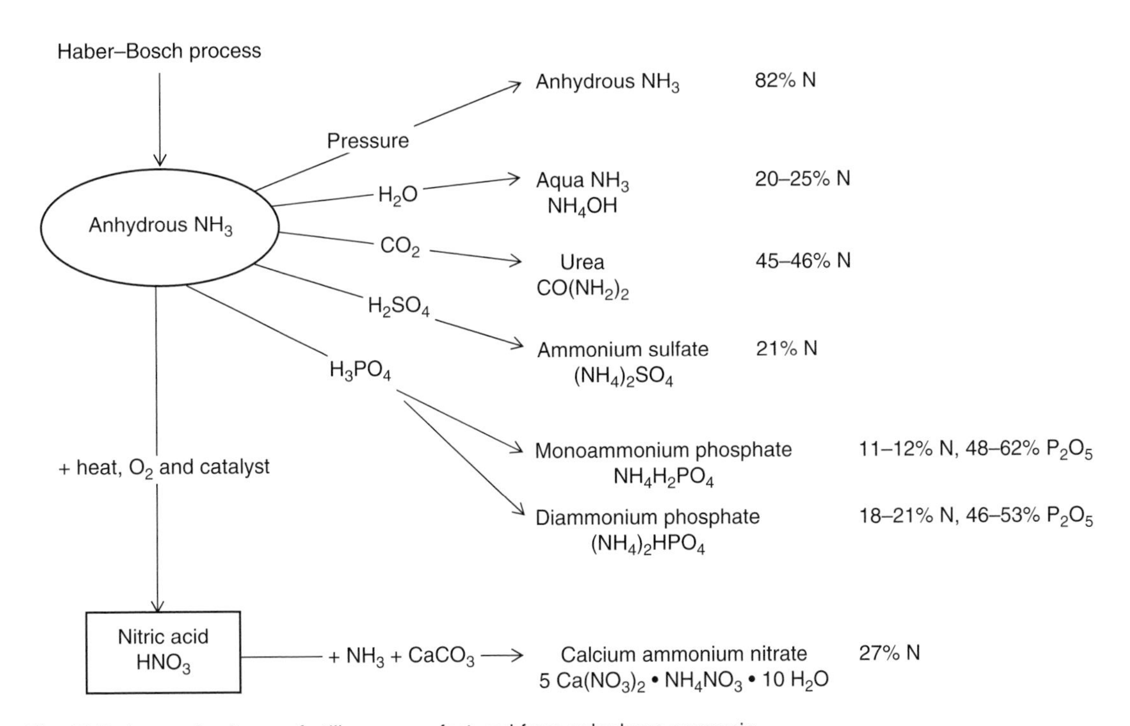

Fig. 12.7. Inorganic nitrogen fertilizers manufactured from anhydrous ammonia.

consider carefully what types of fertilizers could help them to achieve their production goals. There are economic considerations: inorganic fertilizers are more costly than locally available organic residues, some of which may be obtained without charge (e.g. animal manure, municipal wastes, cannery by-products, etc.). There are also environmental consequences to fertilizer use. Inorganic fertilizers are water soluble, so if an application is made in excess of what the crop can use, some of the nutrients may leach below the root zone or be lost in runoff water. However, fields that received high applications of animal manure for many years often exhibit a build-up in nitrogen, phosphorus, potassium and other salts. Excess nitrogen and phosphorus released from these soils is the primary cause of eutrophication and other water quality problems related to agricultural activities in North America and Europe.

Nutrient management programmes have been introduced in many areas to help producers to understand the value of manure and make better choices about fertilizer use on their farms. In the past, some producers applied nutrients in excess of recommended agronomic rates in hopes of obtaining higher yields. This is not a good nutrient management practice if it ignores costs, profits and environmental quality. Neither is applying too few nutrients, since yield and profits will decline. Nutrient management practices can help agricultural producers save money by:

- making better use of on-farm nutrients;
- encouraging the use of rotations, cover crops, residue management and good soil management practices to conserve and recycle nutrients on the farm;
- considering more efficient fertilizer application practices;
- applying the minimum amount of fertilizer needed; and
- identifying opportunities to use low-cost, alternative fertilizer sources.

Fertilizer effects on soil foodweb organisms

In most cases, fertilizer use increases crop production, promoting root and shoot growth during the growing season and leaves more crop residues (roots, stems, leaves) behind when the field is harvested (Fig. 12.8). This suggests that fertilized

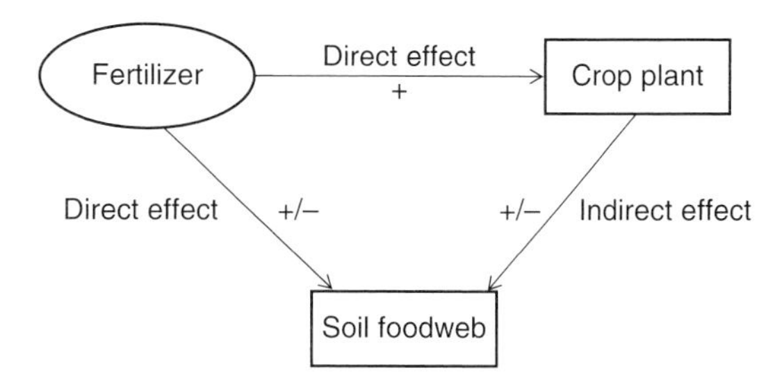

Fig. 12.8. Summary of the effect of fertilizer on the soil foodweb. Fertilizers generally have a strong positive influence on crop production, which leads to a strong indirect effect on the foodweb related to root growth and rhizodeposition (positive interaction). Negative interactions between legumes and N_2-fixing bacteria can occur when nitrogen fertilizers are applied. The direct effect of organic fertilizer on the soil foodweb is generally positive, but some NH_3-rich inorganic and organic fertilizers have a temporary, negative effect on soil organisms.

plants will have greater root growth and exudation during the growing season than unfertilized plants, which will be favourable to free-living microorganisms and other trophic groups in the soil foodweb. However, leguminous plants do not form as many nodules with symbiotic nitrogen-fixing bacteria when nitrogen fertilizers are applied, presumably because they can obtain sufficient nitrogen from soil and do not require the symbiont. For this reason, nitrogen fertilizers are not recommended for any legume or actinorhizal plant.

In greenhouse experiments, the application of phosphorus fertilizers inhibits the development of mycorrhizal associations on crops such as leek and onion, but this is less common in agricultural fields – most plants that are capable of forming a mycorrhizal association are readily infected with the fungus, regardless of the soil phosphorus concentration. This is probably due to the fact that the soil solution contains a low concentration of soluble phosphate, even in highly fertilized soils, due to the rapid adsorption and precipitation of phosphate on soil organo-mineral surfaces. Agronomic guides generally recommend that a small application of phosphorus fertilizer be made with annual crops grown in cold climates to support their early development and root growth before mycorrhizal colonization occurs.

Residues of fertilized crops and the organic matter contained in animal manure, legume residues, compost and other organic fertilizers are a preferred food source for free-living soil biota. Soil microbial biomass and the populations of protists, non-parasitic nematodes, microarthropods and earthworms tend to be larger in organically fertilized soils than in those receiving inorganic fertilizers or no fertilizer. In addition, organic fertilizers contribute to the formation of soil aggregates, which in turn affects the organic matter content, water-holding capacity, bulk density and porosity, important soil physico-chemical characteristics controlling the habitat of free-living soil organisms.

While fertilizers generally have a positive effect on plant growth and consequently on the soil food-web, there are a few situations in which fertilizer applications have negatively affected some groups. Acidic soils that receive ammonia-based fertilizers tend to have lower earthworm populations than unfertilized soils, probably because soil acidification to less than pH 5 was not conducive to earthworm growth (Ma *et al.*, 1990). The application of NH_3-based fertilizers such as anhydrous ammonia and liquid pig slurry affects the soil foodweb because NH_3 is toxic to prokaryotic and eukaryotic cells. Long-term application of these fertilizers seems to have little effect on bacteria and fungi but may affect the actinomycetes and biochemical processes due to the general decline in soil pH resulting from the repeated application of NH_3-based fertilizers (Biederbeck *et al.*, 1996; Deng *et al.*, 2006). High doses of liquid manure can saturate soil pores and inhibit airflow, causing earthworm mortality, but populations generally recover within a few months (Hansen and Engelstad, 1999).

Pesticides

Controlling agricultural pests such as weeds, insects and diseases is essential to achieve economic returns, since crop yields can be reduced or crop failure can occur when pest outbreaks occur. Potato blight, caused by *Phytophthora infestans*, led to potato famines in Ireland (1845) and the Scottish Highlands (1846) that led to human suffering, starvation and mass immigration of about 1.5 million people from the affected areas to North America. While the disease can be partially controlled by planting resistant cultivars, it still poses a risk for potato producers. Plant pathologists and entomologists have made great strides in characterizing and developing control measures for established pests, but new maladies continue to emerge, such as the 'colony collapse disorder' affecting honeybees.This is a major threat to agriculture, as honeybees are the main pollinators of many field and vegetable crops, berries and fruit trees. Colony collapse disorder was reported in the USA in 2006, although the first cases probably occurred as early as 2004, affecting at least one-quarter of an estimated 2.4 million bee colonies in the USA, as well as bees in Canada, Europe and Taiwan. Beekeepers visit a full hive to find that is has been emptied of bees, presumably because they left the hive and died outside, leaving behind a hive with capped brood and food reserves. These observations suggest that colony breakdown may be the final phase of a long process of chronic infection by a silent pathogen. In North America, honeybees have been affected by tracheal mites and varroa mites since the 1980s. When treated with pesticides, the bees can recover from the mite infestation, but they may be weakened and unable to resist other stresses. Other infectious agents affecting honeybees include the parasitic microsporidia *Nosema ceranae* and *Nosema apis*, Kashmir bee virus and Israel acute paralysis virus (Cox-Foster *et al.*, 2007). Research continues to verify the causative agents of colony collapse disorder and develop appropriate treatments to improve bee health and resistance.

Pest control methods

The commercial production of pesticides, which refers to any chemical compound that can slow or prevent the growth of pests, began in the 19th century. The earliest pesticides contained trace elements that were toxic when consumed by pests, or altered the pest's growing conditions in a manner that proved unfavourable. A brief history of three classes of chemical pesticides – fungicides, insecticides and herbicides – is provided below. Additional information on soil fumigants, bacteriocides and nematocides was provided in Chapter 11.

FUNGICIDES Early fungicides included brine, arsenic and copper sulfates, applied as a seed treatment to prevent fungal infection of cereals in the 17th and 18th centuries. Bordeaux Mixture was the first commercial product for fungal disease control, and is still used by home gardeners and in developing countries. The mycologist Pierre Marie Alexis Millardet discovered that a mixture of copper sulfate and lime prevented downy mildew, a fungal disease caused by *Peronospora viticola*, and perfected the formulation so that it did not damage the foliage and fruit of the grape crop. This work was instrumental in saving the wine industry in the Bordeaux region of France, and Millardet was made a Chevalier de la Légion d'Honneur in 1888 (Ayres, 2004).

Advances in organic synthesis from the 1940s led to the development of compounds with fungicidal properties. The dithiocarbamates and the phthalimides were a major improvement over inorganic fungicides in that they were more effective against the fungus, caused less phytotoxicity to the crop and were easily prepared and applied by the user. Further research led to the development of contact fungicides, which kill on contact, and systemic fungicides that are absorbed by the plant and provide global protection against fungal pathogens. The newer systemic fungicides activate the induced resistance pathways of the host to boost the plant's 'immune response' and provide broad protection against fungi, bacteria and viruses. The general classes of fungicides, based on their mode of action, and some representative chemical groups in use since 1970 are described in Tables 12.6 and 12.7.

Table 12.6. The Fungicide Resistance Action Committee (FRAC) classification of fungicide groups, based on their mode of action. (Adapted from Kuck and Gisi, 2007.)

FRAC class	Mode of action/inhibition
A	Fungicides interfering with nucleic acid synthesis
B	Fungicides interfering with mitosis and cell division
C	Fungicides interfering with fungal respiration
D	Fungicides interfering with amino acid and protein synthesis
E	Fungicides interfering with signal transduction
F	Fungicides interfering with lipid and membrane synthesis, cell wall deposition
G	Fungicides interfering with sterol biosynthesis in membranes
H	Fungicides interfering with glucan synthesis
I	Fungicides interfering with melanin synthesis in cell wall
P	Host plant defence inducers
U1	Fungicides with unknown mode of action
U2	Recently introduced fungicides with unknown mode of action and resistance risk
M	Multi-site mode of action

INSECTICIDES One of the first insecticides was the copper acetoarsenite pigment called Paris Green, applied for the control of Colorado potato beetle (*Leptinotarsa decemlineata*) in the USA in 1867. Paris Green sprays were also developed to protect apple against the codling moth (*Cydia pomonella*) until about 1900. However, Paris Green was phytotoxic and damaged plants, leading to the development of lead arsenate sprays to combat the gypsy moth (*Lymantria dispar*) around 1892. Lead arsenates were the principal insecticide used against codling moth on apple trees, insect pests of other fruit trees, garden crops and turfgrasses, for mosquito abatement, in cattle dips and on rubber and coffee trees until the 1940s (Peryea, 1998). Their use was discontinued when the synthetic organic insecticide dichlorodiphenyltrichloroethane (DDT) was introduced.

The insecticide DDT was synthesized by the German scientist Paul Muller in 1939. Initial testing of this chemical showed that it was extremely toxic to flies and other insects. In his laboratory, Mueller showed that flies died when sprayed with DDT; a second group of flies placed in the same cage died before any DDT was added. This demonstrated that DDT persisted in the environment and a small dose could kill a large number of insect pests. It was clear that DDT had the potential to control malaria and other diseases spread by mosquitoes and biting flies, and alleviate human suffering in many parts of the world. For his discovery, Muller received the Nobel Prize in Medicine in 1948. However, the chlorinated benzene ring that made DDT such a persistent and effective insecticide led to unexpected environmental impacts, namely its bioaccumulation in foodwebs, to the detriment of top-level predators such as eagles and raptors. Today, production of insecticides that persist in the environment is generally banned, although limited use of DDT continues in some tropical countries where malaria and other diseases borne by flies pose a significant health risk.

Insecticides can affect the target organism on contact, while systemic insecticides must be ingested before they are effective. The general classes of insecticides that have been used to control insect pests in agroecosystems and for other applications are listed in Table 12.8.

HERBICIDES Before the 1880s, hand-hoeing or repeated cultivation were the only means available to control weeds. The first herbicide was copper

Table 12.7. Fungicide groups and worldwide sales (% of all fungicides sold) in 2004. The general FRAC class and main use of each group is listed. Adapted from: Kuck and Gisi, 2007; Morton and Staub, 2008.

Fungicide group	% of total sales	FRAC class	Main uses
DMI fungicides	28	G	Broad spectrum, including products for cereal, grape and turf diseases
QoI fungicides	19	C	Broad spectrum, effective against cereal diseases and oomycetes
Dithiocarbamates	7	M	Broad spectrum
Copper and sulfur formulations	5	M	Highly effective against powdery mildew and foliar diseases
Benzimidazoles, thiophanates	4	B	Broad spectrum, foliar and seed treatment, postharvest application
Chloronitriles	3	M	Broad spectrum
Dicarboximides	3	F	Control of *Botrytis* and *Monilinia*
Phenylamides	3	A	Control of oomycetes
Amines	3	G	Control of powdery mildew, rusts and Sigatoka
Anilinopyrimidines	3	D	Broad spectrum
MBI fungicides	3	I	Control of rice blast (*Magnaporthe grisea*)
Carboxamides	2	C	Broad spectrum
Uncouplers	2	C	Broad spectrum
Resistance inducers	2	P	Indirect action on fungi, bacteria and viruses
Phosphonates	2	U	Control of oomycetes
Phthalimides	2	M	Broad spectrum, seed treatment
Carboxylic acid amides	2	F	Control of oomycetes
Cyanoacetamide oximes	2	U	Control of oomycetes
Other	5	–	–

Table 12.8. Major groups of insecticides and their mode of action against insect pests. (Adapted from Elbert *et al.*, 2007.)

Insecticide class	Mode of action (representative compound)
Organochlorines	Affect nervous system through sodium channel modulation. Persistent in the environment (DDT)
Organophosphates	Bind to acetylcholinesterase, disrupting nerve impulses. Persists in the environment (Parathion)
Carbamates	Bind to acetylcholinesterases, causing short-term disruption of nervous system. Lower toxicity than organophosphates (Aldicarb)
Phenothiazines	Neuromuscular disruption. Not widely used as an insecticide, but applied to livestock as an anthelmintic drench.
Pyrethroids	Mimic insecticidal activity of the natural compound pyrethrum. Affect sodium channels in nerve axons. Not very persistent, low toxicity (lambda-cyhalothrin)
Neonicotinoids	Synthetic analogues of the natural insecticide nicotine. Bind to nicotinic acetylcholine receptors. Broad-spectrum, systemic insecticides that have low persistence and toxicity (Acetamiprid)
Biological insecticides	*Bacillus thuringiensis* (bacterium) *Metarhizium anisopliae* (entomopathogenic fungus) *Steinernema feltiae* (parasitic nematode) *Cydia pomonella* granulovirus (virus)
Plant insecticides	Chemical defences produced by the plant's induced systemic resistance pathways (caffeine, cinnamon leaf oil, neem, oregano oil and many others)

sulfate, which was found to be as toxic to some broad-leaved weeds as it was to downy mildew. During the 1920s and 1930s, sodium chlorate and benzene derivatives were used as herbicides. Organic synthesis led to the identification of molecules that were bioactive against weeds, and by 1946, many were produced commercially. The first widely used herbicide was 2,4-dichlorophenoxyacetic acid (2,4-D). It is effective at killing broadleaf plants while leaving grasses largely unaffected. Due to its low cost, 2,4-D remains among the most commonly used herbicides in the world.

Contact herbicides kill the plant components with which they come in contact, and thus are not as effective at killing perennial weeds. Systemic herbicides are translocated through the plant, either from the foliage to the rest of the plant if they are applied on leaves or from the soil to the plant if applied to the soil surface. Systemic herbicides can be applied before planting and mechanically incorporated in the soil or as a pre-emergent application, after planting but before the crop has emerged from the soil. Post-emergence applications are generally made with contact herbicides, unless the crop plant has resistance to the herbicide (e.g. Roundup Resistant gene, discussed in Focus Box 12.1). The major mode of action of herbicides is provided in Table 12.9.

Integrated pest management

Because fungi, insects and weeds have short life cycles and are numerically abundant, they can readily develop resistance to pesticides through natural selection. Once resistance develops, the compounds that were previously effective will no longer be able to control the pest. Reproduction of resistant organisms passes the genes for resistance to future generations, rendering the product useless. Agricultural producers, scientists and the agrochemical industry view resistance to insecticides as an extremely serious threat, and the crop protection industry has taken a number of steps to prevent or delay the development of resistance in pests.

Integrated pest management (IPM) is an approach that is consistent with this goal. The IPM concept was first introduced by the entomologist V.M. Stern and collaborators (Stern *et al.*, 1959), who

Table 12.9. Major groups of herbicides and their mode of action against weeds. (Adapted from Menne and Köcher, 2007.)

Herbicide class	Mode of action (representative compound/chemical group)
ACCase inhibitors	Inhibit acetyl coenzyme A carboxylase (ACCase) and affect cell membrane production in the meristems of grasses, but not broadleaf weeds (Diclofop)
ALS inhibitors	Inhibit the acetolactate synthase (ALS) enzyme responsible for producing branched-chain amino acids (valine, leucine and isoleucine). Eventually, DNA synthesis ceases. Both grasses and broadleaf plants are affected. Since this biological pathway exists only in plants (not animals), these herbicides have low toxicity to non-target organisms (Sulfonylureas)
EPSPS inhibitors	Inhibits the enolpyruvylshikimate-3-phosphate synthase enzyme (EPSPS) needed to synthesize amino acids with benzyl side chains (tryptophan, phenylalanine and tyrosine). They affect both grasses and broadleaf weeds (Glyphosate)
Synthetic auxins	Synthetic auxins mimic the natural plant growth regulator. They have several points of action on the cell membrane and are effective at controlling broadleaf weeds (2,4-D)
Photosystem II inhibitors	Reduce the electron flow from water to $NADPH_2^+$ at the photochemical step in photosynthesis. Electrons accumulate on chlorophyll molecules, uncontrolled oxidation reactions occur and the plant dies. Effective against broadleaf weeds and grasses (Atrazine)
PPO inhibitors	Inhibition of the protoporphyrinogen oxidase (PPO) pathway, which is the last step in the biosynthesis of chlorophyll (Phenylpyrazoles)
PDS inhibitors	Inhibition of the phytoene desaturase (PDS) pathway responsible for caretenoid biosynthesis (Pyridazinones)
4-HPPD inhibitors	Inhibition of the 4-hydroxyphenyl-pyruvate dioxygenase (4-HPPD) pathway, responsible for oxidizing the ketoacid of methionine (Pyrazoles)

explained that pest control must be based on a sound understanding of the ecology of pests and their crop hosts. Biological, cultural and chemical control methods must be carefully selected to achieve an ecological and environmentally sound pest management programme. Pesticides were to be used, but only as necessary and with limited applications, based on population monitoring and economic thresholds (Fig. 12.9).

The fundamental principles of IPM – to understand pest-crop ecology, maximize biological and cultural controls and use pesticides as a last resort – has become an accepted philosophy in agriculture. In addition, regulatory measures such as quarantine and certification are essential to prevent the introduction of pests into new areas (Table 12.10).

Pesticide effects on soil foodweb organisms

Certain pesticides applied to agroecosystems are directly toxic to some of the soil foodweb organisms. Soil fumigants affect parasitic organisms as well as the free-living microorganisms. Bacteriocides, fungicides, nematocides and arcaricides applied to kill crop pests will affect some non-target organisms, although the effect will depend on the nature of the pesticide, the dose and the mode of action of the biological active ingredient. Contact pesticides must be absorbed by an organism to be effective, so a contact fungicide such as chlorothalonil, which acts as a protective barrier against fungi on crop foliage, should have limited or no effect on soil fungi. A systemic fungicide such as tricyclazole

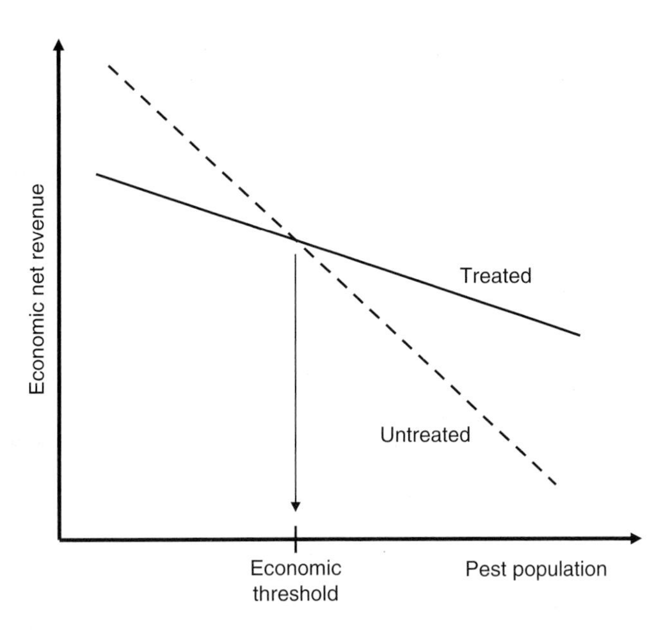

Fig. 12.9. The economic threshold concept for pest control. Net revenue from the crop declines with increasing pest populations. The intersection of the 'Treated' and 'Untreated' lines indicates the economic threshold where the net revenue is equivalent regardless of the treatment. The *y*-intercept of the 'Treated' line is lower than the 'Untreated' line, reflecting that net revenues are lower when producers pay to purchase and apply pesticides. (Adapted from Luna and House, 1990.)

Table 12.10. Approaches for controlling plant diseases and pest outbreaks. (Adapted from Morton and Staub, 2008.)

Regulatory measures	Quarantine, inspection of imported grains, fruits and vegetables, seed certification
Cultural methods	Crop rotation, intercropping, sanitation (burning infected residues, tillage, composting), selecting planting and harvest dates to reduce or avoid pest damage
Biological controls	Breeding of resistant varieties, biological insecticides, biocontrol by soil foodweb organisms
Chemical control	Seed treatment, soil, foliar and postharvest applications of pesticides

is incorporated within rice to control rice blast *Magnaporthe grisea* and could possibly affect other fungi of the Ascomycetes that come into contact with treated rice, but there is no information in the scientific literature to confirm or deny this hypothesis. Some examples of direct toxicity from pesticides approved for use in agroecosystems are:

- The broad-spectrum biocides Kocide (cupric hydroxide) and Agri-Strep (streptomycin sulfate) reduced soil bacterial populations for up to 14 days and fungal populations for up to 7 days. Neither provided consistent or persistent control of soil microbial populations (Donegan *et al.*, 1992).
- The application of the fungicides benomyl and mancozeb to control powdery mildew, rust and leaf spot in apple orchards affected mite populations. Predaceous mite populations declined on leaves of fungicide-treated trees, while phytophagous mites were more numerous (Bower *et al.*, 1995).
- Carbofuran was effective at reducing nematode populations in a tallgrass prairie, but did not reduce macroarthropod (herbivores, detritivores, predators) or earthworm populations (Todd *et al.*, 1992).
- Agricultural soil samples treated with the acaricide bromopropylate had greater bacteria plate counts and more denitrifying bacteria, but there were fewer aerobic diazotrophs and a decline in nitrifying microorganisms (Gonzalez-Lopez *et al.*, 1992).
- Carbaryl and other acethylcholinesterase inhibitors are toxic to earthworms, but did not eliminate earthworms from an arable soil. Perennial hayfields receiving repeated carbaryl applications during the growing season had earthworm populations reduced by 80–98% (Eriksen-Hamel and Whalen, 2007, 2008).

Insecticides that target Lepidoptera, Hemiptera, Chrysomelidae and other common crop pests are not expected to have a direct toxic effect on soil foodweb organisms. Herbicides are effective at killing weeds because they disrupt key biochemical pathways in plants, but should not have any direct toxicity to soil foodweb organisms.

The indirect effects of pesticides on soil foodweb organisms in agroecosystems are not well documented. For example, a fungicide that reduces the fungal population or fungal activity is expected to affect trophic interactions for a period of time, depending on the resiliance of the soil foodweb. Too few studies on fungicides and other pesticides have been conducted to evaluate the ecological impact of such interactions.

In contrast, the indirect effects of herbicides on the soil foodweb have received more attention. Symbiotic soil microorganisms rely on the presence of a healthy host for their survival, whereas the free-living soil decomposers depend on plants for organic substrates (rhizodeposits, residues). Hence, crop protection measures that ensure a high level of primary production are expected to be favourable to the soil foodweb. Indeed, herbicide-induced plant death increases the quantity of dead roots, with a corresponding increase in microbial biomass. Dead weed residues at the soil surface are soon colonized by microorganisms, nematodes, bacterivorous and fungivorous microarthropods, and thereafter consumed by earthworms (Wardle, 1995). The removal of weedy species has a much more profound effect on net primary productivity than on soil decomposers, probably because they can sustain populations on residual soil organic carbon reserves for some time, even when all weeds are removed (Wardle *et al.*, 1999). Weeds appear to be an important alternative host for arbuscular mycorrhizal fungi, which could benefit crop production by maintaining populations of beneficial mycorrhizae (Jordan and Huerd, 2008). Although more work is needed to fully understand these interactions, these findings suggest that weed control with cultural or chemical methods is probably not detrimental to the soil foodweb. Agronomic crops do not require 'weed-free' conditions to achieve good yields, and some weeds could be allowed to grow to maintain mycorrhizal fungi populations in agroecosystems.

Focus Box 12.3. Organic farming systems and the soil foodweb.

Organic farming systems are a type of low-input farming system that relies on crop rotations, biological control and tillage to control pests. Since chemical pesticides and genetically modified crops are not used in organic agriculture, producers rely on bioactive extracts, natural enemies and physical barriers to suppress pathogens, parasites and herbivores. Tillage and hand weeding are the main forms of

Continued

Focus Box 12.3. *Continued*

weed control. Nutrients for crop production are typically supplied from organic fertilizers such as green manure, animal manure, compost and mineral fertilizers from a list of permitted substances. In North America, a certified organic farm has to undergo a 3-year transition period and is regularly inspected to ensure it meets the national standards for organic agriculture (National Standard of Canada, 2006; USDA, 2009).

The DOK trial in Therwil, Switzerland was established in 1978 to evaluate and compare organic and conventional farming systems. The cornerstones of the organic agriculture systems studied in the DOK trial are that they must be low input (i.e. not requiring extensive use of energy to produce crops) and self-sustaining (i.e. relying on nutrient recycling on the farm, rather than external nutrient sources, for crop production). Twenty-seven years after the experiment began, Birkhofer *et al.* (2008) evaluated the effect of diverse farming practices (Table 12.11) on soil quality, nutrient cycling, soil microorganisms and fauna.

The effect of these farming systems on soil food-web organisms is summarized in Table 12.12. In general, bacteria, omnivorous and herbivorous nematodes, enchytraeids and fly larvae were more abundant in the biodynamic and bioorganic plots than the conventional plots. Earthworm populations were larger in systems that received farmyard manure than inorganic fertilizer only. Protists, chilopoda and collembola populations did not differ between farming systems. Surface-dwelling spiders were more abundant in the biodynamic and bio-organic plots, while aphid abundance was highest in the conventional plots that received farmyard manure. Other surface-dwelling macrofauna such as predaceous beetles (Staphylinidae, Carabidae) and herbivorous beetles did not respond to farming practices. Wheat grain and straw yield was greater in the conventional farming systems than in the organic farming systems.

The main conclusion from this study was that farmyard manure increased the resources for soil decomposers, which increased the biomass and activity of bacteria, the omnivorous nematodes, enchytraeids and earthworms. The herbivorous nematodes were also more abundant in the organic agroecosystems, probably because they were feeding on crop roots. At the soil surface, it appeared that spiders were natural enemies capable of controlling aphid populations in the organic farming systems. This supports the assumption of improved pest suppression through biocontrol in organic farming systems. Wheat yield was 23% greater in the conventional than organic farming systems, which may be a reflection of the water-soluble nutrients supplied by inorganic fertilizer compared with farmyard manure.

Table 12.11. Farming systems in the DOK trial in Therwil, Switzerland. (Adapted from Birkhofer *et al.*, 2008.)

Treatment	Biodynamic	Bio-organic	Conventional	Mineral
Abbreviation	BIODYN	BIOORG	CONFYM	CONMIN
Organic fertilizer	Composted farmyard manure	Rotted farmyard manure	Stacked farmyard manure	None
Inorganic fertilizer	None	None	Inorganic fertilizer (NPK)	Inorganic fertilizer (NPK)
Weed control	Mechanical	Mechanical	Mechanical and chemical	Mechanical and chemical
Disease control	Natural enemies	Natural enemies	Chemical*	Chemical*
Pest control	Plant extracts	Plant extracts	Chemical*	Chemical*
Other treatments	Biodynamic preparation	None	Plant growth regulators	Plant growth regulators

*Chemical controls such as fungicides and insecticides were applied when necessary, based on population monitoring and economic thresholds (Fig. 12.9).

Continued

Focus Box 12.3. *Continued*

Table 12.12. Effect of farming systems on the populations of soil microorganisms and fauna (abbreviations for each farming system are given in Table 12.11). Data were collected from the DOK trial in Therwil, Switzerland. (Adapted from Birkhofer *et al.*, 2008.)

Organism	Effect of farming system
Bacteria (direct counts)	BIODYN = BIOORG > CONFYM > CONMIN
Bacteria – phospholipid fatty acids (PLFA)	BIODYN = BIOORG > CONFYM > CONMIN
Fungi (PLFA)	BIODYN = BIOORG = CONFYM > CONMIN
Bacterivorous nematodes	BIODYN = BIOORG = CONFYM > CONMIN
Fungivorous nematodes	BIODYN < BIOORG < CONFYM = CONMIN
Omnivorous nematodes	BIODYN = BIOORG = CONFYM > CONMIN
Herbivorous nematodes	BIODYN = BIOORG > CONFYM > CONMIN
Enchytraeids	BIODYN = BIOORG > CONFYM = CONMIN
Lumbricids	BIODYN = BIOORG > CONFYM = CONMIN
Diptera larvae	BIODYN = BIOORG > CONFYM = CONMIN
Web-building spiders	BIODYN = BIOORG > CONFYM > CONMIN
Cursorial spiders	BIODYN = BIOORG > CONFYM = CONMIN
Aphids	BIODYN = BIOORG < CONFYM = CONMIN
Wheat grain yield	BIODYN = BIOORG < CONFYM = CONMIN
Wheat straw yield	BIODYN = BIOORG < CONFYM = CONMIN

Water management

Water management is a great challenge in agriculture, because too much or too little water can lead to a reduction in crop yields. The Ancient Egyptians constructed water storage basins and an elaborate canal system in the Nile Valley to support crop production more than 6000 years ago. Today, vast expanses of semi-arid and arid lands are irrigated to support food production in all parts of the world. Diverting excess water is also extremely important in humid and semi-humid regions. Dykes, levies and other structures that prevent water from entering agricultural land represent a huge investment to build and maintain, but are critical to agricultural production in the Netherlands and along the Mississipi flood plain of the USA. Artificial drainage of agricultural land is common in humid regions. Subsurface drainage is particularly beneficial on heavy soils (e.g. clays) that remain wet after snowmelt and rainfall events, due to slow natural drainage. Drainage reduces the risk of soil compaction from farm machinery, and crops grow better when the root zone is not waterlogged.

Soil water content is a highly dynamic variable that is affected by infiltration, percolation, runoff and drainage processes. The amount, duration and intensity of water input from rainfall and irrigation have an important control on these processes. Water loss from the soil–plant system through evapotranspiration is strongly controlled by temperature. The daily, weekly and seasonal changes in precipitation and temperature explain a considerable amount of the fluctuations in primary production and trophic interactions in ecosystems.

Global climate change has a tremendous effect on the hydrologic cycle as well as the Earth's temperature. The accelerated melting of glaciers in polar and alpine regions is affecting the water level of oceans and other water bodies, making some parts of the world more susceptible to flooding. The timing, quantity and intensity of rainfall events are also expected to change; some regions are forecast to receive more annual precipitation and heavy rainfall events, while other regions are expected to become drier and more susceptible to drought (IPCC, 2007). In a mesic tallgrass prairie, Knapp *et al.* (2002) found that sporadic, heavy rainfall events reduced the dominance of productive C_4 grasses and net primary production, resulting in less soil organic C storage and soil respiration. This study illustrates the vulnerability of terrestrial ecosystems to changes in hydrologic cycling. Changes in rainfall patterns are expected to affect the survival of soil organisms and plants, perhaps even leading to the local extinction of certain species, thus altering decomposition and nutrient cycling processes.

In some semi-arid and arid regions, salinity is a significant issue for soil biota. An increase in the dissolved salt concentration interferes with the passive osmotic transport processes that normally permit water and nutrients to move from the soil solution into plant roots or into microbial and animal cells. Salts are present naturally in soils, but can migrate and become concentrated in the topsoil where plant roots and soil foodweb organisms are most abundant. The major causes of soil salinity are:

- Clearing of deep-rooted vegetation: when trees and deep-rooted perennial grasses are removed and replaced by shallow-rooting annual species, more water moves through the soil and into the water table. This effectively causes the water table to rise, bringing salts towards the topsoil.
- Overirrigation: irrigating soils in excess of crop needs allows more water to move through the soil, causing the water table to rise.

Once the water table is within 2 m of the soil surface, evapotranspiration and transpiration draw the water containing dissolved salts toward the surface. Water is lost, but the salts remain behind.

Water management effects on soil foodweb organisms

As discussed in Part II, soil organisms are highly sensitive to the soil water content and have developed strategies to tolerate and survive extremely wet or extremely dry soil conditions. As expected, soil organisms respond favourably to irrigation in well-drained soils. In southern Idaho, USA, Entry *et al.* (2008) showed that microbial biomass and DNA concentrations tended to be greater in irrigated agricultural land (lucerne–potato–wheat–bean rotation) and pasture (Kentucky bluegrass–orchardgrass) than in native, non-irrigated sagebrush on the same soil type (Table 12.13). In this semi-arid region, irrigation provided a favourable soil–water balance for microbial activity and also promoted more even, vigorous crop growth than the native conditions, which resulted in a positive feedback for the soil microbial community. Soil fauna, particularly the soft-bodied enchytraeids and earthworms, benefit from irrigation. Prolonged inundation is not beneficial for the aerobic soil biota, although facultative and obligate anaerobes are expected to exhibit higher activity during periods when the soil is waterlogged.

Halophilic archaea and bacteria thrive in saline environments such as the Sargasso Sea due to their ability to exclude salt from the cell. Melanized ascomycetes called black yeasts are also found in hypersaline habitats such as salt marshes. Plants are more sensitive to salts than these microorganisms. In agroecosystems, the salt tolerance of crops ranges from low (0–4 dS/m) for many pulse crops, maize, potatoes, carrots and onions to very high (20 dS/m) for certain grasses. A study from Australia showed two to four times more microbial biomass at 30 dS/m than at 0.5 or 10 dS/m, and more respiration in the most saline conditions. Since the soils in this area tend to be saline and sodic (i.e. have a high concentration of sodium), this suggests that at least some microorganisms have developed a high salt tolerance (Wong *et al.*, 2008). Earthworms are more vulnerable to saline substrates. Earthworms in non-

Table 12.13. Active bacterial biomass, active fungal biomass, microbial biomass and DNA in native sagebrush and irrigated agricultural soils in southern Idaho, USA during summer sampling. (Adapted from Entry *et al.*, 2008.)

Treatment (depth, cm)	Bacterial biomass (mg C/g)	Fungal biomass (mg C/g)	Microbial biomass (mg C/g)	DNA (mg/g)
Native sagebrush (0–5)	6.7	6.1	12.9	41.8
Native sagebrush (5–15)	6.1	4.1	10.3	26.6
Native sagebrush (15–30)	4.8	2.8	7.7	31.6
Irrigated mouldboard ploughing (0–30)	12.0	5.0	17.0	51.0
Irrigated conservation tillage (0–15)	7.9	6.5	14.5	65.2
Irrigated conservation tillage (15–30)	5.6	3.5	9.2	40.0
Irrigated pasture (0–30)	7.9	9.6	17.4	75.3

saline soils survived the application of sludges with salinity levels of about 10 dS/m by avoiding contact with these materials (Schaefer, 2005).

12.2 Grassland and Forest Management

Terrestrial ecosystems classified as forests, woodland and grassland represent more than half of the global land use (Table 12.14). Although less intensively managed than agroecosystems, these ecosystems are modified by natural and anthropogenic processes, including global climate change.

Grasslands are lands dominated by grasses rather than by large shrubs or trees. The most extensive grasslands are savannahs, which cover nearly half of the land area in Africa and large areas of Australia, South America and India. Savannahs are tropical and subtropical grasslands with scattered individual trees. The climate is warm to hot, with annual rainfall of about 500–1300 mm/year. Rainfall is concentrated during 6–8 months, followed by a dry period during which fire may occur. Herbivores such as elephants, giraffes, zebra and other ungulates are abundant.

Temperate grasslands have moderate rainfall, hot summers and cold winters. They are found in South Africa (veldts), Hungary (puszta), Argentina and Uruguay (pampas), the former Soviet Union (steppes) and central North America (prairies). Most of the rainfall generally occurs in the spring to early summer, with additional moisture from snow. Vegetation is adapted to tolerate seasonal drought, occasional fires and grazing by large animals such as bison, antelope and deer, as well as herbivory by grasshoppers and other arthropods. Extensive grazing of livestock such as cattle is common in some areas.

Table 12.14. Global extent of land use categories in 2000. (Adapted from Holmgren, 2006.)

Category	Extent (million km^2)	Extent (% of total land base)
Forest	40	30
Woodland/grassland	34	25
Agricultural crops	15	11
Urban areas	0.4	1
Other land	44	33
Total	134	100

A smaller land area is covered by flooded, montane, polar and xeric grasslands. Flooded grasslands can be in temperate, subtropical or tropical regions and are flooded periodically or year round. Montane and polar grasslands are subject to short, warm summers and cold winters, whereas xeric grasslands have sparse vegetation that can survive in deserts and other arid regions. Each has characteristic vegetation that is well adapted to survive grazing by herbivorous animals and fire, the major natural disturbances of grasslands.

The first forests were dominated by giant horsetails, club mosses and ferns that were over 12 m tall. The gymnosperms that appeared during the Triassic period (245 million years ago) and angiosperms that evolved during the late Cretaceous period (144 million years ago) are the ancestors of trees found in present-day forests. Modern forests contain 70% of the carbon within living biomass, so they are vitally important in the global carbon cycle. Tropical, temperate and boreal forests are the three major types of forests found at different latitudes and altitudes.

Tropical forests occur near the equator (from 23.5°N to 23.5°S).These forests do not experience winter, and generally have only two seasons (rainy and dry). The annual rainfall may exceed 2000 mm. Soils tend to be acidic and nutrient-poor. Litter is decomposed rapidly. Herbivory is widespread. Tropical forests have a complex spatial structure (multilayered canopy) that supports a diverse flora and fauna, perhaps the greatest biodiversity of any terrestrial ecosystem. Initiatives to promote sustainable forest harvesting from these forests are being promoted strongly.

Temperate forests are found in eastern North America, north-eastern Asia, western and central Europe. These forests have moderate to high precipitation (750–1500 mm/year) that is evenly distributed through the year, with hot summers and cold winters. Dominant trees and the herbaceous understorey vegetation support diverse wildlife. Herbivory and fires are common, as is forest harvesting.

The boreal forest grows at latitudes between 50 and 60°N, in Eurasia and North America. The boreal forest is characterized by short, moist, warm summers and long, cold, dry winters. Soils are thin and nutrient-poor. A thick litter layer develops in some areas. However, the understorey vegetation is poorly developed, compared with other forests. Boreal forests are susceptible to herbivory, fire and extensive forest harvesting.

Grazing and forest harvesting

Grazing is a form of herbivory in which plant biomass is removed by animals and invertebrates that cut leaves or suck the sap from the stems. The constitutive and inducible defences that plants invoke in response to herbivory were discussed in Chapter 11. It should be noted that the plant's genetic, physiological and morphological responses to grazing depend on the magnitude of tissue damage in relation to the plant's capacity for regeneration. There is a threshold 'break-even' point beyond which physiological functions are irreversibly impaired and the plant may not be able to survive. When the frequency and intensity of herbivory exceed this break-even point, a change in plant community dynamics and even ecosystem function is expected to occur (Table 12.15).

Humans are responsible for defoliation in natural and managed ecosystems. Free-range grazing by livestock released into grasslands and forests contributes to herbivory on grasses and forbs, woody shrubs and trees. Livestock and other large herbivores cause multiple disturbances in ecosystems, including:

- removal of above-ground biomass;
- removal of below-ground biomass when grazing pressure is high, which affects root productivity and growth;
- return of nutrients to the soil in labile forms (e.g. dung and urine patches);
- physical alteration of the soil environment through compaction and trampling; and
- modulation of plant species composition due to selective grazing of more palatable plants.

While herbivory by livestock or wildlife tends to be heterogeneous and patchy in space and time, this is not generally the case in managed ecosystems. Mechanical forage harvesting or lawn mowing is punctual and uniform across the management zone.

Forest harvesting is an extreme type of 'herbivory', since the entire above-ground biomass is generally removed. Trees such as willow, alder and eucalyptus will regrow after harvest, but most do not. The gaps left in the forest by tree harvest permit the growth of saplings in the understorey that were previously constrained due to insufficient light, space or limiting factors. The regeneration of forests after cutting activities is strongly influenced by the intensity of the harvest and secondary practices such as burning, herbicide applications and replanting the area with selected tree species (e.g. fast-growing pines, spruce and hybrid poplar).

Grazing effects on soil foodweb organisms

Grazing modifies the growth of above- and below-ground plant components, which is expected to influence the activity of soil foodweb organisms. Grasses and forbs often respond to moderate

Table 12.15. Responses of vegetation to herbivory. The capacity of plants to function after tissue removal, and the frequency and intensity of herbivory, determine whether the plant community can accommodate the stress or whether there will be a change in the ecosystem due to herbivory (SOM = soil organic matter). (Adapted from Anderson, 2000.)

System	Dominant species/type	Plant responses	Stressor	Stress intensity	Scale (nominal)	Change in system
Boreal	Spruce Aspen	Slow Slow	Moose graze aspen	Infrequent, severe	10 m^2/year	More spruce, more SOM
Moorland	Heather Grass	Slow Fast	Heather beetle	Infrequent, severe	1 m^2/year	More grass, more SOM
Meadow	Grass	Medium	Foliar invertebrates	Frequent, moderate	10 cm^2/year	Accelerates succession
	Forbs	Medium	Root-eating invertebrates	Frequent, moderate	10 cm^2/year	Retards succession
Pot experiment	Soybean Mycorrhizae	Medium Fast	Nematodes	Frequent, severe	1 mm/h	Reduces N_2 fixation and soybean growth

herbivory by reallocating photosynthates to the roots, which boosts root growth and increases the exudation of labile carbon compounds such as organic acids, sugars and amino acids through their roots (Bardgett and Wardle, 2003). This induces a positive feedback in the rhizosphere, whereby free-living bacteria and associated trophic groups become more active, accelerating decomposition and nutrient cycling processes, especially nitrogen mineralization. The increase of plant-available nutrients in the rhizosphere improves nutrient uptake and boosts plant growth (Fig. 12.10).

Other mechanisms have been suggested to explain the improvement in plant growth following defoliation in grasslands. Dead roots contain organic carbon and other nutrients that can be recycled by root-associated organisms, releasing plant-available nutrients that can be absorbed through the living root system. Herbivores excrete urine, which can be rapidly converted to plant-available NH_4^+ by soil microorganisms, and dung that can be recycled by soil decomposers.

Grazing in forests also exhibits a similar feedback. Moderate defoliation (< 15% foliage removal) of dwarf bamboo (*Sasa nipponica*) by deer stimulated rhizodeposition of carbon and nitrogen-rich compounds, which enhanced bacterial growth and the abundance of bacterivorous nematodes. Fungi and fungivorous nematodes did not respond to defoliation in this study (Niwa *et al.*, 2008).

Forest harvesting is a major disturbance that is expected to affect soil foodweb organisms. The removal of dominant trees alters the litter input, light intensity and temperature of the forest floor. Large trees intercept rainfall, which slowly percolates through the canopy and falls on the soil surface with far less intensity than when these trees are absent. Thus, the resources, habitat and environmental conditions for soil biota are affected by tree harvesting and other forest management practices. Salmon *et al.* (2008) studied the diversity and composition of soil invertebrate communities along a spruce forest chronosequence in the south-eastern Italian Alps. Mineralization rates were used as an indirect measurement of soil microbial community responses. One of the key findings from this study was that the site aspect had an important effect on soil properties and biological diversity (Table 12.16). The Shanon diversity index and species richness of soil invertebrates were lower in the mature stand than in the regeneration stands (intermediate succession state). There was also greater carbon mineralization in clearing and regeneration stands than in the mature stands (Table 12.16). In the clearings, this could be related to the higher temperature and resource availability for microorganisms. Regeneration stands had both a rich soil invertebrate community and high microbial activity, suggesting a positive feedback between these groups within the detrital foodweb. If the forest could be managed as a series of small stands, this would increase the spatial heterogeneity across the forests and leave patches in various states of regrowth, which would support a high level of species diversity and richness (Salmon *et al.*, 2008).

Fire

Many terrestrial biomes worldwide are subject to fire due to natural processes or anthropogenic (accidental or deliberate) intervention. Naturally occurring fires are the result of lightning strikes in areas

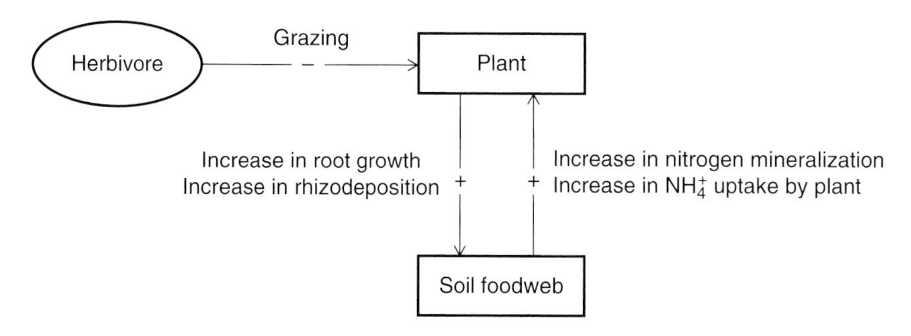

Fig. 12.10. Grazing induces a positive feedback between plants and below-ground communities. The allocation of photosynthates for root growth and exudation stimulates the growth of soil foodweb organisms in the rhizosphere, accelerating nitrogen mineralization.

Table 12.16. Soil properties, invertebrate diversity and richness, and carbon mineralization potential in a spruce forest chronosequence in the south-eastern Italian Alps (Province of Trento, Italy). (Data selected from Salmon *et al.*, 2008.)

Stand	Aspect	Organic carbon (g/kg)	Shannon diversity index (H')	Richness (zoological groups)	Carbon mineralization (mg CO_2/h/g)
Mature	South	337	0.5	5.4	0.14
Regeneration	South	134	1.1	8.6	0.25
Clearing	South	94	1.3	6.4	0.25
Mature	North	318	0.8	5.4	0.08
Regeneration	North	332	1.1	6.6	0.09
Clearing	North	174	1.4	4.6	0.17

that are sufficiently dry and have a fuel source (dead trees or grasses) to sustain the burn. When forest or rangeland managers implement policies of fire suppression for many years, the build-up of dead branches, leaves and other fuel will result in a very high-intensity burn. Yellowstone National Park in Montana, USA, experienced severe wildfires during the summer of 1988, which burned more than 360,000 ha (3213 km^2), representing about 36% of the park area. These started as several smaller fires and were exacerbated by wind and drought. Although firefighters worked all summer to bring the fires under control, it was not until after several rains in the autumn that the last fires were extinguished. Due to fire suppression since the 1960s, Yellowstone was overdue a major fire.

Recovery from the fire began within days after fires ceased, with fireweed appearing and a great number of plants regrowing from sprouts. A profusion of wildflowers was observed in burned areas of Yellowstone in the first 5 years after the fire. Aspen became abundant after the fire, while lodgepole pine declined in areas where the heat and flames had been intense. Most of the large mammals inhabiting the park escaped unharmed and recolonized the burned areas soon after the fires were extinguished. Ecologists surveying the area concluded that the Yellowstone area was specially adapted to large and intense wildfires. The ecological structure and function of the plant and animal communities is maintained naturally by periodic fires (Turner *et al.*, 2003).

In the context of ecological management, fires may either initiate succession or reset the ecosystem to an initial successional state. The effect of fire on plant communities and above-ground organisms has received considerable attention. Plant responses to fire occur fairly quickly, and affect the re-establishment of animals that return to the ecosystem after the fire has been extinguished. It is difficult to predict the responses of plant communities to fire because the temperature and the rate that fire spreads through an ecosystem are not consistent or uniform. The manner in which an ecosystem burns depends on climatic variables such as temperature, wind speed and humidity, as well as on vegetation properties such as fuel loads and the type, density and height of plants in the ecosystem. Wildfires (high-intensity, fast-spreading fires) are particularly favoured by vegetation that contains combustible terpenes, fats, oils and waxes such as chaparral vegetation. Grasses are also highly flammable due to their high surface-to-volume ratio compared with other fuels. Fire intensity is measured as the rate of heat production, which determines the temperatures experienced by organisms and their probability of surviving the fire. Intense fires are not always the most severe – a slow, smouldering fire that consumes more fuel and exposes organisms to high temperatures for a longer duration of time can cause more damage than a fire that moves quickly through the stand. In general, the change in plant community after a fire depends upon:

1. Fire intensity: affects the proportion of plant populations that are burned. In forests, a surface fire burns the understorey plants and litter layer, while a crown fire burns the canopy and a ground fire burns most things. Temperatures can range from 50°C to more than 1500°C, and the heat release can vary between 2.1×10^3 and 2.1×10^6 J/kg of fuel (Neary *et al.*, 1999).
2. Fire frequency: the mean number of fires in an area within a certain time interval. The return interval is the inverse of fire frequency. Fires tend to occur more frequently: (i) when there has been a

drought or period of low rainfall; (ii) in ecosystems that have high fuel accumulation or highly flammable fuel; and (iii) when the burns are of low intensity and leave fuel for subsequent fires. Human interference with the natural fire cycle through fire suppression programmes also affects the fire frequency.
3. Scale of the fire: how much area was affected by fire, and how large are the gaps left for colonization by surviving plants or seeds dispersed during the fire.

Grassland fires and effects on soil foodweb organisms

In grasslands, fire has long been used to remove litter or halt the spread of woody biomass. Native people inhabiting the Great Plains of North America set fires deliberately to remove dead litter and stimulate the growth of prairie grasses, flush out wildlife and to eliminate or reduce the growth of woody trees. In the tallgrass prairie, fire removes much of the above-ground vegetative cover and leads to nitrogen volatilization. Within a few days, forbs and shrubs resprout from buds below the soil surface that survived the fire. Grasses grow back from intercalary meristems and produce new tillers. Exposure of the soil surface increases soil temperature and stimulates the growth of warm-season grasses, which leads to a dense grass canopy in the first season after a burn.

Fire induces changes in the plant community that are expected to affect soil microorganisms and other foodweb organisms, namely:

1. Removal of vegetation and surface litter, which increases soil surface temperature through solar heating. Evapotranspiration is reduced in the short term, which alters the soil moisture balance.
2. Partial or complete combustion of soil organic matter near the surface and input of nutrient-rich ash. This provides a source of inorganic nutrients that can stimulate microbial growth and plant regeneration. Net immobilization of nutrients in microbial biomass after fire is essential for conserving nitrogen in ecosystems.
3. Rapid vegetative regrowth has a positive feedback on symbiotic and free-living microorganisms and higher trophic groups. Competitive pressure for some groups may be alleviated if competitors and predators perished in the fire.

The removal of above-ground biomass stimulates the growth of roots and rhizomes compared with unburned areas (Table 12.17). Plant pathogens, especially those inhabiting the litter layer and associated with shallow-rooted weeds, are killed when soil temperatures reach 50–120°C (Neary *et al.*, 1999). Soil microbial populations tend to show a short-term decline due to fire, probably because of heating and soil drying at the soil surface, but quickly recover. The recovery time is related to fire intensity and the time elapsed between burning and sampling.

Soil microfauna and mesofauna differ in their response to burning. Nematodes are among the most responsive organisms to different fire frequencies in tallgrass prairies. Plant-parasitic nematodes are more numerous after fires, probably a response to the stimulation of root growth. An ecological genomics study by Jones *et al.* (2006) showed that most microbial-feeding nematodes in the Cephalobidae and Plectidae taxa were not affected by the harsh environmental conditions associated with frequent burning, although the *Chiloplacus* sp. (Cephalobidae) may be sensitive to drier soil conditions In contrast, populations of *Oscheius* belonging to the bacterivorous Rhabditidae family were reduced significantly by burning, possibly due to a reduction in preferred food or other resources. Microarthropods and macroarthropods may increase or decrease in abundance after fire (Fig. 12.11). The most susceptible are invertebrates that are not highly mobile or that reside in the litter or surface soil horizons, which are the most vulnerable to the direct effects of intense surface fires or ground fires. Soil physico-chemical conditions after the fire, as well as changes in the plant community and other biotic

Table 12.17. Comparison of root and rhizome biomass of burned and unburned tallgrass prairie. (Adapted from Blair *et al.*, 2000.)

		Root + rhizome biomass (g/m)	
Sample depth (cm)	Study site	Burned prairie	Unburned prairie
35	Illinois	1064	839
100	Illinois	2107	1908
5	Missouri	956	669
20	Kansas	1618	1362
20	Kansas	960	838
30	Kansas	1002	790
30	Kansas	1086	859

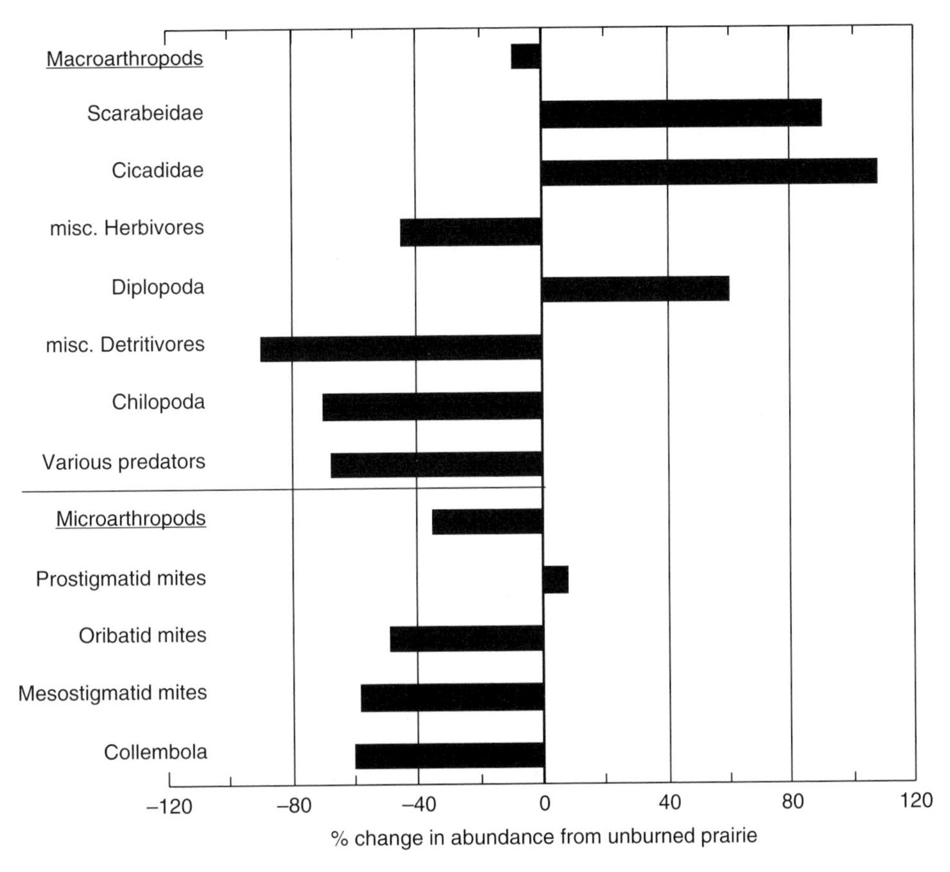

Fig. 12.11. Percentage of change in mean abundance of selected arthropod groups on burned prairie, compared with unburned prairie. (Adapted from Blair *et al.*, 2000.)

interactions, are expected to affect soil foodweb organisms. The effect of fire on below-ground soil organisms is difficult to predict, but it has been suggested that fire severity in grasslands rarely reaches a level that would have a prolonged, negative impact on the soil foodweb.

Forest fires and effects on soil foodweb organisms

Fire is an essential aspect of forest regeneration. The fire return frequency ranges from less than 10 years in the Pine Barrens of New Jersey, to 30–80 years for lodgepole pine forests in the Sierra Nevada and about 50–150 years for boreal forests in North America. The regeneration of lodgepole pine forests that dominate western North America depends on fire, because pine produces serotinous cones that are coated with a sticky resin and seal the seeds inside the cones. The optimum temperature range for melting the resin, allowing the cone to open and disperse seeds is 60–65°C, as higher temperatures reduce seed germination (Knapp and Anderson, 1980).

Fire creates gaps in the forest canopy, permitting seeds to germinate, small seedlings to grow and forest plants to regenerate from sprouts and suckers. Changes in the plant community after fire have a strong influence on plant–microbial symbiosis. Pioneer species such as black alder in deciduous forests and aspen in temperate coniferous forests are actinorhizal (form a symbiosis with the N_2-fixing actinomycetes). Intense fire that burns litter and organic soil layers can lead to a temporary decline in ectomycorrhizal fungi and arbuscular mycorrhizal fungi. Within 1 to 2 years after fire, the roots of newly sprouted vegetation and trees tend to be strongly colonized by mycorrhizal fungi (Cairney and Bastias, 2007).

Fire affects the free-living soil microorganisms and fauna of forests in a similar manner as reported in grasslands. Soil nematode populations fluctuate in the first weeks after fire, due to resource availability and changes in soil physical conditions (temperature, soil moisture content). McSorley (1993) reported that plant-parasitic nematodes were not affected or declined after fire in a pine forest. Bacterivorous nematodes of the *Acrobeles* genus increase in abundance after burning, but other bacterivorous nematodes, including individuals of *Rhabiditis* sp., were less abundant at one site. Fungivorous nematodes declined, and the numbers of omnivorous nematodes and predators increased at one of two sites, suggesting that nematode responses to fire are site specific. A census of 99 forested sites in southern Mississippi, USA, confirmed that sites with a history of frequent fire had similar soil properties and nematode communities to undisturbed sites along a 60-year chronosequence. These results indicate that above-ground disturbances due to fire probably have transient effects on soil nematode communities (Matlack, 2001).

Malmström *et al.* (2008) evaluated microarthropod populations in a boreal forest site before and after burning. The short-term effect of forest fire on soil microarthropods is related to fire- and heat-induced mortality, rather than starvation or habitat alteration. Collembola, oribatid mite and protura populations were lower 1 week after fire than pre-burn, whereas mesostigmatid mites were less affected. Within 6 months of the fire, the collembola and mesostigmatid mite populations recovered to pre-burn conditions, but oribatid mites and protura populations remained low for at least 3 years. The long-term effects of fire on these soil microarthropods are probably due to habitat modification (i.e. destruction of the litter layer and changes in the above-ground vegetation cover).

Forest fire also affects biogeochemical cycling. Burning leads to some nutrient volatilization and leaves behind nutrient-rich ash. Some of these nutrients will be retained and stabilized through chemical processes and immobilized by soil microorganisms, but nutrient losses through accelerated erosion, leaching and denitrification can also occur. The effects of fire on biogeochemical cycling in a chaparral shrubland are partially explained by fire severity (Table 12.18).

In tropical areas, the deliberate cutting and burning of forests to clear the land for agriculture and recycle nutrients from the woody biomass is referred to as 'slash and burn agriculture'. A study by Kennard and Gholz (2001) indicated that soil organic matter content was 72% lower in a burned plot (high-intensity fire) than unburned forest, and a similar quantity of organic nitrogen was probably lost from the topsoil. Burning increased the concentration of plant-available nitrogen (NH_4^+ and NO_3^-) and other essential nutrients (phosphorus, potassium, calcium and magnesium). Base cations contributed from plant ash were effective at buffering soil pH, which is an important constraint for agricultural production in the strongly weathered, acidic soils of the tropics. Soil fertility was improved

Table 12.18. Fire effects on soil organic matter, soil microorganisms and roots at the surface and in soil (5 cm depth) of chaparral, in relation to fire severity. (Adapted from Neary *et al.*, 1999.)

	Fire severity		
Parameter	Light	Moderate	High
Surface temperature (°C)	250	400	675
Soil temperature (°C)	< 50	50	75
Surface litter	Partially scorched	Mostly burned	Totally burned
Soil organic matter	Not affected	Pyrolysis begins	Pyrolysis begins
Surface roots	Dead	Dead	Dead
Soil roots	Live	Live	Dead
Surface microorganisms	Dead	Dead	Dead
Soil microorganisms	Live	Some dead	Some dead
Surface nutrient volatilization	N	N and P	N, P, K and S
Soil nutrient volatilization	None	None	None

after fire, as shown by the more rapid growth of *Anadenanthera colubrina*, a shade-intolerant tree species. Measurements after 18 months showed that seedlings of *A. colubrina* were 80–100 cm tall in burned areas, and generally less than 40 cm tall in forest gaps and clear-cut areas. There are few reports of how slash and burn agriculture affects soil biota. However, the reliance on controlled, low-intensity fires to clear excess vegetation in tropical regions is expected to have minor, short-term impacts on soil foodweb organisms.

Further Reading and Web Sites

National Sustainable Agriculture Information Service
http://attra.ncat.org/

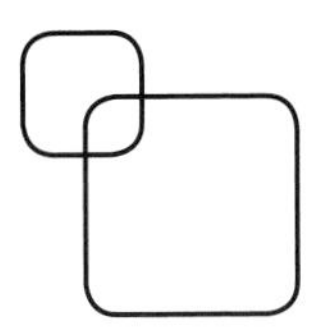

PART V

Conclusions and Future Directions

13 Sustaining the Soil Foodweb in a Changing Environment

Soil organisms are a prominent component of terrestrial ecosystems due to their ability to influence primary productivity, energy flow, nutrient cycling and soil pedogenesis. At a global scale, a diverse soil foodweb is essential for maintaining vital ecosystem functions and providing ecosystem services. Soils are the foundation to maintaining ecosystems that can support plants and animals, clean and recycle water, capture and exchange gases with the atmosphere and meet human demands for food and fibre. There is no doubt that human well-being depends on understanding and supporting ecological management decisions that preserve the rich biodiversity of soils.

Soil microorganisms and fauna have evolved to survive in a dynamic and complex environment. Within a particular ecosystem, some disturbance of the spatial fabric of the soil environment and temporal changes in moisture, temperature and other variables are generally tolerated by the soil organisms (Table 13.1). Small-scale disturbances that occur within soil aggregates or around fine roots have a minimal impact on organisms living in these niches, while a large-scale disturbance such as a fire that burns many square kilometres of forest will impact soil organisms living in a broad geographical region that may comprise several soil types with varying topography and land use.

Few people would dispute that we are in the midst of a period of unprecedented global change. The exponential rise in human population, coupled with more intense fossil fuel use and other environmental impacts, is causing dramatic and rapid changes at a global scale (Table 13.2). Soil organisms are not immune to the effects of global change – even if there is no direct effect on the soil microorganisms and animals of a particular environmental stress, the biota could be impacted through feedback from terrestrial plant and animal communities within the ecosystem (Fig. 13.1).

The consequences of global changes on soil biota are not entirely known or understood. Further research is needed to determine how the soil foodweb responds to changing environmental conditions, particularly factors that could cause biological stress. Some relevant questions might be:

- Do the soil organisms living in a particular ecosystem have sufficient adaptive plasticity in their genome and life history to tolerate more stressful conditions such as infrequent rainfall or hotter soil temperatures?
- Will carbon dioxide and nutrient enrichment make terrestrial plants less or more dependent on microbial symbionts?
- How do soil animals allocate energy and what are the reproductive trade-offs in a changing environment?
- Is loss of diversity in the soil foodweb an inevitable consequence of global changes?

The answers to these and other questions will provide insight into the ability of the soil foodweb to persist in a changing environment. Gene expression and biochemical reaction rates may be early indicators of organisms under stress. The emerging field of functional genomics is a promising tool for detecting individuals, populations and communities that are sensitive to environmental changes. This could lead to the selection of indicator species or ecological groups for further monitoring, to determine whether certain ecosystems are at risk of losing biodiversity or having their ecological functions impaired.

This chapter provides an overview of how large-scale environmental change may alter terrestrial ecosystems irrevocably, and the potential effects on the diversity and functional capabilities of the soil foodweb.

13.1 Climate Change

Any long-term, significant change in the expected weather pattern for a region is described as climate change. Some variation in weather from year to

©CAB International 2010. *Soil Ecology and Management* (J.K. Whalen and L. Sampedro)

Table 13.1. Temporal variation in environmental factors that affect the abundance and activity of biota in the soil foodweb, and the impact at different spatial scales.

Temporal scale	Environmental factors affecting the soil foodweb
Hours	Water percolation through soils, root exudation, input and depletion of organic substrates and nutrients
Days	Diurnal fluxes in temperature and soil water (evapotranspiration, dew, rainfall/irrigation)
Months/years	Weather pattern, seasonal fluctuations in climate, freezing/thawing, wetting/drying, detrital inputs, management practices (more frequent in agricultural systems than in natural ecosystems)
Decades/centuries	Climate change, large-scale disturbances (wildfires, floods, hurricanes, volcanic eruptions)
Spatial scale	**Extent of disturbance of the soil foodweb**
Micro- to millimetres	Soil aggregates, fine roots
Milli- to centimetres	Macropores, earthworm burrows
Centimetres to metres	Fungal hyphal networks, root architecture and rooting depth, soil profile
Metres to kilometres	Large-scale changes across soil series, topographical locations and different land use/ management

Table 13.2. Examples of human activities and environmental impacts that contribute to global change.

Human activity	Environmental impact(s)
Fossil fuel combustion	↑ carbon dioxide (CO_2) in atmosphere, global warming due to greenhouse effect
Land use change from grassland/forests to agricultural use	↑ carbon dioxide (CO_2) in atmosphere, global warming due to greenhouse effect; ↓ biodiversity, especially due to habitat loss
Fossil fuel combustion, industrial activities	Nutrient enrichment of terrestrial and aquatic systems due to ↑ nitrogen and sulfur deposition
Agricultural production	↑ nitrous oxide (N_2O) in atmosphere from nitrogen fertilizer and manure, global warming
Intensive livestock production	Nutrient enrichment of aquatic systems due to ↑ phosphorus application and transport from agricultural land; ↑ methane (CH_4) generation from animals and manure storage facilities, global warming
Natural resource extraction	↑ trace metals in the environment
Organic synthesis	↑ novel chemical substances in the environment
Import of non-native species	Biological invasions can reduce the diversity of native taxa through competition, predation or habitat modification
Urbanization	Change in land use and hydrological cycle, waste management, pollution from transportation and industrial activities

year is normal, but there seems to be a trend of global warming, supported by disappearance of ice caps and melting of glaciers. The World Glacier Monitoring Service, which reports on glacier retreat and glacier mass balance, reports that strong glacier retreats have occurred from the mid-1980s to present. There has been a net loss in glacier ice mass every year since 1985 (Zemp *et al.*, 2009). At the same time, the Earth's temperature has risen at a rate of about 0.2°C per decade. The IPCC (2007) predicts that the average surface temperature could increase by 1.1–6.4°C by the end of the 21st century. The increase in surface temperature also influences precipitation and storm patterns, and will probably lead to an increase in sea levels.

The magnitude of climate change during the next century will be affected by a number of factors, and many questions remain to be answered:

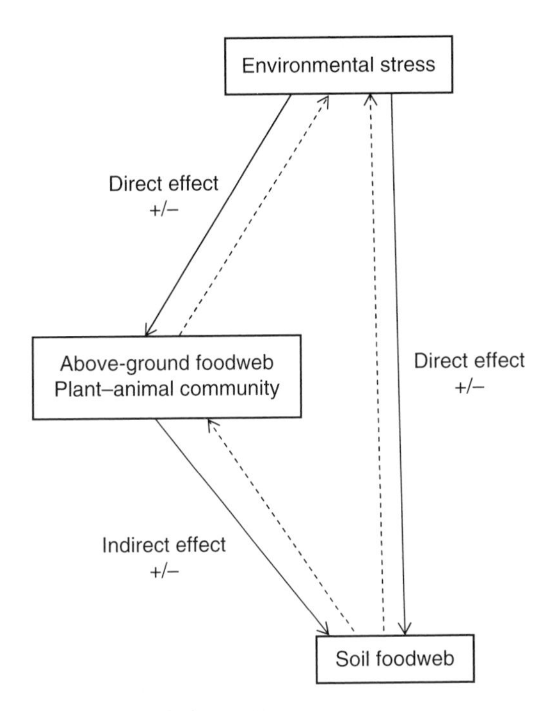

Fig. 13.1. Effects of environmental stress on terrestrial ecosystems, which may affect soil organisms directly or indirectly through the above-ground foodweb (terrestrial plant and animal communities). The dotted lines indicate feedbacks between the compartments. For example, an increase in the atmospheric CO_2 concentration could stimulate plant growth (+ direct effect), which releases more root exudates that benefit soil microbial growth (+ indirect effect). Respiration from plants, animals and the soil foodweb would release CO_2 back to the atmosphere (dotted lines from the above-ground foodweb and soil foodweb to the environmental stress compartment).

- Will greenhouse gas concentrations in the atmosphere increase, stay the same or decrease?
- How strongly will weather patterns (e.g. temperature fluctuations, the amount and intensity of precipitation events) respond to changes in greenhouse gas concentrations?
- Are there other natural phenomena (e.g. volcanic eruptions, changes in the sun's intensity) and internal variability within the planet (e.g. random changes in the circulation of the atmosphere and oceans) that will affect the Earth's climate?

Even if the emissions of greenhouse gases (summarized in Table 13.3) were reduced dramatically to steady-state conditions, it is predicted that the Earth's surface temperature would continue to warm because of heat stored in the oceans and other positive feedbacks (IPCC, 2007). For example, more solar radiation is absorbed by sea water than by ice, and there is a decline in ice covering polar and alpine regions that could reflect this solar radiation.

Soil organisms respond to changes in temperature and soil water content, so they will probably be affected by climate change. This can be tested experimentally by measuring the populations at field sites where the temperature has risen over a number of years or in the laboratory under precisely controlled temperature and moisture conditions. Many researchers rely on mathematical models to extrapolate results from a few sites or laboratory experiments to broader geographical areas. Mathematical models are also helpful in

Table 13.3. Greenhouse gases and their potential for trapping radiant energy in the Earth's stratosphere, expressed as the global warming potential compared with CO_2 over a 100-year period (IPCC, 2007).

Compound(s)	Global warming potential, compared with CO_2	Comments
Carbon dioxide (CO_2)	1	Most abundant greenhouse gas, mostly from fossil fuel combustion
Methane (CH_4)	25	From anaerobic environments: swamps and ruminant animals' digestive systems
Nitrous oxide (N_2O)	298	70% comes from agriculture
Sulfur hexafluoride (SF_6)	22,200	Fire suppressant, inert gas
Hydrofluorocarbons (CFCs)	100–12,200	Production stopped in 1994, stocks to be eliminated by 2020
Perfluorocarbons (PFCs)	5,800–12,000	Use in medicine and coolant systems is closely regulated

testing hypotheses about decomposition, nitrogen mineralization/immobilization and other soil processes. They can also be used to make predictions and provide estimates about soil processes controlled by the soil biota. A simple example is illustrated in Fig. 13.2. Imagine that an organism (soil biota) is responsible for converting cellulose (state 1) into biomass and release CO_2 through respiration (state 2). Imagine also that with global change, the environment is more favourable for the organism and the reaction occurs more efficiently, i.e. the biochemical reaction produces more CO_2 during a period of time (scenario a). Another possibility is that several physiological processes in the organism are affected, in addition to the biochemical reaction that produces CO_2 (scenario b). In this case, the CO_2 production may be slightly, moderately or highly stimulated by global change.

Nitrogen deposition

In addition to greenhouse gases, the Earth's atmosphere has become enriched with other gaseous substances. Nitrogen oxides (NO_x) and ammonia (NH_3) emitted from fossil fuel combustion, industrial activities and agricultural operations enter the atmosphere and are sometimes transported over long distances before they are deposited in natural ecosystems that would not normally receive fertilizer. While most of the concern with nitrogen enrichment focuses on terrestrial ecosystems, impacts may also be seen in freshwater and marine ecosystems (nitrogen-induced eutrophication).

Nitrogen oxides are produced from high-temperature combustion. In the atmosphere, NO_x is transformed to a range of secondary pollutants,

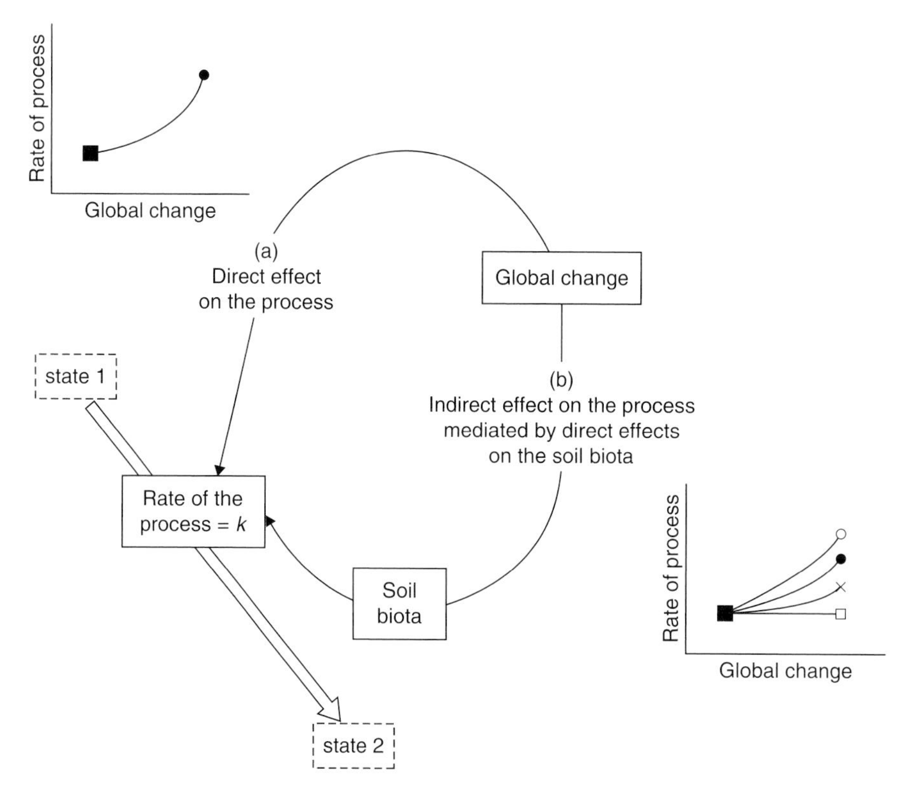

Fig. 13.2. Schematic diagram of a soil process converting a component from state 1 to state 2 at rate *k*. It is assumed that the process rate is mediated by the soil biota present. It is postulated that global change may affect the process rate directly (scenario a) or through an effect on soil community function (scenario b). If global change acts only through scenario a, the future process rate should be predictable as long as the change is within the limits of model calibration. If global change also acts through scenario b, and the mediating effect of the soil biota on the process rate is altered, then the future process rate may not be predictable. (Adapted from Smith *et al.*, 1998.)

including nitric acid (HNO_3), nitrates (NO_3^-) and organic nitrogen compounds such as peroxyacetyle nitrate. Ammonia is emitted from vehicles and livestock production (manure storages and land spreading), transported to the atmosphere and converted to ammonium (NH_4^+). The primary or secondary pollutants leave the atmosphere through wet deposition (with precipitation) or by dry deposition (nitrogen-rich particulate materials). Consequences of nitrogen deposition are unintentional fertilization and acidification of natural ecosystems. Soil acidification is associated with the leaching of base cations, in association with NO_3^-, through the profile. Nitrate is susceptible to leaching due to its small molecular size and negative charge; it attracts positively charged cations as it leaches through the profile. In moderately acidic soils (pH > 4.0), soil bacteria transform NH_4^+ to NO_3^-.

Nitrogen deposition in terrestrial ecosystems affects plant community dynamics. Bryophytes are sensitive to excessive nitrogen inputs. Slow-growing plants are susceptible to competition from invasive grasses, which benefit from nitrogen fertilization. The biodiversity of low-alpine *Calluna–Caldonia* heathlands was reduced by nitrogen deposition, with *Calluna* declining with nitrogen loads of 5–10 kg N/ha/year (Helliwell *et al.*, 2008). An increase in plant foliar N concentration increases the risk of damage from pests and pathogens both above and below ground. The absence of some ectomycorrhizal species in tree roots is associated with high net nitrogen mineralization rates, suggesting that nitrogen deposition might also trigger changes in the mycorrhizal community of forest soils (Parrent *et al.*, 2006). The unintentional fertilization of ecosystems with excess nitrogen also affects decomposition, as discussed for peatlands in Focus Box 13.1.

Focus Box 13.1. Global changes and the conservation of peat bogs.

Peatlands as long-term carbon sinks

Peatlands are extensive sinks of atmospheric carbon, meaning that more carbon is retained than emitted from these ecosystems. Peatlands cover approximately 3% of the land surface and contain about 20–30% of the global soil carbon. The long-term accumulation and sequestration of organic carbon as peat derives from the extremely slow decomposition of decaying detritus in peatlands. Several factors are thought to be responsible for slow decomposition, including low oxygen availability due to frequent waterlogging, low soil temperature, acidic pH conditions, low nutrient and high polyphenol concentrations in plant litter. All of these factors reduce microbial activities that lead to decomposition.

Upland grasslands and northern peatlands have been long-term sinks for atmospheric CO_2 since the last Ice Age. Recent research showed that temperate northern soil carbon stocks are declining by 2% per year, due to increased C mineralization and loss of soil organic carbon (Bellamy *et al.*, 2005). Some factors other than global warming have been proposed as explanatory factors for the observed trend in soil organic carbon content. Changes in land use and agricultural practices such as animal manure application, deeper ploughing depths and less litter input due to the harvest of agricultural products are thought to be important (Smith *et al.*, 2007). But the fact is that in the current context of global change, peatlands could be an ecosystem close to extinction. Besides their relevance as huge global carbon stocks, they harbour great microbial, faunal and plant biodiversity.

This case study demonstrates the importance of mathematical models describing global warming, nitrogen deposition and rain patterns to understand the current trends in peatland carbon dynamics. The accuracy of our predictions about the behaviour of soil carbon stocks over a wide geographical area depends greatly on our knowledge of the feedbacks between temperature, soil biota and recalcitrant-carbon respiration. Another important point is that biochemical constraints affecting a single extracellular enzyme could have a major control on the decomposition pathway, with ecosystem-level consequences.

Extracellular enzymes, phenolics and waterlogging

The perceptive research of Freeman *et al.* (2001a) showed that a high concentration of polyphenolic compounds is actually limiting the activity of several extracellular enzymes involved in recalcitrant organic matter breakdown in peatlands. However, according to these authors, the major limiting factor of soil respiration in peatlands is the anaerobic conditions due to waterlogging, which prevent the enzyme phenol oxidase from depleting the phenolic compounds that inhibit biodegradation.

Continued

Focus Box 13.1. *Continued*

Under experimental conditions, they found that moderate aerobic conditions boosted phenol oxidase activity by sevenfold in comparison with waterlogged conditions, effectively reducing the concentrations of phenolics in the soil solution, and so increasing the activity of accompanying hydrolases.

They suggested that increased peat aeration as a result of drought (due to a global change in rainfall patterns) could switch on a critical mechanism, leading to the release of the buried organic carbon to the atmosphere. This is an example of a positive feedback constrained by a single extracellular enzyme. Such a positive feedback on soil carbon respiration could have profound implications in the context of global change.

Nitrogen deposition

As atmospheric deposition is the major source of nutrients for peatlands, Bragazza *et al.* (2006) hypothesized that the global increase of atmospheric nitrogen fertilization could alter the litter biochemistry of peats. Analysing the biochemistry of litter decomposition in nine peatlands along a gradient of increased atmospheric N deposition, they showed that decomposition rates, and thus CO_2 emission and dissolved organic carbon release, were significantly enhanced by N supply derived from human activity (Fig. 13.3).

DOC exportation

Another study by Freeman *et al.* (2001b) reported that dissolved organic carbon concentration in fresh waters draining from uplands and peatlands in the UK (11 lakes and 11 streams) increased by 65% in 12 years (1988–2000). They speculated that rising temperatures, changing rainfall patterns and other human-induced environmental changes could be responsible for an increase in the exportation of formerly buried organic terrestrial carbon to the oceans.

Feedbacks between temperature, soil fauna, DOC and recalcitrant-carbon respiration

Besides altered litter biochemistry, interactions due to the soil fauna could also have a relevant role in releasing ancient preserved soil organic carbon. Several recent papers (Briones *et al.*, 2004, 2007) have shown that enchytraeid populations respond to small increases in soil temperature increasing biomass and activity, resulting in greater dissolved organic carbon concentration in the soil solution, greater soil respiration and the simultaneous exploitation of 'older' soil carbon than at lower temperatures. Thus, the interaction of soil biota with temperature and litter chemistry could be a major factor in determining the sensitivity of different soil organic carbon pools to rising global temperature.

If atmospheric temperatures continue to rise and precipitation increases in the summer months across the northern hemisphere as some models predict, Briones *et al.* (2007) predict that invertebrate populations will increase in size and activity. This could lead to greater consumption and processing of organic carbon, and progressive respiration of old, previously unused soil carbon to the atmosphere.

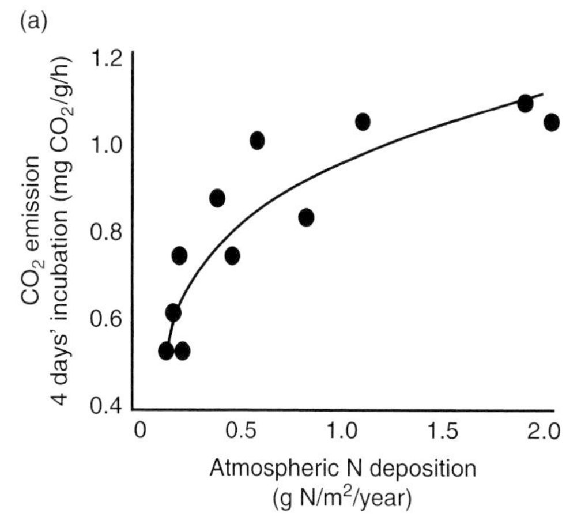

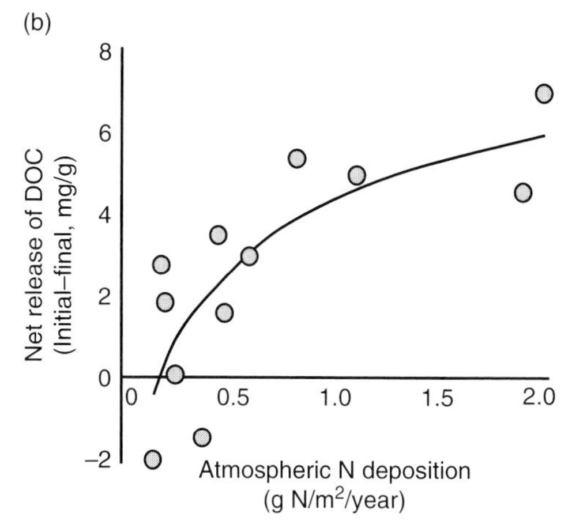

Fig. 13.3. CO_2 emission from litter peat samples after 4 days of incubation (a) and dissolved organic carbon (DOC) release from the peat samples (b), in relation to atmospheric N deposition in 12 peat bogs. Both relationships were explained by logarithmic regressions. Each dot is the mean value of three to six peat samples. (Modified from Bragazza *et al.*, 2006.)

Mitigation strategies

Despite continuous improvements in energy efficiency and public awareness of the need for energy conservation, global energy use and supply are projected to continue to grow, especially as developing countries pursue industrialization. If there is no change in energy policies, it is projected that more than 80% of the energy supply will continue to be based on fossil fuels. As a consequence, greenhouse gas emissions will continue to rise, exacerbating the ongoing effects of climate change.

The IPCC (2007) recognizes that no single sector or technology can address the entire mitigation challenge, so a concerted effort will be required (Table 13.4). The soil foodweb has a role in greenhouse gas mitigation efforts. Soil organisms such as plant growth-promoting rhizobacteria can be harnessed to improve the growth of energy crops with lower external nutrient inputs. Symbiotic and free-living microorganisms living in the rhizosphere could be beneficial in this respect (also see Focus Box 13.2). The process of carbon sequestration relies on the conversion of organic residues into stable soil carbon by microorganisms and fauna. Additional information on the stability and longevity of carbon-rich organo-mineral complexes generated by these organisms could help achieve target soil organic carbon levels. A reduction in nitrous oxide from agricultural soils could be achieved by manipulating the communities of nitrifying and denitrifying microorganisms, which may include bacteria, fungi and archaea. Soil ecology research is extremely relevant and fundamental to the problem of mitigating global climate change.

Table 13.4. Key mitigation technologies and practices by sector. (Adapted from IPCC, 2007.)

Sector	Key mitigation technologies and practices currently available
Energy supply	Improved supply and distribution efficiency; fuel switching from coal to gas; nuclear power; renewable heat and power (hydropower, solar, wind, geothermal and bioenergy); combined heat and power; early applications of carbon capture and storage (e.g. storage of removed CO_2 from natural gas)
Transport	More fuel-efficient vehicles; hybrid vehicles; cleaner diesel vehicles; biofuels; modal shifts from road transport to rail and public transport systems; non-motorized transport (cycling, walking); land-use and transport planning
Buildings	Efficient lighting; more efficient electrical appliances, heating and cooling devices; improved cooking stoves; improved insulation; passive and active solar design for heating and cooling; alternative refrigeration fluids; recovery and recycling of fluorinated gases
Industry	More efficient electrical equipment; heat and power recovery; material recycling and substitution; control of non-CO_2 gas emissions; a wide array of process-specific technologies
Agriculture	Improved crop and grazing land management to increase soil carbon storage; restoration of cultivated peaty soils and degraded lands; improved rice cultivation techniques, livestock and manure management to reduce CH_4 emissions; improved nitrogen fertilizer application techniques to reduce N_2O emissions; dedicated energy crops to replace fossil fuel use; improved energy efficiency
Forestry/forests	Afforestation; reforestation; forest management; reduced deforestation; harvested wood product management; use of forestry products for bioenergy to replace fossil fuel use
Waste	Landfill methane recovery; waste incineration with energy recovery; composting of organic waste; controlled wastewater treatment; recycling and waste reduction

Focus Box 13.2. Biomass energy crops and soil biota.

Biomass energy crops are plants that are burned to produce heat for homes and industries or transformed into biofuel (ethanol/biodiesel) before they are combusted in automobiles, homes and industries. Biomass energy crops are considered by many to be a 'green' energy source because they recycle the CO_2 recently fixed through photosynthesis, whereas fossil fuels release CO_2 that has been stored for millennia.

Continued

Focus Box 13.2. *Continued*

Biofuels are categorized as first-generation and second-generation products. The first-generation biofuels come from grain crops that are high in oil or starch (Fig. 13.4). Oilseed crops such as sunflower undergo transesterification, which produces biodiesel that can then be mixed with the diesel oil used in buses, cars, trucks and trains and to heat homes (furnace oil). In starchy crops such as maize, enzymes are used to catalyse the degradation of starch into sugars, which are then fermented to produce ethanol that can be mixed with gasoline (Fig. 13.4a). Gasoline containing 10% to 20% ethanol can be burned in vehicles without engine modification. Certain vehicles have engines that can burn 85% ethanol and 15% gasoline (USDOE, 2008a).

The second-generation biofuels are derived from non-food parts (ligno-cellulosic residues) such as straw, leaves and tree branches. It is more complicated to release the energy contained in these plant parts. One approach is thermochemical (Fig. 13.4b). The residue is placed in a furnace at > 700°C and combusted to produce a carbon-rich material called char (represented below as 'C'), which undergoes the following reactions:

$$\text{Combustion: } C + 0.5\, O_2 \rightarrow CO + \text{energy} \quad [13.1]$$

$$\text{Gasification: } C + H_2O \rightarrow H_2 + CO \quad [13.2]$$

The energy comes from the combustion reaction, while gasification produces a mixture of CO and H_2 gases called syngas. The syngas can be combusted or further processed by the chemical industry to produce biodiesel (USDOE, 2008b).

The biochemical approach to transform ligno-cellulosic residues into ethanol is very complicated (Fig. 13.4b). In nature, at least three enzymes are necessary for the hydrolysis reaction. Research is still under way to choose bacteria and fungi that will produce large quantities of these enzymes for full-scale, commercial production (USDOE, 2008b).

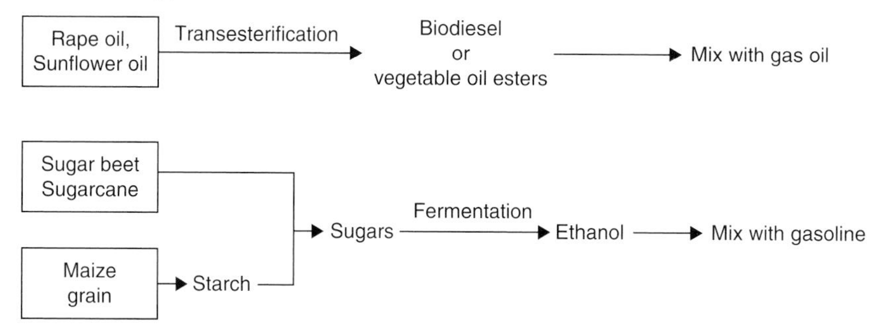

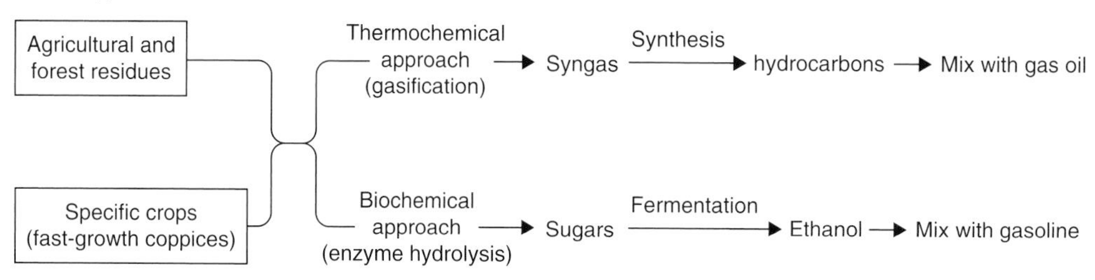

Fig. 13.4. (a) First-generation biofuels are transformed into biodiesel through a chemical transesterification process or biochemically degraded from starch to yield sugars, which are then fermented to produce ethanol. (b) Second-generation biofuels are ligno-cellulosic residues that can undergo gasification to produce hydrocarbons. They can also be subjected to enzymatic degradation, which yields sugars that can be fermented to produce ethanol. (Modified from: IFP – Innovation, Energy, Environment (http://www.ifp.com, accessed on 14 March 2009).)

Continued

Focus Box 13.2. *Continued*

Biofuels have been accepted as an appropriate substitute for fossil fuels and many countries are already passing legislation that will compel manufacturers to add ethanol or biodiesel to fuel sources. There remain many questions about the environmental, economic and energetic costs and benefits of biofuels. Hill *et al.* (2006) wrote that a biofuel should have clear environmental benefits over fossil fuels, be economically competitive and be produced in sufficient quantities to have a meaningful impact on energy demands. Ideally, there would be a net energy gain during the life cycle of the biofuel production and use (Fig. 13.5).

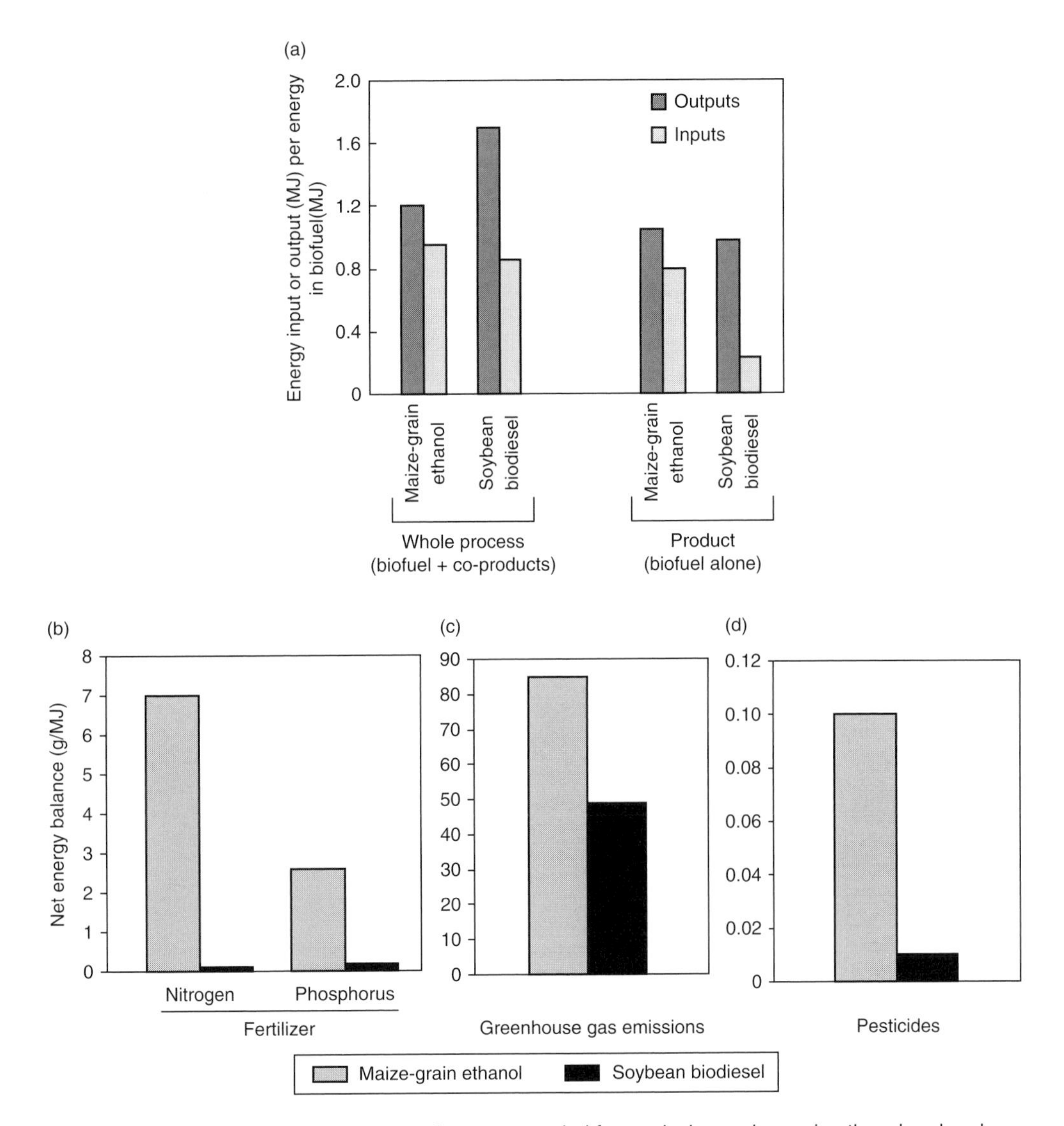

Fig. 13.5. (a) Summary of energy inputs and outputs needed for producing maize-grain ethanol and soybean biodiesel. The advantage of biodiesel over ethanol comes from lower agricultural inputs and more efficient conversion of feedstocks to fuel. Environmental costs associated with the production of maize-grain ethanol and soybean biodiesel include: (b) fertilizer, (c) greenhouse gas emissions and (d) pesticides. (Modified from Hill *et al.*, 2006.)

Continued

Focus Box 13.2. *Continued*

After considering the current industrial efficiencies and practices, Hill *et al.* (2006) found that ethanol yields 25% more energy than the energy used to produce this material, whereas biodiesel yields 93% more. In terms of collateral environmental effects, they report that biodiesel releases just 1.0, 8.3 and 13% of the agricultural nitrogen, phosphorus and pesticide pollutants, respectively, per net energy gain. Relative to the fossil fuels they displace, greenhouse gas emissions are reduced 12% by the production and combustion of ethanol and 41% by biodiesel. Biodiesel also releases less air pollutants per net energy gain than ethanol. It should be noted that the advantage of biodiesel over ethanol comes from lower agricultural inputs and more efficient conversion of feedstocks to fuel, and of fuel to energy.

The study of Hill *et al.* (2006) was based on the net energy yield and environmental impacts of first-generation biofuels. There are two major problems with converting oilseed and starchy grains into biofuel. First, these crops are food crops and it is not ethical to convert food into biofuel, given that a considerable portion of the world's population is malnourished or barely has the means to purchase food at the present time (FAO, 2008). Reducing the food supply would drive up the price and cause unnecessary hardship to millions of people. Secondly, there is not enough food produced to meet the demand for biofuel. Only 12% of gasoline demand and 6% of diesel demand could be met by transforming all of the maize and soybean produced in the USA (Hill *et al.*, 2006).

Second-generation biofuels produced from non-food crops such as switchgrass, reed canarygrass, miscanthus, willow and hybrid poplars are seen as a better alternative to meet future energy. However, it is not clear where these plants should be grown. Since the most productive agricultural land is needed for food production, the marginal lands could be transformed for the production of biomass energy crops. This could be appropriate on marginal lands that were cleared for agricultural use but are not farmed intensively due to low soil fertility and other production constraints. Low-productivity grasslands and forests could also be considered marginal lands. Indeed, converting some of these lands to produce short-rotation woody or herbaceous biomass crops contributes to sustainable land use. Mann and Tolbert (2000) compiled data from several European countries to show that such perennial crops provide wildlife habitat, reduce nitrate pollution of water and act as a potential carbon sink. For example, willow is an excellent filter crop that sequesters metal contaminants from wastewater. Plant biomass is a carbon sink and contributes residues (roots, leaf litter, branches, etc.) that become stabilized in the soil organic carbon pool, thus contributing to carbon sequestration.

The environmental benefits of second-generation biofuels may not be as apparent in heavily populated areas such as South America and East Asia where land use pressure is high. There is a risk that the economic benefits from biofuel production may lead some producers in these areas to convert some 'unproductive' lands, such as tropical forests, flooded forests and wetlands and other valuable, undisturbed ecosystems. Due to the potential loss of ecosystem services resulting from such activities, clear policies and incentives need to be put in place so that these areas are properly valued and protected.

A major challenge in producing second-generation biofuels is to achieve a high yield with minimal inputs so the net energy gain is as favourable as possible. Organic fertilizers such as animal manure and green manure are preferred to inorganic nitrogen fertilizers. Biofertilizers, which are essentially formulations of rhizosphere-compatible microorganisms that promote nutrient recycling and crop growth, are being investigated by many research groups. Arbuscular mycorrhizae hold promise for enhancing the uptake of nitrogen, phosphorus and other essential nutrients while also improving plant–water relations. Bredja *et al.* (1998) reported that arbuscular mycorrhizal fungi and associated rhizosphere microflora isolated from switchgrass stands were highly effective in enhancing seedling growth and nutrient uptake of switchgrass under greenhouse conditions. Replicating these results under field conditions remains difficult due to abiotic factors such as former soil use, climate and soil properties and biotic interactions, including competition from indigenous soil microorganisms and predation by higher trophic groups in the soil foodweb.

Scientists and policy makers continue of work on questions related to the appropriate method of incorporating biomass energy crops into our current land-use framework. Clearly, there are socio-economic and environmental considerations that must be addressed if we hope to harvest energy-rich crops to attain our global objective of reducing greenhouse gas emissions and fossil fuel combustion while promoting sustainable land use.

13.2 Conclusions

Our concept and understanding of soil ecology has expanded remarkably in the past 50 years. The ready accessibility of molecular tools and isotopic tracers is allowing us to probe more deeply into the hidden soil microbial world and make discoveries about the dynamic and complex relations within the soil foodweb. The recent discovery of non-culturable soil archaea that may be important in nitrogen cycling reveals that we still have much to learn about soil diversity and functioning. Further work is needed on how these and other organisms can provide benefits to humans, whether it be by improving food production, by purifying water, by storing carbon and buffering climate change, by degrading potentially harmful pollutants or by supplying new antimicrobial compounds for use in medicine.

It is in our best interest to preserve soil resources and manage them to the best of our ability to protect their physical, chemical and biological integrity. In doing so, we ensure that the soil will continue to supply the ecological services that are essential for our survival on this planet. Still much remains to be learned about the dynamics of soil systems, but the emergence of scientific tools such as GIS and remote sensing that permit us to visualize the landscape and changes due to land use, soil properties, erosion, desertification, etc. can be combined with dynamic predictive models that can estimate the rates at which past and future changes in the terrestrial ecosystem are occurring. Such information provides guidance on how soil ecology should be managed in the future to sustain life on Earth.

Further Reading and Web Sites

United States Department of Agriculture – Farm Service Agency (USDA-FSA). Conservation Reserve Program http://www.fsa.usda.gov/FSA/webapp?area=home&subject=copr&topic=crp/

World Glacier Monitoring Service http://www.geo.uzh.ch/wgms/

References

Abe, T., Bignell, D.E. and Higsahi, H. (2000) *Termites: Evolution, Sociality, Symbioses, Ecology.* Kluwer Academic Publishers, Dordrecht, The Netherlands, 488 pp.

Abiven, S., Menasseri, S. and Chenu, C. (2009) The effects of organic inputs over time on soil aggregate stability – a literature analysis. *Soil Biology and Biochemistry* 41, 1–12.

Adamczyk, B., Kitunen, V. and Smolander, A. (2008) Protein precipitation by tannins in soil organic horizon and vegetation in relation to tree species. *Biology and Fertility of Soils* 45, 55–64.

Adams, J.D.W. and Frostick, L.E. (2008) Analysis of bacterial activity, biomass and diversity during windrow composting. *Waste Management* 29, 598–605.

Adl, M.S., Simpson, A.G.B., Farmer, M.A., Andersen, R.A., Anderson, O.R., Barta, J.R. *et al.* (2005) The new classification of eukaryotes with emphasis on the taxonomy of protists. *Journal of Eukaryotic Microbiology* 52, 399–451.

Aira M., Dominguez, J., Monroy, F. and Velando, A. (2007) Stress promotes changes in resource allocation to growth and reproduction in a simultaneous hermaphrodite with indeterminate growth. *Biological Journal of the Linnean Society* 91, 593–600.

Al-Deeb, M.A., Wilde, G.E., Blair, J.M. and Todd, T.C. (2003) Effect of Bt corn for corn rootworm control on nontarget soil microarthopods and nematodes. *Environmental Entomology* 32, 859–865.

Allen, M.F. (2007) Mycorrhizal fungi: highways for water and nutrients in arid soils. *Vadose Zone Journal* 6, 291–297.

Altieri, M.A. (1999) The ecological role of biodiversity in agroecosystems. *Agriculture, Ecosystems and Environment* 74, 19–31.

Anderson, J.M. (2000) Food web functioning and ecosystem processes: problems and perceptions of scaling. In: Coleman, D.C. and Hendrix, P.F. (eds) *Invertebrates as Webmasters in Ecosystems.* CAB International, Wallingford, UK, pp. 3–24.

Angers, D.A., Recous, S. and Aita, C. (1997) Fate of carbon and nitrogen in water-stable aggregates during decomposition of $^{13}C^{15}N$-labelled wheat straw in situ. *European Journal of Soil Science* 48, 295–300.

Arias, T. (2004) Arquitectura de raíces y vástagos de *Vismia baccifera* y de raíces de *Vismia macrophylla* (Clusiaceae). *Caldasia* 26, 333–358.

Ashraf, R., Shahid, F. and Ali, T.A. (2007) Association of fungi, bacteria and actinomycetes with different composts. *Pakistan Journal of Botany* 39, 2141–2151.

Attwood, S.J., Maron, M., House, A.P.N. and Zammit, C. (2008) Do arthropod assemblages display globally consistent responses to intensified agricultural land use and management? *Global Ecology and Biogeography* 17, 585–599.

Ayres, P.G. (2004) Alexis Millardet: France's forgotten mycologist. *Mycologist* 18, 23–26.

Badenes-Perez, F.R., Nault, B.A. and Shelton, A.M. (2005) Manipulating the attractiveness and suitability of hosts for Diamondback moth (Lepidoptera: Plutellidae). *Journal of Economic Entomology* 98, 836–844.

Balestrini, R. and Lanfranco, L. (2006) Fungal and plant gene expression in arbuscular mycorrhizal symbiosis. *Mycorrhiza* 16, 509–524.

Barbercheck, M.E., Neher, D.A., Anas, O., El-Allaf, S.M. and Weicht, T.R. (2009) Response of soil invertebrates to disturbance across three resource regions in North Carolina. *Environmental Monitoring and Assessment* 152, 283–298.

Bardgett, R.D. and Wardle, D.A. (2003) Herbivore mediated linkages between above-ground and below-ground communities. *Ecology* 84, 2258–2268.

Bartelink, H.H. (1998) A model of dry matter partitioning in trees. *Tree Physiology* 18, 91–101.

Bellamy, P.H., Loveland, P.J., Bradley, R.I., Lark, R.M. and Kirk, G.J.D. (2005) Carbon losses from all soils across England and Wales 1978–2003. *Nature* 437, 245–248.

Benton Jones J., Jr (2003) *Agronomic Handbook: Management of Crops, Soils and Their Fertility.* CRC Press, Boca Raton, Florida.

Bi, H.H., Song, Y.Y. and Zeng, R.S. (2007) Biochemical and molecular responses of host plants to mycorrhizal infection and their roles in plant defence. *Allelopathy Journal* 20, 15–27.

Biederbeck, V.O., Campbell, C.A., Ukrainetz, H., Curtin, D. and Bouman, O.T. (1996) Soil microbial and biochemical properties after ten years of fertilization with urea and anhydrous ammonia. *Canadian Journal of Soil Science* 76, 7–14.

Birkhofer, K., Bezemer, T.M., Bloem, J., Bonkowski, M., Christensen, S., Dubois, D. *et al.* (2008) Long-term organic farming fosters below and aboveground biota: implications for soil quality, biological control and productivity. *Soil Biology and Biochemistry* 40, 2297–2308.

Blair, J.M., Todd, T.C. and Callaham, M.A., Jr (2000) Responses of grassland soil invertebrates to natural and anthropogenic disturbances. In: Coleman, D.C. and Hendrix, P.F. (eds) *Invertebrates as Webmasters in Ecosystems*. CAB International, Wallingford, UK, pp. 43–71.

Blanchart, E., Lavelle, P., Bradueau, E., Le Bissonnais, Y. and Valentin, C. (1997) Regulation of soil structure by geophagous earthworm activities in humid savannas of Cote d'Ivoire. *Soil Biology and Biochemistry* 29, 431–439.

Blouin, M., Zuily-Fodil, Y., Pham-Thi, A.T., Laffray, D., Reversat, G., Pando, A. *et al.* (2005) Belowground organism activities affect plant aboveground phenotype, inducing plant tolerance to parasites. *Ecology Letters* 8, 202–208.

Bonanomi, G., Antignani, V., Pane, C. and Scala, F. (2007) Suppression of soilborne fungal diseases with organic amendments. *Journal of Plant Pathology* 89, 311–324.

Bossio, D.A., Scow, K.M., Gunapala, N. and Graham, K.J. (1998) Determinants of soil microbial communities: effects of agricultural management, season, and soil type on phospholipid fatty acid profiles. *Microbial Ecology* 36, 1–12.

Boström, U. (1995) Earthworm populations (Lumbricidae) in ploughed and undisturbed leys. *Soil and Tillage Research* 35, 125–133.

Boulter, J.I., Trevors, J.T. and Boland, G.J. (2002) Microbial studies of compost: bacterial identification, and their potential for turfgrass pathogen suppression. *World Journal of Microbiology and Biotechnology* 18, 661–671.

Bourne, J.M., Kerry, B.R. and Leij, F.A. (1996) The importance of the host plant on the interaction between root-knot nematodes (*Meloidogyne* spp.) and the nematophagous fungus, *Verticillium chlamydosporium*. *Biocontrol Science and Technology* 6, 539–548.

Bower, K.N., Berkett, L.P. and Costante, J.F. (1995) Nontarget effect of a fungicide spray program on phytophagous and predacious mite populations in a scab-resistant apple orchard. *Environmental Entomology* 24, 423–430.

Brady, N.C. (1984) *The Nature and Properties of Soils*, 9th edn. Macmillan Publishing Co., New York.

Brady, N.C. and Weil, R.R. (2008) *The Nature and Properties of Soils*, 14th edn. Pearson Education, Upper Saddle River, New Jersey, 905 pp.

Bragazza, L., Freeman, C., Jones, T., Rydin, H., Limpens, J., Fenner, N. *et al.* (2006) Atmospheric nitrogen deposition promotes carbon loss from peat bogs. *Proceedings of the National Academy of Sciences* 103, 19386–19389.

Brandes, J.A., Devol, A.H. and Deutsch, C. (2007) New developments in the marine nitrogen cycle. *Chemical Reviews* 107, 577–589.

Bredja, J.J., Moser, L.E. and Vogel, K.P. (1998) Evaluation of switchgrass rhizosphere microflora for enhancing seedling yield and nutrient uptake. *Agronomy Journal* 90, 753–758.

Bridge, J. and Starr, J.L. (2007) *Plant Nematodes of Agricultural Importance: a Colour Handbook*. Manson, London, 152 pp.

Briones, M.J.I., Ostle, N. and Poskitt, J. (2004) Influence of warming and enchytraeid activities on soil CO_2 and CH_4 fluxes. *Soil Biology and Biochemistry* 36, 1851–1859.

Briones, M.J.I., Ostle, N. and Garnett, M.H. (2007) Invertebrates increase the sensitivity of non-labile soil carbon to climate change. *Soil Biology and Biochemistry* 39, 816–818.

Broadbent, P. and Baker, K.F. (1974) Behaviour of *Phytophthora cinnamomi* in soils as suppressive and conducive to root rot. *Australian Journal of Agricultural Research* 25, 121–137.

Brune, A. (1998) Termite guts: the world's smaller bioreactors. *Trends in Biotechnology* 16, 16–21.

Cairney, J.W.G. and Bastias, B.A. (2007) Influences of fire on forest soil fungal communities. *Canadian Journal of Forest Research* 37, 207–215.

Canfield, D.E. (2005) The early history of atmospheric oxygen. *Annual Reviews of Earth and Planetary Sciences* 33, 1–36.

Carlsson, G. and Huss-Danell, K. (2003) Nitrogen fixation in perennial forage legumes in the field. *Plant and Soil* 253, 353–372.

Casper, B.B., Schenk, H.J. and Jackson, R.B. (2003) Defining a plant's belowground zone of influence. *Ecology* 84, 2313–2321.

Cerri, C.E.P., Paustian, K., Bernoux, M.A.L., Victoria, R.A.L., Melillo, J.M. and Cerri, C.C. (2004) Modeling changes in soil organic matter in Amazon forest to pasture conversion with the Century model. *Global Change Biology* 10, 815–832.

Christensen, N.L., Bartuska, A.M., Brown, J.H., Carpenter, S., D'Antonio, C., Francis, R. *et al.* (1996) The report of the Ecological Society of America committee on the scientific basis for ecosystem management. *Ecological Applications* 6, 665–691.

Clough, A. and Skjemstad, J.O. (2000) Physical and chemical protection of soil organic carbon in three agricultural soils with different contents of calcium carbonate. *Australian Journal of Soil Research* 38, 1005–1016.

Coleman, D.C. and Crossley, D.A., Jr (2003) *Fundamentals of Soil Ecology*. Academic Press, London, 205 pp.

Cox-Foster, D.L., Conlan, S., Holmes, E.C., Palacios, G., Evans, J.D., Moran, N.A. *et al.* (2007) A megagenomic survey of microbes in honey bee colony collapse disorder. *Science* 318, 283–287.

Cross, J.V., Solomon, M.G., Chandler, D., Jarrett, P., Richardson, P.N., Winstanley, D. *et al.* (1999) Biocontrol of pests of apples and pears in Northern and Central Europe: 1. Microbial agents and nematodes. *Biocontrol Science and Technology* 9, 125–149.

Dalsgaard, T., Canfield, D.E., Petersen, J., Thandrup, B. and Acuña-Gonzalez, J. (2003) N_2 production by annammox reaction in the anoxic water column of Golfo Dulce, Costa Rica. *Nature* 422, 606–608.

Darwin, C. (1881) *The Formation of Vegetable Mould Through the Action of Worms*. John Murray, London.

De Bona, F.D., Bayer, C., Dieckow, J. and Bergamaschi, H. (2008) Soil quality assessed by carbon management index in a subtropical Acrisol subjected to tillage systems and irrigation. *Australian Journal of Soil Research* 46, 469–475.

Deng, S.P., Parham, J.A., Hattey, J.A. and Babu, D. (2006) Animal manure and anhydrous ammonia amendment alter microbial carbon use efficiency, microbial biomass, and activities of dehydrogenase and amidohydrolases in semiarid agroecosystems. *Applied Soil Ecology* 33, 258–268.

Dieckow, J., Mielniczuk, J., Knicker, H., Bayer, C., Dick, D.P. and Kögel-Knabner, I. (2005) Organic N forms of a subtropical Acrisol under no-till cropping systems as assessed by acid hydrolysis and solid-state NMR spectroscopy. *Biology and Fertility of Soils* 42, 153–158.

Donegan, K., Fieland, V., Fowles, N., Ganio, L. and Seidler, R. (1992) Efficacy of burning, tillage, and biocides in controlling bacteria released at field sites and effects on indigenous bacteria and fungi. *Applied and Environmental Microbiology* 58, 1207–1214.

Doran, J.W., Sarrantonio, M. and Liebig, M.A. (1996) Soil health and sustainability. *Advances in Agronomy* 56, 1–54.

Dorioz, J.M., Robert, M. and Chenu, C. (1993) The role of roots, fungi and bacteria on clay particle organization. An experimental approach. *Geoderma* 56, 179–194.

Drenovsky, R.E., Feris, K.P., Batten, K.M. and Hristova, K. (2008) New and current microbiological tools for ecosystem ecologists: towards a goal of linking structure and function. *American Midland Naturalist* 160, 140–159.

Eggleton, P. (2001) Termites and trees: a review of recent advances in termite phylogenetics. *Insectes Sociaux* 48, 187–193.

Eggleton, P. (2006) The termite gut habitat: its evolution and co-evolution. In: Konig, H. and Varma, A. (eds) *Intestinal Microorganisms of Termites and Other Invertebrates*. Soil Biology vol. 6. Springer-Verlag, Berlin, pp. 373–404.

Elbert, A., Nauen, R. and McCaffery, A. (2007) IRAC, insecticide resistance and mode of action classification of insecticides. In: Krämer, W. and Schirmer, U. (eds) *Modern Crop Protection Compounds*. Wiley-VCH, Verlag GmbH and Co., Weinheim, Germany, pp. 753–771.

Elliott, E.T. and Coleman, D.C. (1988) Let the soil work for us. *Ecological Bulletins* 39, 23–32.

Entry, J.A., Mills, D., Mathee, K., Jayachandran, K., Sojka, R.E. and Narasimhan, G. (2008) Influence of irrigated agriculture on soil microbial diversity. *Applied Soil Ecology* 40, 146–154.

Eriksen-Hamel, N.S. and Whalen, J.K. (2007) Impacts of earthworms on soil nutrients and plant growth in soybean and maize agroecosystems. *Agriculture, Ecosystems and Environment* 120, 442–448.

Eriksen-Hamel, N.S. and Whalen, J.K. (2008) Earthworms, soil mineral nitrogen and forage production in grass-based hayfields. *Soil Biology and Biochemistry* 40, 1004–1010.

Esteban, G.F., Clarke, K.J., Olmo, J.L. and Finlay, B.J. (2006) Soil protozoa – an intensive study of population dynamics and community structure in an upland grassland. *Applied Soil Ecology* 33, 137–151.

Ettema, C.H. and Wardle, D.A. (2002) Spatial soil ecology. *Trends in Ecology & Evolution* 17, 177–183.

European Commission (2006) *Soil Protection: the Story Behind the Strategy*. Office for Official Publications of the European Communities, Luxembourg City, Luxembourg.

Fierer, N., Schimel, J.P. and Holden, P.A. (2003) Variations in microbial community composition through two soil depth profiles. *Soil Biology and Biochemistry* 35, 167–176.

Food and Agriculture Organization of the United Nations (FAO) (2008) *High Food Prices to Blame – Economic Crisis Could Compound Woes*. Available at: http://www.fao.org/news/story/en/item/8836/icode/ (accessed on 3 March 2009).

Foth, H.D. (1990) *Fundamentals of Soil Science*, 8th edn. Wiley, New York, 360 pp.

Foth, H.D. and Turk, L.M. (1972) *Fundamentals of Soil Science*, 5th edn. John Wiley & Sons, New York.

Fox, O., Vetter, S., Ekschmitt, K. and Wolters, V. (2006) Soil fauna modifies the recalcitrance-persistence relationship of soil carbon pools. *Soil Biology and Biochemistry* 38, 1353–1363.

Fraser, P.M., Williams, P.H. and Haynes, R.J. (1996) Earthworm species, population size and biomass under different cropping systems across the Canterbury Plains, New Zealand. *Applied Soil Ecology* 3, 49–57.

Freeman, C., Evans, C.D., Monteith, D.T., Reynolds, B. and Fenner, N. (2001a) Export of organic carbon from peat soils. *Nature* 412, 785.

Freeman, C., Ostle, N. and Kang, H. (2001b) An enzymic 'latch' on a global carbon store. *Nature* 409, 149.

Friberg, H., Lagerlo, J. and Rämert, B. (2005) Influence of soil fauna on fungal plant pathogens in agricultural and horticultural systems. *Biocontrol Science and Technology* 15, 641–658.

Friedel, J.K. and Scheller, D. (2002) Composition of hydrolysable amino acids in soil organic matter and soil microbial biomass. *Soil Biology and Biochemistry* 34, 315–325.

Frouz, J., Elhottová, D., Helingerová, M. and Kocourek, F. (2008) The effect of Bt-corn on soil invertebrates, soil microbial community and decomposition rates of corn post-harvest residues under field and laboratory conditions. *Journal of Sustainable Agriculture* 32, 645–655.

Gadd, G.M. (2005) Microorganisms in toxic metal-polluted soils. In: Buscot, F. and Varma, A. (eds) *Microorganisms in Soils: Roles in Genesis and Functions*. Springer, Heidelberg, Germany, pp. 325–356.

Gallardo, A. (2003) Effect of tree canopy on the spatial distribution of soil nutrients in a Mediterranean Dehesa. *Pedobiologia* 47, 117–125.

Gallardo, A., Rodriguez-Saucedo, J.J., Covelo, F. and Fernandez-Alés, R. (2000) Soil nitrogen heterogeneity in a Dehesa ecosystem. *Plant and Soil* 222, 71–82.

Gams, W. and Seifert, K.A. (2008) Anamorphic fungi. In: *Encyclopedia of Life Sciences*. John Wiley & Sons, Chichester, UK. DOI:10.1002/9780470015902.a0000351.pub2

Giller, K.E., Beare, M.H., Lavelle, P., Izac, A.M.N. and Swift, M.J. (1997) Agricultural intensification, soil biodiversity and agroecosystem function. *Applied Soil Ecology* 6, 3–16.

Giraud, T., Pedersen, J.S. and Keller, L. (2002) Evolution of supercolonies: the Argentine ants of southern Europe. *Proceedings of the National Academy of Sciences* 99, 6075–6079.

Giri, B., Giang, P.H., Kumari, R., Prasad, R., Sachdev, M., Garg, A.P. *et al.* (2005) Mycorrhizosphere: strategies and functions. In: Buscot, F. and Varma, A. (eds) *Microorganisms in Soils: Roles in Genesis and Functions*. Springer-Verlag, Berlin, pp. 213–252.

Gitelson, A.A., Viña, A., Masek, J.G., Verma, S.B. and Suyker, A.E. (2008) Synoptic monitoring of gross primary productivity of maize using Landsat data. *IEEE Geoscience and Remote Sensing Letters* 5, 133–137.

Gongalsky, K.B., Pokarzhevskii, A.D., Filimonova, Z.V. and Savin, F.A. (2004) Stratification and dynamics of bait-lamina perforation in three forest soils along a north-south gradient in Russia. *Applied Soil Ecology* 25, 111–122.

Gonzalez-Lopez, J., Martinez-Toledo, M.V. and Salmeron, V. (1992) Effect of the acaricide bromopropylate on agricultural soil microflora. *Soil Biology and Biochemistry* 24, 815–817.

Gracia-Garza, J.A., Neumann, S., Vyn, T.J. and Boland, G.J. (2002) Influence of crop rotation and tillage on production of apothecia by *Sclerotinia sclerotiorum*. *Canadian Journal of Plant Pathology* 24, 137–143.

Graham, J.H. and Mitchell, D.J. (2005) Biological control of soilborne plant pathogens and nematodes. In: Sylvia, D.M., Fuhrmann, J.J., Hartel, P.G. and Zuberer, D.D. (eds) *Principles and Applications of Soil Microbiology*. Prentice Hall, Upper Saddle River, New Jersey, pp. 427–446.

Haas, D. and Défago, G. (2005) Biological control of soil-borne pathogens by fluorescent pseudomonads. *Nature Reviews Microbiology* 3, 307–319.

Hansen, S. and Engelstad, F. (1999) Earthworm populations in a cool and wet district as affected by tractor traffic and fertilization. *Applied Soil Ecology* 13, 237–250.

Havlin, J.L., Beaton, J.D., Tisdale, S.L. and Nelson, W.L. (2005) *Soil Fertility and Fertilizers: an Introduction to Nutrient Management*, 7th edn. Macmillian Publishing Co., New York.

Hayatsu, M., Tago, K. and Saito, M. (2008) Various players in the nitrogen cycle: diversity and functions of the microorganisms involved in nitrification and denitrification. *Soil Science and Plant Nutrition* 54, 33–45.

Helliwell, R., Britton, A., Gibbs, S., Fisher, J. and Aherne, J. (2008) Who put the N in pristiNe? Impacts of nitrogen enrichment in fragile mountain environments. *Mountain Research and Development* 28, 210–215.

Herridge, D.F., Peoples, M.B. and Boddey, R.M. (2008) Global inputs of biological nitrogen fixation in agricultural systems. *Plant and Soil* 311, 1–18.

Hill, J., Nelson, E., Tilman, D., Polasky, S. and Tiffany, D. (2006) Environmental, economic, and energetic costs and benefits of biodiesel and ethanol biofuels. *Proceedings of the National Academy of Sciences* 103, 11206–11210.

Hoitink, H.A.J. and Fahy, P.C. (1986) Basis for the control of soilborne plant pathogens with composts. *Annual Review of Phytopathology* 24, 93–144.

Holmgren, P. (2006) *Global Land Use Area Change Matrix*. Working Paper 134, Forestry Department, Food and Agricultural Organization of the United Nations, Rome.

Hooks, C.R.R. and Johnson, M.W. (2003) Impact of agricultural diversification on the insect community of cruciferous crops. *Crop Protection* 22, 223–238.

Houghton, R.A. (2008) Carbon flux to the atmosphere from land-use changes: 1850–2005. In: *TRENDS: A Compendium of Data on Global Change*. Carbon Dioxide Information Analysis Center, Oak Ridge National Laboratory, US Department of Energy, Oak Ridge, Tennessee.

Hu, H.Y., Liu, X.X., Zhao, Z.W., Sun, J.G., Zhang, Q.W., Liu, X.Z. *et al.* (2009) Effects of repeated cultivation of transgenic Bt cotton on functional bacterial populations in rhizosphere soil. *World Journal of Microbiology and Biotechnology* 25, 357–366.

Hunt, H.W., Coleman, D.C., Ingham, E.R., Ingham, R.E., Elliott, E.T., Moore, J.C. *et al.* (1987) The detrital food web in a shortgrass prairie. *Biology and Fertility of Soils* 3, 57–68.

Hwang, S.C. and Ko, W.L. (1977) Biology of chlamydospores, sporangia, and zoospores of *Phytophthora cinnamomi* in soil. *Phytopathology* 68, 726–731.

Intergovernmental Panel on Climate Change (IPCC) (2007) *Climate Change 2007 Synthesis Report, Contribution of Working Groups I, II and III to the Fourth Assessment Report of the Intergovernmental Panel on Climate Change.* Core Writing Team, Pachauri, R.K. and Reisinger, A. (eds). IPCC, Geneva, Switzerland, 104 pp. (available at: http://www.ipcc.ch/ipccreports/assessments-reports.htm).

Iritani, W.M. and Arnold, C.Y. (1960) Nitrogen release of vegetable crop residues during incubation as related to their chemical composition. *Soil Science* 89, 74–82.

James, E.K. and Olivares, F.L. (1998) Infection and colonization of sugar cane and other graminaceous plants by endophytic diazotrophs. *Critical Reviews in Plant Sciences* 17, 77–119.

Jansson, H.B., Persson, C. and Odeslius, R. (2000) Growth and capture activities of nematophagous fungi in soil visualized by low temperature scanning electron microscopy. *Mycologia* 92, 10–15.

Jastrow, J.D., Miller, R.M. and Lussenhop, J. (1998) Contributions of interacting biological mechanisms to soil aggregate stabilization in restored prairies. *Soil Biology and Biochemistry* 30, 905–916.

Jenny, H. (1941) *Factors of Soil Formation; a System of Quantitative Pedology.* McGraw-Hill, New York.

Jenny, H. (1980) *The Soil Resource: Origin and Behavior.* Springer-Verlag, New York, 377 pp.

Johnson, J.M.F., Allmaras, R.R. and Reicosky, D.C. (2006) Estimating source carbon from crop residues, roots and rhizodeposits using the national grain-yield database. *Agronomy Journal* 98, 622–636.

Johnson, J.M.F., Barbour, N.W. and Weyers, S.L. (2007) Chemical composition of crop biomass impacts its decomposition. *Soil Science Society of America Journal* 71, 155–162.

Jones, K.L., Todd, T.C., Wall-Beam, J.L., Coolon, J.D., Blair, J.M. and Herman, M.A. (2006) Molecular approach for assessing responses of microbial-feeding nematodes to burning and chronic nitrogen enrichment in a native grassland. *Molecular Ecology* 15, 2601–2609.

Jönsson, K.I., Rabbow, E., Schill, R.O., Harms-Ringdahl, M. and Rettberg, P. (2008) Tardigrades survive exposure to space in low Earth orbit. *Current Biology* 18, R729–R731.

Jordan, N. and Huerd, S. (2008) Effects of soil fungi on weed communities in a corn-soybean rotation. *Renewable Agriculture and Food Systems* 23, 108–117.

Kennard, D.K. and Gholz, H.L. (2001) Effects of high- and low-intensity fires on soil properties and plant growth in a Bolivian dry forest. *Plant and Soil* 234, 119–129.

Khan, A., Williams, K.L. and Nevalainen, H.K. (2006) Control of plant-parasitic nematodes by *Paecilomyces lilacinus* and *Monacrosporium lysipagum* in pot trials. *BioControl* 51, 643–658.

Kimura, M., Jia, Z.-J., Nakayama, N. and Asakawa, S. (2008) Ecology of viruses in soils: past, present and future perspectives. *Soil Science and Plant Nutrition* 54, 1–32.

Kleber, M., Sollins, P. and Sutton, R. (2007) A conceptual model of organo-mineral interactions in soils: self-assembly of organic molecular fragments into zonal structures on mineral surfaces. *Biogeochemistry* 85, 9–24.

Klok, C., Zorn, M., Koolhaas, J.E., Eijsackers, H.J.P. and van Gestel, C.A.M. (2006) Does reproductive plasticity in *Lumbricus rubellus* improve the recovery of populations in frequently inundated river floodplains? *Soil Biology and Biochemistry* 38, 611–618.

Knapp, A.K. and Anderson, J.E. (1980) Effect of heat on germination of seeds from serotinous lodgepole pine cones. *American Midland Naturalist* 104, 370–372.

Knapp, A.K., Fay, P.A., Blair, J.M., Collins, S.L., Smith, M.D., Carlisle, J.D. *et al.* (2002) Rainfall variability, carbon cycling, and plant species diversity in a mesic grassland. *Science* 298, 2202–2205.

Knicker, H. (2000) Biogenic nitrogen in soils as revealed by solid-state carbon-13 and nitrogen-15 nuclear magnetic resonance spectroscopy. *Journal of Environmental Quality* 29, 715–723.

Knicker, H., Almendros, G., Gonzàlez-Vila, F.J., Lüdemann, H.D. and Martin, F. (1995) ^{13}C and ^{15}N NMR analysis of some fungal melanins in comparison with soil organic matter. *Organic Geochemistry* 23, 1023–1028.

Konig, H. and Varma, A. (eds) (2006) *Intestinal Microorganisms of Termites and Other Invertebrates.* Soil Biology Series vol. 6. Springer-Verlag, Berlin, 483 pp.

Kraulis, P.J. (1991) MOLSCRIPT: a program to produce both detailed and schematic plots of protein structure. *Journal of Applied Crystallography* 24, 946–950.

Kuck, K.H. and Gisi, U. (2007) FRAC mode of action classification and resistance risk of fungicides. In: Krämer, W. and Schirmer, U. (eds) *Modern Crop Protection Compounds.* Wiley-VCH, Verlag GmbH and Co., Weinheim, Germany, pp. 416–432.

Kulkarni, M.V., Groffman, P.M. and Yavitt, J.B. (2008) Solving the global nitrogen problem: it's a gas! *Frontiers in Ecology and the Environment* 6, 199–206.

Lavelle, P. and Spain, A.V. (2001) *Soil Ecology.* Kluwer Academic Publishers, Dordrecht, The Netherlands, 654 pp.

Lee, K.E. (1985) *Earthworms: Their Ecology and Relationship with Soils and Land Use.* Academic Press, Sydney, 411 pp.

Liedgens, M. and Richner, W. (2001) Minirhizotron observations of the spatial distribution of the maize root system. *Agronomy Journal* 93, 1097–1104.

Lo, N., Bandi, C., Watanabe, H., Nalepa, C. and Beninati, T. (2003) Evidence for cocladogenesis between diverse dictyopteran lineages and their intracellular endosymbionts. *Molecular Biology and Evolution* 20, 907–913.

Lœgreid, M., Bøckman, O.C. and Kaarstad, O. (1999) *Agriculture, Fertilizers and the Environment.* CAB International/Norsk Hydro ASA, Wallingford, UK, 294 pp.

Löhnis, F. (1910) *Handbuch der Landwirtschtliche Bakteriologie.* Borntraeger, Berlin.

Lowe, C.N. and Butt, K.R. (2005) Culture techniques for soil dwelling earthworms: a review. *Pedobiologia* 49, 401–413.

Lowe, C.N. and Butt, K.R. (2007) Earthworm culture, maintenance and species selection in chronic ecotoxicological studies: a critical review. *European Journal of Soil Biology* 43, S281–S288.

Lubbock, J. (1873) *Monograph of the Collembola and Thysanura.* The Ray Society, London.

Luna, J.M. and House, G.J. (1990) Pest management in sustainable agricultural systems. In: Edwards, C.A., Lal, R., Madden, P., Miller, R.H. and House, G. (eds) *Sustainable Agricultural Systems.* Soil and Water Conservation Society, Ankeny, Iowa, pp. 157–173.

Ma, W.C., Brussard, L. and deRidder, J.A. (1990) Long-term effects of nitrogenous fertilizers on grassland earthworms (Oligochaeta: Lumbricidae): their relation to soil acidification. *Agriculture, Ecosystems and Environment* 30, 71–80.

Maene, L.M. (2007) International fertilizer supply and demand. *Proceedings of the Australian Fertilizer Industry Conference*, Hamilton Island, Queensland, 6–10 August, 2007.

Malmström, A., Persson, T. and Ahlström, K. (2008) Effects of fire intensity on survival and recovery of soil microarthropods after a clearcut burning. *Canadian Journal of Forest Research* 38, 2465–2475.

Mann, L. and Tolbert, V. (2000) Soil sustainability in renewable biomass plantings. *Ambio* 29, 492–498.

Marschner, B., Brodowski, S., Dreves, A., Gleixner, G., Gude, A., Grootes, P.M. *et al.* (2008) How relevant is recalcitrance for the stabilization of organic matter in soils? *Journal of Plant Nutrition and Soil Science* 171, 91–110.

Matlack, G.R. (2001) Factor determining the distribution of soil nematodes in a commercial forest landscape. *Forest Ecology and Management* 146, 129–143.

McSorley, R. (1993) Short-term effect of fire on the nematode community in a pine forest. *Pedobiologia* 37, 39–48.

Menne, H. and Köcher, H. (2007) HRAC classification of herbicides and resistance development. In: Krämer, W. and Schirmer, U. (eds) *Modern Crop Protection Compounds.* Wiley-VCH, Verlag GmbH and Co., Weinheim, Germany, pp. 5–26.

Mieth, A. and Bork, H.R. (2005) History, origin and extent of soil erosion on Easter Island (Rapa Nui). *Catena* 63, 244–260.

Mokany, K., Raison, R.J. and Prokushkin, A.S. (2006) Critical analysis of root:shoot ratios in terrestrial biomes. *Global Change Biology* 12, 84–96.

Monnier, G. (1965) Action des matières organiques sur la stabilité structural des sols. Thèse de la faculté des sciences de Paris, 140 pp.

Moore, J.C. and de Ruiter, P.C. (1991) Temporal and spatial heterogeneity of trophic interactions within below-ground food webs. *Agriculture Ecosystems and Environment* 34, 371–394.

Moore, J.C., Walter, D.E. and Hunt, H.W. (1988) Arthropod regulation of micro- and mesobiota in below-ground detrital food webs. *Annual Review of Entomology* 33, 419–439.

Moore, J.C., Berlow, E.L., Coleman, D.C., de Ruiter, P.C., Dong, Q., Hastings, A. *et al.* (2004) Detritus, trophic dynamics and biodiversity. *Ecology Letters* 7, 584–600.

Mort, A.J. and Bauer, W.D. (1980) Composition of the capsular and extracellular polysaccharides of *Rhizobium japonicum. Plant Physiology* 66, 158–163.

Morton, V. and Staub, T. (2008) *A Short History of Fungicides.* Available at: http://www.apsnet.org/online/feature/fungi/

Murillo-Williams, A. and Pedersen, P. (2008) Arbuscular mycorrhizal colonization response to three seed-applied fungicides. *Agronomy Journal* 100, 795–800.

Myrold, D.D. (1998) Transformations of nitrogen. In: Sylvia, D., Fuhrmann, J., Hatrel, P. and Zuberer, D. (eds) *Principles and Applications of Soil Microbiology.* Prentice Hall, Upper Saddle River, New Jersey, pp. 259–294.

Nalepa, C.A., Bignell, D.E. and Bandi, C. (2001) Detritivory, coprophagy, and the evolution of digestive mutualisms in Dictyoptera. *Insectes Sociaux* 48, 194–201.

National Standard of Canada (2006) *Organic Production Systems: General Principles and Management Standards.* CAN/CGSB-32.310-2006. Canadian General Standards Board, Gatineau, Quebec, 40 pp.

Neary, D.G., Klopatek, C.C., DeBano, L.F. and Ffolliott, P.F. (1999) Fire effects on belowground sustainability: a review and synthesis. *Forest Ecology and Management* 122, 51–71.

Nel, B., Steinberg, C., Labuschagne, N. and Vilijoen, A. (2007) Evaluation of fungicides and sterilants for potential application in the management of *Fusarium* wilt of banana. *Crop Protection* 26, 697–705.

Nelson, D.W. and Sommers, L.E. (1996) Total carbon, organic carbon, and organic matter. In: Page, A.L. *et al.* (eds) *Methods of Soil Analysis*, Part 2, 2nd edn. *Agronomy* 9, 961–1010. American Society of Agronomy, Madison, Wisconsin.

Nguyen, C. (2003) Rhizodeposition of organic carbon by plants: mechanisms and controls. *Agronomie* 23, 375–396.

Niwa, S., Kaneko, N., Okada, H. and Sakamoto, K. (2008) Effects of fine-scale simulation of deer browsing on soil micro-foodweb structure and N mineralization rate in a temperate forest. *Soil Biology and Biochemistry* 40, 699–708.

Northup, R.R., Yu, Z., Dahlgren, R.A. and Vogt, K.A. (1995) Polyphenol control of nitrogen release from pine litter. *Nature* 377, 227–229.

Oades, J.M. and Waters, A.G. (1991) Aggregate hierarchy in soils. *Australian Journal of Soil Research* 29, 815–828.

O'Hara, G.W. (2001) Nutritional constraints on root nodule bacteria affecting symbiotic nitrogen fixation: a review. *Australian Journal of Experimental Agriculture* 41, 417–433.

Ohkuma, M., Noda, S. and Kudo, T. (1999) Phylogenetic relationships of symbiotic methanogens in diverse termites. *FEMS Microbiology Letters* 171, 147–153.

Paré, T., Dinel, H., Moulin, A.P. and Townley-Smith, L. (1999) Organic matter quality and structural stability of a Black Chernoemic soil under different manure and tillage practices. *Geoderma* 91, 311–326.

Parrent, J.L., Morris, W.F. and Vilgalys, R. (2006) CO_2-enrichment and nutrient availability alter ectomycorrhizal fungal communities. *Ecology* 87, 2278–2287.

Parton, W.J., Schimel, D.S., Cole, C.V. and Ojima, D.S. (1987) Analysis of factors controlling soil organic matter levels in Great Plains grasslands. *Soil Science Society of America Journal* 51, 1173–1179.

Paterson, E., Gebbing, T., Abel, C., Sim, A. and Telfer, G. (2007) Rhizodeposition shapes rhizosphere microbial community structure in organic sol. *New Phytologist* 173, 600–610.

Pattey, E., Blackburn, L.G., Strachan, I.B., Desjardins, R. and Dow, D. (2008) Spring thaw and growing season N_2O emissions from a field planted with edible peas and a cover crop. *Canadian Journal of Soil Science* 88, 241–249.

Paul, E.A. and Clark, F.E. (1989) *Soil Microbiology and Biochemistry*, Academic Press, San Diego, California, 275 pp.

Paul, J.P. and Williams, B.L. (2005) Contribution of α-amino N to extractable organic nitrogen (DON) in three soil types from the Scottish uplands. *Soil Biology and Biochemistry* 37, 801–803.

Perrichot, V., Lacau, S., Neraudeau, D. and Nel, A. (2008) Fossil evidence for the early ant evolution. *Naturwissenschaften* 95, 85–90.

Peryea, F.J. (1998) Historical use of lead arsenate insecticides, resulting soil contamination and implications for soil remediation. *Proceedings 16th World Congress of Soil Science*, Montepellier, France, no. 274.

Popa, V.I., Dumitru, M., Volf, I. and Anghel, N. (2008) Lignin and polyphenols as allelochemicals. *Industrial Crops and Products* 27, 144–149.

Post, W.M., Peng, T.H., Emanuel, W.R., King, A.W., Dale, V.H. and deAngelis, D.L. (1990) The global carbon cycle. *American Scientist* 78, 310–326.

Poulton, A.J., Adey, T.R., Balch, W.M. and Holligan, P.M. (2007) Relating coccolithophore calcification rates to phytoplankton community dynamics: regional differences and implications for carbon export. *Deep-Sea Research II* 54, 538–557.

Prasad, A.K., Chai, L., Singh, R.P. and Kafatos, M. (2006) Crop yield estimation model for Iowa using remote sensing and surface parameters. *International Journal of Applied Earth Observation and Geoinformation* 8, 26–33.

Prasifka, J.R., Schmidt, N.P., Kohler, K.A., O'Neal, M.E., Hellmich, R.L. and Singer, J.W. (2006) Effects of living mulches on predator abundance and sentinel prey in a corn-soybean-forage rotation. *Environmental Entomology* 35, 1423–1431.

Prietzel, J., Thieme, J., Salomé, M. and Knicker, H. (2007) Sulfur K-edge XANES spectroscopy reveals differences in sulfur speciation of bulk soils, humic acid, fulvic acid, and particle size separates. *Soil Biology and Biochemistry* 39, 877–890.

Prihoda, K.R. and Coats, J.R. (2008) Fate of *Bacillus thuringiensis* (bt) Cry3Bb1 protein in a soil microcosm. *Chemosphere* 73, 1102–1107.

Prince, S.D., Haskett, J., Steininger, M., Strand, H. and Wright, R. (2001) Net primary production of US Midwest croplands from agricultural harvest yield data. *Ecological Applications* 11, 1194–1205.

Quiquampoix, H. and Mousain, D. (2005) Enzymatic hydrolysis of organic phosphorus. In: Turner, B.L., Frossard, E. and Baldwin, D.S. (eds) *Organic Phosphorus in the Environment.* CAB International, Wallingford, UK, pp. 88–112.

Radek, R. (1999) Flagellates, bacteria and fungi associated with termites: diversity and function in nutrition – a review. *Ecotropica* 5, 183–196.

Rahman, L., Chan, K.Y. and Heenan, D.P. (2007) Impact of tillage, stubble management and crop rotation on nematode populations in a long-term field experiment. *Soil and Tillage Research* 95, 110–119.

Rasse, D.P., Rumpel, C. and Dignac, M.F. (2005) Is soil carbon mostly root carbon? Mechanisms for specific stabilization. *Plant and Soil* 269, 341–356.

Rayner, M.C. (1927) Mycorrhiza. An account of non-pathogenic infection by fungi in vascular plants and bryophytes. Wheldon and Wesley, London.

Read, D.S., Sheppard, K., Bruford, M.W., Glen, D.M. and Symondson, W.O.C. (2006) Molecular detection of predation by soil micro-arthropods on nematodes. *Molecular Ecology* 15, 1963–1972.

Rees, R.M., Bingham, I.J., Baddeley, J.A. and Watson, C.A. (2005) The role of plants and land management in sequestering soil carbon in temperate arable and grassland ecosystems. *Geoderma* 128, 130–154.

Rillig, M.C., Wright, S.F., Nichols, K.A., Schmidt, W.F. and Torn, M.S. (2001) Large contribution of arbuscular mycorrhizal fungi to soil carbon pools in tropical forest soils. *Plant Soil* 233, 167–177.

Robinson, J.M. (1990) Lignin, land plants, and fungi: biological evolution affecting Phanerozoic oxygen balance. *Geology* 15, 607–610.

Rodriguez, J. (1999) *Ecología*. Ed. Pirámide, Madrid, Spain, 411 pp.

Rossi, J.-P., Delaville, L. and Quénéhervé, P. (1996) Microspatial structure of a plant-parasitic nematode community in a sugarcane field in Martinique. *Applied Soil Ecology* 3, 17–26.

Rousseau, J.V.D., Sylvia, D.M. and Fox, A.J. (1994) Contribution of ecotomycorrhiza to the potential nutrient-absorbing surface of pine. *New Phytologist* 128, 639–644.

Ruess, L., Schütz, K., Migge-Kleian, S., Häggblom, M.M., Kandeler, E. and Scheu, S. (2007) Lipid composition of *Collembola* and their food resources in deciduous forest stands – implications for feeding strategies. *Soil Biology and Biochemistry* 39, 1990–2000.

Salmon, S., Frizzera, L. and Camaret, S. (2008) Linking forest dynamics to richness and assemblage of soil zoological groups and to soil mineralization processes. *Forest Ecology and Management* 256, 1612–1623.

Sampedro, L. (1999) Biodegradación de residuos cellulósicos mediante lombrices y microorganismos. PhD thesis. Universidade de Vigo, Spain.

Sampedro, L. and Whalen, J.K. (2007) Changes in the fatty acid profiles through the digestive tract of the earthworm *Lumbricus terrestris*. *Applied Soil Ecology* 35, 226–236.

Sampedro, L., Jeannotte, R. and Whalen, J.K. (2006) Trophic transfer of fatty acids from gut microbiota to the earthworm *Lumbricus terrestris*. *Soil Biology and Biochemistry* 38, 2188–2198.

Sánchez-Moreno, S., Alonso-Prados, E., Alonso-Prados, J.L. and García-Baudín, J.M. (2009) Multivariate analysis of toxicological and environmental properties of soil nematicides. *Pest Management Science* 65, 82–92.

Saxena, D. and Stotzky, G. (2001) *Bacillus thuringiensis* (Bt) toxin released from root exudates and biomass of Bt corn has no apparent effect on earthworms, nematodes, protozoa, bacteria and fungi in soil. *Soil Biology and Biochemistry* 33, 1225–1230.

Schaefer, M. (2005) The landfill of TBT contaminated harbor sludge on rinsing fields – a hazard for the soil fauna? Risk assessment with earthworms. *Water, Air and Soil Pollution* 165, 265–278.

Schindler Wessells, M.L., Bohlen, P.J., McCartney, D.A., Subler, S. and Edwards, C.A. (1997) Earthworm effects on soil respiration in corn agroecosystems receiving different nutrient inputs. *Soil Biology and Biochemistry* 29, 409–412.

Schleper, C., Jurgens, G. and Jonuscheit, M. (2005) Genomic studies of uncultivated archaea. *Nature Reviews: Microbiology* 3, 479–488.

Schmidt, O., Clements, R.O. and Donaldson, G. (2003) Why do cereal-legume intercrops support large earthworm populations? *Applied Soil Ecology* 22, 181–190.

Schulze, J. (2004) How are nitrogen fixation rates regulated in legumes? *Journal of Plant Nutrition and Soil Science* 167, 125–137.

Seixas, C.D.S., Barreto, R.W., Freitas, L.G., Monteiro, F.T. and Oliveira, R.D.L. (2004) *Ditylenchus drepanocercus* rediscovered in the Neotropics causing angular leaf spots on *Miconia calvescens*. *Journal of Nematology* 36, 481–486.

Sexstone, A.J., Revsbech, N.P., Parkin, T.B. and Tiedje, J.M. (1985) Direct measurement of oxygen profiles and denitrification rates in soil aggregates. *Soil Science Society of America Journal* 49, 645–651.

Silvester, W. and Musgrave, D. (1991) Free-living diazotrophs. In: Dilworth, M. and Glenn, A. (eds) *Biology and Biochemistry of Nitrogen Fixation*. Elsevier, Amsterdam, The Netherlands, pp. 162–186.

Six, J., Bossuyt, H., Degryze, S. and Denef, K. (2004) A history of research on the link between (micro) aggregates, soil biota, and soil organic matter dynamics. *Soil and Tillage Research* 79, 7–31.

Smernik, R.J. and Baldock, J.A. (2005) Does solid-state ^{15}N NMR spectroscopy detect all soil organic nitrogen? *Biogeochemistry* 75, 507–528.

Smil, V. (2001) *Enriching the Earth*. MIT Press, Cambridge, Massachusetts.

Smith, P., Andren, O., Brussaard, L., Dangerfield, M., Ekschmitt, K., Lavelle, P. *et al.* (1998) Soil biota and global change at the ecosystem level: describing soil biota in mathematical models. *Global Change Biology* 4, 773–784.

Smith, P., Chapman, S.J., Scott, W.A., Black, H.I.J., Wattenbach, M.N., Milne, R. *et al.* (2007) Climate change cannot be entirely responsible for soil carbon loss observed in England and Wales, 1978–2003. *Global Change Biology* 13, 2605–2609.

Soil Survey Laboratory (1992) *Soil Survey Laboratory Methods Manual.* Soil Survey Investigations Report No. 42. US Department of Agriculture, Washington, DC.

Steinaker, D.F. and Wilson, S.D. (2008) Phenology of fine roots and leaves in forest and grassland. *Journal of Ecology* 96, 1222–1229.

Stern, V.M., Smith, R.F., van den Bosch, R. and Hagen, K. (1959) The integrated control concept. *Hilgardia* 29, 81–101.

Stinner, B.R. and Blair, J.M. (1990) Ecological and agronomic characteristics of innovative cropping systems. In: Edwards, C.A., Lal, R., Madden, P., Miller, R.H. and House, G. (eds) *Sustainable Agricultural Systems*. Soil and Water Conservation Society, Ankeny, Iowa, pp. 123–140.

Stirling, G.R. (1984) Biological control of *Meloidogyne javanica* with *Bacillus penetrans*. *Phytopathology* 74, 55–60.

Stringer, L. (2008) Can the UN Convention to Combat Desertification guide sustainable use of world's soils? *Frontiers in Ecology and Environment* 6, 138–144.

Sylvia, D.M., Fuhrmann, P.G., Hartel, P.G. and Zuberer, D.A. (eds) (2005) *Principles and Applications of Soil Microbiology*. Prentice Hall, Upper Saddle River, New Jersey.

Tapp, H. and Stotzky, G. (1998) Persistence of the insecticidal toxin from *Bacillus thuringiensis* subsp. *Kurstaki* in soil. *Soil Biology and Biochemistry* 30, 471–476.

Tate, R.L. (1992) *Soil Organic Matter: Biological and Ecological Effects*. Krieger Publishing Co., Malabar, Florida, 291 pp.

Tisdall, J.M. and Oades, J.M. (1982) Organic matter and water-stable aggregates in soils. *Journal of Soil Science* 62, 141–163.

Todd, T.C., James, S.W. and Seastedt, T.R. (1992) Soil invertebrate and plant responses to mowing and carbofuran application in a North American tallgrass prairie. *Plant and Soil* 144, 11–124.

Torsvik, V., Goksoyr, J. and Daae, F.L. (1990) High diversity in DNA of soil bacteria. *Applied and Environmental Microbiology* 56, 782–787.

Tremblay, L. and Brenner, R. (2006) Microbial contributions to N-immobilization and organic matter preservation in decaying plant detritus. *Geochimica et Cosmochimica Acta* 70, 133–146.

Turner, B.L. (2007) Inositol phosphates in soil: amounts, forms and significance of the phosphorylated inositol stereoisomers. In: Turner, B.L., Richardson, A.E. and Mullaney, E.J. (eds) *Inositol Phosphates: Linking Agriculture and the Environment*. CAB International, Wallingford, UK, pp. 186–206.

Turner, M., Romme, W.H. and Tinker, D.B. (2003) Surprises and lessons from the 1988 Yellowstone fires. *Frontiers in Ecology and the Environment* 1, 351–358.

United States Department of Agriculture (USDA) (2009) *Agricultural Marketing Program – National Organic Program*. Available at: http://www.ams.usda.gov/ (accessed on 12 February 2009).

United States Department of Energy (USDOE) (2008a) *Alternative Fuels and Advanced Vehicles Data Center: E85 Vehicle Toolkit*. Available at: http://www.afdc.energy.gov/afdc/e85toolkit/ (accessed on 3 March 2009).

United States Department of Energy (USDOE) (2008b) *Biomass Program Technologies: Processing and Conversion*. Available at: http://www1.eere.energy.gov/biomass/processing_conversion.html (accessed on 3 March 2009).

University of Michigan Extension (2008) *Bt Corn and European Corn Borer: Long-term Success Through Resistance Management*. Available at: www.extension.umn.edu/distribution/cropsystems/DC7055.html

Upchurch, R., Chiu, C.Y., Everett, K., Dyszynski, G., Coleman, D.C. and Whitman, W.B. (2008) Differences in the composition and diversity of bacterial communities from agricultural and forest soils. *Soil Biology and Biochemistry* 40, 1294–1305.

vanVliet, P.C.J., Coleman, D.C. and Hendrix, P.F. (1997) Population dynamics of Enchytraeidae (Oligochaeta) in different agricultural systems. *Biology and Fertility of Soils* 25, 123–129.

Varma, A. and Oelmuller, R. (eds) (2007) *Advanced Techniques in Soil Microbiology*. Soil Biology Series vol. 11. Springer Berlin, Heidelberg, Germany, 427 pp.

Velando, A., Domínguez, J. and Ferreiro, A. (2006) Inbreeding and outbreeding reduces cocoon production in the earthworm *Eisenia andrei*. *European Journal of Soil Biology* 42, 354–357.

Velando, A., Eiroa, J. and Domínguez, J. (2008) Brainless but not clueless: earthworms boost their ejaculates when they detect fecund non-virgin partners. *Proceedings of the Royal Society B* 275, 1067–1072.

Venter, J.C., Remington, K., Heidelberg, J.F., Halpern, A.L., Rusch, D., Eisen, J.A. *et al.* (2004) Environmental genome shotgun sequencing of the Sargasso Sea. *Science* 304, 66–74.

Vessey, J.K., Pawlowski, K. and Bergman, B. (2005) Root-based N_2-fixing symbioses: legumes, actinorhial plants, *Parasponia* sp. and cycads. *Plant and Soil* 274, 51–78.

Vetter, S., Fox, O., Ekschmitt, K. and Wolters, V. (2004) Limitations of faunal effects on soil carbon flow: density dependence, biotic regulation and mutual inhibition. *Soil Biology and Biochemistry* 36, 387–397.

Visser, S. and Parkinson, D. (1975) Fungal succession on aspen poplar leaf litter. *Canadian Journal of Botany* 53, 1640–1651.

Wagner, G.H. and Wolf, D.C. (1998) Carbon transformations and soil organic matter formation. In: Sylvia, D.M., Fuhrmann, J.J., Hartel, P.G. and Zuberer, D.A. (eds) *Principles and Applications of Soil Microbiology*. Prentice Hall, Upper Saddle River, New Jersey, pp. 218–258.

Wall, D.H. and Moore, J.C. (1999) Interactions underground: soil biodiversity, mutualism and ecosystem processes. *Bioscience* 49, 109–117.

Wall, D.H., Fitter, A. and Paul, E.A. (2005) Developing new perspectives from advances in soil biodiversity research. In: Bardgett, R.D., Usher, M.B. and Hopkins, D.W. (eds) *Biological Diversity and Function in Soils*. British Ecological Society, Cambridge University Press, Cambridge, UK, pp. 3–30.

Wallwork, J.A. (1976) *The Distribution and Diversity of Soil Fauna*. Academic Press, London.

Wardle, D.A. (1995) Impacts of disturbance on detritus food-webs in agroecosystems of contrasting tillage and weed management practices. *Advances in Ecological Research* 26, 105–185.

Wardle, D.A., Bonner, K.I., Barker, G.M., Yeates, G.W., Nicholson, K.S., Bardgett, R.D. *et al.* (1999) Plant removals in perennial grassland: vegetation dynamics, decomposer, soil biodiversity, and ecosystem properties. *Ecological Monographs* 69, 535–568.

Weiner, R., Langille, S. and Quintero, E. (1995) Structure, function and immunochemistry of bacterial exopolysaccharides. *Journal of Industrial Microbiology* 15, 339–346.

Weller, D.M. (1988) Biological control of soilborne plant pathogens in the rhizosphere with bacteria. *Annual Review of Phytopathology* 26, 379–407.

Weller, D.M., Raaijmakers, J.M., McSpadden Gardener, B.B. and Thomashow, L.S. (2002) Microbial populations responsible for specific soil suppressiveness to plant pathogens. *Annual Review of Phytopathology* 40, 309–348.

Whalen, J.K., Chang, C. and Clayton, G.W. (2002) Cattle manure and lime amendments to improve crop production of acidic soils in Northern Alberta. *Canadian Journal of Soil Science* 82, 227–238.

Wienhold, B.J., Andrews, S.S. and Karlen, D.L. (2004) Soil quality: a review of the science and experiences in the USA. *Environmental Geochemistry and Health* 26, 89–95.

Woese, C.R., Kandler, O. and Wheelis, M.L. (1990) Towards a natural system of organisms: Proposal for the domains Archaea, Bacteria, and Eucarya. *Proceedings of the National Academy of Sciences* 87, 4576–4579.

Wong, V.N.L., Dalal, R.C. and Greene, R.S.B. (2008) Salinity and sodicity effects on respiration and microbial biomass of soil. *Biology and Fertility of Soils* 44, 943–953.

Yeates, G.W. (2007) Diversity of nematodes. In: Benckiser, G. and Schnell, S. (eds) *Biodiversity in Agricultural Production Systems*. Taylor and Francis Group, CRC Press, Boca Raton, Florida, pp. 215–235.

Zas, R. (2008) Autocorrelación espacial y el diseño y análisis de experimentos. In: Maestre, F., Escudero, A. and Bonet, A. (eds) *Introduccion al Análisis Espacial de Datos en Ecología y Ciencias Ambientales: Metodos y Aplicaciones*. Servicio de Publicaciones Universidad Rey Juan Carlos, Madrid, pp. 542–590.

Zemp, M., Hoelzle, M. and Haeberli, W. (2009) Six decades of glacier mass-balance observations: a review of the worldwide monitoring network. *Annals of Glaciology* 50, 101–111.

Zhou, Z., Takaya, N., Nakamura, A., Yamaguchi, M., Takeo, K. and Shoun, H. (2002) Ammonia fermentation, a novel anoxic metabolism of nitrate by fungi. *Journal of Biological Chemistry* 277, 1892–1896.

Zhu, X.G., Long, S.P. and Ort, D.R. (2008) What is the maximum efficiency with which photosynthesis can convert solar energy into biomass? *Current Opinion in Biotechnology* 19, 153–159.

Zibilske, L.M. (2005) Biological control of soilborne plant pathogens and nematodes. In: Sylvia, D.M., Fuhrmann, J.J., Hartel, P.G. and Zuberer, D.D. (eds) *Principles and Applications of Soil Microbiology*. Prentice Hall, Upper Saddle River, New Jersey, pp. 482–497.

Zuberer, D. (1998) Biological nitrogen fixation: introduction and non-symbiotic. In: Sylvia, D.M., Fuhrmann, J.J., Hartel, P.G. and Zuberer, D.D. (eds) *Principles and Applications of Soil Microbiology*. Prentice Hall, Upper Saddle River, New Jersey, pp. 295–321.

Index

Page numbers in **bold** type refer to figures, tables and Focus Box material.